NUMERIC METHODS
FOR
LEAST SQUARES PROBLEMS

NUMERICAL METHODS FOR LEAST SQUARES PROBLEMS

ÅKE BJÖRCK

Linköping University
Linköping, Sweden

siam.
Society for Industrial and Applied Mathematics
Philadelphia

Copyright © 1996 by the Society for Industrial and Applied Mathematics.

10 9 8 7 6 5 4 3 2

All rights reserved. Printed in the United States of America. No part of this book may be reproduced, stored, or transmitted in any manner without the written permission of the publisher. For information, write to the Society for Industrial and Applied Mathematics, 3600 University City Science Center, Philadelphia, PA 19104-2688.

Library of Congress Cataloging-in-Publication Data

Björck, Åke, 1934-
    Numerical methods for least squares problems / Åke Björck.
       p. cm.
    Includes bibliographic references (p. -) and index.
    ISBN 0-89871-360-9 (pbk.)
       1. Equations, Simultaneous—Numerical solutions. 2. Least squares. I Title.
QA214.B56   1996
512.9'42—dc20   96-3908

Portions were adapted with permission from *Handbook of Numerical Analysis, Volume I, Least Squares Methods* by Åke Björck, © 1990, North-Holland, Amsterdam.

siam. is a registered trademark.

Dedicated to

Germund Dahlquist
and
Gene H. Golub

Contents

Preface   xv

1. Mathematical and Statistical Properties of Least Squares Solutions   1
    1.1 Introduction 1
        1.1.1 Historical remarks. 2
        1.1.2 Statistical preliminaries. 2
        1.1.3 Linear models and the Gauss–Markoff theorem. 3
        1.1.4 Characterization of least squares solutions. 5
    1.2 The Singular Value Decomposition 9
        1.2.1 The singular value decomposition. 9
        1.2.2 Related eigenvalue decompositions. 11
        1.2.3 Matrix approximations. 12
        1.2.4 The sensitivity of singular values and vectors. 13
        1.2.5 The SVD and pseudoinverse. 15
        1.2.6 Orthogonal projectors and angles between subspaces. . . 17
    1.3 The QR Decomposition 19
        1.3.1 The full rank case. 19
        1.3.2 Rank revealing QR decompositions. 21
        1.3.3 The complete orthogonal decomposition. 23
    1.4 Sensitivity of Least Squares Solutions 24
        1.4.1 Vector and matrix norms. 24
        1.4.2 Perturbation analysis of pseudoinverses. 26
        1.4.3 Perturbation analysis of least squares solutions. 27
        1.4.4 Asymptotic forms and derivatives. 32
        1.4.5 Componentwise perturbation analysis. 32
        1.4.6 A posteriori estimation of errors. 34

2. Basic Numerical Methods   37
    2.1 Basics of Floating Point Computation 37
        2.1.1 Rounding error analysis. 37
        2.1.2 Running rounding error analysis. 39

		2.1.3	Stability of algorithms.	40
2.2	The Method of Normal Equations			42
		2.2.1	Forming the normal equations.	42
		2.2.2	The Cholesky factorization.	44
		2.2.3	Conditioning and scaling.	49
2.3	Elementary Orthogonal Transformations			51
		2.3.1	Householder transformations.	51
		2.3.2	Givens transformation.	53
		2.3.3	Fast Givens transformations.	56
2.4	Methods Based on the QR Decomposition			58
		2.4.1	Householder and Givens QR decomposition.	58
		2.4.2	Gram–Schmidt orthogonalization.	60
		2.4.3	Least squares by Householder QR decomposition.	63
		2.4.4	Least squares problems by MGS.	64
		2.4.5	Gram–Schmidt with reorthogonalization.	66
		2.4.6	Hybrid algorithms.	69
		2.4.7	Block algorithms.	71
2.5	Methods Based on Gaussian Elimination			73
		2.5.1	The Peters–Wilkinson method.	73
		2.5.2	Pseudoinverse solutions from LU decompositions.	76
		2.5.3	The augmented system method.	77
2.6	Computing the SVD			81
		2.6.1	SVD and least squares problems.	81
		2.6.2	Transformation to bidiagonal form.	81
		2.6.3	The QR algorithm for real symmetric matrices.	83
		2.6.4	The QR algorithm for the SVD.	85
		2.6.5	Zero shift QR algorithm.	90
		2.6.6	Jacobi methods for the SVD.	92
		2.6.7	Singular values by spectrum slicing.	96
2.7	Rank Deficient and Ill-Conditioned Problems			99
		2.7.1	SVD and numerical rank.	99
		2.7.2	Truncated SVD solutions and regularization.	100
		2.7.3	QR decompositions with column pivoting.	103
		2.7.4	Pseudoinverse solutions from QR decompositions.	106
		2.7.5	Rank revealing QR decompositions.	108
		2.7.6	Complete orthogonal decompositions.	110
		2.7.7	Subset selection by SVD and RRQR.	113
2.8	Estimating Condition Numbers and Errors			114
		2.8.1	The LINPACK condition estimator.	114
		2.8.2	Hager's condition estimator.	116
		2.8.3	Computing the variance-covariance matrix.	118
2.9	Iterative Refinement			120
		2.9.1	Iterative refinement for linear systems.	120
		2.9.2	Extended precision iterative refinement.	121

| | | 2.9.3 Fixed precision iterative refinement. . . . . 124 |

3. Modified Least Squares Problems   **127**
   3.1 Introduction 127
       3.1.1 Updating problems. 127
       3.1.2 Modified linear systems. 128
       3.1.3 Modifying matrix factorizations. 129
       3.1.4 Recursive least squares. 131
   3.2 Modifying the Full QR Decomposition 132
       3.2.1 Introduction. 132
       3.2.2 General rank one change. 132
       3.2.3 Deleting a column. 133
       3.2.4 Appending a column. 135
       3.2.5 Appending a row. 136
       3.2.6 Deleting a row. 137
       3.2.7 Modifying the Gram–Schmidt decomposition. 138
   3.3 Downdating the Cholesky Factorization 140
       3.3.1 Introduction. 140
       3.3.2 The Saunders algorithm. 141
       3.3.3 The corrected seminormal equations. 142
       3.3.4 Hyperbolic rotations. 143
   3.4 Modifying the Singular Value Decomposition 145
       3.4.1 Introduction. 145
       3.4.2 Appending a row. 145
       3.4.3 Deleting a row. 147
   3.5 Modifying Rank Revealing QR Decompositions 149
       3.5.1 Appending a row. 149
       3.5.2 Deleting a row. 152

4. Generalized Least Squares Problems   **153**
   4.1 Generalized QR Decompositions 153
       4.1.1 Introduction. 153
       4.1.2 Computing the GQR and PQR. 153
   4.2 The Generalized SVD 155
       4.2.1 The CS decomposition. 155
       4.2.2 The generalized SVD. 157
       4.2.3 Computing the GSVD. 159
   4.3 General Linear Models and Generalized Least Squares 160
       4.3.1 Gauss–Markoff linear models. 160
       4.3.2 Generalized linear least squares problems. 162
       4.3.3 Paige's method. 164
   4.4 Weighted Least Squares Problems 165
       4.4.1 Introduction. 165
       4.4.2 Methods based on Gaussian elimination. 166
       4.4.3 QR decompositions for weighted problems. 168

		4.4.4	Weighted problems by updating.	171
	4.5	Minimizing the l_p Norm	172
		4.5.1	Introduction.	172
		4.5.2	Iteratively reweighted least squares.	173
		4.5.3	Robust linear regression.	175
		4.5.4	Algorithms for l_1 and l_∞ approximation.	175
	4.6	Total Least Squares	176
		4.6.1	Errors-in-variables models.	176
		4.6.2	Total least squares problem by SVD.	177
		4.6.3	Relationship to the least squares solution.	180
		4.6.4	Multiple right-hand sides.	181
		4.6.5	Generalized TLS problems.	182
		4.6.6	Linear orthogonal distance regression.	184

5. Constrained Least Squares Problems   187

	5.1	Linear Equality Constraints	187
		5.1.1	Introduction.	187
		5.1.2	Method of direct elimination.	188
		5.1.3	The nullspace method.	189
		5.1.4	Problem LSE by generalized SVD.	191
		5.1.5	The method of weighting.	192
		5.1.6	Solving LSE problems by updating.	194
	5.2	Linear Inequality Constraints	194
		5.2.1	Classification of problems.	194
		5.2.2	Basic transformations of problem LSI.	196
		5.2.3	Active set algorithms for problem LSI.	198
		5.2.4	Active set algorithms for BLS.	201
	5.3	Quadratic Constraints	203
		5.3.1	Ill-posed problems.	203
		5.3.2	Quadratic inequality constraints.	205
		5.3.3	Problem LSQI by GSVD.	206
		5.3.4	Problem LSQI by QR decomposition.	208
		5.3.5	Cross-validation.	211

6. Direct Methods for Sparse Problems   215

	6.1	Introduction	215
	6.2	Banded Least Squares Problems	217
		6.2.1	Storage schemes for banded matrices.	218
		6.2.2	Normal equations for banded problems.	219
		6.2.3	Givens QR decomposition for banded problems.	221
		6.2.4	Householder QR decomposition for banded problems.	222
	6.3	Block Angular Least Squares Problems	224
		6.3.1	Block angular form.	224
		6.3.2	QR methods for block angular problems.	225
	6.4	Tools for General Sparse Problems	227

		6.4.1	Storage schemes for general sparse matrices. 227
		6.4.2	Graph representation of sparse matrices. 230
		6.4.3	Predicting the structure of A^TA. 231
		6.4.4	Predicting the structure of R. 232
		6.4.5	Block triangular form of a sparse matrix. 234
	6.5	Fill Minimizing Column Orderings 237
		6.5.1	Bandwidth reducing ordering methods. 237
		6.5.2	Minimum degree ordering. 238
		6.5.3	Nested dissection orderings. 240
	6.6	The Numerical Cholesky and QR Decompositions 242
		6.6.1	The Cholesky factorization. 242
		6.6.2	Row sequential QR decomposition. 242
		6.6.3	Row orderings for sparse QR decomposition. 244
		6.6.4	Multifrontal QR decomposition. 245
		6.6.5	Iterative refinement and seminormal equations. 250
	6.7	Special Topics 252
		6.7.1	Rank revealing sparse QR decomposition. 252
		6.7.2	Updating sparse least squares solutions. 254
		6.7.3	Partitioning for out-of-core solution. 255
		6.7.4	Computing selected elements of the covariance matrix. . . . 256
	6.8	Sparse Constrained Problems 257
		6.8.1	An active set method for problem BLS. 257
		6.8.2	Interior point methods for problem BLS. 262
	6.9	Software and Test Results 264
		6.9.1	Software for sparse direct methods. 264
		6.9.2	Test results. 266

7. Iterative Methods For Least Squares Problems   269

	7.1	Introduction 269
		7.1.1	Iterative versus direct methods. 270
		7.1.2	Computing sparse matrix-vector products. 270
	7.2	Basic Iterative Methods 274
		7.2.1	General stationary iterative methods. 274
		7.2.2	Splittings of rectangular matrices. 276
		7.2.3	Classical iterative methods. 276
		7.2.4	Successive overrelaxation methods. 279
		7.2.5	Semi-iterative methods. 280
		7.2.6	Preconditioning. 283
	7.3	Block Iterative Methods 284
		7.3.1	Block column preconditioners. 284
		7.3.2	The two-block case. 286
	7.4	Conjugate Gradient Methods 288
		7.4.1	CGLS and variants. 288
		7.4.2	Convergence properties of CGLS. 290

		7.4.3	The conjugate gradient method in finite precision. 292
		7.4.4	Preconditioned CGLS. 293
	7.5	Incomplete Factorization Preconditioners 294	
		7.5.1	Incomplete Cholesky preconditioners. 294
		7.5.2	Incomplete orthogonal decompositions. 297
		7.5.3	Preconditioners based on LU factorization. 299
	7.6	Methods Based on Lanczos Bidiagonalization 303	
		7.6.1	Lanczos bidiagonalization. 303
		7.6.2	Best approximation in the Krylov subspace. 306
		7.6.3	The LSQR algorithm. 307
		7.6.4	Convergence of singular values and vectors. 309
		7.6.5	Bidiagonalization and total least squares. 310
	7.7	Methods for Constrained Problems 312	
		7.7.1	Problems with upper and lower bounds. 312
		7.7.2	Iterative regularization. 314

8. Least Squares Problems with Special Bases   317

	8.1	Least Squares Approximation and Orthogonal Systems 317	
		8.1.1	General formalism. 317
		8.1.2	Statistical aspects of the method of least squares. 318
	8.2	Polynomial Approximation 319	
		8.2.1	Triangle family of polynomials. 319
		8.2.2	General theory of orthogonal polynomials. 320
		8.2.3	Discrete least squares fitting. 321
		8.2.4	Vandermonde-like systems. 323
		8.2.5	Chebyshev polynomials. 325
	8.3	Discrete Fourier Analysis 328	
		8.3.1	Introduction. 328
		8.3.2	Orthogonality relations. 329
		8.3.3	The fast Fourier transform. 330
	8.4	Toeplitz Least Squares Problems 332	
		8.4.1	Introduction. 332
		8.4.2	QR decomposition of Toeplitz matrices. 333
		8.4.3	Iterative solvers for Toeplitz systems. 334
		8.4.4	Preconditioners for Toeplitz systems. 335
	8.5	Kronecker Product Problems 336	

9. Nonlinear Least Squares Problems   339

	9.1	The Nonlinear Least Squares Problem 339	
		9.1.1	Introduction. 339
		9.1.2	Necessary conditions for local minima. 340
		9.1.3	Basic numerical methods. 341
	9.2	Gauss–Newton-Type Methods 342	
		9.2.1	The damped Gauss–Newton method. 343
		9.2.2	Local convergence of the Gauss–Newton method. 345

		9.2.3	Trust region methods.	346
9.3	Newton-Type Methods			348
	9.3.1		Introduction.	348
	9.3.2		A hybrid Newton method.	348
	9.3.3		Quasi-Newton methods.	349
9.4	Separable and Constrained Problems			351
	9.4.1		Separable problems.	351
	9.4.2		General constrained problems.	353
	9.4.3		Orthogonal distance regression.	354
	9.4.4		Least squares fit of geometric elements.	357

Bibliography   359

Index   401

Chapter 1
Preface

A basic problem in science is to fit a model to observations subject to errors. It is clear that the more observations that are available the more accurately will it be possible to calculate the parameters in the model. This gives rise to the problem of "solving" an overdetermined linear or nonlinear system of equations. It can be shown that the solution which minimizes a weighted sum of the squares of the residual is optimal in a certain sense. Gauss claims to have discovered the method of least squares in 1795 when he was 18 years old. Hence this book also marks the bicentennial of the use of the least squares principle.

The development of the basic modern numerical methods for solving linear least squares problems took place in the late sixties. The QR decomposition by Householder transformations was developed by Golub and published in 1965. The implicit QR algorithm for computing the singular value decomposition (SVD) was developed about the same time by Kahan, Golub, and Wilkinson, and the final algorithm was published in 1970. These matrix decompositions have since been developed and generalized to a high level of sophistication. Great progress has been made in the last decade in methods for generalized and modified least squares problems and in direct and iterative methods for large sparse problems. Methods for total least squares problems, which allow errors also in the system matrix, have been systematically developed.

Applications of least squares of crucial importance occur in many areas of applied and engineering research such as statistics, geodetics, photogrammetry, signal processing, and control. Because of the great increase in the capacity for automatic data capturing, least squares problems of large size are now routinely solved. Therefore, sparse direct methods as well as iterative methods play an increasingly important role. Applications in signal processing have created a great demand for stable and efficient methods for modifying least squares solutions when data are added or deleted. This has led to renewed interest in rank revealing QR decompositions, which lend themselves better to updating than the singular value decomposition. Generalized and weighted least squares problems and problems of Toeplitz and Kronecker structure are becoming increasingly important.

Chapter 1 gives the basic facts and the mathematical and statistical background of least squares methods. In Chapter 2 relevant matrix decompositions and basic numerical methods are covered in detail. Although most proofs are omitted, these two chapters are more elementary than the rest of the book and essentially self-contained. Chapter 3 treats modified least squares problems and includes many recent results. In Chapter 4 generalized QR and SVD decompositions are presented, and methods for generalized and weighted problems surveyed. Here also, robust methods and methods for total least squares are treated. Chapter 5 surveys methods for problems with linear and quadratic constraints. Direct and iterative methods for large sparse least squares problems are covered in Chapters 6 and 7. These methods are still subject to intensive research, and the presentation is more advanced. Chapter 8 is devoted to problems with special bases, including least squares fitting of polynomials and problems of Toeplitz and Kronecker structures. Finally, Chapter 9 contains a short survey of methods for nonlinear problems.

This book will be of interest to mathematicians working in numerical linear algebra, computational scientists and engineers, and statisticians, as well as electrical engineers. Although a solid understanding of numerical linear algebra is needed for the more advanced sections, I hope the book will be found useful in upper-level undergraduate and beginning graduate courses in scientific computing and applied sciences.

I have aimed to make the book and the bibliography as comprehensive and up-to-date as possible. Many recent research results are included, which were only available in the research literature before. Inevitably, however, the content reflects my own interests, and I apologize in advance to those whose work has not been mentioned. In particular, work on the least squares problem in the former Soviet Union is, to a large extent, not covered.

The history of this book dates back to at least 1981, when I wrote a survey entitled "Least Squares Methods in Physics and Engineering" for the Academic Training Programme at CERN in Geneva. In 1985 I was invited to contribute a chapter on "Least Squares Methods" in the *Handbook of Numerical Analysis*, edited by P. G. Ciarlet and J. L. Lions. This chapter [95] was finished in 1988 and appeared in Volume 1 of the *Handbook*, published by North-Holland in 1990. The present book is based on this contribution, although it has been extensively updated and made more complete.

The book has greatly benefited from the insight and knowledge kindly provided by many friends and colleagues. In particular, I have been greatly influenced by the work of Gene H. Golub, Nick Higham, and G. W. Stewart. Per-Åke Wedin gave valuable advice on the chapter on nonlinear problems. Part of the *Handbook* chapter was written while I had the benefit of visiting the Division of Mathematics and Statistics at CSIRO in Canberra and the Chr. Michelsen Institute in Bergen.

Thanks are due to Elsevier Science B.V. for the permission to use part of the material from the *Handbook* chapter. Finally, I thank Beth Gallagher and Vickie Kearn at SIAM for the cheerful and professional support they have given throughout the copy editing and production of the book.

Åke Björck
Linköping, February 1996

Chapter 1
Mathematical and Statistical Properties of Least Squares Solutions

De tous les principes qu'on peut proposer pour cet objet, je pense qu'il n'en est pas de plus general, de plus exact, ni d'une application plus facile que celui qui consiste à rendre *minimum* la somme de carrés des erreurs.[1]
Adrien Marie Legendre, Nouvelles méthodes pour la détermination des orbites des comètes. Appendice. Paris, 1805.

1.1. Introduction

The linear least squares problem is a computational problem of primary importance, which originally arose from the need to fit a linear mathematical model to given observations. In order to reduce the influence of errors in the observations one would then like to use a greater number of measurements than the number of unknown parameters in the model. The resulting problem is to "solve" an **overdetermined** linear system of equations. In matrix terms, given a vector $b \in \mathbf{R}^m$ and a matrix $A \in \mathbf{R}^{m \times n}$, $m > n$, we want to find a vector $x \in \mathbf{R}^n$ such that Ax is the "best" approximation to b.

EXAMPLE 1.1.1. Consider a model described by a scalar function $y(t) = f(x,t)$, where $x \in \mathbf{R}^n$ is a parameter vector to be determined from measurements (y_i, t_i), $i = 1, \ldots, m$, $m > n$. In particular, let $f(x,t)$ be *linear* in x:

$$f(x,t) = \sum_{j=1}^{n} x_j \phi_j(t).$$

Then the equations $y_i = \sum_{j=1}^{n} x_j \phi_j(t_i)$, $i = 1, \ldots, m$ form an overdetermined linear system $Ax = b$, where $a_{ij} = \phi_j(t_i)$ and $b_i = y_i$. ∎

There are many possible ways of defining the "best" solution. A choice which can often be motivated for statistical reasons (see below) and which also leads to a simple computational problem is to let x be a solution to the minimization problem

(1.1.1) $$\min_{x} \|Ax - b\|_2, \quad A \in \mathbf{R}^{m \times n}, \quad b \in \mathbf{R}^m,$$

---
[1] Of all the principles that can be proposed, I think there is none more general, more exact, and more easy of application, than that which consists of rendering the sum of the squares of the errors a minimum.

where $\|\cdot\|_2$ denotes the Euclidean vector norm. We call this a **linear least squares problem** and x a linear least squares solution of the system $Ax = b$. We refer to $r = b - Ax$ as the residual vector. A least squares solution minimizes $\|r\|_2^2 = \sum_{i=1}^m r_i^2$, i.e., the sum of the squared residuals. If rank $(A) < n$, then the solution x to (1.1) is not unique. However, among all least squares solutions there is a unique solution which minimizes $\|x\|_2$; see Theorem 1.2.10.

1.1.1. Historical remarks. Laplace in 1799 used the principle of minimizing the sum of the absolute errors $\sum_{i=1}^m |r_i|$, with the added condition that the sum of the errors be equal to zero; see Goldstine [363, 1977]. He showed that the solution x must then satisfy exactly n out of the m equations. Gauss argued that since, by the principles of probability, greater or smaller errors are equally possible in all equations, it is evident that a solution which satisfies precisely n equations must be regarded as less consistent with the laws of probability. He was then led to the principle of least squares. The algebraic procedure of the method of least squares was first published by Legendre [523, 1805]. It was justified as a statistical procedure by Gauss [320, 1809], where he (much to the annoyance of Legendre) claimed to have discovered the method of least squares in 1795.[2]

Most historians agree that Gauss was right in his claim. Gauss used the least squares principle for analyzing survey data and in astronomical calculations. A famous example is when Gauss successfully predicted the orbit of the asteroid Ceres in 1801. The method of least squares quickly became the standard procedure for analysis of astronomical and geodetic data. There are several good accounts of the history of the invention of least squares and the dispute between Gauss and Legendre; see Placket [660, 1972], Stigler [757, 1977], [758, 1981], and Goldstine [363, 1977].

Gauss gave the method a sound theoretical basis in "Theoria Combinationis" [322, 1821], [323, 1823]. These two memoirs of Gauss, which contain his definitive treatment of the area, have recently been collected for the first time in an English translation by Stewart [325, 1995]. Gauss proves here the optimality of the least squares estimate without any assumptions that the random variables follow a particular distribution. This contribution of Gauss was somehow neglected until being rediscovered by Markoff [566, 1912]; see Theorem 1.1.1.

1.1.2. Statistical preliminaries. Let y be a random variable having the distribution function $F(y)$, where $F(y)$ is nondecreasing, right continuous, and

$$0 \leq F(y) \leq 1, \quad F(-\infty) = 0, \quad F(\infty) = 1.$$

The expected value and the variance of y is then defined as

$$\mathcal{E}(y) = \mu = \int_{-\infty}^{\infty} y \, dF(y), \quad \mathcal{E}((y-\mu)^2) = \sigma^2 = \int_{-\infty}^{\infty} (y-\mu)^2 \, dF(y).$$

---

[2] "Our principle, which we have made use of since 1795, has lately been published by Legendre...," C. F. Gauss, *Theory of the Motion of the Heavenly Bodies Moving about the Sun in Conic Sections*, Hamburg [320, 1809].

1.1. Introduction

Let $y = (y_1, \ldots, y_n)^T$ be a vector of random variables and let $\mu = (\mu_1, \ldots, \mu_n)$, where $\mu_i = \mathcal{E}(y_i)$. Then we write $\mu = \mathcal{E}(y)$. If y_i and y_j have the joint distribution $F(y_i, y_j)$ the **covariance** σ_{ij} between y_i and y_j is defined by

$$\mathrm{cov}(y_i, y_j) = \mathcal{E}[(y_i - \mu_i)(y_j - \mu_j)] = \sigma_{ij} = \int_{-\infty}^{\infty} (y_i - \mu_i)(y_j - \mu_j) dF(y_i, y_j).$$

Note that $\sigma_{ij} = \mathcal{E}(y_i y_j) - \mu_i \mu_j$. The variance-covariance matrix $V \in \mathbf{R}^{n \times n}$ of y is defined by

$$\mathcal{V}(y) = V = \mathcal{E}[(y - \mu)(y - \mu)^T] = \mathcal{E}(yy^T) - \mu\mu^T.$$

We now prove some properties which will be useful in the remainder of the book.

LEMMA 1.1.1. *Let $z = Fy$, where $F \in \mathbf{R}^{r \times n}$ is a given matrix and y a random vector with $\mathcal{E}(y) = \mu$ and covariance matrix V. Then*

$$\mathcal{E}(z) = F\mu, \qquad \mathcal{V}(z) = FVF^T.$$

Proof. The first property follows directly from the definition of expected value. The second is proved as

$$\mathcal{V}(Fy) = \mathcal{E}[F(y - \mu)(y - \mu)^T F^T] = F\mathcal{E}[(y - \mu)(y - \mu)^T]F^T = FVF^T. \quad \blacksquare$$

In the special case when $F = f^T$ is a row vector, then $z = f^T y$ is a linear functional of y and $\mathcal{V}(z) = \mu \|b\|_2^2$. The following lemma is given without proof.

LEMMA 1.1.2. *Let $A \in \mathbf{R}^{n \times n}$ be a symmetric matrix and consider the quadratic form $y^T A y$, where y is a random vector with expected value μ and covariance matrix V. Then*

$$\mathcal{E}(y^T A y) = \mu^T A \mu + \mathrm{trace}\,(AV),$$

where $\mathrm{trace}\,(AV)$ *denotes the sum of diagonal elements of AV.*

1.1.3. Linear models and the Gauss–Markoff theorem.

In linear statistical models one assumes that the vector $b \in \mathbf{R}^m$ of observations is related to the unknown parameter vector $x \in \mathbf{R}^n$ by a linear relation

(1.1.2) $$Ax = b + \epsilon,$$

where $A \in \mathbf{R}^{m \times n}$ is a known matrix and ϵ is a vector of random errors. In the **standard linear model** we have

(1.1.3) $$\mathcal{E}(\epsilon) = 0, \qquad \mathcal{V}(\epsilon) = \sigma^2 I,$$

i.e., the random variables ϵ_i are uncorrelated and all have zero means and the same variance. We also assume that $\mathrm{rank}\,(A) = n$.

We make the following definitions.

DEFINITION 1.1.1. *A function $g(y)$ of the random vector y is an unbiased estimate of a parameter θ if $\mathcal{E}(g(y)) = \theta$. When such a function exists, then θ is called an estimable parameter.*

DEFINITION 1.1.2. *Let c be a constant vector. Then the linear function $g = c^T y$ is called a minimum variance unbiased estimate of θ if $\mathcal{E}(g) = \theta$, and $\mathcal{V}(g)$ is minimized over all linear estimators.*

The following theorem by Gauss placed the method of least squares on a sound theoretical basis without any assumption that the random errors follow a normal distribution.

THEOREM 1.1.1. *The Gauss–Markoff theorem. Consider the linear model (1.1.2), where $A \in \mathbf{R}^{m \times n}$ is a known matrix of rank n, $\hat{b} = b + \epsilon$, where ϵ is a random vector with mean and variance given by (1.1.3). Then the best linear unbiased estimator of any linear function $c^T x$ is $c^T \hat{x}$, where $\hat{x}$ is the least squares estimator, obtained by minimizing the sum of squares $\|Ax - \hat{b}\|_2^2$. Furthermore, $\mathcal{E}(s^2) = \sigma^2$, where s^2 is the quadratic form*

$$(1.1.4) \qquad s^2 = \frac{1}{m-n}(b - A\hat{x})^T(b - A\hat{x}) = \frac{1}{m-n}\|b - A\hat{x}\|_2^2.$$

Proof. For a modern proof see Zelen [850, 1962, pp. 560–561]. ∎

COROLLARY 1.1.1. *The variance-covariance matrix of the least squares estimate $\hat{x}$ is*
$$(1.1.5) \qquad \mathcal{V}(\hat{x}) = \sigma^2 (A^T A)^{-1}.$$

Proof. Since $\hat{x} = (A^T A)^{-1} A^T \hat{b}$ it follows from Lemma 1.1.1 that

$$\mathcal{V}(\hat{x}) = (A^T A)^{-1} A^T \mathcal{V}(\hat{b}) A (A^T A)^{-1} = \sigma^2 (A^T A)^{-1}. \qquad \blacksquare$$

The residual vector $\hat{r} = b - A\hat{x}$ satisfies $A^T \hat{r} = 0$, and hence there are n linear relations among the m components of $\hat{r}$. It can be shown that the residuals $\hat{r}$, and therefore also the quadratic form s^2, are uncorrelated with $\hat{x}$, i.e.,

$$\operatorname{cov}(\hat{r}, \hat{x}) = 0, \qquad \operatorname{cov}(s^2, \hat{x}) = 0.$$

In the **general univariate linear model** the covariance matrix is $\mathcal{V}(\epsilon) = \sigma^2 W$, where $W \in \mathbf{R}^{m \times m}$ is a positive semidefinite symmetric matrix. If A has rank n and W is positive definite then the best unbiased linear estimate for x was shown by Aiken [4, 1934] to be the solution of

$$(1.1.6) \qquad \min_x (Ax - b)^T W^{-1} (Ax - b).$$

General linear models are considered in Section 4.3.1, and special methods for the corresponding generalized least squares problems are treated in Sections 4.3 and 4.4. It is important to note that for singular W the best unbiased linear estimate of x can *not* always be obtained by replacing W^{-1} in (1.1.6) by the Moore–Penrose pseudoinverse $W^\dagger$!

1.1. Introduction

In some applications it might be more adequate to consider the more general minimization problem

$$\min_x \|Ax - b\|_p, \tag{1.1.7}$$

where the Hölder vector p-norms $\|\cdot\|_p$ are defined by

$$\|x\|_p = \left(\sum_{i=1}^n |x_i|^p\right)^{1/p}, \qquad 1 \le p < \infty. \tag{1.1.8}$$

The Euclidian norm corresponds to $p = 2$, and in the limiting case

$$\|x\|_\infty = \max_{1 \le i \le n} |x_i|. \tag{1.1.9}$$

EXAMPLE 1.1.2. To illustrate the effect of using a Hölder norm with $p \ne 2$, we consider the problem of estimating the scalar γ from m observations $y \in \mathbf{R}^m$. This is equivalent to minimizing $\|A\gamma - y\|_p$, with $A = (1, 1, \ldots, 1)^T$. It is easily verified that if $y_1 \ge y_2 \ge \cdots \ge y_m$, then the solution for some different values p are

$$\begin{aligned}
\gamma_1 &= y_{(m+1)/2}, \quad (m \text{ odd}), \\
\gamma_2 &= (y_1 + y_2 + \cdots + y_m)/m, \\
\gamma_\infty &= (y_1 + y_m)/2.
\end{aligned}$$

These estimates correspond to the **median**, **mean**, and **midrange**, respectively. Note that the estimate γ_1 is insensitive to extreme values of y_i. This property carries over to more general problems, and a small number of isolated large errors will usually not change the l_1 solution. For a treatment of problem (1.1.7) when $p \ne 2$ see Section 4.5.

1.1.4. Characterization of least squares solutions. We begin by characterizing the set of all solutions to the least squares problem (1.1.1).

THEOREM 1.1.2. *Denote the set of all solutions to* (1.1.1) *by*

$$\mathcal{S} = \{x \in \mathbf{R}^n \mid \|Ax - b\|_2 = \min\}. \tag{1.1.10}$$

Then $x \in \mathcal{S}$ if and only if the following orthogonality condition holds:

$$A^T(b - Ax) = 0. \tag{1.1.11}$$

Proof. Assume that $\hat{x}$ satisfies $A^T \hat{r} = 0$, where $\hat{r} = b - A\hat{x}$. Then for any $x \in \mathbf{R}^n$ we have $r = b - Ax = \hat{r} + A(\hat{x} - x) \equiv \hat{r} + Ae$. Squaring this we obtain

$$r^T r = (\hat{r} + Ae)^T(\hat{r} + Ae) = \hat{r}^T \hat{r} + \|Ae\|_2^2,$$

which is minimized when $x = \hat{x}$.

On the other hand suppose $A^T \hat{r} = z \ne 0$, and take $x = \hat{x} + \epsilon z$. Then $r = \hat{r} - \epsilon A z$, and

$$r^T r = \hat{r}^T \hat{r} - 2\epsilon z^T z + \epsilon^2 (Az)^T Az < \hat{r}^T \hat{r}$$

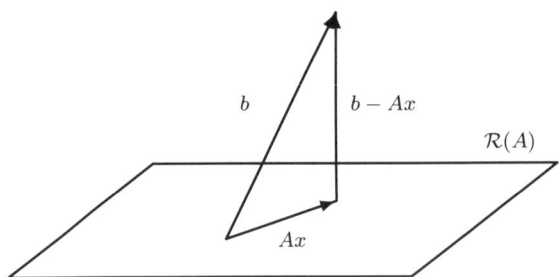

FIG. 1.1.1. *Geometric interpretation of least squares property.*

for sufficiently small ϵ. Hence $\hat{x}$ is not a least squares solution. ∎

The **range** (or column space) of a matrix $A \in \mathbf{R}^{m \times n}$ is defined to be

$$\mathcal{R}(A) = \{\, z = Ax \mid x \in \mathbf{R}^n \,\}. \tag{1.1.12}$$

The set of solutions to $A^T y = 0$ is a subspace called the **nullspace** of A^T and denoted by

$$\mathcal{N}(A^T) = \{\, y \in \mathbf{R}^m \mid A^T y = 0 \,\}, \tag{1.1.13}$$

and is the orthogonal complement in $\mathbf{R}^m$ to the space $\mathcal{R}(A)$. These are two of the four fundamental subspaces of the matrix A; see Section 1.2. Theorem 1.1.2 asserts that the residual vector $r = b - Ax$ of a least squares solution lies in $\mathcal{N}(A^T)$. Hence any least squares solution x uniquely decomposes the right-hand side b into two orthogonal components

$$b = Ax + r, \quad Ax \in \mathcal{R}(A), \quad r \in \mathcal{N}(A^T). \tag{1.1.14}$$

This geometric interpretation is illustrated for $n = 2$ in Figure 1.1.1.

From (1.1.11) it follows that a least squares solution satisfies the **normal equations**

$$A^T A x = A^T b. \tag{1.1.15}$$

The matrix $A^T A \in \mathbf{R}^{n \times n}$ is symmetric and nonnegative definite. The normal equations are always consistent since

$$A^T b \in \mathcal{R}(A^T) = \mathcal{R}(A^T A).$$

Furthermore we have the following theorem.

THEOREM 1.1.3. *The matrix $A^T A$ is positive definite if and only if the columns of A are linearly independent, i.e., $\operatorname{rank}(A) = n$.*

Proof. If the columns of A are linearly independent, then $x \neq 0 \Rightarrow Ax \neq 0$ and therefore $x \neq 0 \Rightarrow x^T A^T A x = \|Ax\|_2^2 > 0$. Hence $A^T A$ is positive definite. On the other hand, if the columns are linearly dependent then for some $x_0 \neq 0$ we have $Ax_0 = 0$ and so $x_0^T A^T A x_0 = 0$, and $A^T A$ is not positive definite. ∎

1.1. Introduction

From Theorem 1.1.3 it follows that if $\operatorname{rank}(A) = n$, then the unique least squares solution x and the corresponding residual $r = b - Ax$ are given by

$$(1.1.16) \qquad x = (A^T A)^{-1} A^T b, \qquad r = b - A(A^T A)^{-1} A^T b.$$

If $S \subset \mathbf{R}^m$ is a subspace, then $P_S \in \mathbf{R}^{m \times m}$ is the **orthogonal projector** onto S if $\mathcal{R}(P_S) = S$, and

$$(1.1.17) \qquad P_S^2 = P_S, \qquad P_S^T = P_S.$$

Moreover,
$$(I - P_S)^2 = (I - P_s), \qquad (I - P_S) P_S = 0,$$

and $(I - P_S)$ is the projector for the subspace complementary to that of S.

Let P_1 and P_2 be orthogonal projectors onto S. Then using (1.1.17) we have for all $z \in \mathbf{R}^m$

$$\|(P_1 - P_2) z\|_2^2 = z^T P_1 (I - P_2) z + z^T P_2 (I - P_1) z = 0.$$

It follows that $P_1 = P_2$, and hence the orthogonal projector is unique.

From the geometric interpretation (see Figure 1.1.1) Ax is the orthogonal projection of b onto $\mathcal{R}(A)$. We have $r = (I - P_{\mathcal{R}(A)}) b$, and in the full rank case

$$(1.1.18) \qquad P_{\mathcal{R}(A)} = A(A^T A)^{-1} A^T.$$

If $\operatorname{rank}(A) < n$ then A has a nontrivial nullspace and the least squares solution is not unique. If $\hat{x}$ is a particular least squares solution then the set of all least squares solutions is

$$\mathcal{S} = \{ x = \hat{x} + z \mid z \in \mathcal{N}(A) \}.$$

If $\hat{x} \perp \mathcal{N}(A)$ then $\|x\|_2^2 = \|\hat{x}\|_2^2 + \|z\|_2^2$, and therefore $\hat{x}$ is the unique least squares solution of minimum norm.

The problem of computing the minimum norm solution $y \in \mathbf{R}^m$ to an underdetermined system of linear equations

$$(1.1.19) \qquad \min \|y\|_2, \qquad A^T y = c,$$

where $A \in \mathbf{R}^{m \times n}$, occurs as a subproblem in optimization algorithms. If $\operatorname{rank}(A) = n$, then the system $A^T y = c$ is consistent and the unique solution of (2.5) is given by the **normal equations of the second kind**

$$(1.1.20) \qquad A^T A z = c, \qquad y = Az,$$

that is, $y = A(A^T A)^{-1} c$.

The classical method for solving the normal equations is based on the following matrix factorization.

THEOREM 1.1.4. **Cholesky Decomposition.** *Let the matrix $C \in R^{n \times n}$ be symmetric and positive definite. Then there is a unique upper triangular matrix R with positive diagonal elements such that*

$$C = R^T R. \tag{1.1.21}$$

R is called the Cholesky factor of C and (1.1.21) is called the Cholesky factorization.

Proof. The proof is by induction on the order n of C. The result is trivial for $n = 1$. Assume that (1.1.21) holds for all positive definite matrices of order n. Consider the positive definite matrix $\bar{C}$ of order $n + 1$, and seek a factorization

$$\bar{C} = \begin{pmatrix} C & c \\ c^T & \gamma \end{pmatrix} = \begin{pmatrix} R^T & 0 \\ r^T & \rho \end{pmatrix} \begin{pmatrix} R & r \\ 0 & \rho \end{pmatrix}. \tag{1.1.22}$$

C is a principal minor of $\bar{C}$ and hence positive definite. By the induction hypothesis the factorization $C = R^T R$ exists and thus (1.1.22) holds provided r and $\rho > 0$ satisfy

$$R^T r = m, \quad \rho^2 = \gamma - r^T r. \tag{1.1.23}$$

Since R^T has positive diagonal elements and is lower triangular, $r = R^{-T} m$ is uniquely determined. Now, from the positive definiteness of $\bar{C}$ it follows that

$$0 < (r^T R^{-T} \quad -1) \begin{pmatrix} C & c \\ c^T & \gamma \end{pmatrix} \begin{pmatrix} R^{-1} r \\ -1 \end{pmatrix} = r^T R^{-T} C R^{-1} r - 2 r^T R^{-T} m + \gamma$$
$$= r^T r - 2 r^T r + \gamma = \gamma - r^T r.$$

Hence also $\rho = (\gamma - r^T r)^{1/2}$ is uniquely determined. ∎

Another characterization of the least squares solution is given in the following theorem.

THEOREM 1.1.5. *Assume that $A \in \mathbf{R}^{m \times n}$ has rank n. Then the symmetric linear system*

$$\begin{pmatrix} I & A \\ A^T & 0 \end{pmatrix} \begin{pmatrix} y \\ x \end{pmatrix} = \begin{pmatrix} b \\ c \end{pmatrix} \tag{1.1.24}$$

is nonsingular and gives the condition for the solution of both the primal and dual least squares problem

$$\min_x \left\{ \|Ax - b\|_2^2 + 2 c^T x \right\}, \tag{1.1.25}$$

$$\min_y \|y - b\|_2^2, \quad A^T y = c. \tag{1.1.26}$$

Proof. The system (1.1.24), often called the **augmented system**, can be obtained by differentiating (1.1.25) to give $A^T(b - Ax) = c$, and setting y to be the residual $y = b - Ax$. The system can also be obtained by differentiating the Lagrangian

$$L(x, y) = \|y - b\|_2^2 + 2 x^T (A^T y - c)$$

of (1.1.26), and equating to zero. Here x is the vector of Lagrange multipliers. ∎

1.2. The Singular Value Decomposition

Setting $c = 0$ in (1.1.25) gives the linear least squares problem (1.1.2). Setting $b = 0$ in (1.1.26) gives the problem of minimum 2-norm solution of an underdetermined linear system $A^T y = c$; see (1.1.19).

1.2. The Singular Value Decomposition

1.2.1. The singular value decomposition. The singular value decomposition (SVD) of a matrix $A \in \mathbf{R}^{m \times n}$ is a matrix decomposition of great theoretical and practical importance for the treatment of least squares problems. It provides a diagonal form of A under an orthogonal equivalence transformation. The history of this matrix decomposition goes back more than a century; see the very interesting survey of the early history of the SVD by Stewart [750, 1993]. However, only recently has the SVD been as much used as it should. Now it is a main tool in numerous application areas such as signal and image processing, control theory, pattern recognition, time-series analysis, etc.

Because applications exist also for complex matrices we state the theorem below for matrices with complex elements. (The matrix A^H will denote the matrix formed by conjugating each element and taking the transpose.)

THEOREM 1.2.1. *Singular Value Decomposition. Let $A \in \mathbf{C}^{m \times n}$ be a matrix of rank r. Then there exist unitary matrices $U \in \mathbf{C}^{m \times m}$ and $V \in \mathbf{C}^{n \times n}$ such that*

$$(1.2.1) \qquad A = U \Sigma V^H, \quad \Sigma = \begin{pmatrix} \Sigma_1 & 0 \\ 0 & 0 \end{pmatrix},$$

where $\Sigma \in \mathbf{R}^{m \times n}$, $\Sigma_1 = \mathrm{diag}\,(\sigma_1, \sigma_2, \ldots, \sigma_r)$, and

$$\sigma_1 \geq \sigma_2 \geq \cdots \geq \sigma_r > 0.$$

The σ_i are called the singular values of A, and if we write

$$(1.2.2) \qquad U = (u_1, \ldots, u_m), \quad V = (v_1, \ldots, v_n),$$

the u_i and v_i are, respectively, the left and right singular vectors associated with σ_i, $i = 1, \ldots, r$.

Proof. (See Golub and Van Loan [389, 1989].) Let $v_1 \in \mathbf{C}^n$ be a vector such that

$$\|v_1\|_2 = 1, \quad \|Av_1\|_2 = \|A\|_2 = \sigma,$$

where σ is real and positive. The existence of such a vector follows from the definition of a matrix subordinate norm $\|A\|$. If $\sigma = 0$, then $A = 0$, and we can take $\Sigma = 0$ and U and V arbitrary unitary matrices. Therefore assume that $\sigma > 0$, and take $u_1 = (1/\sigma) A v_1 \in \mathbf{C}^m$, $\|v_1\|_2 = 1$. Let the matrices

$$V = (v_1, V_1) \in \mathbf{C}^{n \times n}, \quad U = (u_1, U_1) \in \mathbf{C}^{m \times m}$$

be unitary. (Recall that it is always possible to extend a unitary set of vectors to a unitary basis for the whole space.) Since $U_1^H A v_1 = \sigma U_1^H u_1 = 0$ it follows that $U^H A V$ has the following structure:

$$A_1 \equiv U^H A V = \begin{pmatrix} \sigma & w^H \\ 0 & B \end{pmatrix},$$

where $w^H = u_1^H A V_1$, and $B = U_1^H A V_1 \in \mathbf{C}^{(m-1) \times (n-1)}$. From the two inequalities

$$\|A_1\|_2(\sigma^2 + w^H w)^{1/2} \geq \left\|A_1 \begin{pmatrix} \sigma \\ w \end{pmatrix}\right\|_2 = \left\|\begin{pmatrix} \sigma^2 + w^H w \\ Bw \end{pmatrix}\right\|_2 \geq \sigma^2 + w^H w,$$

it follows that $\|A_1\|_2 \geq (\sigma^2 + w^H w)^{1/2}$. But since U and V are unitary, $\|A_1\|_2 = \|A\|_2 = \sigma$, and thus $w = 0$. The proof can now be completed by an induction argument on the smallest dimension $\min(m, n)$. ∎

A rectangular matrix $A \in \mathbf{R}^{m \times n}$ represents a linear mapping from $\mathbf{C}^n$ to $\mathbf{C}^m$. The significance of Theorem 1.2.1 is that it shows that there is an orthogonal basis in each of these spaces, with respect to which this mapping is represented by a generalized diagonal matrix Σ with real elements. Methods for computing the SVD are described in Section 2.6.

The SVD of A can be written

$$(1.2.3) \qquad A = U_1 \Sigma_1 V_1^H = \sum_{i=1}^{r} \sigma_i u_i v_i^H,$$

where

$$(1.2.4) \qquad U_1 = (u_1, \ldots, u_r), \qquad V_1 = (v_1, \ldots, v_r).$$

By this a matrix A of rank r is decomposed into a sum of $r = \mathrm{rank}\,(A)$ matrices of rank one.

The singular values of A are unique. The singular vector v_j, $j \leq r$, will be unique only when σ_j^2 is a simple eigenvalue of $A^H A$. For multiple singular values, the corresponding singular vectors can be chosen as any orthonormal basis for the unique subspace that they span. Once the singular vectors v_j, $1 \leq j \leq r$, have been chosen, the vectors u_j, $1 \leq j \leq r$, are uniquely determined from

$$(1.2.5) \qquad A v_j = \sigma_j u_j, \qquad j = 1, \ldots, r.$$

Similarly, given u_j, $1 \leq j \leq r$, the vectors v_j, $1 \leq j \leq r$, are uniquely determined from

$$(1.2.6) \qquad A^H u_j = \sigma_j v_j, \qquad j = 1, \ldots, r.$$

The SVD gives complete information about the four fundamental subspaces associated with A. It is easy to verify that

$$(1.2.7) \quad \mathcal{N}(A) = \mathrm{span}[v_{r+1}, \ldots, v_n], \qquad \mathcal{R}(A) = \mathrm{span}[u_1, \ldots, u_r],$$
$$(1.2.8) \qquad \mathcal{R}(A^H) = \mathrm{span}[v_1, \ldots, v_r], \qquad \mathcal{N}(A^H) = \mathrm{span}[u_{r+1}, \ldots, u_m],$$

and we find the well-known relations

$$\mathcal{N}(A)^\perp = \mathcal{R}(A^H), \quad \mathcal{R}(A)^\perp = \mathcal{N}(A^H).$$

Note that with $V = (V_1, V_2)$ and $z \in \mathbf{C}^{n-r}$ an arbitrary vector,

$$(1.2.9) \qquad x = V_2 z = \sum_{j=r+1}^{n} z_j v_j$$

gives the general solution to the homogeneous linear system $Ax = 0$. This result is often useful in optimization problems.

1.2.2. Related eigenvalue decompositions.

There is a close relationship between the SVD and the Hermitian (or real symmetric) eigenvalue problem from (1.2.1) it follows that

$$(1.2.10) \qquad A^H A = V \Sigma^T \Sigma V^H, \qquad AA^H = U \Sigma \Sigma^T U^H.$$

Here,

$$\Sigma^T \Sigma = \begin{pmatrix} \Sigma_1^2 & 0 \\ 0 & 0 \end{pmatrix} \in \mathbf{R}^{n \times n}, \qquad \Sigma \Sigma^T = \begin{pmatrix} \Sigma_1^2 & 0 \\ 0 & 0 \end{pmatrix} \in \mathbf{R}^{m \times m},$$

and thus $\sigma_1^2, \ldots, \sigma_r^2$, are the nonzero eigenvalues of the Hermitian and positive semidefinite matrices $A^H A$ and AA^H, and v_j and u_j are the corresponding eigenvectors. For a proof of the SVD using this relationship, see Stewart [729, 1973, p. 319].

A matrix $A \in \mathbf{C}^{n \times n}$ is Hermitian if $A^H = A$. A Hermitian matrix A has real eigenvalues $\lambda_1, \ldots, \lambda_n$, and then $A^H A = A^2$ as real nonnegative eigenvalues equal to λ_i^2, $i = 1, \ldots, n$. Hence, (1.2.10) shows that for a Hermitian matrix the singular values are given by $\sigma_i = |\lambda_i|$, $i = 1, \ldots, n$.

In principle, the SVD can be found from the eigenvalue decomposition of the two Hermitian matrices $A^H A$ and AA^H. However, this does not lead to a stable algorithm for computing the SVD.

EXAMPLE 1.2.1. Consider the case $n = 2$,

$$A = (a_1, a_2) \in \mathbf{R}^{m \times 2}, \qquad a_1^T a_2 = \cos \gamma,$$

and $\|a_1\|_2 = \|a_2\|_2 = 1$. Here γ is the angle between the vectors a_1 and a_2. The matrix

$$A^T A = \begin{pmatrix} 1 & \cos \gamma \\ \cos \gamma & 1 \end{pmatrix}$$

has eigenvalues $\lambda_1 = 2\cos^2(\gamma/2)$, $\lambda_2 = 2\sin^2(\gamma/2)$, and so,

$$\sigma_1 = \sqrt{2} \cdot \cos \frac{\gamma}{2}, \qquad \sigma_2 = \sqrt{2} \sin \frac{\gamma}{2}.$$

The eigenvectors of $A^T A$,

$$v_1 = \frac{1}{\sqrt{2}} \begin{pmatrix} 1 \\ 1 \end{pmatrix}, \qquad v_2 = \frac{1}{\sqrt{2}} \begin{pmatrix} -1 \\ 1 \end{pmatrix},$$

are the right singular vectors of A. The left singular vectors can be determined from (1.2.5).

Numerically, if γ is less than the square root of machine precision then $\cos \gamma \approx 1 - \gamma^2/2 = 1$, and $A^T A$ has only one nonzero eigenvalue equal to 2. Thus the smallest singular value of A has been lost! ∎

The following relationship between the SVD and a Hermitian eigenvalue problem, which can easily be verified, was exploited by Lanczos [513, 1961, Chap. 3].

THEOREM 1.2.2. *Let the SVD of $A \in \mathbf{C}^{m \times n}$ be $A = U\Sigma V^H$, where $\Sigma = \text{diag}(\Sigma_1, 0)$,*

$$U = (U_1, U_2), \quad U_1 \in \mathbf{C}^{m \times r}, \quad V = (V_1, V_2), \quad V_1 \in \mathbf{C}^{n \times r}.$$

Then

(1.2.11) $$C = \begin{pmatrix} 0 & A \\ A^H & 0 \end{pmatrix} = P^H \begin{pmatrix} \Sigma_1 & 0 & 0 \\ 0 & -\Sigma_1 & 0 \\ 0 & 0 & 0 \end{pmatrix} P,$$

where P is unitary

(1.2.12) $$P = \frac{1}{\sqrt{2}} \begin{pmatrix} U_1 & U_1 & \sqrt{2}\,U_2 & 0 \\ V_1 & -V_1 & 0 & \sqrt{2}\,V_2 \end{pmatrix}^H.$$

Hence the eigenvalues of C are $\pm\sigma_1, \pm\sigma_2, \ldots, \pm\sigma_r$, and zero repeated $(m+n-2r)$ times, where $r = \text{rank}(A)$.

1.2.3. Matrix approximations. The singular value decomposition plays an important role in a number of matrix approximation problems. In the theorem below we consider the approximation of one matrix by another of lower rank. Several other results can be found in Golub [365, 1968] and in Golub and Van Loan [389, 1989, Chap. 12.4].

THEOREM 1.2.3. *Let $A \in \mathbf{C}^{m \times n}$ have $\text{rank}(A) = r$, and the SVD*

$$A = U\Sigma V^H = \sum_{i=1}^{r} \sigma_i u_i v_i^H.$$

Let $B \in \mathcal{M}_k^{m \times n}$, where $\mathcal{M}_k^{m \times n}$ is the set of matrices in $\mathbf{C}^{m \times n}$ of rank $k < r$. Then

$$\min \|A - X\|_2, \quad X \in \mathcal{M}_k^{m \times n},$$

is obtained for $X = B$, where

$$B = \sum_{i=1}^{k} \sigma_i u_i v_i^H, \qquad \|A - B\|_2 = \sigma_{k+1}.$$

Proof. See Golub and Van Loan [389, 1989, Chap. 2.5.4] and Mirsky [578, 1960]. ∎

As a special case of this theorem it follows that if $\text{rank}(A) = n$, then σ_n is the shortest distance from A to the set of singular matrices in the spectral norm.

REMARK 1.2.1. The theorem was originally proved for the Frobenius norm (see (1.4.7)). For this norm the minimum distance is

$$\|A - B\|_F = (\sigma_{k+1}^2 + \cdots + \sigma_r^2)^{1/2},$$

and the solution is unique; see Eckhart and Young [261, 1936]. A generalization of the Eckhart–Young theorem is given by Golub, Hoffman, and Stewart [369, 1987]. ∎

1.2. THE SINGULAR VALUE DECOMPOSITION

Closely related to the singular value decomposition is the **polar decomposition**.

THEOREM 1.2.4. *Polar Decomposition. Let $A \in \mathbf{C}^{m \times n}$, $m \geq n$. Then there exist a matrix $Q \in \mathbf{C}^{m \times n}$ and a unique Hermitian positive semidefinite matrix $H \in \mathbf{C}^{n \times n}$ such that*

(1.2.13) $$A = QH, \quad Q^H Q = I.$$

If rank $(A) = n$ then H is positive definite and Q is uniquely determined.

Proof. Let A have the singular value decomposition

$$A = U \begin{pmatrix} \Sigma_1 \\ 0 \end{pmatrix} V^H, \quad \Sigma_1 = \operatorname{diag}(\sigma_1, \sigma_2, \ldots, \sigma_n),$$

where U and V are unitary and $\sigma_1 \geq \sigma_2 \geq \cdots \geq \sigma_n \geq 0$. It follows that $A = QH$, where

$$Q = U_1 V^H, \quad H = V \Sigma V^H,$$

and $U_1 = (u_1, \ldots, u_n)$. ∎

The polar decomposition can be regarded as a generalization to matrices of the complex number representation $z = re^{i\theta}$, $r \geq 0$. Since $H^2 = V\Sigma^2 V^H = A^H A$ it follows that H equals the unique Hermitian positive semidefinite square root of $A^H A$,

$$H = (A^H A)^{1/2}.$$

The unitary factor U in the polar decomposition possesses a best approximation property described in the following theorem from Higham [453, 1986].

THEOREM 1.2.5. *Let $A, B \in \mathbf{C}^{m \times n}$ and let $B^H A \in \mathbf{C}^{n \times n}$ have the polar decomposition $B^H A = UH$. Then, for any unitary $Z \in \mathbf{C}^{n \times n}$,*

(1.2.14) $$\|A - BU\|_F \leq \|A - BZ\|_F \leq \|A + BU\|_F,$$

where $\|\cdot\|_F$ denotes the Frobenius norm. In the special case in which $m = n$ and $B = I$ we have

(1.2.15) $$\|A - U\|_F \leq \|A - Z\|_F \leq \|A + U\|_F,$$

and the minimum is

$$\|A - U\|_F = \left(\sum_{i=1}^{n} (\sigma_i - 1)^2 \right)^{1/2},$$

where $\sigma_i = \sigma_i(A)$.

Hence the nearest unitary matrix to $A \in \mathbf{C}^{n \times n}$ is the unitary factor of the polar decomposition. Fan and Hoffman [286, 1955] showed that (1.2.15) holds for any unitarily invariant norm. Higham [453, 1986] also discusses the approximation properties of the Hermitian factor H.

1.2.4. The sensitivity of singular values and vectors. Like the eigenvalues of a real Hermitian matrix, the singular values of a general matrix have a minmax characterization.

THEOREM 1.2.6. *Let $A \in \mathbf{R}^{m \times n}$ have singular values*

$$\sigma_1 \geq \sigma_2 \geq \cdots \geq \sigma_p \geq 0, \qquad p = \min(m, n),$$

and S be a linear subspace of $\mathbf{R}^n$. Then

$$(1.2.16) \qquad \sigma_i = \min_{\dim(S) = n - i + 1} \max_{\substack{x \in S \\ x \neq 0}} \frac{\|Ax\|_2}{\|x\|_2}.$$

Proof. The result is established in almost the same way as for the corresponding eigenvalue theorem, the Courant–Fischer theorem; see Wilkinson [836, 1965, pp. 99–101]. ∎

The minmax characterization of the singular values may be used to establish results on the sensitivity of the singular values of A to perturbations.

THEOREM 1.2.7. *Let A and $\tilde{A} = A + E \in \mathbf{R}^{m \times n}$, $m \geq n$, have singular values $\sigma_1 \geq \sigma_2 \geq \cdots \geq \sigma_n$ and $\tilde{\sigma}_1 \geq \tilde{\sigma}_2 \geq \cdots \geq \tilde{\sigma}_n$, respectively. Then*

$$(1.2.17) \qquad |\sigma_i - \tilde{\sigma}_i| \leq \|E\|_2,$$

$$(1.2.18) \qquad \sum_{i=1}^{n} |\sigma_i - \tilde{\sigma}_i|^2 \leq \|E\|_F^2.$$

Proof. See Stewart [729, 1973, pp. 321–322]. ∎

The result (1.2.18) is known as the Wielandt–Hoffman theorem for singular values. The theorem shows the important fact that the singular values of a matrix A are well-conditioned with respect to perturbations of A. Perturbations of the elements of a matrix produce perturbations of the same, or smaller, magnitude in the singular values. This is of great importance for the use of the SVD to determine the "numerical rank" of a matrix; see Section 2.7.1.

The next result gives a perturbation result for singular vectors.

THEOREM 1.2.8. *Let $A \in \mathbf{R}^{m \times n}$, $m \geq n$, have singular values $\sigma_1 \geq \sigma_2 \geq \cdots \geq \sigma_n$ and singular vectors u_i, v_i, $i = 1, \ldots, n$. Let $\tilde{\sigma}_i$, $\tilde{u}_i$, and $\tilde{v}_i$ be the corresponding values for $\tilde{A} = A + E$. Then if $\|E\|_2 < \gamma_i$ it holds*

$$(1.2.19) \qquad \max\left(\sin\theta(u_i, \tilde{u}_i),\ \sin\theta(v_i, \tilde{v}_i)\right) \leq \frac{\|E\|_2}{\gamma_i - \|E\|_2},$$

*where γ_i is the **absolute gap** between σ_i and the other singular values,*

$$(1.2.20) \qquad \gamma_i = \min_{j \neq i} |\sigma_i - \sigma_j|.$$

Proof. A more general result is given in Golub and Van Loan [389, 1989, Thm. 8.3.5]. ∎

Sharper perturbation results can be given for singular values and vectors of bidiagonal matrices; see Theorem 2.6.2.

It is well known that the eigenvalues of the leading principal minor of order $(n-1)$ of a Hermitian matrix $A \in \mathbf{R}^{n \times n}$ interlace the eigenvalues of A; see Wilkinson [836, 1965, p. 103]. A similar theorem holds for singular values.

1.2. THE SINGULAR VALUE DECOMPOSITION

THEOREM 1.2.9. *Let A be bordered by a column $u \in \mathbf{R}^m$,*

$$\hat{A} = (A, u) \in \mathbf{R}^{m \times n}, \quad m \geq n.$$

Then the ordered singular values σ_i of A separate the ordered singular values $\hat{\sigma}_i$ of $\hat{A}$ as follows:

$$\hat{\sigma}_1 \geq \sigma_1 \geq \hat{\sigma}_2 \geq \sigma_2 \geq \cdots \geq \hat{\sigma}_{n-1} \geq \sigma_{n-1} \geq \hat{\sigma}_n.$$

Similarly, if A is bordered by a row $v \in \mathbf{R}^n$,

$$\hat{A} = \begin{pmatrix} A \\ v^H \end{pmatrix} \in \mathbf{R}^{m \times n}, \quad m \geq n,$$

$$\hat{\sigma}_1 \geq \sigma_1 \geq \hat{\sigma}_2 \geq \sigma_2 \geq \cdots \geq \hat{\sigma}_{n-1} \geq \sigma_{n-1} \geq \hat{\sigma}_n \geq \sigma_n.$$

Proof. The theorem is a consequence of the minmax characterization of the singular values in Theorem 1.2.6; cf. Lawson and Hanson [520, 1974, p. 26]. ∎

1.2.5. The SVD and pseudoinverse. The SVD is a powerful tool for solving the linear least squares problem. This is because the unitary matrices that transform A to diagonal form (1.2.1) do not change the l_2-norm of vectors. We have the following fundamental result, which applies to both overdetermined and underdetermined linear systems.

THEOREM 1.2.10. *Consider the general linear least squares problem*

$$(1.2.21) \qquad \min_{x \in S} \|x\|_2, \quad S = \{x \in \mathbf{R}^n \mid \|b - Ax\|_2 = \min\},$$

where $A \in \mathbf{C}^{m \times n}$ and $\mathrm{rank}\,(A) = r \leq \min(m,n)$. This problem always has a unique solution, which can be written in terms of the SVD of A as

$$(1.2.22) \qquad x = V \begin{pmatrix} \Sigma_r^{-1} & 0 \\ 0 & 0 \end{pmatrix} U^H b.$$

Proof. Let

$$z = V^H x = \begin{pmatrix} z_1 \\ z_2 \end{pmatrix}, \quad c = U^H b = \begin{pmatrix} c_1 \\ c_2 \end{pmatrix},$$

where $z_1, c_1 \in \mathbf{C}^r$. Then

$$\begin{aligned} \|b - Ax\|_2 &= \|U^H(b - AVV^H x)\|_2 \\ &= \left\| \begin{pmatrix} c_1 \\ c_2 \end{pmatrix} - \begin{pmatrix} \Sigma_1 & 0 \\ 0 & 0 \end{pmatrix} \begin{pmatrix} z_1 \\ z_2 \end{pmatrix} \right\|_2 = \left\| \begin{pmatrix} c_1 - \Sigma_1 z_1 \\ c_2 \end{pmatrix} \right\|_2. \end{aligned}$$

Thus, the residual norm will be minimized for z_2 arbitrary and $z_1 = \Sigma_r^{-1} c_1$. The choice $z_2 = 0$ minimizes $\|z\|_2$, and therefore $\|x\|_2 = \|Vz\|_2$ as well. ∎

DEFINITION 1.2.1. *We write (1.2.22) as $x = A^\dagger b$, where*

$$(1.2.23) \qquad A^\dagger = V \begin{pmatrix} \Sigma_r^{-1} & 0 \\ 0 & 0 \end{pmatrix} U^H \in \mathbf{C}^{n \times m}$$

*is called the **pseudoinverse** of A, and the solution (1.2.22) is called the pseudoinverse solution.*

It follows easily from Theorem 1.2.10 that $A^\dagger$ minimizes $\|AX - I\|_F$. For computing the pseudoinverse solution it suffices to compute Σ, V, and the vector $c = U^H b$. The pseudoinverse of a scalar is

$$(1.2.24) \qquad \sigma^\dagger = \begin{cases} 1/\sigma, & \text{if } \sigma \neq 0, \\ 0, & \text{if } \sigma = 0. \end{cases}$$

This shows the important fact that the pseudoinverse $A^\dagger$ is not a continuous function of A, unless we allow only perturbations which do not change the rank of A. The pseudoinverse can also be uniquely characterized by the two geometrical conditions

$$(1.2.25) \qquad A^\dagger b \perp \mathcal{N}(A), \quad (I - AA^\dagger)b \perp \mathcal{R}(A) \quad \forall b \in \mathbf{R}^m.$$

The matrix $A^\dagger$ is often called the **Moore–Penrose** inverse. E. H. Moore introduced the general reciprocal in 1920. It was rediscovered by Bjerhammar [83, 1951] and Penrose [655, 1955], who gave the following elegant algebraic characterization.

THEOREM 1.2.11. *Penrose's conditions. The pseudoinverse $X = A^\dagger$ is uniquely determined by the following four conditions.*

$$(1.2.26) \qquad \begin{array}{ll} (1)\ AXA = A, & (2)\ XAX = X, \\ (3)\ (AX)^H = AX, & (4)\ (XA)^H = XA. \end{array}$$

It follows in particular that $A^\dagger$ in (1.2.23) does not depend on the particular choice of U and V in the SVD. It can be directly verified that $A^\dagger$ given by (1.2.23) satisfies these four conditions. If only part of the Penrose conditions hold, the corresponding matrix X is called a generalized inverse. Such inverses have been extensively analyzed; see Nashed [596, 1976].

The pseudoinverse can be shown to have the following properties.

THEOREM 1.2.12.
1. $(A^\dagger)^\dagger = A$;
2. $(A^\dagger)^H = (A^H)^\dagger$;
3. $(\alpha A)^\dagger = \alpha^\dagger A^\dagger$;
4. $(A^H A)^\dagger = A^\dagger (A^\dagger)^H$;
5. *if U and V are unitary* $(UAV^H)^\dagger = VA^\dagger U^H$;
6. *if $A = \sum_i A_i$, where $A_i A_j^H = 0$, $A_i^H A_j = 0$, $i \neq j$, then $A^\dagger = \sum_i A_i^\dagger$*;
7. *if A is normal $(AA^H = A^H A)$ then $A^\dagger A = AA^\dagger$ and $(A^n)^\dagger = (A^\dagger)^n$*;
8. A, A^H, $A^\dagger$, *and* $A^\dagger A$ *all have rank equal to* $\text{trace}(A^\dagger A)$.

Proof. The statements easily follow from (1.2.23). See also Penrose [655, 1955]. ∎

The pseudoinverse does not share some other properties of the ordinary inverse. For example, in general

$$(AB)^\dagger \neq B^\dagger A^\dagger \quad \text{and} \quad AA^\dagger \neq A^\dagger A.$$

1.2. THE SINGULAR VALUE DECOMPOSITION

EXAMPLE 1.2.2. If we take $A = \begin{pmatrix} 1 & 0 \end{pmatrix}$ and $B = \begin{pmatrix} 1 & 1 \end{pmatrix}^T$, then $AB = 1$,

$$1 = (AB)^\dagger \neq B^\dagger A^\dagger = \frac{1}{2}\begin{pmatrix} 1 & 1 \end{pmatrix}\begin{pmatrix} 1 \\ 0 \end{pmatrix} = \frac{1}{2},$$

and

$$AA^\dagger = \begin{pmatrix} 1 & 0 \end{pmatrix}\begin{pmatrix} 1 \\ 0 \end{pmatrix} = 1, \quad A^\dagger A = \begin{pmatrix} 1 \\ 0 \end{pmatrix}\begin{pmatrix} 1 & 0 \end{pmatrix} = \begin{pmatrix} 1 & 0 \\ 0 & 0 \end{pmatrix}. \quad \blacksquare$$

In the special case in which $A \in \mathbf{C}^{m \times n}$, rank$(A) = n$,

$$(1.2.27) \qquad A^\dagger = (A^H A)^{-1} A^H, \qquad (A^H)^\dagger = A(A^H A)^{-1}.$$

The first expression follows from Theorems 1.1.2 and 1.1.3 and relates to the least squares solution in the case of full column rank. The second expression follows using property 2 in Theorem 1.2.12 and relates to the minimum norm solution to an underdetermined system of full row rank.

Necessary and sufficient conditions for the relation $(AB)^\dagger = B^\dagger A^\dagger$ to hold have been given by Greville [399, 1966]. The following theorem gives useful *sufficient* conditions.

THEOREM 1.2.13. *Assume that $A \in \mathbf{C}^{m \times r}$, $B \in \mathbf{C}^{r \times n}$, where* rank$(A) =$ rank$(B) = r$. *Then it holds that*

$$(1.2.28) \qquad (AB)^\dagger = B^\dagger A^\dagger = B^H (BB^H)^{-1} (A^H A)^{-1} A^H.$$

Proof. The last equality follows from (1.2.27). The first equality is verified by showing that the four Penrose conditions are satisfied. $\blacksquare$

1.2.6. Orthogonal projectors and angles between subspaces. An important property of the pseudoinverse is that it gives simple expressions for the orthogonal projections onto the four fundamental subspaces of A:

$$(1.2.29) \qquad \begin{aligned} P_{\mathcal{R}(A)} &= AA^\dagger, & P_{\mathcal{N}(A^H)} &= I - AA^\dagger, \\ P_{\mathcal{R}(A^H)} &= A^\dagger A, & P_{\mathcal{N}(A)} &= I - A^\dagger A. \end{aligned}$$

These expressions are easily verified using the Penrose conditions (1.2.26).

If the columns of a matrix U are orthonormal then $U^H U = I$, and $P_{\mathcal{R}(U)} = UU^H$ satisfies (1.1.17). Using (1.2.10) we can therefore express the projections (1.2.29) in terms of the singular vectors of A as

$$(1.2.30) \qquad \begin{aligned} P_{\mathcal{R}(A)} &= U_1 U_1^H, & P_{\mathcal{N}(A^H)} &= U_2 U_2^H, \\ P_{\mathcal{R}(A^T)} &= V_1 V_1^H, & P_{\mathcal{N}(A)} &= V_2 V_2^H, \end{aligned}$$

where $U_1 = (u_1, \ldots, u_r)$ and $V_1 = (v_1, \ldots, v_r)$.

DEFINITION 1.2.2. *Let $S_A = \mathcal{R}(A)$ and $S_B = \mathcal{R}(B)$ be two subspaces of $\mathbf{C}^m$ where without restriction we assume that $p = \dim(S_A) \geq \dim(S_B) = q \geq 1$. The **principal angles** θ_k, between S_A and S_B and the corresponding **principal vectors** u_k, v_k, $k = 1, \ldots, q$, are recursively defined by*

$$(1.2.31) \qquad \cos\theta_k = \max_{u \in S_A} \max_{v \in S_B} u^H v, \quad \|u\|_2 = \|v\|_2 = 1,$$

subject to the constraints

$$(1.2.32) \qquad u \perp u_j, \quad v \perp v_j, \quad j = 1, \ldots, k-1.$$

Note that for $k = 1$, the constraints are empty, and θ_1 is the smallest principal angle between S_A and S_B. The principal vectors need not be uniquely defined, but the principal angles always are. Principal angles and vectors have important applications, e.g., in statistics.

If $p = q$ the subspaces have the same dimension. In this case the distance between the subspaces S_A and S_B is defined to be

$$(1.2.33) \qquad \operatorname{dist}(S_A, S_B) = |\sin(\theta_p)| = (1 - \sigma_p^2)^{1/2},$$

where θ_p is the *largest* principal angle.

The relationship between principal angles and the SVD is given in the following theorem.

THEOREM 1.2.14. *Assume that $Q_A \in \mathbf{R}^{m \times p}$ and $Q_B \in \mathbf{R}^{m \times q}$ form unitary bases for the two subspaces S_A and S_B. Consider the SVD*

$$(1.2.34) \qquad M = Q_A^H Q_B = YCZ^H, \quad C = \operatorname{diag}(\sigma_1, \ldots, \sigma_q),$$

where $\sigma_1 \geq \sigma_2 \geq \cdots \geq \sigma_q$, $Y^H Y = Z^H Z = I_q$. Then the principal angles and principal vectors are given by

$$(1.2.35) \qquad \cos\theta_k = \sigma_k, \quad U = Q_A Y, \quad V = Q_B Z.$$

Proof. The proof follows from the minmax characterization of the singular values and vectors; see Theorem 1.2.6.

It can also be shown that the nonzero singular values of $(P_{S_A} - P_{S_B})$, where $P_{S_A} = Q_A Q_A^H$ and $P_{S_B} = Q_B Q_B^H$ are the orthogonal projectors, equal $\sin(\theta_k)$, $k = 1, \ldots, q$. This gives the alternative definition

$$\operatorname{dist}(S_A, S_B) = \|P_{S_A} - P_{S_B}\|_2.$$

Methods for computing principal angles and vectors, and applications are discussed in Björck and Golub [111, 1973] and Golub and Zha [394, 1994].

1.3. The QR Decomposition

1.3.1. The full rank case. The SVD of A gives the solution of the general rank deficient least squares problem (1.2.21). However, in many applications it is too expensive to compute the SVD, and one has to use simpler decompositions. Among these the most important are the QR and related decompositions.

Let $A \in \mathbf{R}^{m \times n}$ and $b \in \mathbf{R}^m$, and let $Q \in \mathbf{R}^{m \times m}$ be an orthogonal matrix. Since orthogonal transformations preserve the Euclidean length it follows that the linear least squares problem

$$\min_x \|Q^T(Ax - b)\|_2 \tag{1.3.1}$$

is equivalent to (1.1.1). We now show how to choose Q so that the problem (1.3.1) becomes simple to solve.

THEOREM 1.3.1. QR Decomposition. *Let $A \in R^{m \times n}, m \geq n$. Then there is an orthogonal matrix $Q \in R^{m \times m}$ such that*

$$A = Q \begin{pmatrix} R \\ 0 \end{pmatrix}, \tag{1.3.2}$$

where R is upper triangular with nonnegative diagonal elements. The decomposition (1.3.2) is called the QR decomposition of A, and the matrix R will be called the R-factor of A.

Proof. The proof is by induction on n. Let A be partitioned in the form $A = (a_1, A_2)$, $a_1 \in R^m$, and put $\rho = \|a_1\|_2$. Let $U = (y, U_1)$ be an orthogonal matrix with $y = a_1/\rho$ if $a_1 \neq 0$, and $y = e_1$ otherwise. Since $U_1^T y = 0$ it follows that

$$U^T A = \begin{pmatrix} \rho & r^T \\ 0 & B \end{pmatrix}, \qquad B = U_1^T A_2 \in R^{(m-1) \times (n-1)},$$

where $\rho = \|a_1\|_2$, $r = A_2^T y$.

For $n = 1$, A_2 is empty and the theorem holds with $Q = U$ and $R = \rho$, a scalar. Assume now that the induction hypothesis holds for $n - 1$. Then there is an orthogonal matrix $\bar{Q}$ such that $\bar{Q}^T B = \begin{pmatrix} \bar{R} \\ 0 \end{pmatrix}$, and (1.3.2) will hold if we define

$$Q = U \begin{pmatrix} 1 & 0 \\ 0 & \bar{Q} \end{pmatrix}, \qquad R = \begin{pmatrix} \rho & r^T \\ 0 & \bar{R} \end{pmatrix}. \qquad \blacksquare$$

The proof of Theorem 1.3.1 gives a way to compute Q and R, provided we can construct an orthogonal matrix $U = (y, U_1)$ given its first column. Several ways to perform this construction using elementary orthogonal transformations are given in Section 2.2.1. The systematic use of orthogonal transformations to reduce matrices to simpler form was initiated by Givens [361, 1958] and Householder [475, 1958]. The application to linear least squares problems is due to Golub [364, 1965], although Householder [475] discussed least squares.

Note that from the form of the decomposition (1.3.2) it follows immediately that R has the same singular values and right singular vectors as A. A relationship between the Cholesky factorization of $A^T A$ and the QR decomposition of A is given next.

THEOREM 1.3.2. *Let $A \in \mathbf{R}^{m \times n}$ have rank n. Then if the R-factor in the QR decomposition of A has positive diagonal elements it equals the Cholesky factor of $A^T A$.*

Proof. If rank $(A) = n$, then by Theorem 1.1.4 the Cholesky factor of $A^T A$ is unique. Now from (1.3.2) it follows that

$$A^T A = (R^T\ 0) Q^T Q \begin{pmatrix} R \\ 0 \end{pmatrix} = R^T R,$$

which concludes the proof. ∎

Assume that rank $(A) = n$, and partition Q in the form

(1.3.3) $\qquad Q = (Q_1, Q_2), \quad Q_1 \in \mathbf{R}^{m \times n}, \quad Q_2 \in \mathbf{R}^{m \times (m-n)}.$

Then by (1.3.2) and nonsingularity of R we have

(1.3.4) $\qquad A = (Q_1, Q_2) \begin{pmatrix} R \\ 0 \end{pmatrix} = Q_1 R, \quad Q_1 = A R^{-1}.$

Hence we can express Q_1 uniquely in terms of A and R. However the matrix Q_2 will not, in general, be uniquely determined.

From (1.3.4) it follows that

(1.3.5) $\qquad \mathcal{R}(A) = \mathcal{R}(Q_1), \quad \mathcal{R}(A)^\perp = \mathcal{R}(Q_2),$

which shows that the columns of Q_1 and Q_2 form orthonormal bases for $\mathcal{R}(A)$ and its complement. It follows that the corresponding orthogonal projections are

(1.3.6) $\qquad P_{\mathcal{R}(A)} = Q_1 Q_1^T, \quad P_{\mathcal{R}(A)^\perp} = Q_2 Q_2^T.$

We now show how to use the QR decomposition (1.3.2) to solve the augmented system (1.1.24). As shown in Theorem 1.1.5, this includes as special cases both the solution of the linear least squares problem ($b = 0$) and the minimum norm solution of an underdetermined system ($c = 0$).

THEOREM 1.3.3. *Let $A \in R^{m \times n}, m \geq n, b \in R^m$, and $c \in R^n$ be given. Assume that rank $(A) = n$, and let the QR decomposition of A be given by (1.3.2). Then the solution to the augmented system*

$$\begin{pmatrix} I & A \\ A^T & 0 \end{pmatrix} \begin{pmatrix} y \\ x \end{pmatrix} = \begin{pmatrix} b \\ c \end{pmatrix}$$

can be computed from

(1.3.7) $\qquad z = R^{-T} c, \quad \begin{pmatrix} d_1 \\ d_2 \end{pmatrix} = Q^T b,$

(1.3.8) $\qquad x = R^{-1}(d_1 - z), \quad y = Q \begin{pmatrix} z \\ d_2 \end{pmatrix}.$

Proof. The augmented system can be written $y + Ax = b$, $A^T y = c$, and using the factorization (1.3.2),

1.3. THE QR DECOMPOSITION

$$y + Q \begin{pmatrix} R \\ 0 \end{pmatrix} x = b, \quad (R^T \ \ 0) Q^T y = c.$$

Multiplying the first equation with Q^T and the second with R^{-T} we get

$$Q^T y + \begin{pmatrix} R \\ 0 \end{pmatrix} x = Q^T b, \quad (I_n \ \ 0) Q^T y = R^{-T} c.$$

Using the second equation to eliminate the first n components of $Q^T y$ in the first equation, we can solve for x. The last $m - n$ components of $Q^T y$ are obtained from the last $m - n$ equations in the first block. ∎

Taking $c = 0$ and $r = y = b - Ax$ in (1.3.7)–(1.3.8) it follows that the solution to the least squares problem $\min_x \|Ax - b\|_2$ is obtained from

$$(1.3.9) \qquad \begin{pmatrix} d_1 \\ d_2 \end{pmatrix} = Q^T b, \quad Rx = d_1, \quad r = Q \begin{pmatrix} 0 \\ d_2 \end{pmatrix}.$$

In particular, $\|r\|_2 = \|d_2\|_2$. Taking $b = 0$, we find that the solution to the problem $\min \|y\|_2$ such that $A^T y = c$ is obtained from

$$(1.3.10) \qquad R^T z = c, \quad y = Q \begin{pmatrix} z \\ 0 \end{pmatrix}.$$

It follows that when A has full rank n the pseudoinverses of A and A^T are given by the expressions

$$(1.3.11) \qquad A^\dagger = R^{-1} Q_1^T, \quad (A^T)^\dagger = Q_1 R^{-T}.$$

1.3.2. Rank revealing QR decompositions. According to Theorem 1.3.1 any matrix $A \in R^{m \times n}$ has a QR decomposition. However, as illustrated in the following example, if $\operatorname{rank}(A) < n$, then the decomposition is not unique.

EXAMPLE 1.3.1. For any c and s such that $c^2 + s^2 = 1$ we have

$$A = \begin{pmatrix} 0 & 0 \\ 0 & 1 \end{pmatrix} = \begin{pmatrix} c & -s \\ s & c \end{pmatrix} \begin{pmatrix} 0 & s \\ 0 & c \end{pmatrix} = QR.$$

Here $\operatorname{rank}(A) = 1 < 2 = n$. Note that the columns of Q no longer provide orthogonal bases for $R(A)$ and its complement. ∎

We now show how the QR decomposition can be modified for the case when $\operatorname{rank}(A) < n$. (Note that this includes the case when $m < n$.)

THEOREM 1.3.4. *Given $A \in R^{m \times n}$ with $\operatorname{rank}(A) = r$ there is a permutation matrix Π and an orthogonal matrix $Q \in R^{m \times m}$ such that*

$$(1.3.12) \qquad A\Pi = Q \begin{pmatrix} R_{11} & R_{12} \\ 0 & 0 \end{pmatrix} \begin{matrix} \}r \\ \}m-r \end{matrix},$$

where $R_{11} \in R^{r \times r}$ is upper triangular with positive diagonal elements.

Proof. Since $\text{rank}(A) = r$, we can always choose a permutation matrix Π such that $A\Pi = (A_1, A_2)$, where $A_1 \in R^{m \times r}$ has linearly independent columns. Let
$$Q^T A_1 = \begin{pmatrix} R_{11} \\ 0 \end{pmatrix}, \qquad Q = (Q_1, Q_2),$$
be the QR decomposition of A_1, where $Q_1 \in \mathbf{R}^{m \times r}$. By Theorem 1.3.2, Q_1 and R_{11} are uniquely determined, and R_{11} has positive diagonal elements. Put
$$Q^T A\Pi = (Q^T A_1, Q^T A_2) = \begin{pmatrix} R_{11} & R_{12} \\ 0 & R_{22} \end{pmatrix}.$$
From $\text{rank}(Q^T A\Pi) = \text{rank}(A) = r$ it follows that $R_{22} = 0$, since otherwise $Q^T A\Pi$ would have more than r linearly independent rows. Hence the decomposition must have the form (1.3.12). ∎

The decomposition (1.3.12) is not in general unique. Several strategies for determining a suitable column permutation Π are described in Section 2.7. When Π has been chosen, Q_1, R_{11}, and R_{12} are uniquely determined.

Also, when A has full column rank, but is close to a rank deficient matrix, QR decompositions with column permutations are of interest.

THEOREM 1.3.5. (See H. P. Hong and C. T. Pan [473, 1992].) *Let $A \in \mathbf{R}^{m \times n}$, $(m \geq n)$, and r be any integer $0 < r < n$. Then there exists a permutation matrix Π such that the QR factorization has the form*

$$(1.3.13) \qquad A\Pi = Q \begin{pmatrix} R_{11} & R_{12} \\ 0 & R_{22} \end{pmatrix},$$

with $R_{11} \in \mathbf{R}^{r \times r}$ upper triangular, $c = \sqrt{r(n-r) + \min(r, n-r)}$, and

$$(1.3.14) \qquad \sigma_r(R_{11}) \geq \frac{1}{c}\sigma_r(A), \qquad \|R_{22}\|_2 \leq c\sigma_{r+1}(A).$$

A QR decomposition of the form (1.3.13)–(1.3.14) is called a **rank revealing QR (RRQR) decomposition**. If $\sigma_{r+1} = 0$ we recover the decomposition (1.3.12). Although the existence of a column permutation so that the corresponding QR decomposition satisfies (1.3.14) has been proved, it is still an open question if an algorithm of polynomial complexity exists for finding such a permutation; see Section 2.7.5. (Note that an exhaustive search for Π has combinatorial complexity!)

REMARK 1.3.1. From (1.3.13) it follows that
$$A\Pi \begin{pmatrix} R_{11}^{-1} R_{12} \\ -I \end{pmatrix} = Q \begin{pmatrix} R_{11} & R_{12} \\ 0 & R_{22} \end{pmatrix} \begin{pmatrix} R_{11}^{-1} R_{12} \\ -I \end{pmatrix} = Q \begin{pmatrix} 0 \\ R_{22} \end{pmatrix}.$$

Hence if $R_{22} = 0$, a dimensional argument shows that the nullspace of $A\Pi$ is given by

$$(1.3.15) \qquad \mathcal{N}(A\Pi) = \mathcal{R}\begin{pmatrix} R_{11}^{-1} R_{12} \\ -I_{n-r} \end{pmatrix}.$$

1.3.3. The complete orthogonal decomposition.

For some applications it will be useful to carry the reduction in (1.3.12) one step further, using orthogonal transformations from the right as well. By performing a QR decomposition of the transpose of the triangular factor, the off-diagonal block can be eliminated:

$$\begin{pmatrix} R_{11}^T & 0 \\ R_{12}^T & 0 \end{pmatrix} = \hat{Q} \begin{pmatrix} \hat{R}_{11} & 0 \\ 0 & 0 \end{pmatrix}. \tag{1.3.16}$$

We then obtain a decomposition of the following form.

DEFINITION 1.3.1. *A* **complete orthogonal decomposition** *of $A \in R^{m \times n}$ with* $\operatorname{rank}(A) = r$ *is a decomposition of the form*

$$A = Q \begin{pmatrix} T & 0 \\ 0 & 0 \end{pmatrix} V^T, \tag{1.3.17}$$

where $Q \in R^{m \times m}$ and $V \in R^{n \times n}$ are orthogonal matrices and $T \in R^{r \times r}$ is upper or lower triangular with positive diagonal elements.

Obviously, a decomposition of the form (1.3.17) is not unique. For example, the SVD of A is one example of a complete orthogonal decomposition. The form closest to the SVD that can be achieved by a finite algorithm is the **bidiagonal decomposition**

$$A = Q \begin{pmatrix} B & 0 \\ 0 & 0 \end{pmatrix} V^T, \tag{1.3.18}$$

where B is a bidiagonal matrix with nonnegative diagonal elements; see Section 2.6.2.

From Theorem 1.2.12 it follows that the pseudoinverse of A can be expressed in terms of the decomposition (1.3.17) as

$$A^\dagger = V \begin{pmatrix} T^{-1} & 0 \\ 0 & 0 \end{pmatrix} Q^T. \tag{1.3.19}$$

Further, partitioning the orthogonal matrices in (1.3.17) by rows we have

$$A = (Q_1, Q_2) \begin{pmatrix} T & 0 \\ 0 & 0 \end{pmatrix} \begin{pmatrix} V_1^T \\ V_2^T \end{pmatrix}.$$

It follows that the complete orthogonal decomposition, like the SVD, provides orthogonal bases for the fundamental subspaces of A.

For many computational purposes the complete QR decomposition (1.3.17) is as useful as the SVD. An important advantage over the SVD is that the complete QR decomposition can be updated much more efficiently than the SVD when A is subject to a change of low rank; see Section 3.5. Different methods for computing the various QR decompositions are described in Sections 2.4 and 2.7.

1.4. Sensitivity of Least Squares Solutions

In this section we give results on the sensitivity of pseudoinverses and least squares solutions to perturbations in A and b. Many of the results below were first given by Wedin [824, 1973]. Stewart in [731, 1977] gives a unified treatment with interesting historical comments on the perturbation theory for pseudoinverses and least squares solutions. A more recent and excellent source of information is Stewart and Sun [754, 1990, Chap. 3].

1.4.1. Vector and matrix norms.
In perturbation and error analyses it is useful to have a measure of the size of a vector or a matrix. Such measures are provided by vector and matrix norms, which can be regarded as generalizations of the absolute value.

A vector norm is a function $\|\cdot\| : \mathbf{C}^n \to \mathbf{R}$ that satisfies the following three conditions:

1. $\|x\| > 0 \quad \forall x \in \mathbf{C}^n, \quad x \neq 0$ (definiteness);
2. $\|\alpha x\| = |\alpha| \|x\| \quad \forall \alpha \in \mathbf{C}, \quad x \in \mathbf{C}^n$ (homogeneity);
3. $\|x + y\| \leq \|x\| + \|y\| \quad \forall x, y \in \mathbf{C}^n$ (triangle inequality).

The most common vector norms are the Hölder p-norms

$$(1.4.1) \qquad \|x\|_p = (|x_1|^p + |x_2|^p + \cdots + |x_n|^p)^{1/p}, \quad 1 \leq p < \infty.$$

The l_p-norms have the property that $\|x\|_p = \| |x| \|_p$. Vector norms with this property are said to be **absolute**. The three most important particular cases are $p = 1, 2$ and the limit when $p \to \infty$:

$$(1.4.2) \qquad \begin{aligned} \|x\|_1 &= |x_1| + \cdots + |x_n|, \\ \|x\|_2 &= (|x_1|^2 + \cdots + |x_n|^2)^{1/2} = (x^H x)^{1/2}, \\ \|x\|_\infty &= \max_{1 \leq i \leq n} |x_i|. \end{aligned}$$

The vector 2-norm is the Euclidean length of the vector, and is invariant under unitary transformations, i.e.,

$$\|Qx\|_2^2 = x^H Q^H Q x = x^H x = \|x\|_2^2$$

if Q is unitary. Another important property is the Hölder inequality

$$|x^H y| \leq \|x\|_p \|y\|_q, \quad \frac{1}{p} + \frac{1}{q} = 1.$$

The special case with $p = q = 2$ is called the Cauchy–Schwarz inequality.

A matrix norm is a function $\|\cdot\| : \mathbf{C}^{m \times n} \to \mathbf{R}$ that satisfies analogues of the three vector norm properties. A matrix norm can be constructed from any vector norm by defining

$$(1.4.3) \qquad \|A\| = \max_{x \neq 0} \frac{\|Ax\|}{\|x\|} = \max_{\|x\|=1} \|Ax\|.$$

1.4. SENSITIVITY OF LEAST SQUARES SOLUTIONS

This norm is called the matrix norm **subordinate** to the vector norm. From the definition it follows directly that

$$\|Ax\| \leq \|A\|\,\|x\|, \qquad x \in \mathbf{C}^n.$$

It is an easy exercise to show that subordinate matrix norms are submultiplicative, i.e., whenever the product AB is defined it satisfies the condition $\|AB\| \leq \|A\|\|B\|$.

The matrix norms subordinate to the vector p-norms are especially important. For these it holds that $\|I_n\|_p = 1$. Formulas for $\|A\|_p$ are known only for $p = 1, 2, \infty$. It can be shown that

$$(1.4.4) \qquad \|A\|_1 = \max_{1 \leq j \leq n} \sum_{i=1}^{m} |a_{ij}|, \qquad \|A\|_\infty = \max_{1 \leq i \leq m} \sum_{j=1}^{n} |a_{ij}|,$$

respectively. Hence these norms are easily computable, and it holds that $\|A\|_1 = \|A^H\|_\infty$. The 2-norm, also called the **spectral norm**, is given by

$$(1.4.5) \qquad \|A\|_2 = \max_{\|x\|_2 = 1} \|Ax\|_2 = \sigma_1(A).$$

Since the nonzero singular values of A and A^H are the same it follows that $\|A\|_2 = \|A^H\|_2$. The spectral norm is expensive to compute, but a useful upper bound is

$$(1.4.6) \qquad \|A\|_2 \leq (\|A\|_1 \|A\|_\infty)^{1/2}.$$

It is well known that on a finite-dimensional space two norms differ by at most a positive constant, which only depends on the dimension. For the vector p-norms it holds that

$$\|x\|_{p_2} \leq \|x\|_{p_1} \leq n^{\left(\frac{1}{p_1} - \frac{1}{p_2}\right)} \|x\|_{p_2}, \qquad p_1 \leq p_2.$$

Another way to proceed in defining norms for matrices is to regard $\mathbf{C}^{m \times n}$ as an mn-dimensional vector space and apply a vector norm over that space. With the exception of the **Frobenius norm** derived from the vector 2-norm,

$$(1.4.7) \qquad \|A\|_F = \left(\sum_{i=1}^{m} \sum_{j=1}^{n} |a_{ij}|^2 \right)^{1/2},$$

such norms are not much used. Note that $\|A^H\|_F = \|A\|_F$. Useful alternative characterizations of the Frobenius norm are

$$(1.4.8) \qquad \|A\|_F^2 = \operatorname{trace}(A^H A) = \sum_{i=1}^{k} \sigma_i^2(A), \qquad k = \min(m, n).$$

The Frobenius norm is submultiplicative, but is often larger than necessary, e.g., $\|I\|_F = n^{1/2}$. This tends to make bounds derived in terms of the Frobenius norm not as sharp as they might be. From (1.4.8) we also get lower and upper bounds for the matrix 2-norm

$$\frac{1}{\sqrt{n}}\|A\|_F \leq \|A\|_2 \leq \|A\|_F.$$

An important property of the Frobenius norm and the 2-norm is that they are invariant with respect to orthogonal transformations, i.e., for all unitary matrices Q and P ($Q^H Q = I$ and $P^H P = I$) of appropriate dimensions we have $\|QAP^H\| = \|A\|$. We finally remark that the 1-,∞-, and Frobenius norms satisfy

$$\| |A| \| = \|A\|, \qquad |A| = (|a_{ij}|),$$

but for the 2-norm the best result is that $\| |A| \|_2 \leq n^{1/2}\|A\|_2$.

1.4.2. Perturbation analysis of pseudoinverses. We first give some perturbation bounds for the pseudoinverse. We consider a matrix $A \in \mathbf{R}^{m \times n}$ and let $B = A + E$ be the perturbed matrix. The theory is complicated by the fact that $A^\dagger$ varies discontinuously when the rank of A changes; cf. (1.2.24).

THEOREM 1.4.1. *If* rank $(A + E) \neq$ rank (A) *then*

$$\|(A+E)^\dagger - A^\dagger\|_2 \geq 1/\|E\|_2.$$

Proof. See Wedin [824, 1973]. ∎

EXAMPLE 1.4.1. By Theorem 1.4.1, when the rank changes the perturbation in $A^\dagger$ may be unbounded when $\|E\|_2 \to 0$. A trivial example of this is obtained by taking

$$A = \begin{pmatrix} \sigma & 0 \\ 0 & 0 \end{pmatrix}, \quad E = \begin{pmatrix} 0 & 0 \\ 0 & \epsilon \end{pmatrix},$$

where $\sigma > 0$, $\epsilon \neq 0$. Then $1 = \text{rank}(A) \neq \text{rank}(A+E) = 2$,

$$A^\dagger = \begin{pmatrix} \sigma^{-1} & 0 \\ 0 & 0 \end{pmatrix}, \quad (A+E)^\dagger = \begin{pmatrix} \sigma^{-1} & 0 \\ 0 & \epsilon^{-1} \end{pmatrix},$$

and $\|(A+E)^\dagger - A^\dagger\|_2 = |\epsilon|^{-1} = 1/\|E\|_2$. ∎

In case the perturbation E does not change the rank of A, such unbounded growth of $(A + E)^\dagger$ cannot occur.

THEOREM 1.4.2. *If* rank $(A + E) =$ rank $(A) = r$, *and* $\eta = \|A^\dagger\|_2 \|E\|_2 < 1$, *then*

(1.4.9) $$\|(A+E)^\dagger\|_2 \leq \frac{1}{1-\eta}\|A^\dagger\|_2.$$

Proof. From the assumption and Theorem 1.2.7 it follows that

$$1/\|(A+E)^\dagger\|_2 = \sigma_r(A+E) \geq \sigma_r(A) - \|E\|_2 = 1/\|A^\dagger\|_2 - \|E\|_2 > 0,$$

which implies (1.4.9). ∎

We now characterize perturbations for which the pseudoinverse is well behaved. An **acute** perturbation of A is a perturbation such that the column and row spaces of A do not alter fundamentally.

1.4. Sensitivity of Least Squares Solutions

DEFINITION 1.4.1. *The subspaces $\mathcal{R}(A)$ and $\mathcal{R}(B)$ are said to be acute if the corresponding orthogonal projections satisfy*

$$\|P_{\mathcal{R}(A)} - P_{\mathcal{R}(B)}\|_2 < 1.$$

Further, the matrix $B = A + E$ is said to be an acute perturbation of A if $\mathcal{R}(A)$ and $\mathcal{R}(B)$ a well as $\mathcal{R}(A^T)$ and $\mathcal{R}(B^T)$ are acute.

Acute perturbations can be characterized by the following theorem.

THEOREM 1.4.3. *The matrix B is an acute perturbation of A if and only if*

$$(1.4.10) \qquad \operatorname{rank}(A) = \operatorname{rank}(B) = \operatorname{rank}(P_{\mathcal{R}(A)} B P_{\mathcal{R}(A^T)}).$$

Proof. See Stewart [731, 1977]. ∎

Let A and $B = A + E$ be square nonsingular matrices. Then, from the well-known identity $B^{-1} - A^{-1} = -B^{-1}EA^{-1}$, it follows that

$$\|A^{-1} - B^{-1}\| \le \|A^{-1}\|\,\|B^{-1}\|\,\|E\|.$$

The following generalization of this result can be proved by expressing the projections in terms of pseudoinverses using the relations in (1.2.29):

$$(1.4.11) \quad B^\dagger - A^\dagger = -B^\dagger E A^\dagger + (B^T B)^\dagger E^T P_{\mathcal{N}(A^T)} - P_{\mathcal{N}(B)} E^T (AA^T)^\dagger.$$

This identity can be used to obtain bounds for $\|B^\dagger - A^\dagger\|$ in the general case. For the case when $\operatorname{rank}(B) = \operatorname{rank}(A)$ the following theorem applies.

THEOREM 1.4.4. *If $B = A + E$ and $\operatorname{rank}(B) = \operatorname{rank}(A)$, then*

$$(1.4.12) \qquad \|B^\dagger - A^\dagger\| \le \mu \|B^\dagger\|\,\|A^\dagger\|\,\|E\|$$

where $\mu = 1$ for the Frobenius norm $\|\cdot\|_F$, and for the spectral norm $\|\cdot\|_2$,

$$\mu = \begin{cases} \frac{1+\sqrt{5}}{2} & \text{if } \operatorname{rank}(A) < \min(m,n), \\ \sqrt{2} & \text{if } \operatorname{rank}(A) = \min(m,n). \end{cases}$$

Proof. For the $\|\cdot\|_2$ norm, see Wedin [824, 1973]. The result that $\mu = 1$ for the Frobenius norm is due to van der Sluis and Veltkamp [784, 1979]. ∎

From the results above we deduce the following corollary.

COROLLARY 1.4.1. *A necessary and sufficient condition that*

$$(1.4.13) \qquad \lim_{E \to 0} (A + E)^\dagger = A^\dagger$$

is that $\lim_{E \to 0} \operatorname{rank}(A + E) = \operatorname{rank}(A)$.

1.4.3. Perturbation analysis of least squares solutions. We now consider the effect of perturbations of A and b upon the pseudoinverse solution $x = A^\dagger b$. In this analysis the **condition number** of a rectangular matrix $A \in \mathbf{R}^{m \times n}$ plays a significant role. The following definition generalizes the condition number of a square nonsingular matrix.

DEFINITION 1.4.2. *The condition number of $A \in \mathbf{R}^{m \times n}$ ($A \neq 0$) is*

(1.4.14) $$\kappa(A) = \|A\|_2 \|A^\dagger\|_2 = \sigma_1/\sigma_r,$$

where $0 < r = \text{rank}(A) \leq \min(m,n)$, and $\sigma_1 \geq \sigma_2 \geq \cdots \geq \sigma_r > 0$, are the nonzero singular values of A.

The last equality in (1.4.14) follows from the relations $\|A\|_2 = \sigma_1$, $\|A^\dagger\|_2 = \sigma_r^{-1}$.

In the following we denote the perturbed A and b by

(1.4.15) $$\tilde{A} = A + \delta A, \quad \tilde{b} = b + \delta b,$$

and the perturbed solution $\tilde{x} = \tilde{A}^\dagger \tilde{b} = x + \delta x$. We start by deriving the first-order perturbation estimate for least squares solutions when $\text{rank}(A) = n$. Then if $\|\delta A\|_2 < \sigma_n$ we have $\text{rank}(A + \delta A) = n$, and the perturbed solution $x + \delta x$ satisfies the normal equations

$$(A + \delta A)^T \Big((A + \delta A)(x + \delta x) - (b + \delta b) \Big) = 0.$$

Subtracting $A^T(Ax - b) = 0$, and neglecting second-order terms, we get

(1.4.16) $$\delta x = A^\dagger(\delta b - \delta A x) + (A^T A)^{-1} \delta A^T r,$$

where $A^\dagger = (A^T A)^{-1} A^T$ and $r = b - Ax$.

In the special case when only the right-hand side is perturbed, i.e., $\delta A = 0$, no second-order terms occur, and the exact perturbation equals

(1.4.17) $$\delta x = A^\dagger \delta b = A^\dagger A A^\dagger \delta b = A^\dagger A_{\mathcal{R}(A)} \delta b = A^\dagger \delta b_1,$$

where we have split δb into orthogonal components

$$\delta b = \delta b_1 + \delta b_2, \quad \delta b_1 = A_{\mathcal{R}(A)} \delta b, \quad \delta b_2 = A_{\mathcal{N}(A^T)} \delta b.$$

Hence the perturbation δx depends only on the component of δb in $\mathcal{R}(A)$.

From the SVD of A we have $A^\dagger = V \Sigma^\dagger U^T$, $(A^T A)^{-1} = V(\Sigma^T \Sigma)^{-1} V^T$, and it follows that

$$\|(A^T A)^{-1} A^T\|_2 = \frac{1}{\sigma_n}, \quad \|(A^T A)^{-1}\|_2 = \frac{1}{\sigma_n^2}.$$

Using (1.4.17) and taking norms in (1.4.16) we obtain the first-order result

(1.4.18) $$\|\delta x\|_2 \leq \frac{1}{\sigma_n}(\|\delta b_1\|_2 + \|\delta A\|_2 \|x\|_2) + \frac{1}{\sigma_n^2} \|\delta A\|_2 \|r\|_2.$$

Since $1/\sigma_n = \kappa(A)/\|A\|_2$ the last term here is proportional to $\kappa^2(A)$. Golub and Wilkinson [393, 1966] were the first to note that such a term occurs when $r \neq 0$. In van der Sluis [781, 1975] a geometrical explanation for the occurrence of this term is given, and lower bounds for the worst perturbation are also derived.

EXAMPLE 1.4.2. (See van der Sluis [781, 1975] and Figure 1.4.1.) Let $A = (a_1, a_2)$ be the matrix in Example 1.2.1, and assume that the angle

1.4. Sensitivity of Least Squares Solutions

$\gamma = \arccos(a_1^T a_2)$ is small. Choose perturbations δa_1 and δa_2 of size $\|\delta a_1\|_2 = \|\delta a_2\|_2 = \epsilon$, so that the plane $\hat{S} = \text{span}(a_1 + \delta a_1, a_2 + \delta a_2)$ is obtained by rotation of the plane $S = \text{span}(a_1, a_2)$ around the bisector $u_1 = \frac{1}{2}(a_1 + a_2)$, which according to Example 1.2.1 is an approximate left singular vector. If δa_1 and δa_2 are orthogonal to S and of opposite direction, then the angle of rotation will be $\theta \approx \epsilon/(\frac{1}{2}\gamma)$. Now let $c = P_S b$ be the orthogonal projection of b onto S and assume that the approximate direction of c is along u_1. Then $\hat{c} = P_{\hat{S}} b$ is obtained by rotating the residual vector r through the angle θ and hence

$$\|\hat{c} - c\|_2 \approx \sin\theta \|r\|_2 \approx 2\epsilon \|r\|_2/\gamma.$$

Further, the direction of $\hat{c} - c$ will be approximately along $u_2 = \frac{1}{\gamma}(a_2 - a_1)$. Since $\delta a_1 + \delta a_2 = 0$ we have $\delta A x \approx 0$ and hence

$$A\delta x \approx \hat{c} - c, \qquad \|\delta x\|_2 \approx \frac{1}{\sigma_2} \|\hat{c} - c\|_2.$$

It follows that

$$\|\delta x\|_2 \approx \epsilon 2\sqrt{2} \|r\|_2/\gamma^2 = \frac{1}{\sqrt{2}} \cdot \epsilon \|r\|_2 \kappa^2,$$

which is what we wished to show. This example illustrates that the occurrence of κ^2 is due to two coinciding events: rotation of the projection plane around a dominant left singular vector produces a large change in r, and this has the direction of the minimal left singular vector. ∎

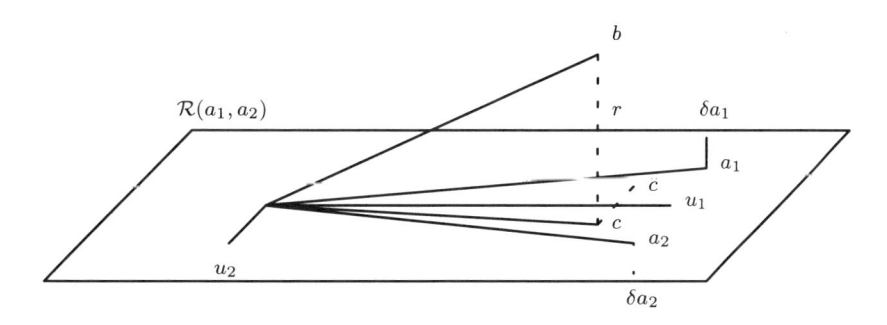

FIG. 1.4.1. *Exhibiting the squaring of $\kappa(A)$.*

We now give a more refined perturbation analysis, which follows that of Wedin [824, 1973] and applies to both overdetermined and underdetermined systems. In order to be able to prove any meaningful result we assume that the two conditions

(1.4.19) $\qquad \text{rank}(A + \delta A) = \text{rank}(A), \qquad \eta = \|A^\dagger\|_2 \|\delta A\|_2 < 1$

are satisfied. Note that if $\text{rank}(A) = \min(m, n)$ then the condition $\eta < 1$ suffices to guarantee that $\text{rank}(A + \delta A) = \text{rank}(A)$.

In the analysis we will need an estimate for the largest principal angle between the fundamental subspaces of $\tilde{A}$ and A. (For a definition of the principal angles between two subspaces see Definition 1.2.2.)

THEOREM 1.4.5. *Let $\tilde{A} = A + \delta A$ and assume that the conditions in (1.4.19) are satisfied. Then if $\chi(\cdot)$ denotes any of the four fundamental subspaces,*

(1.4.20) $$\sin\theta_{\max}(\chi(\tilde{A}), \chi(A)) \leq \eta < 1.$$

Proof. The result follows from Lemma 4.1 in Wedin [824, 1973]. ∎

We decompose the error δx as follows:
$$\delta x = \tilde{x} - x = \tilde{A}^\dagger \tilde{b} - x = \tilde{A}^\dagger(Ax + r + \delta b) - x,$$
or, using $P_{\mathcal{N}(\tilde{A})} = I - \tilde{A}^\dagger \tilde{A}$,

(1.4.21) $$\delta x = \tilde{A}^\dagger P_{\mathcal{R}(\tilde{A})}(\delta b - \delta A x) + \tilde{A}^\dagger r - P_{\mathcal{N}(\tilde{A})} x.$$

We separately estimate each of the three terms in this decomposition of δx. Using Theorem 1.4.2 and the assumption (1.4.25) it follows that

(1.4.22) $$\|\tilde{A}^\dagger(\delta b - \delta A x)\|_2 \leq \frac{1}{1-\eta} \|A^\dagger\|_2 \cdot (\|\delta A\|_2 \|x\|_2 + \|\delta b\|_2).$$

(We remark that a sharper estimate can be obtained by substituting for $\|\delta b\|_2$ in (1.4.22) $\|\delta b_1\|_2 + \eta \|\delta b_2\|_2$.)

Since $r \perp \mathcal{R}(A)$ we have $r = P_{\mathcal{N}(A^T)} r$, and, using (1.2.29), we can write the second term as

(1.4.23) $$\tilde{A}^\dagger r = \tilde{A}^\dagger \tilde{A} \tilde{A}^\dagger r = \tilde{A}^\dagger P_{\mathcal{R}(\tilde{A})} P_{\mathcal{N}(A^T)} r.$$

Now, by definition, $\|P_{\mathcal{R}(\tilde{A})} P_{\mathcal{N}(A^T)}\|_2 = \sin\theta_{\max}(\mathcal{R}(\tilde{A}), \mathcal{R}(A))$, where $\theta_{\max}$ is the largest principal angle between the two subspaces $\mathcal{R}(\tilde{A})$ and $\mathcal{R}(A)$.

Similarly, since $x = P_{\mathcal{R}(A^T)} x$, we can write the third term

(1.4.24) $$P_{\mathcal{N}(\tilde{A})} x = P_{\mathcal{N}(\tilde{A})} P_{\mathcal{R}(A^T)} x$$

and
$$\|P_{\mathcal{N}(\tilde{A})} P_{\mathcal{R}(A^T)}\|_2 = \sin\theta_{\max}(\mathcal{N}(\tilde{A}), \mathcal{N}(A)).$$

Using Theorem 1.4.5 to estimate (1.4.23) and (1.4.24) we arrive at the following result.

THEOREM 1.4.6. *Assume that* $\operatorname{rank}(A + \delta A) = \operatorname{rank}(A)$, *and let*

(1.4.25) $$\|\delta A\|_2/\|A\|_2 \leq \epsilon_A, \quad \|\delta b\|_2/\|b\|_2 \leq \epsilon_b.$$

Then if $\eta = \kappa \epsilon_A < 1$ the perturbations δx and δr in the least squares solution x and the residual $r = b - Ax$ satisfy

(1.4.26) $$\|\delta x\|_2 \leq \frac{\kappa}{1-\eta} \left(\epsilon_A \|x\|_2 + \epsilon_b \frac{\|b\|_2}{\|A\|_2} + \epsilon_A \kappa \frac{\|r\|_2}{\|A\|_2} \right) + \epsilon_A \kappa \|x\|_2$$

and

(1.4.27) $$\|\delta r\|_2 \leq \epsilon_A \|x\|_2 \|A\|_2 + \epsilon_b \|b\|_2 + \epsilon_A \kappa \|r\|_2.$$

1.4. SENSITIVITY OF LEAST SQUARES SOLUTIONS

Proof. The estimate (1.4.26) follows from above, and (1.4.27) is proved using the decomposition

$$\begin{aligned}
\tilde{r} - r &= P_{\mathcal{N}(\tilde{A}^T)}(b + \delta b) - P_{\mathcal{N}(A^T)} b \\
&= P_{\mathcal{N}(\tilde{A}^T)} \delta b + P_{\mathcal{N}(\tilde{A}^T)} P_{\mathcal{R}(A)} b - P_{\mathcal{R}(\tilde{A}^T)} P_{\mathcal{N}(A^T)} r
\end{aligned}$$

and using $P_{\mathcal{N}(\tilde{A}^T)} P_{\mathcal{R}(A)} b = P_{\mathcal{N}(\tilde{A}^T)} Ax = -P_{\mathcal{N}(\tilde{A}^T)} \delta A x$. ∎

REMARK 1.4.1. The last term in (1.4.26) (and therefore also in (1.4.27)) vanishes if rank$(A) = n$, since then $\mathcal{N}(\tilde{A}) = \{0\}$. If the system is compatible, e.g., if rank$(A) = m$, then $r = 0$ and the term involving κ^2 in (1.4.26) vanishes. For rank$(A) = n$, and $\epsilon_b = 0$, the condition number of the least squares problem can be written as

$$(1.4.28) \qquad \kappa_{LS}(A, b) = \kappa(A) \left(1 + \kappa(A) \frac{\|r\|_2}{\|A\|_2 \|x\|_2} \right).$$

Note that the conditioning depends on r and therefore on the right-hand side!

REMARK 1.4.2. When rank$(A) = n$ there are perturbations δA and δb such that the estimates in Theorem 1.4.6 can almost be attained for an arbitrary matrix A and vector b. This can be shown by considering first-order approximations of the terms (see Wedin [824, 1973]).

REMARK 1.4.3. It should be stressed that the perturbation analysis above is based on the class of perturbations defined by (1.4.25) and relevant only if the errors in the components of A and b are roughly of equal magnitude. For example, if the columns of A have widely different norms, then a more relevant class of perturbations often is

$$(1.4.29) \qquad \tilde{a}_j = a_j + \delta a_j, \quad \|\delta a_j\|_2 \leq \epsilon_j \|a_j\|_2, \quad j = 1, 2, \ldots, n.$$

Similarly, if the norm of the perturbation bounds for the rows in A differ widely, then (1.4.26) and (1.4.27) may considerably overestimate the perturbations. However, scaling the rows in (A, b) will change the least squares problem. ∎

A sharper estimate is usually obtained by scaling the columns of A so that the relative perturbation bound in all columns is the same:

$$Ax = (AD^{-1})(Dx) = \bar{A}\bar{x}, \quad D = \text{diag}(\epsilon_1 \|a_1\|_2, \ldots, \epsilon_n \|a_n\|_2).$$

In particular, if $\epsilon_j = \epsilon$, $\forall j$, then this scales the matrix so that all column norms are equal. The following result by van der Sluis [780, 1969, Thm. 4.3] shows that this scaling approximately minimizes $\kappa(\bar{A})$ over all diagonal scalings. (Note, however, that scaling the columns changes the norm in which the error in x is measured.)

THEOREM 1.4.7. *Let $C \in \mathbf{R}^{n \times n}$ be symmetric and positive definite, and denote by $\mathcal{D}$ the set of $n \times n$ nonsingular diagonal matrices. Then if in C all diagonal elements are equal, and C has at most q nonzero elements in any row, it holds that*

$$\kappa(C) \leq q \min_{D \in \mathcal{D}} \kappa(DCD).$$

If this result is applied with $q = n$ to the matrix of normal equations $A^T A$, it follows that if all columns in A have unit length, then

$$\kappa(A) \leq \sqrt{n} \min_{D \in \mathcal{D}} \kappa(AD). \tag{1.4.30}$$

1.4.4. Asymptotic forms and derivatives. Derivatives of orthogonal projectors and pseudoinverses were first considered by Golub and Pereyra [378, 1973]. Stewart [731, 1977] gives asymptotic forms and derivatives for orthogonal projectors, pseudoinverses, and least squares solutions.

If $A = A(\tau)$ is differentiable and varies without changing rank, then the projection $P_{\mathcal{R}(A)}$ is differentiable and

$$\frac{dP_{\mathcal{R}(A)}}{d\tau} = P_{\mathcal{N}(A^T)} \frac{dA}{d\tau} P_{\mathcal{R}(A^T)} A^\dagger + (A^\dagger)^T P_{\mathcal{R}(A^T)} \frac{dA^T}{d\tau} P_{\mathcal{N}(A^T)}.$$

For the pseudoinverse,

$$\frac{dA^\dagger}{d\tau} = -A^\dagger P_{\mathcal{R}(A)} \frac{dA}{d\tau} P_{\mathcal{R}(A^T)} A^\dagger + (A^T A)^\dagger P_{\mathcal{R}(A^T)} \frac{dA^T}{d\tau} P_{\mathcal{N}(A^T)}$$
$$- P_{\mathcal{N}(A)} \frac{dA^T}{d\tau} P_{\mathcal{R}(A)} (AA^T)^\dagger.$$

Finally, for the least squares solution $x = A^\dagger b$, we obtain

$$\frac{dx}{d\tau} = -A^\dagger P_{\mathcal{R}(A)} \frac{dA}{d\tau} P_{\mathcal{R}(A^T)} x + (A^T A)^\dagger P_{\mathcal{R}(A^T)} \frac{dA^T}{d\tau} P_{\mathcal{N}(A^T)} b$$
$$- P_{\mathcal{N}(A)} \frac{dA^T}{d\tau} P_{\mathcal{R}(A)} (A^T)^\dagger x.$$

1.4.5. Componentwise perturbation analysis. There are several drawbacks with a normwise perturbation analysis. As already mentioned, it can give huge overestimates when the corresponding problem is badly scaled. Using norms we ignore how the perturbation is distributed among the elements of the matrix and vector. For these reasons *componentwise perturbation analysis* is of interest. An excellent survey of the theory and history behind such an analysis is given by Higham [466, 1994].

In this section we derive perturbation results and condition numbers corresponding to componentwise errors in A and b for the least squares problem. A similar analysis is given in Arioli et al. in [21, 1989]. We assume that we have componentwise bounds on the perturbations in the data

$$|\delta a_{ij}| \leq \omega e_{ij}, \quad |\delta b_i| \leq \omega f_i, \quad i,j = 1, \ldots, n,$$

where $e_{ij} > 0$ and $f_i > 0$ are known. In order to write such componentwise bounds in a simple way we define the absolute value of a matrix A and vector b by

$$|A|_{ij} = (|a_{ij}|), \quad |b|_i = (|b_i|).$$

1.4. Sensitivity of Least Squares Solutions

We introduce the partial ordering "$\leq$" for matrices A, B and vectors x, y, which is to be interpreted *componentwise*:

$$A \leq B \iff a_{ij} \leq b_{ij}, \quad x \leq y \iff x_i \leq y_i.$$

It is easy to show that if $C = AB$, then

$$|c_{ij}| \leq \sum_{k=1}^{n} |a_{ik}| \, |b_{kj}|,$$

and hence $|C| \leq |A| \, |B|$. A similar rule holds for matrix-vector multiplication.

With these notations we can write the componentwise bounds above as

(1.4.31) $$|\delta A| \leq \omega E, \quad |\delta b| \leq \omega f,$$

where $E > 0$, $f > 0$. Taking $E = |A|$ and $f = |b|$ in (1.4.31) corresponds to **componentwise relative error bounds** for A and b.

We first derive estimates for the perturbations in the solution of a nonsingular square system $Ax = b$. The basic identity for this perturbation analysis is

$$\delta x = (I + A^{-1}\delta A)^{-1} A^{-1}(\delta A x + \delta b).$$

Assuming that $|A^{-1}||\delta A| < 1$, taking absolute values gives the inequality

$$|\delta x| \leq (I - |A^{-1}||\delta A|)^{-1} |A^{-1}|(|\delta A||x| + |\delta b|).$$

The matrix $(I - |A^{-1}||\delta A|)$ is guaranteed to be nonsingular if $\| \, |A^{-1}| \, |\delta A| \, \| < 1$. For perturbations satisfying (1.4.31) we obtain

(1.4.32) $$|\delta x| \leq \omega (I - \omega |A^{-1}|E)^{-1} |A^{-1}|(E|x| + f).$$

Provided that $\omega \kappa_E(A) < 1$, $\kappa_E(A) = \| \, |A^{-1}|E\|$, we get from (1.4.32) for any absolute norm the perturbation bound

(1.4.33) $$\|\delta x\| \leq \frac{\omega}{1 - \omega \kappa_E(A)} \| \, |A^{-1}|(E\,|x| + f)\|.$$

For the special case of componentwise relative error bounds ($E = |A|$),

(1.4.34) $$\kappa_{|A|}(A) = \| \, |A^{-1}||A| \, \|$$

is the **Bauer–Skeel condition number** of A (also denoted by $\mathrm{cond}\,(A)$). It is possible for $\kappa_{|A|}(A)$ to be much smaller than $\kappa(A)$. It can be shown that $\kappa_{|A|}(A)$ and the bound (1.4.33) for $E = |A|$ are *invariant under row scalings*.

We now consider a componentwise error analysis for the linear least squares problem. For simplicity we will neglect error terms of order ω^2. From (1.4.16) we obtain

(1.4.35) $$|\delta x| = \omega |A^{\dagger}|(f + E|x|) + \omega |(A^T A)^{-1}|E^T |r|,$$

and for the special case of componentwise relative perturbations,
$$|\delta x| = \omega |A^\dagger|(|x| + |A||x|) + \omega |(A^T A)^{-1}||A^T||r|.$$
It follows that
(1.4.36) $\quad \|\delta x\|_\infty \leq \omega \Big(\| |A^\dagger|(|A||x| + |b|) \|_\infty + \| |(A^T A)^{-1}||A|^T |r| \|_\infty \Big).$

Hence
(1.4.37) $\quad\quad\quad\quad\quad\quad\quad \text{cond}(A) = \| |A^\dagger| |A| \|_\infty$

can be taken as an approximate condition number for componentwise relative perturbations. In the general case when $m > n$, cond(A) often depends only weakly on the row scaling D, but in a way which is complicated to describe. For stiff problems, where some rows are scaled with a large weight w, cond(A) usually tends to a limit value when $w \to \infty$, whereas $\kappa(A)$ grows linearly with w; see Björck [97, 1991].

1.4.6. A posteriori estimation of errors. Let $\bar{x}$ be an approximate solution of the least squares problem $\min_x \|Ax - b\|_2$, where $A \in \mathbf{R}^{m \times n}$, $m \geq n$. Consider the problem of finding the smallest perturbation E such that $\bar{x}$ *exactly* solves the problem $\min_x \|(A + E)x - b\|_2$. For a consistent linear system $Ax = b$ Rigal and Gaches [686, 1967] showed that the perturbation E of smallest l_2-norm is given by the rank one perturbation

(1.4.38) $\quad\quad\quad\quad\quad\quad\quad E_0 = \bar{r}\bar{x}/\|\bar{x}\|_2^2 = \bar{r}\bar{x}^\dagger.$

The corresponding norm $\|E\|_2 = \|\bar{r}\|_2 \|\bar{x}\|_2$ is called the **normwise backward error**.

How to find the normwise backward error for an inconsistent least squares problem was an open problem for a long time. Stewart [733, 1977] showed that for the two perturbations

(1.4.39) $\quad\quad\quad\quad E_1 = -\bar{r}\bar{r}^\dagger A, \quad\quad E_2 = (r - \bar{r})\bar{x}^\dagger,$

$\bar{x}$ solves the perturbed least squares problems exactly. The corresponding norms are
$$\|E_1\|_2 = \|A^T \bar{r}\|_2/\|\bar{r}\|_2, \quad\quad E_2 = \|r - \bar{r}\|_2/\|\bar{x}\|_2.$$
The first is small when the residual $\bar{r}$ is almost orthogonal to the column space of A. The second is small when $\bar{r}$ is almost equal to the exact residual r. However, it is possible for $\bar{x}$ to be a solution of a slightly perturbed least squares problem and yet for both $\|E_1\|_2$ and $\|E_2\|_2$ to be orders of magnitude larger than the norm of the perturbation.

Recently Waldén, Karlsson, and Sun [811, 1995] gave an explicit representation for the set $\mathcal{E}$ of all perturbation matrices E such that $\bar{x}$ exactly solves
$$\min_x \|(A + E)x - b\|_2.$$
They also found an expression for the $E \in \mathcal{E}$ which minimizes $\|E\|_F$. The corresponding solution when perturbations in both A and b are allowed is given in the following theorem.

1.4. Sensitivity of Least Squares Solutions

THEOREM 1.4.8. *Let $A \in \mathbf{R}^{m \times n}$, $m \geq n$, $b \in \mathbf{R}^m$, and $\bar{x}$ be an approximate least squares solution. The normwise backward error*

$$\eta_F(\bar{x}) = \min\{\|(E, \tau e)\|_F \mid \|(A + E)x - (b + e)\|_2 = \min\}$$

is given by

(1.4.40) $$\eta_F(\bar{x}) = \begin{cases} \gamma\sqrt{\mu}, & \lambda_* \geq 0, \\ (\gamma^2 \mu + \lambda_*)^{1/2}, & \lambda_* < 0, \end{cases}$$

where $\bar{r} = b - A\bar{x}$, $\gamma = \|\bar{r}\|_2/\|\bar{x}\|_2$, and

$$\lambda_* = \lambda_{\min}\left(AA^T - \mu \frac{\bar{r}\bar{r}^T}{\|\bar{x}\|_2^2}\right), \qquad \mu = \frac{\tau^2 \|\bar{x}\|_2^2}{1 + \tau^2 \|\bar{x}\|_2^2} \leq 1.$$

The parameter τ in Theorem 1.4.8 allows some flexibility. For example, taking the limit $\tau \to \infty$ gives the case when only A is perturbed. Then $\mu = 1$, and (1.4.40) becomes

$$\eta_F(\bar{x}) = \begin{cases} \gamma, & \lambda_* \geq 0, \\ (\gamma^2 + \lambda_*)^{1/2}, & \lambda_* < 0. \end{cases}$$

Note that the required backward error is no larger than the backward error $\|E_0\|_F$ for a consistent system, where E_0 is given by (1.4.38). It is strictly smaller if $\lambda_* < 0$. Note that a sufficient condition for $\lambda_* < 0$ is $\hat{r} \notin \mathcal{R}(A)$.

The expressions for η_F in the theorem are elegant but unsuitable for computation since they can suffer from cancellation when $\lambda_* < 0$. Higham [467, 1996, Chap. 15] has suggested the alternative formula,

$$\eta_F(\bar{x}) = \min\left(\eta_1, \sigma_{\min}(A, \eta_1 B)\right),$$

where

$$\eta_1 = \gamma\sqrt{\mu}, \qquad B - (I - \bar{r}\bar{r}^T/\|\bar{r}\|_2^2).$$

This is more computationally reliable, but still expensive to compute. Simpler lower and upper bounds are given in Waldén, Karlsson, and Sun [811, 1995]. Optimal backward error bounds for linear least squares problems with multiple right-hand sides have been given by Sun [767, 1995].

Given an arbitrary approximate least squares solution $\bar{x}$ the **componentwise backward error** is the smallest $\omega \geq 0$ in

$$|\delta A| \leq \omega |A|, \qquad |\delta b| \leq \omega |b|,$$

such that $\bar{x}$ is the *exact* solution of the perturbed problem

$$\min_x \|(A + \delta A)x - (b + \delta b)\|_2.$$

For a consistent linear system $b \in \mathcal{R}(A)$, Oettli and Prager [603, 1964] showed that

(1.4.41) $$\omega_B = \max_{1 \leq i \leq n} \frac{|A\bar{x} - b|_i}{(|A||\bar{x}| + |b|)_i},$$

where 0/0 should be interpreted as 0, and $\zeta/0$ ($\zeta \neq 0$) as infinity. (The latter case means that no finite ω satisfying (1.4.41) exists.) Together with the perturbation result (1.4.36) this can be used to compute an a posteriori bound on the error in a given approximate solution $\bar{x}$.

Unfortunately there is no similar result for the inconsistent linear least squares problem. One approach could be to apply the Oettli–Prager bound to the augmented system (1.1.24). Here there are no perturbations in the diagonal blocks of the augmented system matrix and in the last zero vector in the augmented right-hand side. However, a result by Kiełbasiński and Schwetlick [505, 1988, Lemma 8.2.11] shows that allowing *unsymmetric* perturbations in the blocks A and A^T has little effect on the backward error. Also, by a slight modification of the perturbation analysis in the previous section we can accommodate a perturbation to the first diagonal block; see also Higham [461, 1990]. Hence, for an a posteriori error analysis, it makes sense to take the relative backward error of a computed solution $\bar{r}, \bar{x}$ to be the smallest nonnegative number ω such that

$$|\delta I| \leq \omega I, \quad |\delta A_i| \leq \omega E, \quad i = 1, 2, \quad |\delta b| \leq \omega f,$$

and

(1.4.42)
$$\begin{pmatrix} I + \delta I & A + \delta A_1 \\ A^T + \delta A_2 & 0 \end{pmatrix} \begin{pmatrix} \bar{r} \\ \bar{x} \end{pmatrix} = \begin{pmatrix} b + \delta b \\ 0 \end{pmatrix}.$$

Using the result of Oettli and Prager, $\omega(\bar{r}, \bar{x}) = \max(\omega_1, \omega_2)$, where

(1.4.43)
$$\omega_1 = \max_{1 \leq i \leq m} \frac{|\bar{r} + A\bar{x} - b|_i}{(|\bar{r}| + E|\bar{x}| + f)_i}, \quad \omega_2 = \max_{1 \leq i \leq n} \frac{|A^T \bar{r}|_i}{(E^T |\bar{r}|)_i}$$

gives the backward error for a computed solution $\bar{r}$ and $\bar{x}$.

If we only have a computed $\bar{x}$ it may be feasible to put $\bar{r} = b - A\bar{x}$ and apply the result above. With this choice we have $\omega_1 = 0$ (exactly) and hence

$$\omega(\bar{r}, \bar{x}) = \omega_2 = \max_{1 \leq i \leq n} \frac{|A^T(b - A\bar{x})|_i}{(E^T |b - A\bar{x}|)_i}.$$

However, in the case of a nearly consistent least squares problem, $fl(b - A\bar{x})$ will mainly consist of roundoff, and will not be accurately orthogonal to the range of A. Hence, although $\bar{x}$ may have a small relative backward error, ω_2 will not be small. This illustrates a fundamental problem in computing the backward error: for $\bar{x}$ to have a small backward error it is sufficient that *either* $(b - A\bar{x})$ or $A^T(b - A\bar{x})$ be small, but neither of these conditions is *necessary*.

Chapter 2
Basic Numerical Methods

2.1. Basics of Floating Point Computation

2.1.1. Rounding error analysis. A floating point number system consists of numbers x which can be represented as

$$(2.1.1) \qquad x = m \cdot \beta^e, \qquad m = \pm d_0.d_1 d_2 \ldots d_t,$$

where m is the mantissa, e the exponent, and t the number of digits carried in the mantissa, and $0 \leq d_i < \beta$. The integer β is the base (usually $\beta = 2$). The mantissa is usually normalized so that $1 \leq |m| < \beta$, and the exponent satisfies $L \leq e \leq U$. Hence, the floating point number system is characterized by the set (β, t, L, U).

The result of arithmetic operations on floating point numbers cannot generally be represented exactly as floating point numbers. Rounding errors will arise because the computer can only represent a subset F of the real numbers. The elements of this subset are referred to as floating point numbers. Error estimates can be expressed in terms of the **unit roundoff** u, which for the floating point system (2.1.1) may be defined as

$$u = \frac{1}{2}\beta^{1-t}$$

and equals the maximum relative error in storing a number. (Sometimes u is defined as the smallest floating point number ϵ such that $fl(1 + \epsilon) > 1$. However, with this definition the precise value of u varies even among different implementations of the standard IEEE arithmetic.)

To derive error estimates of matrix computations the **standard model** of floating point arithmetic is used. We denote the stored result of any floating point calculation C as $fl(C)$. The assumption in the standard model is that for any $x, y \in F$,

$$(2.1.2) \qquad fl(x \text{ op } y) = (x \text{ op } y)(1 + \delta), \qquad |\delta| \leq u,$$

where op $= +, -, *, /$. Normally it is assumed that (2.1.2) also holds for the square root operation. A thorough presentation of floating point arithmetic is given by Higham [467, 1996, Chap. 2.1.1].

Consider a finite algebraic algorithm which from the data $(a_1, \ldots, a_r)$, through intermediate values $(c_1, \ldots, c_s)$, computes a solution $(w_1, \ldots, w_t)$. There are two basic forms of error analysis for such an algorithm, both of which are useful.

1. In **classical forward error analysis** one attempts to find bounds for the errors in the intermediate values $|\hat{c}_i - c_i|$, and finally bounds for $|\hat{w}_i - w_i|$.

2. In **backward error analysis** one attempts to determine a modified set of data $\tilde{a}_i$ such that the computed solution $\hat{w}_i$ is the *exact solution*, and give bounds for $|\tilde{a}_i - a_i|$. There may be an infinite number of such sets; sometimes there is just one and it can happen, even for very simple algorithms, that no such set exists. Notice that no reference is made to the exact solution for the original data.

Backward error analysis was pioneered by J. H. Wilkinson in the late fifties. When it applies, it tends to be very markedly superior to forward analysis. It usually also gives better insight into the stability (or lack thereof) of the algorithm, which often is the primary purpose of a rounding error analysis.

To yield error bounds for the solution, the backward error analysis has to be complemented with a perturbation analysis. The **condition number** $\kappa(a)$ for the problem of computing w from data a is defined as

$$\kappa(a) = \limsup_{\epsilon \to 0} \frac{1}{\epsilon} \frac{\|\hat{w} - w\|_{out}}{\|w\|_{out}}, \quad \|\hat{a} - a\|_{in} \leq \epsilon \|a\|_{in}.$$

Note that $\kappa(a)$ is a function of the input data a and depends on the choice of norms in the data space and solution space. For sufficiently small ϵ we have the estimate for the perturbations in the output

$$\|\hat{w} - w\|_{out} \leq \epsilon \kappa(a) \|w\|_{out} + O(\epsilon^2).$$

If $\kappa(a)$ is "large" the problem is said to be **ill-conditioned**. The definition of large may differ from problem to problem and depends on the accuracy of the data and the accuracy needed in the solution.

By repeatedly using the formula for floating point multiplication one can show that the computed product $fl(x_1 x_2 \cdots x_n)$ is exactly equal to

$$x_1 x_2 (1 + \epsilon_2) x_3 (1 + \epsilon_3) \cdots x_n (1 + \epsilon_n),$$

where $|\epsilon_i| \leq u$, $i = 2, 3, \ldots, n$. This can be interpreted as a *backward error analysis*; we have shown that the computed product is the exact product of factors x_1, $\tilde{x}_i = x_i(1 + \epsilon_i)$, $i = 2, \ldots, n$. It also follows from this analysis that

$$|fl(x_1 x_2 \cdots x_n) - x_1 x_2 \cdots x_n| \leq \delta |x_1 x_2 \cdots x_n|$$

where

$$\delta = (1 + u)^{n-1} - 1 < 1.06(n-1)u,$$

2.1. Basics of Floating Point Computation

and the last inequality holds if the condition $(n-1)u < 0.1$ is satisfied. This bounds the *forward error* in the computed result.

Similar results can easily be derived for basic vector and matrix operations; see Wilkinson [836, 1965, pp. 114–118]. For an inner product $x^T y$ we have

$$fl(x^T y) = x_1 y_1 (1+\delta_1) + x_2 y_2 (1+\delta_2) + \cdots + x_n y_n (1+\delta_n),$$

where

$$|\delta_1| < 1.06nu, \qquad |\delta_r| < 1.06(n+2-r)u, \quad r = 2, \ldots, n.$$

Hence, the computed result is the exact inner product of x and $\hat{y}$, where $\hat{y}_i = y_i(1+\delta_i)$. Useful consequences of this result are

$$(2.1.3) \qquad |fl(x^T y) - x^T y| \leq \sum |x_i||y_i||\delta_i| \leq 1.06nu|x^T||y|$$
$$\leq 1.06nu\|x\|_2 \|y\|_2.$$

This result is easily generalized to yield a *forward* error analysis of matrix-matrix multiplication. However, for this case there is no backward error analysis, since the rows and columns of the two matrices participate in many inner products!

2.1.2. Running rounding error analysis. A different approach to rounding error analysis is to perform the analysis automatically, for *each particular computation*, i.e., to perform a **running error** analysis. This gives an a posteriori error analysis, as compared to the a priori error analysis discussed above. An example is the use of **interval analysis** for which special-purpose hardware and software now exist.

A simple form of a posteriori analysis, called running error analysis, was used in the early days of computing by Wilkinson; see [842, 1986]. To illustrate his idea we rewrite the basic model for floating point arithmetic as

$$x \text{ op } y = fl(x \text{ op } y)(1+\epsilon), \quad |\epsilon| \leq u.$$

These relations are satisfied for most implementations of floating point arithmetic. Then the actual error can be estimated as $|fl(x \text{ op } y) - x \text{ op } y| \leq u|fl(x \text{ op } y)|$. Note that the error is now given in terms of the *computed* result and is available in the computer at the time the operation is performed. A running error analysis can often be easily implemented by taking an existing program and modifying it so that as each arithmetic operation is performed, the absolute value of the computed results is added into the accumulating error bound.

EXAMPLE 2.1.1. The inner product $fl(x^T y)$ and a running error bound is computed by the program

$$\hat{s}_0 = 0; \quad \eta = 0;$$
$$\textbf{for } i = 1, 2, \ldots, n$$
$$\hat{t}_i = fl(x_i y_i); \quad \eta = \eta + |\hat{t}_i|;$$
$$\hat{s}_i = fl(\hat{s}_{i-1} + \hat{t}_i);$$
$$\eta = \eta + |\hat{s}_i|;$$
$$\textbf{end}$$

For the final error we have the estimate $|fl(x^T y) - x^T y| \leq \eta u$. ∎

Note that a running error analysis takes advantage of cancellations in the sum. This is in contrast to the previous a priori error estimates, where the error estimate is the same for all distributions of signs of the elements x_i and y_i.

2.1.3. Stability of algorithms. We say that an algorithm for computing the result w is **forward stable** for the data a if the computed solution $\bar{w}$ is close to the exact solution,

$$|\bar{w} - w|/|w| \leq c_1 u,$$

where c_1 is a not-too-large constant. Similarly, we say that an algorithm is **backward stable** for the data a if the computed solution $\bar{w}$ is the exact solution of a slightly perturbed problem with data $\tilde{a}$ such that

$$|\tilde{a} - a|/|a| < c_2 u,$$

where c_2 is a not-too-large constant and u is a measure of the precision used in the computations. We would like these estimates to hold not only for a single problem, but for some *class of input data*. We are usually satisfied if we can prove forward or backward stability for the norm $\|\cdot\|_2$ or $\|\cdot\|_\infty$, although we may like the estimates to hold elementwise,

$$|\tilde{a}_i - a_i|/|a_i| < c_2 u, \qquad i = 1, \ldots, r.$$

Forward stability guarantees an accurate solution. However, very few algorithms are forward stable. For a backward stable algorithm we are not guaranteed to get an accurate solution. We can only show that

$$\|\bar{w} - w\| \leq c_1 u \kappa \|w\|,$$

where $\kappa(X)$ is the condition number for the problem. However, if the perturbations $\bar{a} - a$ are within the uncertainties of the given data, *the computed solution is as good as our data warrants*! Therefore, backward stability is usually very satisfactory. As we shall see in the next section, although many important algorithms are not backward stable, they are **mixed stable**. By this we mean that the computed solution $\bar{w}$ is close to the exact solution $\tilde{w}$ of a slightly perturbed problem with data $\tilde{a}$, where $\tilde{a}$ satisfies (2.1.2) and

$$\|\tilde{w} - \bar{w}\|/\|w\| \leq c_1 u.$$

Stability of algorithms for solving systems of linear equations $Ax = b$ is defined similarly. The following definition of backward stability is the one introduced by Wilkinson [836, 1965]. Forward and mixed stability are defined analogously.

DEFINITION 2.1.1. *A numerical algorithm for solving systems of linear equations $Ax = b$ is **backward stable** for a class of matrices $\mathcal{A}$ if for each $A \in \mathcal{A}$ and for each b the computed solution $\bar{x}$ satisfies $\bar{A}\bar{x} = \bar{b}$ where $\bar{A}$ and $\bar{b}$ are close to A and b.*

2.1. BASICS OF FLOATING POINT COMPUTATION

For example, the Cholesky algorithm is backward stable for the class of symmetric positive definite matrices. An important property of backward stable algorithms is given in the following theorem.

THEOREM 2.1.1. *An algorithm for solving $Ax = b$ is backward stable according to Definition 2.1.1 if and only if the computed solution has a small residual*

(2.1.4) $$\|b - A\bar{x}\| \leq c_2 u \|A\| \|\bar{x}\|.$$

Proof. If (2.1.4) holds we define for the 2-norm

$$\delta A = r\bar{x}^T / \|\bar{x}\|_2^2, \qquad r = b - A\bar{x}.$$

Then it holds exactly that $(A + \delta A)\bar{x} = A\bar{x} + r = b$, where

$$\|\delta A\|_2 = \|r\|_2 / \|\bar{x}\|_2 \leq c_2 u \|A\|_2.$$

We can take $\delta b = 0$, and hence the algorithm is stable by Definition 2.1.1. On the other hand, if $\bar{A}\bar{x} = \bar{b}$, we have $b - A\bar{x} = (\bar{A} - A)\bar{x} + b - \bar{b}$, and if

$$\|\bar{A} - A\|_2 \leq c_3 u \|A\|_2, \qquad \|\bar{b} - b\|_2 \leq c_4 u \|b\|_2,$$

then an estimate of the form (2.1.4) for the norm of the residual holds. ∎

Note that in Definition 2.1.1 it is not required that $\bar{A}$ be in class $\mathcal{A}$. If the system comes from a physical problem such that A always belongs to some class $\mathcal{A}$, then we have not solved a nearby physical problem unless the perturbed $\bar{A}$ also is in $\mathcal{A}$. For example, if A is Toeplitz we would also like $\bar{A}$ to be Toeplitz. Therefore, the following alternative definition of stability has been suggested (Bunch [134, 1987]).

DEFINITION 2.1.2. *A numerical algorithm for solving linear equations is* **strongly backward stable** *for a class of matrices $\mathcal{A}$ if for each $A \in \mathcal{A}$ and for each b the computed solution $\bar{x}$ satisfies $\bar{A}\bar{x} = \bar{b}$, where $\bar{A}$ and $\bar{b}$ are close to A and b, and $\bar{A}$ is in $\mathcal{A}$.*

Strong forward stability does not seem to be a useful concept, since any known structure of the solution can usually be imposed. Although many algorithms are known to be strongly backward stable, it may be difficult to prove strong stability for other important algorithms. For example, only recently did Bunch, Demmel, and Van Loan [135, 1989] prove that Gaussian elimination with partial pivoting is strongly backward stable for the class of symmetric positive definite systems, the difficulty being to show that $\bar{A}$ is symmetric. In this particular case one can also argue that since the condition number for unsymmetric perturbations of a symmetric system is the same as for symmetric perturbations, backward stability is sufficient. Hence Definition 2.1.1 often seems the more useful one.

Many important algorithms for solving linear systems (for example, most iterative methods) are not backward stable. Therefore, the following weaker form of stability is needed.

DEFINITION 2.1.3. *An algorithm for solving $Ax = b$ is **acceptable-error stable** for a class of matrices $\mathcal{A}$ if the computed solution $\bar{x}$ satisfies*

$$\|\bar{x} - x\| \leq c_3 u \kappa(A) \|x\|.$$

It is straightforward to show that backward stability implies stability, but the converse is not true.

Many users are satisfied if the algorithm they use produces accurate solutions for well-conditioned problems. The following definition of weak stability has been suggested by Bunch [134, 1987].

DEFINITION 2.1.4. *An algorithm for solving $Ax = b$ is **weakly stable** for a class of matrices $\mathcal{A}$ if for each well-conditioned A in $\mathcal{A}$ the computed solution $\bar{x}$ is such that $\|\bar{x} - x\|/\|x\|$ is small.*

Bunch notes that weak stability "may be sufficient for giving confidence that an algorithm may be used successfully."

2.2. The Method of Normal Equations

2.2.1. Forming the normal equations.

In Section 1.1.4 it was shown that when A has full column rank the solution x to the least squares problem (1.1.1) satisfies the normal equations

$$(2.2.1) \qquad A^T A x = A^T b.$$

Similarly the solution y to the minimum norm problem (1.1.19) can be obtained from the normal equations of the second kind

$$(2.2.2) \qquad A^T A z = c, \qquad y = Az.$$

In this section we discuss numerical methods, which date back to Gauss, based on forming and solving the normal equations. We assume here that rank $(A) = n$, and defer treatment of rank deficient problems to Section 2.7. Then by Theorem 1.2.2 the matrix $A^T A$ is positive definite and the least squares problem has a unique solution.

The first step in the method of normal equations for the least squares problem is to form the matrix and vector

$$(2.2.3) \qquad C = A^T A \in \mathbf{R}^{n \times n}, \quad d = A^T b \in \mathbf{R}^n.$$

Since the cross-product matrix C is symmetric it is only necessary to compute and store its upper triangular part. If A is partitioned by columns

$$A = (a_1, a_2, \ldots, a_n),$$

the relevant elements in C and d are given by

$$(2.2.4) \qquad \begin{aligned} c_{jk} &= a_j^T a_k, & 1 \leq j \leq k \leq n, \\ d_j &= a_j^T b, & 1 \leq j \leq n. \end{aligned}$$

2.2. The Method of Normal Equations

The column-oriented scheme is not suitable for large problems, where the matrix A is held in secondary storage, since each column needs to be accessed many times. By sequencing the operations differently we can obtain a row-oriented algorithm for forming the normal equations, which uses only one pass through the data and no more storage than that needed for $A^T A$ and $A^T b$. Partitioning A by rows, i.e.,

$$A^T = (\tilde{a}_1, \tilde{a}_2, \ldots, \tilde{a}_m),$$

where $\tilde{a}_i^T$ denotes the ith row of A, we can write

(2.2.5) $$C = \sum_{i=1}^{m} \tilde{a}_i \tilde{a}_i^T, \quad d = \sum_{i=1}^{m} b_i \tilde{a}_i.$$

Here C is expressed as the sum of m matrices of rank one, and d as a linear combination of the rows of A. Again, only the upper triangular part of A is computed and stored.

We call (2.2.4) the inner product form and (2.2.5) the outer product form of the normal equations. Row-wise accumulation of C and d using (2.2.5) is advantageous to use if the data A and b are stored row by row on secondary storage. The outer product form is also preferable if the matrix A is sparse; see Section 6.6.1.

REMARK 2.2.1. If $m \gg n$, then the number of elements in the upper triangular part of C, which is $\frac{1}{2}n(n+1)$, is much smaller than mn, the number of elements in A. In this case the formation of C and d can be viewed as a data compression.

The number of multiplicative operations needed to form the matrix $A^T A$ is $\frac{1}{2}n(n+1)m$. In the following, to quantify the operation counts in matrix algorithms we will use the concept of a **flop**. This will be used to mean roughly the amount of work associated with the statement

$$s := s + a_{ik} \cdot b_{kj},$$

i.e., it comprises a floating point add, a floating point multiply, and some subscripting.[1] Thus we will say that forming $A^T A$ and $A^T b$ requires

$$\frac{1}{2}n(n+1)m + mn \text{ flops},$$

or approximately $\frac{1}{2}n^2 m$ flops if $n \gg 1$. It should be stressed that on modern computers a flop may not be an adequate measure of work.

We now consider rounding errors made in the formation of the system of normal equations and their effect on the computed solution. For the computed elements in the matrix $C = A^T A$ we have

$$\bar{c}_{ij} = fl\left(\sum_{k=1}^{m} a_{ik} a_{jk}\right) = \sum_{k=1}^{m} a_{ik} a_{jk}(1 + \delta_k),$$

---

[1] In several recent textbooks (e.g., Golub and Van Loan [389, 1989]) a flop is instead defined as a floating point add *or* multiply, doubling all the flop counts.

where $|\delta_k| < 1.06(m+2-k)u$, and u is the unit roundoff. It follows (cf. 2.1.3) that the computed matrix satisfies $\bar{C} = C + E$, where

(2.2.6) $$|e_{ij}| < 1.06mu|a_i|^T|a_j| \leq 1.06mu\|a_i\|_2\|a_j\|_2.$$

A similar estimate holds for the rounding errors in the computed vector A^Tb. Note that it is *not* generally possible to show that $\bar{C} = (A+E)^T(A+E)$ for some error matrix E unless we allow E to be proportional to $\kappa(A)$. That is, the rounding errors in forming the matrix A^TA are not generally equivalent to small perturbations of the initial data matrix A. We deduce that the solution computed by the method of normal equations does not generally equal the exact solution to a problem where the data A and b have been perturbed by small amounts. Hence the method of normal equations is *not backward stable* in the sense of Definition 2.1.1.

It is possible to accumulate the inner products in (2.2.4) in double precision. Then the only rounding errors in the computation of C and d will be when the double precision results are rounded to single precision. However, even this might lead to a serious loss of accuracy. As the following example illustrates, when A^TA is ill-conditioned it might be necessary to use double precision in both forming and solving the normal equations in order to avoid loss of significant information.

EXAMPLE 2.2.1. It is important to note that information in the data matrix A may be lost when A^TA is formed unless sufficient precision is used. As a simple example, consider the matrix in Läuchli [517, 1961]:

$$A = \begin{pmatrix} 1 & 1 & 1 \\ \epsilon & & \\ & \epsilon & \\ & & \epsilon \end{pmatrix}, \quad A^TA = \begin{pmatrix} 1+\epsilon^2 & 1 & 1 \\ 1 & 1+\epsilon^2 & 1 \\ 1 & 1 & 1+\epsilon^2 \end{pmatrix}.$$

Assume that $\epsilon = 10^{-3}$ and that six decimal digits are used for the elements of A^TA. Then, since $1 + \epsilon^2 = 1 + 10^{-6}$ is rounded to 1 we lose all information contained in the last three rows of A. ∎

More generally, a similar loss of accuracy will occur in forming the normal matrix

$$A^TA = \begin{pmatrix} A_1^T & \epsilon A_2^T \end{pmatrix} \begin{pmatrix} A_1 \\ \epsilon A_2 \end{pmatrix} = A_1^T A_1 + \epsilon^2 A_2^T A_2,$$

when $\epsilon \leq \sqrt{u}$, where u is the unit roundoff.

2.2.2. The Cholesky factorization. We now consider the solution of the symmetric positive definite system of normal equations (2.2.1). Gauss solved the normal equation by elimination, preserving symmetry, and solving for x by back-substitution. This method is closely related to computing the **Cholesky factorization**

(2.2.7) $$C = R^TR,$$

2.2. The Method of Normal Equations

where R is upper triangular with positive diagonal elements, and solving two triangular systems for the least squares solution x

$$(2.2.8) \qquad R^T z = d, \quad Rx = z.$$

Similarly the solution to the normal equations of second kind (2.2.2) can be computed from $y = Az$, where

$$(2.2.9) \qquad R^T w = c, \quad Rz = w.$$

A constructive proof of the existence of this factorization was given in Theorem 1.1.4. The Cholesky factorization was developed to solve least squares problems in geodetic survey problems, and was first published by Benoit [61, 1924].[2] (In statistical applications it is often known as the square root method, although the proper square root of A should satisfy $B^2 = A$.)

When the Cholesky factor R of the matrix $A^T A$ has been computed, the solution of (2.2.8) requires a forward and a backward substitution which takes about $2 \cdot 1/2 n^2 - n^2$ flops. The total work required to solve (1.1.1) by the method of normal equations is $\frac{1}{2}mn^2 + \frac{1}{6}n^3 + O(mn)$ flops.

The Cholesky factorization can be used in a slightly different way to solve the least squares problem.

THEOREM 2.2.1. *Consider the bordered matrix*

$$(2.2.10) \qquad \bar{A} = (A, b) \in R^{m \times (n+1)},$$

and let the corresponding cross-product matrix and its Cholesky factor be

$$\bar{C} = \bar{A}^T \bar{A} = \begin{pmatrix} C & d \\ d^T & b^T b \end{pmatrix} = \bar{R}^T \bar{R}, \qquad \bar{R} = \begin{pmatrix} R & z \\ 0 & \rho \end{pmatrix}.$$

Then the least squares solution and the residual norm are obtained from

$$(2.2.11) \qquad Rx = z, \quad \|b - Ax\|_2 = \rho.$$

Proof. By equating $\bar{A}^T \bar{A}$ and $\bar{R}^T \bar{R}$ it follows that R is the Cholesky factor of $A^T A$ and that

$$R^T z = d, \quad b^T b = z^T z + \rho^2.$$

From (2.2.8) it follows that x satisfies the first equation in (2.2.11). Further, since $r = b - Ax$ is orthogonal to Ax,

$$\|Ax\|_2^2 = (r + Ax)^T Ax = b^T Ax = b^T A R^{-1} R^{-T} A^T b = z^T z,$$

and hence $\|r\|_2^2 = b^T b - z^T z$. ∎

Working with the bordered matrix (2.2.10) often simplifies other methods for solving least squares problems as well. Note that if $b \in R(A)$, then $\rho = 0$ and $\bar{R}$ is singular.

---
[2]Sometimes this factorization is attributed to Banachiewicz [34, 1938].

We now describe several variants of an algorithm for computing the Cholesky factorization. Here and in the following we express our algorithms in a programming-like language, which is precise enough to express important algorithmic concepts, but permits suppression of unimportant details. The notations should be self-explanatory.

ALGORITHM 2.2.1. CHOLESKY FACTORIZATION (COLUMNWISE VERSION). Given a symmetric positive definite matrix $M \in R^{n \times n}$ the following algorithm computes the unique upper triangular matrix R with positive diagonal elements such that $C = R^T R$:

$$\text{for } j = 1:n$$
$$\quad \text{for } i = 1:j-1$$
$$\quad\quad r_{ij} = \left(c_{ij} - \sum_{k=1}^{i-1} r_{ki} r_{kj}\right)/r_{ii}$$
$$\quad \text{end}$$
$$\quad r_{jj} = \left(c_{jj} - \sum_{k=1}^{j-1} r_{kj}^2\right)^{1/2}$$
$$\text{end}$$

The algorithm requires about $n^3/6$ flops. Note that R is computed column by column and that the elements r_{ij} can overwrite the elements c_{ij}. At any stage we have the Cholesky factor of a leading principal minor of A. It is also possible to sequence the algorithm so that R is computed row by row.

ALGORITHM 2.2.2. CHOLESKY FACTORIZATION (ROW-WISE VERSION).

$$\text{for } i = 1:n$$
$$\quad r_{ii} = \left(c_{ii} - \sum_{k=1}^{i-1} r_{ki}^2\right)^{1/2}$$
$$\quad \text{for } j = i+1:n$$
$$\quad\quad r_{ij} = \left(c_{ij} - \sum_{k=1}^{i-1} r_{ki} r_{kj}\right)/r_{ii}$$
$$\quad \text{end}$$
$$\text{end}$$

The two versions of the Cholesky algorithm are numerically equivalent, i.e., they will compute the same factor R, even taking rounding errors into account. We remark that from

$$r_{ji}^2 \leq \sum_{k=1}^{i} r_{ki}^2 = c_{ii}, \quad j \leq i,$$

it follows that the elements in R are bounded in terms of the diagonal elements in C. This is essential for the numerical stability of the Cholesky factorization. The different sequencing in these two versions of the Cholesky factorization is illustrated in Figure 2.2.1. For a further discussion of the different ways of

2.2. THE METHOD OF NORMAL EQUATIONS

sequencing the operations in the Cholesky factorization, see George and Liu [336, 1981, pp. 15–20].

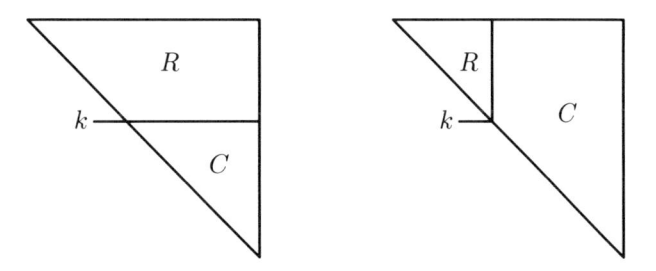

FIG. 2.2.1. *Computed part of the Cholesky factor R after k steps in the row-wise and columnwise algorithm.*

We finally describe a version of Cholesky factorization, which has the advantage that it can be adapted to use diagonal pivoting. Although the Cholesky factorization is stable without pivoting, this is needed when C is only positive semidefinite, i.e., $x^T C x \geq 0$ whenever $x \neq 0$. An important application of pivoting in the Cholesky factorization is to rank deficient least squares problems; see Section 2.7.

ALGORITHM 2.2.3. CHOLESKY FACTORIZATION (OUTER PRODUCT VERSION).

$$
\begin{aligned}
&\text{for } k = 1 : n - 1 \\
&\quad d_k = c_{kk}; \\
&\quad \text{for } i = k + 1 : n \\
&\quad\quad r_{ki} = c_{ki}/d_k; \\
&\quad\quad \text{for } j = k + 1 : i \\
&\quad\quad\quad c_{ji} = c_{ji} - r_{ki} d_k r_{kj}; \\
&\quad\quad \text{end} \\
&\quad \text{end} \\
&\text{end}
\end{aligned}
$$

Algorithm 2.2.3 also differs in that it produces a factorization of the form $C = R^T D R$, where D is diagonal and R is *unit* upper triangular, i.e., has unit elements on its diagonal. Since this algorithm does not require square roots, it may be slightly faster. The Cholesky factorization can be obtained by taking $d_k = (c_{kk}^{(k)})^{1/2}$ in Algorithm 2.2.3, which has the advantage of retaining compatibility with the QR decomposition of A.

Note that the elements in R and D can overwrite the elements in the upper triangular part of C. Pivoting can be incorporated in this algorithm by choosing, at each stage, the pivot as the maximal diagonal element in the current reduced matrix. In the kth step we compute the quantities

$$(2.2.12) \qquad s_j^{(k)} = c_{jj} - \sum_{i=1}^{k-1} r_{ij}^2, \quad j = k, \ldots, n,$$

select an index p such that

$$s_p^{(k)} \geq s_j^{(k)}, \quad j = k, \ldots, n,$$

and interchange rows and columns k and p. Obviously, this will maximize the diagonal element r_{kk}. The quantities $s_j^{(k)}$ can be recursively updated, which lowers the overhead of pivoting to $O(n^2)$ flops. Note that

$$(A\Pi)^T A\Pi = \Pi^T A^T A\Pi = \Pi^T C \Pi, \quad C = A^T A,$$

i.e., a permutation of the columns of A is equivalent to a symmetric permutation of rows and columns of the matrix C.

Block versions of the Cholesky algorithm can be developed, which are more efficient in high performance computing since the main work is performed as matrix-matrix multiplications. Such algorithms are also useful when the matrix C is too large to be stored in the main memory. If, for simplicity, we assume that C can be partitioned into $N \times N$ blocks with square diagonal blocks, we get the following algorithm using a block row-wise order.

ALGORITHM 2.2.4. BLOCK CHOLESKY FACTORIZATION.

$$\begin{aligned}
&\text{for } I = 1 : N \\
&\qquad R_{II}^T R_{II} = C_{II} - \sum_{K=1}^{I-1} R_{KI}^T R_{KI}; \\
&\qquad \text{for } J = I+1 : N \\
&\qquad\qquad R_{IJ}^T = \left(C_{JI} - \sum_{K=1}^{I-1} R_{KJ}^T R_{KI} \right) R_{II}^{-1}; \\
&\qquad \text{end} \\
&\text{end}
\end{aligned}$$

Note that the diagonal block R_{II} is obtained by computing the Cholesky factorization of the matrix

$$C_{II} - \sum_{K=1}^{I-1} R_{KI}^T R_{KI}.$$

The multiplication with R_{II}^{-1} in the computation of $X = R_{IJ}^T$ is performed by solving the triangular systems of the form $R_{II}^T X = C^T$ by forward substitution:

$$R_{II}^T R_{IJ} = \left(C_{JI} - \sum_{K=1}^{I-1} R_{KJ}^T R_{KI} \right)^T.$$

An operation count shows that for large N the main work in the block Cholesky algorithm is in the matrix-matrix multiplications $R_{KJ}^T R_{KI}$.

2.2. THE METHOD OF NORMAL EQUATIONS

2.2.3. Conditioning and scaling. A detailed error analysis for the Cholesky factorization is carried out in Wilkinson [838, 1968]. We state the main result below.

THEOREM 2.2.2. *Let $C \in R^{n \times n}$ be a symmetric positive definite matrix. Provided that*

(2.2.13) $$2n^{3/2}u\kappa(C) < 0.1$$

the Cholesky factor of C can be computed without breakdown, and the computed $\bar{R}$ will satisfy

(2.2.14) $$\bar{R}^T \bar{R} = C + E, \quad \|E\|_2 < 2.5n^{3/2}u\|R\|_2^2.$$

Hence, $\bar{R}$ is the exact Cholesky factor of a matrix close to C.

Proof. See Wilkinson [838, 1968]. ∎

For large n the constants 2 and 2.5 in (2.2.13) and (2.2.14) can be improved to 1 and $\frac{2}{3}$, respectively.

Let $C = A^T A$ be positive definite and $C = R^T R$ its Cholesky factorization. Then

(2.2.15) $$\|C\|_2 = \max_{\|x\|_2=1} x^T C x = \max_{\|x\|_2=1} \|Rx\|_2^2 = \|R\|_2^2 = \|A\|_2^2,$$

which bounds the elements in R without requiring any pivoting. Hence by Theorem 2.2.2 the Cholesky algorithm for computing R from C is backward stable. However, since $\kappa(C) = \kappa^2(A)$, the Cholesky algorithm may fail, i.e., square roots of negative numbers may arise, even when $\kappa(A)$ is of the order of $\sqrt{u}$. This squaring of the condition number will cause a loss of accuracy in the method of normal equations when A is ill-conditioned.

To assess the errors in the least squares solution $\bar{x}$ computed by the method of normal equations, we make the simplifying assumption that no rounding errors occur during the formation of $A^T A$ and $A^T b$. We also neglect errors coming from the solution of the triangular systems (2.2.8), which usually are small. (For an analysis of rounding errors in the solution of triangular systems, see Higham [459, 1989].) Then from Theorem 2.2.2 the computed solution $\bar{x}$ satisfies

$$(A^T A + E)\bar{x} = A^T b, \quad \|E\|_2 < 2.5n^{3/2}u\|A\|_2^2.$$

A perturbation analysis then shows that

(2.2.16) $$\|\bar{x} - x\|_2 \leq 2.5n^{3/2}u\kappa^2(A)\|x\|_2.$$

Examples are sometimes given where the errors in the method of normal equations are much smaller than the bounds in (2.2.16). Usually these results are a consequence of the fact that the error bounds can often be sharpened significantly by a scaling argument. By carefully studying the proof of Theorem 2.2.2 it is seen that if $C = A^T A$ the requirement (2.2.13) can be replaced with $2n^{3/2}u(\kappa')^2 < 0.1$, where

(2.2.17) $$\kappa' = \min_D \kappa(AD), \quad D > 0,$$

is the minimum condition number under a diagonal scaling. Similarly, the bound (2.2.16) can be improved to

$$\|\bar{x} - x\|_2 \leq 2.5 n^{3/2} u \kappa' \kappa(A) \|x\|_2. \tag{2.2.18}$$

Note that these improvements are achieved *without* explicitly carrying out any scaling in the algorithm; see Björck [88, 1978]. Moreover, from Theorem 1.4.7 it follows that choosing D so that the columns in AD have unit length is nearly optimal.

EXAMPLE 2.2.2. The matrix $A \in R^{21 \times 6}$ with elements

$$a_{ij} = (i-1)^j, \quad 1 \leq i \leq 21, \quad 1 \leq j \leq 6,$$

arises when fitting a fifth-degree polynomial $p(t) = x_0 + x_1 t + x_2 t^2 + \cdots + x_5 t^5$ to observations at points $t_i = 0, 1, \ldots, 20$. The condition numbers are

$$\kappa(A) = 6.40 \cdot 10^6, \quad \kappa(AD_1) = 2.22 \cdot 10^3.$$

Thus, by scaling, the condition number is reduced by more than three orders of magnitude. ∎

Sometimes ill-conditioning is caused by an unsuitable formulation of the problem. Then a different choice of parametrization can significantly reduce the condition number. For example, in approximation problems one should try to use orthogonal, or nearly orthogonal, base functions; see Chapter 7. An important example occurs in **linear regression**.

EXAMPLE 2.2.3. In linear regression we want to fit the linear model $y(t) = \alpha + \beta t$ to given data (y_i, t_i), $i = 1, \ldots, m$. This leads to an overdetermined linear system

$$\begin{pmatrix} 1 & t_1 \\ 1 & t_2 \\ \vdots & \vdots \\ 1 & t_m \end{pmatrix} \begin{pmatrix} \alpha \\ \beta \end{pmatrix} = \begin{pmatrix} y_1 \\ y_2 \\ \vdots \\ y_m \end{pmatrix}.$$

From the normal equations

$$\begin{pmatrix} m & \sum_{i=1}^m t_i \\ \sum_{i=1}^m t_i & \sum_{i=1}^m t_i^2 \end{pmatrix} \begin{pmatrix} \alpha \\ \beta \end{pmatrix} = \begin{pmatrix} \sum_{i=1}^m y_i \\ \sum_{i=1}^m y_i t_i \end{pmatrix}$$

we obtain the solution

$$\beta = \left(\sum_{i=1}^m y_i t_i - m \bar{y} \bar{t} \right) / \left(\sum_{i=1}^m t_i^2 - m \bar{t}^2 \right), \qquad \alpha = \bar{y} - \beta \bar{t},$$

where $\bar{y}$ and $\bar{t}$ are the mean values $\bar{y} = (\sum_{i=1}^m y_i)/m$, $\bar{t} = (\sum_{i=1}^m t_i)/m$. Note that the point $(\bar{y}, \bar{t})$ lies on the fitted line.

A more accurate formula for β is obtained by making the change of variable $\tilde{t} = t - \bar{t}$, and writing the model as $y(t) = \tilde{\alpha} + \tilde{\beta} \tilde{t}$. Then $\sum_{i=1}^m \tilde{t}_i = 0$, i.e., the two columns in A are orthogonal and the matrix of normal equations is diagonal. Using the identity $\sum_{i=1}^{m}(y_i - \bar{y})(t_i - \bar{t}) = \sum_{i=1}^{m} y_i(t_i - \bar{t})$ we find

$$(2.2.19) \qquad \beta = \sum_{i=1}^{m}(y_i - \bar{y})(t_i - \bar{t}) / \sum_{i=1}^{m}(t_i - \bar{t})^2, \qquad \alpha = \bar{y} - \beta\bar{t}.$$

When the elements in A and b are the original data, ill-conditioning cannot be avoided by choosing another parametrization. The accuracy of the computed normal equation solution may then depend on the square of the condition number of A. In view of the perturbation result in Theorem 1.4.6 this is not consistent with the mathematical sensitivity of small residual problems. We conclude that the normal equations approach can introduce errors much greater than those of a backward stable algorithm. For mildly ill-conditioned problems this can be offset by fixed precision iterative refinement; see Section 2.9.2.

In the next several sections we review methods for solving least squares problems which work directly with A. In particular, the methods based on orthogonalization will be shown to have very satisfactory stability properties.

2.3. Elementary Orthogonal Transformations

The QR decomposition of $A \in R^{m \times n}$, $m \geq n$, is a decomposition of the form

$$(2.3.1) \qquad A = Q \begin{pmatrix} R \\ 0 \end{pmatrix},$$

where $Q \in R^{m \times m}$ is orthogonal and $R \in R^{n \times n}$ upper triangular. The matrix R is called the R-factor of A. To develop algorithms for computing Q and R we first consider the following standard task.

Given a nonzero vector $a \in R^m$, find an orthogonal matrix U such that

$$(2.3.2) \qquad U^T a = \sigma e_1, \qquad \sigma = \|a\|_2,$$

where e_1 is the first unit vector. From (2.3.2) it follows that $a = \sigma U e_1$, i.e., the first column in U is given by $y = a/\|a\|_2$; compare the proof of Theorem 1.3.1. The construction of such a matrix U plays an important part in developing numerical methods in linear algebra.

2.3.1. Householder transformations. We now show how to solve the construction in (2.3.2) using certain elementary orthogonal matrices. A matrix P of the form
$$(2.3.3) \qquad P = I - \frac{1}{\gamma} u u^T, \qquad \gamma = \frac{1}{2} u^T u,$$

is called a **Householder transformation**, and u a Householder vector. The name derives from A. S. Householder, who in [475, 1958] initiated their use in numerical linear algebra. It is easily verified that P is orthogonal and symmetric,

$$P^T P = I, \qquad P^T = P,$$

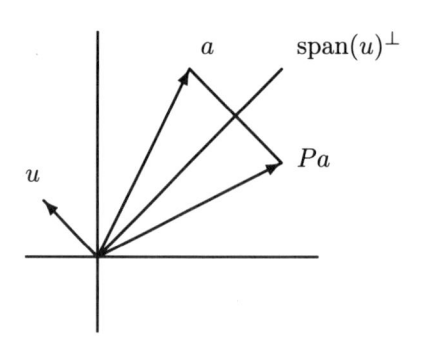

FIG. 2.3.1. *Reflection of a vector a in a hyperplane with normal u.*

and hence $P^2 = I$. The product of P with a given vector x can easily be found without explicitly forming P itself, since

$$Pa = \left(I - \frac{1}{\gamma}uu^T\right)a = a - \frac{1}{\gamma}(u^Ta)u.$$

Note that $Pu = -u$ so that P reverses u and $Pa \in \text{span}\{a, u\}$. Hence the effect of the transformation is to reflect the vector x in the hyperplane with normal vector u; see Figure 2.3.1. Therefore, P is also called a Householder **reflector**.

We now show how to choose u to find a Householder transformation P which solves the standard task. From Figure 2.3.1 it is seen that if we take

(2.3.4) $$u = a \mp \sigma e_1, \quad \sigma = \|a\|_2,$$

then $Pa = \pm \sigma e_1$. Further, with $\alpha_1 = a^T e_1$ we have

(2.3.5) $$u^T u = (a \mp \sigma e_1)^T (a \mp \sigma e_1) = \sigma^2 \mp 2\sigma\alpha_1 + \sigma^2 = 2\sigma(\sigma \mp \alpha_1),$$

so that $\gamma = \sigma(\sigma \mp \alpha_1)$. If a is close to a multiple of e_1, then $\sigma \approx |\alpha_1|$ and cancellation may occur in (2.3.6). This can lead to a large relative error in γ, and to avoid this, one usually takes

(2.3.6) $$u = a + \text{sign}(\alpha_1)\sigma e_1, \quad \gamma = \sigma(\sigma + |\alpha_1|).$$

ALGORITHM 2.3.1. Let a be a vector with $\|a\|_2 = \sigma$ and $a^T e_1 = \alpha_1$. Construct a Householder transformation $P = (I - uu^T/\gamma)$ such that $Pa = \text{sign}(\alpha_1)\sigma e_1 = \hat{\sigma}$, where $\hat{\sigma} = -\text{sign}(\alpha_1)\sigma$.

$$[u, \gamma, \hat{\sigma}] = \text{house}(a)$$
$$\sigma = (a^T a)^{1/2};$$
$$\alpha_1 = a^T e_1;$$
$$\hat{\sigma} = -\text{sign}(\alpha_1)\sigma;$$
$$u = a - \hat{\sigma} e_1;$$
$$\gamma = \sigma(\sigma + |\alpha_1|);$$

2.3. Elementary Orthogonal Transformations

REMARK 2.3.1. The choice of sign in Algorithm 2.3.1 has the small drawback that the vector $a = e_1$ will be mapped onto $-e_1$. The other choice, $u = a - \text{sign}(\alpha_1)\sigma e_1$, may give rise to numerical cancellation in $(\sigma - |\alpha_1|)$. It has been pointed out by Parlett [651, 1980, p. 91] that this can be avoided by rewriting the formulas in the form

$$\gamma = \sigma\tau/(\sigma + |\alpha_1|), \quad \tau = \sum_{i=2}^{m} \alpha_i^2.$$

If $A = (a_1, \ldots, a_n) \in R^{m \times n}$ is a matrix, the product PA is computed as

(2.3.7) $$PA = (Pa_1, \ldots, Pa_n), \quad Pa_j = a_j - \frac{1}{\gamma}(u^T a_j)u.$$

Thus P need not be formed explicitly, and the product can be computed in $2mn$ flops. Writing the product as

$$PA = \left(I - \frac{1}{\gamma}uu^T\right)A = A - \frac{1}{\gamma}u(A^T u)^T$$

shows that A is altered by a matrix of rank one, when premultiplied by a Householder transformation. An analogous algorithm exists for postmultiplying A by a Householder matrix

$$AP = A\left(I - \frac{1}{\gamma}uu^T\right) = A - \frac{1}{\gamma}(Au)u^T.$$

We define a *complex* Householder transformation to have the form

(2.3.8) $$P = I - \frac{1}{\gamma}uu^H, \quad \gamma = \frac{1}{2}u^H u.$$

It can be verified that P is unitary and Hermitian, $P^H P = I$, $P^H = P$. To determine $u \in \mathbf{C}^n$ such that $Px = ke_1$, where

$$x \in \mathbf{C}^n, \quad |k| = \sigma = \|x\|_2, \quad x_1 = e^{i\alpha_1}|x_1|,$$

we take

(2.3.9) $$u = x - ke_1, \quad k = -\sigma e^{i\alpha_1}, \quad \gamma = \sigma(\sigma + |x_1|).$$

Hence the unitary matrix $e^{-i\alpha_1}P$ reduces the arbitrary vector x to real form, $e^{-i\alpha_1}Px = -\sigma e_1$.

2.3.2. Givens transformations. We next consider orthogonal matrices representing plane rotations. These are also called **Givens rotations** after W. Givens [361, 1958]. In two dimensions the matrix representing a rotation clockwise through an angle θ is

(2.3.10) $$R(\theta) = \begin{pmatrix} c & s \\ -s & c \end{pmatrix}, \quad c = \cos\theta, \quad s = \sin\theta.$$

In n dimensions the matrix representing a rotation in the plane spanned by the unit vectors e_i and $e_j, i < j$, is a rank two modification of the unit matrix I_n:

$$
(2.3.11) \qquad R_{ij}(\theta) = \begin{pmatrix} 1 & & & & & & \\ & \ddots & & & & & \\ & & c & & s & & \\ & & & \ddots & & & \\ & & -s & & c & & \\ & & & & & \ddots & \\ & & & & & & 1 \end{pmatrix} \begin{matrix} \\ \\ i \\ \\ j \\ \\ \end{matrix}.
$$

When a column vector $a = (\alpha_1, \ldots, \alpha_n)^T$ is premultiplied by $R_{ij}(\theta)$, we obtain $R_{ij}(\theta) a = b = (\beta_1, \ldots, \beta_n)^T$, where $\beta_k = \alpha_k$, $k \neq i, j$, and

$$
(2.3.12) \qquad \begin{aligned} \beta_i &= c\alpha_i + s\alpha_j, \\ \beta_j &= -s\alpha_i + c\alpha_j. \end{aligned}
$$

Thus a plane rotation may be multiplied into a vector at a cost of two additions and four multiplications. If in (2.3.12) we set

$$
(2.3.13) \qquad c = \alpha_i/\sigma, \quad s = \alpha_j/\sigma, \quad \sigma = (\alpha_i^2 + \alpha_j^2)^{1/2} \neq 0,
$$

then

$$
\begin{pmatrix} c & s \\ -s & c \end{pmatrix} \begin{pmatrix} \alpha_i \\ \alpha_j \end{pmatrix} = \begin{pmatrix} \sigma \\ 0 \end{pmatrix},
$$

i.e., we have introduced a zero in the jth component of the vector a. A more robust algorithm for computing the Givens rotation is given below. This computes c, s, and σ to nearly full machine precision, barring underflow and overflow, which can only occur if the true value of σ itself would overflow.

ALGORITHM 2.3.2. Construct c, s, σ in a Givens rotation such that $-s\alpha + c\beta = 0$:

$$
\begin{aligned}
&[c, s, \sigma] = \text{givrot}(\alpha, \beta) \\
&\quad \text{if } \beta \equiv 0 \\
&\qquad c = 1.0; \quad s = 0.0; \quad \sigma = \alpha; \\
&\quad \text{else if } |\beta| > |\alpha| \\
&\qquad t = \alpha/\beta; \quad tt = \sqrt{1 + t^2}; \\
&\qquad s = 1/tt; \quad c = ts; \quad \sigma = tt\beta; \\
&\quad \text{else} \\
&\qquad t = \beta/\alpha; \quad tt = \sqrt{1 + t^2}; \\
&\qquad c = 1/tt; \quad s = tc; \quad \sigma = tt\alpha; \\
&\quad \text{end}
\end{aligned}
$$

2.3. ELEMENTARY ORTHOGONAL TRANSFORMATIONS

Premultiplication of a matrix $A \in R^{m \times n}$ with a Givens rotation R_{ij} will only affect the two rows i and j in A, which are transformed according to

$$a_{ik} := ca_{ik} + sa_{jk},$$
$$a_{jk} := -sa_{ik} + ca_{jk}, \quad k = 1, 2, \ldots, n.$$

The product requires $4n$ multiplications and $2n$ additions. Similarly, postmultiplying A with R_{ij} will only affect columns i and j.

Givens rotations can be used in several different ways to solve the standard task (2.3.2). We can take

$$R_{1m} \cdots R_{13} R_{12} a = \sigma e_1,$$

where the Givens rotation R_{1k} is determined to zero the kth component in the current vector. Note that R_{1k} only affects the components 1 and k and hence previously introduced zeros will not be destroyed. Another possibility is to take

$$R_{12} R_{23} \cdots R_{n-1,n} a = \sigma e_1,$$

where $R_{k-1,k}$ is chosen to zero the kth component. This demonstrates the greater flexibility of Givens rotations compared to Householder reflections.

It is essential to note that the matrix R_{ij} need never be explicitly formed, but can be represented by the numbers c and s. When a large number of rotations need to be stored it is more economical to store just a single number. This can be done in a numerically stable way using a scheme devised by Stewart [730, 1976]. The idea is to save c or s, whichever is smaller. To distinguish between the two cases we store the reciprocal of c and treat $c = 0$ as a special case. Thus, for the matrix (2.3.10) we define the scalar

$$(2.3.14) \quad \rho = \begin{cases} 1 & \text{if } c = 0, \\ \text{sign}(c)s & \text{if } |s| < |c|, \\ \text{sign}(s)/c & \text{if } |c| \leq |s|. \end{cases}$$

Given ρ the numbers c and s can be retrieved up to a common factor ± 1 by

$$\text{if } \rho = 1, \text{ then } c = 0; \quad s = 1;$$
$$\text{if } |\rho| < 1, \text{ then } s = \rho; \quad c = (1 - s^2)^{1/2};$$
$$\text{if } |\rho| > 1, \text{ then } c = 1/\rho; \quad s = (1 - c^2)^{1/2}.$$

We use this scheme because the formula $\sqrt{1 - x^2}$ gives poor accuracy when $|x|$ is close to unity.

It is also possible to construct unitary plane rotations; see Wilkinson [836, 1965, Sec. 43–44]. We define the matrix R by

$$(2.3.15) \quad R(\theta) = \begin{pmatrix} \bar{c} & \bar{s} \\ -s & c \end{pmatrix}, \quad c = e^{i\alpha} \cos\theta, \quad s = e^{i\beta} \sin\theta,$$

where $|c|^2 + |s|^2 = 1$. When $\alpha = \beta = 0$ the matrix is real and orthogonal. For any vector $x \in \mathbf{C}^n$, given i and $j > i$, we can choose a unitary plane rotation so that $(Rx)_i$ is real and nonnegative and $(Rx)_j = 0$. From

$$(Rx)_j = -sx_i + cx_j = 0,$$

we obtain

(2.3.16) $\qquad c = x_i/\sigma, \quad s = x_j/\sigma, \quad \sigma = (|x_i|^2 + |x_j|^2)^{1/2} \neq 0.$

Then we have

$$(Rx)_i = \bar{c}x_i + \bar{s}x_j = \frac{|x_i|^2}{\sigma} + \frac{|x_j|^2}{\sigma} = \sigma > 0.$$

If we do not require $(Rx)_i$ to be real, we can drop the parameter α in (2.3.15).

2.3.3. Fast Givens transformations. Gentleman [329, 1973] and Hammarling [422, 1974] showed how to reduce the arithmetic when using Givens transformations. We first describe the version used in the Basic Linear Algebra Subprograms (BLAS); see Lawson et al. [522, 1979].

Assume that we want to perform the Givens transformation

$$QA = \begin{pmatrix} \gamma & \sigma \\ -\sigma & \gamma \end{pmatrix} A, \qquad A = \begin{pmatrix} \alpha & a_2 & \dots & a_n \\ \beta & b_2 & \dots & b_n \end{pmatrix},$$

where $\gamma^2 + \sigma^2 = 1$, and the transformation is constructed to zero the element β in A. In fast rotations, the number of multiplications is reduced by keeping the matrix A in the scaled form

(2.3.17) $\qquad A = DA', \qquad D = \begin{pmatrix} d_1 & 0 \\ 0 & d_2 \end{pmatrix}.$

These two factors are then updated separately. The rotation may then be represented in the factored form

$$QA = QDA' = \tilde{D}PA', \qquad QD = \tilde{D}P.$$

In actual computation, in order to avoid square roots, D^2 is stored rather than D. There are several ways to choose the diagonal matrix $\tilde{D}$ so that two elements in P exactly equal unity. This will eliminate $2n$ multiplications when forming the product PA'.

Consider first the case $|\gamma| \geq |\sigma|$, i.e., $|\theta| \leq \pi/4$. Then

$$QD = \begin{pmatrix} d_1\gamma & d_2\sigma \\ -d_1\sigma & d_2\gamma \end{pmatrix} = \gamma D \begin{pmatrix} 1 & (\frac{d_2}{d_1})\frac{\sigma}{\gamma} \\ -(\frac{d_1}{d_2})\frac{\sigma}{\gamma} & 1 \end{pmatrix} = \tilde{D}P,$$

and $\tilde{D}^2 = \gamma^2 D^2$. Since

$$\frac{\sigma}{\gamma} = \frac{\beta}{\alpha} = \frac{d_2}{d_1}\frac{\beta'}{\alpha'},$$

2.3. ELEMENTARY ORTHOGONAL TRANSFORMATIONS

we have

$$(2.3.18) \quad P = \begin{pmatrix} 1 & p_{12} \\ -p_{21} & 1 \end{pmatrix}, \quad p_{21} = \frac{\beta'}{\alpha'}, \quad p_{12} = \left(\frac{d_2}{d_1}\right)^2 p_{21}.$$

Hence we only need the squares of the scale factors d_1 and d_2. These may be updated using the identity $\gamma^2 = (1 + \sigma^2/\gamma^2)^{-1}$, giving

$$(2.3.19) \quad \tilde{d}_1^2 = \frac{d_1^2}{t}, \quad \tilde{d}_2^2 = \frac{d_2^2}{t}, \quad t = 1 + \frac{\sigma^2}{\gamma^2} = 1 + p_{12}p_{21}.$$

Thus we have also managed to eliminate the square root in the Givens transformation.

Similar formulas are easily derived for the other case $|\gamma| < |\sigma|$, i.e., $|\theta| > \pi/4$. We write

$$QD = \begin{pmatrix} d_1\gamma & d_2\sigma \\ -d_1\sigma & d_2\gamma \end{pmatrix} = \sigma \begin{pmatrix} d_2 & 0 \\ 0 & d_1 \end{pmatrix} \begin{pmatrix} (\frac{d_1}{d_2})\frac{\gamma}{\sigma} & 1 \\ -1 & (\frac{d_2}{d_1})\frac{\gamma}{\sigma} \end{pmatrix} = \tilde{D}P.$$

This gives

$$(2.3.20) \quad P = \begin{pmatrix} p_{11} & 1 \\ -1 & p_{22} \end{pmatrix}, \quad p_{22} = \frac{\alpha'}{\beta'}, \quad p_{11} = \left(\frac{d_1}{d_2}\right)^2 p_{22},$$

and further we have

$$(2.3.21) \quad \tilde{d}_1^2 = \frac{d_2^2}{t}, \quad \tilde{d}_2^2 = \frac{d_1^2}{t}, \quad t = 1 + \frac{\gamma^2}{\sigma^2} = p_{11}p_{22} + 1.$$

When using these modified Givens transformations the square of the scale factors is always updated by a factor in the interval $[\frac{1}{2}, 1]$. Thus after many transformations the elements in D may underflow. Therefore the size of the scale factors has to be monitored and occasional rescalings done. This can decrease the efficiency substantially, and although we have eliminated half the multiplications, the fast Givens transformations are not twice as fast in practice. Depending on the computer they tend to be more efficient by factors varying between 1.2 and 1.6.

Recently, self-scaling fast rotations, which obviate the rescalings, have been developed by Anda and Park [14, 1994]. They write the fast rotation in a modified form such that the diagonal scalings are updated by

$$\tilde{d}_1^2 = c^2 d_1^2, \quad \tilde{d}_2^2 = c^{-2} d_2^2,$$

where $c = \cos(\theta)$. Thus the value c^2 is multiplied into one diagonal element and divided into the other, and the diagonal elements no longer decrease monotonically. In [14] four variations of the modified fast rotation are developed for the case of large and small rotation angles and the cases when the ordering of the rows is reversed. The decision among these four variants is made to diminish the larger diagonal element while decreasing the smaller.

2.4. Methods Based on the QR Decomposition

In Theorem 1.3.3 it was shown how the solutions of the primal and dual least squares problems

(2.4.1) $$\min_x \{\|Ax - b\|_2^2 + 2c^T x\},$$

(2.4.2) $$\min_y \|y - b\|_2, \qquad A^T y = c,$$

where $A \in \mathbf{R}^{m \times n}$ has full column rank, could be obtained from the QR decomposition of A. In this chapter we give several algorithms for computing the QR decomposition (QRD) and the resulting algorithms for solving least squares problems.

2.4.1. Householder and Givens QR decomposition. We now describe how the QR decomposition of a matrix $A \in \mathbf{R}^{m \times n}$ of rank n can be computed using a sequence of Householder or Givens transformations. We let $A^{(1)} = A$, and compute a sequence of matrices

$$A^{(k+1)} = P_k A^{(k)}, \qquad k = 1, \ldots, n.$$

Here P_k is an orthogonal matrix chosen to zero the elements below the main diagonal in the kth column of the reduced matrix $A^{(k)}$. Hence the matrix $A^{(k+1)}$ will be triangular in its first k columns, i.e., it has the form

(2.4.3) $$A^{(k+1)} = P_k \cdots P_1 A = \begin{pmatrix} R_{11} & R_{12} \\ 0 & \tilde{A}^{(k+1)} \end{pmatrix}$$

with $R_{11} \in \mathbf{R}^{k \times k}$ upper triangular.

If we put

$$\tilde{A}^{(k)} = (\tilde{a}_k^{(k)}, \ldots, \tilde{a}_n^{(k)}),$$

then we should choose $P_k = \operatorname{diag}(I_{k-1}, \tilde{P}_k)$, where

(2.4.4) $$\tilde{P}_k \tilde{a}_k^{(k)} = \sigma_k e_1, \qquad \sigma_k = r_{kk} = \|\tilde{a}_k^{(k)}\|_2.$$

Note that in this step only the matrix $\tilde{A}^{(k)}$ is transformed, and (R_{11}, R_{12}) are the first $(k-1)$ rows in the final matrix R. It follows that

$$P_n \cdots P_2 P_1 A = A^{(n+1)} = \begin{pmatrix} R \\ 0 \end{pmatrix},$$

and hence

$$Q = (P_n \cdots P_2 P_1)^T = P_1 P_2 \cdots P_n.$$

The key construction is to find an orthogonal matrix $\tilde{P}_k$ which satisfies (2.4.4). However, this is just the standard task (2.3.2) with $a = \tilde{a}_k^{(k)}$. Using (2.3.4) and (2.3.6) to construct $\tilde{P}_k$ as a Householder matrix we get the following algorithm.

2.4. Methods Based on the QR Decomposition

ALGORITHM 2.4.1. QRD BY HOUSEHOLDER TRANSFORMATIONS. Given a matrix $A^{(1)} = A \in \mathbf{R}^{m \times n}$ of rank n, the following algorithm computes R and Householder matrices:

$$P_k = \text{diag}\,(I_{k-1}, \tilde{P}_k), \quad \tilde{P}_k = I - u_k u_k^T/\gamma_k, \quad k = 1, 2, \ldots, n$$

so that $Q = P_1 P_2 \cdots P_n$.

$$\begin{aligned}
&\text{for } k = 1, 2, \ldots, n \\
&\quad [u_k, \gamma_k, r_{kk}] = \text{house}(\tilde{a}_k^{(k)}); \\
&\quad \text{for } j = k+1, \ldots, n \\
&\quad\quad \beta_{jk} = u_k^T \tilde{a}_j^{(k)}/\gamma_k; \\
&\quad\quad \begin{pmatrix} r_{kj} \\ \tilde{a}_j^{(k+1)} \end{pmatrix} = \tilde{a}_j^{(k)} - \beta_{jk} u_k; \\
&\quad \text{end} \\
&\text{end}
\end{aligned}$$

This factorization requires $n^2(m - \frac{1}{3}n)$ flops, which can be compared to the method of normal equations which uses $\frac{1}{2}n^2(m + \frac{1}{3}n)$ flops. Hence, for $m = n$ both methods require the same work but for $m \gg n$ the QR method is twice as expensive.

REMARK 2.4.1. Note that the vectors $\tilde{u}^{(k)}$, $k = 1, 2, \ldots, n$, can overwrite the elements on and below the main diagonal of A. Thus, all information associated with the factors Q and R can be kept in A and two extra vectors of length n for $(r_{11}, \ldots, r_{nn})$ and $(\gamma_1, \ldots, \gamma_n)$.

REMARK 2.4.2. Let $\bar{R}$ denote the computed R. It can be shown that there exists an exactly orthogonal matrix $\tilde{Q}$ (*not* the computed Q) such that

(2.4.5) $$A + E = \tilde{Q} \cdot \bar{R}, \quad \|E\|_F \leq c_1 u \|A\|_F,$$

where the error constant $c_1 = c_1(m, n)$ is a polynomial in m and n, and $\|\cdot\|_F$ denotes the Frobenius norm. Golub and Wilkinson [393, 1966] show that $c_1 = 12.5n$ if inner products are accumulated in double precision. ∎

Normally it is more economical to keep Q in factored form and access it through β_k and $\tilde{u}^{(k)}$, $k = 1, 2, \ldots, n$, than to compute Q explicitly. If Q is explicitly required it can be computed by taking $Q^{(1)} = I$ and computing $Q = Q^{(n+1)}$ by

$$Q^{(k+1)} = P_k Q^{(k)}, \quad k = 1, 2, \ldots, n.$$

If we take advantage of the property that $P_k = \text{diag}(I_{k-1}, \tilde{P}_k)$ this requires $2n(m(m-n) + \frac{1}{3}n^2)$ flops. Alternatively, we can compute

$$Q_1 = Q \begin{pmatrix} I_n \\ 0 \end{pmatrix}, \quad Q_2 = Q \begin{pmatrix} 0 \\ I_{m-n} \end{pmatrix}$$

separately in $n^2(m - \frac{1}{3}n)$ and $n(m-n)(2m-n)$ flops, respectively.

An algorithm similar to Algorithm 2.4.1 but using Givens rotations can easily be developed. In this algorithm the matrix $\tilde{P}_k$ in (2.4.4) is constructed as a product of $(m-k)$ Givens rotations.

ALGORITHM 2.4.2. QRD BY GIVENS TRANSFORMATIONS. Given a matrix $A \in \mathbf{R}^{m \times n}$, $m \geq n$, of rank n the following algorithm overwrites A with $Q^T A = \begin{pmatrix} R \\ 0 \end{pmatrix}$:

$$
\begin{aligned}
&\text{for } k = 1, 2, \ldots, n \\
&\quad \text{for } i = k+1, \ldots, m \\
&\quad\quad [c, s] = \text{givrot}(a_{kk}, a_{ik}); \\
&\quad\quad \text{for } j = k, \ldots, n \\
&\quad\quad\quad t = c a_{k,j} + s a_{i,j}; \\
&\quad\quad\quad a_{i,j} = -s a_{k,j} + c a_{i,j}; \\
&\quad\quad\quad a_{k,j} = t \\
&\quad\quad \text{end} \\
&\quad \text{end} \\
&\text{end}
\end{aligned}
$$

The algorithm uses the procedure in Algorithm 2.3.2 to construct the Givens rotations, and requires a total of $2n^2(m - \frac{1}{3}n)$ multiplications.

REMARK 2.4.3. Using Stewart's storage scheme (2.3.14) for the rotations $R_{ij}(\theta)$ we can store the information defining Q in the zeroed part of the matrix A. As for the Householder algorithm it is advantageous to keep Q in factored form.

REMARK 2.4.4. The error properties of Algorithm 2.4.2 are as good as for the algorithm based on Householder transformations. Wilkinson [836, 1965, p. 240] showed that for $m = n$ the bound (2.4.5) holds with $c_1 = 3n^{3/2}$. Gentleman [330, 1975] has improved this error bound to $c_1 = 3(m+n-2)$, $m \geq n$, and notes that actual errors are observed to grow even more slowly.

2.4.2. Gram–Schmidt orthogonalization. Let $A \in \mathbf{R}^{m \times n}$, $m \geq n = \text{rank}(A)$. The Gram–Schmidt orthogonalization produces $Q_1 \in \mathbf{R}^{m \times n}$ and $R \in \mathbf{R}^{n \times n}$ in the factorization

(2.4.6) $\qquad A = (a_1, \ldots, a_n) = Q_1 R, \qquad Q_1 = (q_1, \ldots, q_n),$

where Q_1 has orthogonal columns and R is upper triangular. The columns of Q_1 in the factorization are obtained by successively orthogonalizing the columns of A. In this paper we survey a number of numerical properties of Gram–Schmidt orthogonalization. We show that in spite of a sometimes bad reputation the Gram–Schmidt algorithm has a number of remarkable properties that make it the algorithm of choice in a variety of applications.

In this section we give several computational variants of Gram–Schmidt orthogonalization. These different versions have an interesting history. The

2.4. Methods Based on the QR Decomposition

"modified" Gram–Schmidt (MGS) algorithm was already derived by Laplace [515, 1816, §2] as an elimination method using weighted row sums; see Farebrother [288, 1988, Chap. 4]. However, Laplace did not interpret his algorithm in terms of orthogonalization, nor did he use it for computing least squares solutions! Bienaymé [73, 1853] gave a similar derivation of a slightly more general algorithm; see Björck [101, 1994]. The idea of elimination with weighted row combinations also appears in Bauer [57, 1965], but without reference to earlier sources. What is now called the "classical" Gram–Schmidt (CGS) algorithm first appeared explicitly *later* in a paper by Schmidt [711, 1908, p. 61], which treats the solution of linear systems with infinitely many unknowns. The orthogonalization is used here as a theoretical tool rather than a computational procedure.

The bad reputation of Gram–Schmidt orthogonalization as a numerical algorithm has arisen mostly because of the (sometimes catastrophic) loss of orthogonality which can occur in the classical algorithm. However, for MGS the loss of orthogonality can be shown to occur in a predictable manner, and be directly proportional to the $\kappa(A)$.

In the **modified Gram–Schmidt** (MGS) algorithm a sequence of matrices, $A = A^{(1)}, A^{(2)}, \ldots, A^{(n)} = Q_1 \in \mathbf{R}^{m \times n}$, is computed, where $A^{(k)}$ has the form

$$A^{(k)} = (q_1, \ldots, q_{k-1}, a_k^{(k)}, \ldots, a_n^{(k)}),$$

and $a_k^{(k)}, \ldots, a_n^{(k)}$ have been made orthogonal to $q_1, \ldots, q_{k-1}$, which are final columns in Q_1. In the kth step we first obtain q_k by normalizing the vector $a_k^{(k)}$:

$$\tilde{q}_k = a_k^{(k)}, \quad r_{kk} = (\tilde{q}_k^T \tilde{q}_k)^{1/2}, \quad q_k = \tilde{q}_k / r_{kk}. \tag{2.4.7}$$

We then orthogonalize $a_{k+1}^{(k)}, \ldots, a_n^{(k)}$ against q_k:

$$a_j^{(k+1)} = a_j^{(k)} - r_{kj} q_k, \quad r_{kj} = q_k^T a_j^{(k)}, \quad j = k+1, \ldots, n. \tag{2.4.8}$$

Equivalently, this can be written

$$a_j^{(k+1)} = P_k a_j^{(k)}, \quad P_k = I - q_k q_k^T, \tag{2.4.9}$$

where P_k is the orthogonal projection onto the complement of q_k. The unnormalized vector $\tilde{q}_k$ is just the orthogonal projection of a_k onto the complement of $\mathrm{span}[a_1, a_2, \ldots, a_{k-1}] = \mathrm{span}[q_1, q_2, \ldots, q_{k-1}]$. After k steps we have obtained the first k columns of Q and the first k rows of R in the QR factorization. After n steps we have obtained the factorization $A = Q_1 R$, where the columns of Q_1 are orthonormal by construction. Since R has positive diagonal elements it equals, in exact computation, the unique upper triangular Cholesky factor of $A^T A$. The MGS algorithm requires mn^2 flops, and is summarized below.

ALGORITHM 2.4.3. MGS (ROW-ORIENTED VERSION). Given $A \in R^{m \times n}$ with $\mathrm{rank}(A) = n$ the following algorithm computes the factorization $A = Q_1 R$:

for $k = 1 : n$
$\quad \hat{q}_k = a_k^{(k)}$;
$\quad r_{kk} = (\hat{q}_k^T \hat{q}_k)^{1/2}$;
$\quad q_k = \hat{q}_k / r_{kk}$;
$\quad$ for $j = k+1 : n$
$\quad\quad r_{kj} = q_k^T a_j^{(k)}$;
$\quad\quad a_j^{(k+1)} = a_j^{(k)} - r_{kj} q_k$;
$\quad$ end
end

It is possible to get a column-oriented version of the MGS algorithm by interchanging the two loops in Algorithm 2.4.3 so that the elements of R are instead computed column by column.

ALGORITHM 2.4.4. MGS (COLUMN-ORIENTED VERSION). Given $A^{(1)} = A \in R^{m \times n}$ with rank$(A) = n$ the following algorithm computes Q_1 and R in the factorization $A = Q_1 R$.

for $k = 1 : n$
$\quad$ for $i = 1 : k-1$
$\quad\quad r_{ik} = q_i^T a_k^{(i)}$;
$\quad\quad a_k^{(i+1)} = a_k^{(i)} - r_{ik} q_i$;
$\quad$ end
$\quad \hat{q}_k = a_k^{(k)}$;
$\quad r_{kk} = (\hat{q}_k^T \hat{q}_k)^{1/2}$;
$\quad q_k = \hat{q}_k / r_{kk}$;
end

The column-oriented version of the MGS Algorithm 2.4.4 was used by Rutishauser [694, 1967]; see also Gander [316, 1980]. It was independently derived by Longley [543, 1981].

REMARK 2.4.5. There is no numerical difference between the column- and row-oriented versions of the MGS algorithm. Since the operations and rounding errors are the same they will produce exactly the same numerical results. However, for treating rank deficient problems column pivoting is necessary (see Section 2.7.3) and then the row-oriented version is preferable.

REMARK 2.4.6. A square root-free version of MGS orthogonalization results if the normalization of the vectors $\hat{q}_k$ is omitted. In this version one computes $\hat{Q}_1$ and $\hat{R}$ so that $A = \hat{Q}_1 \hat{R}$ and $\hat{R}$ *unit* upper triangular. The changes in Algorithm 2.4.4 are to put

$$\hat{r}_{ii} = 1, \quad d_i = \hat{q}_i^T \hat{q}_i, \quad \hat{r}_{ik} = \hat{q}_i^T a_k^{(i)} / d_i$$

and subtract $\hat{r}_{ik} \hat{q}_i$ instead of $r_{ik} q_i$. ∎

Another way to derive the Gram–Schmidt factorization is as follows. Assume that $q_1, \ldots, q_k$ have been determined. Then, by the orthonormality of the q_i, we have $r_{ik} = q_i^T a_k$, $k = 1, \ldots, k-1$, and we can solve for q_k:

$$q_k = \frac{1}{r_{kk}} \hat{q}_k, \qquad \hat{q}_k = a_k - \sum_{i=1}^{k-1} r_{ik} q_i.$$

This leads to the CGS algorithm, where the factors Q_1 and R are generated column by column.

ALGORITHM 2.4.5. CLASSICAL GRAM–SCHMIDT. Given $A \in R^{m \times n}$ with rank$(A) = n$ the following algorithm computes for $k = 1, 2, \ldots, n$ the column q_k of Q_1 and the elements $r_{1k}, \ldots, r_{kk}$ of R in the factorization $A = Q_1 R$:

$$\begin{aligned}
&\text{for } k = 1 : n \\
&\quad \text{for } i = 1 : k - 1 \\
&\quad\quad r_{ik} = q_i^T a_k; \\
&\quad \text{end} \\
&\quad \hat{q}_k = a_k - \sum_{i=1}^{k-1} r_{ik} q_i; \\
&\quad r_{kk} = (\hat{q}_k^T \hat{q}_k)^{1/2}; \\
&\quad q_k = \hat{q}_k / r_{kk}; \\
&\text{end}
\end{aligned}$$

REMARK 2.4.7. In CGS the main work can be performed as a matrix-vector multiplication
$$r_k = Q_{k-1}^T a_k, \qquad \hat{q}_k = a_k - Q_{k-1} r_k,$$
where $Q_{k-1} = (q_1, \ldots, q_{k-1})$ and r_k is the kth column in R (excluding the diagonal element). Hence CGS is better adapted to parallel computing than MGS. ∎

For $n > 2$ the CGS algorithm is numerically different from the modified versions 2.4.3 and 2.4.4. The important difference between the algorithms is that in the modified Algorithm 2.4.3 the projections $r_{ik} q_i$ are subtracted from a_k as soon as they are computed. CGS should not be used numerically without reorthogonalization; see Section 2.4.5. The superiority of the MGS algorithm over CGS for solving least squares problems was experimentally established by Rice [684, 1966], and proved by Björck [85, 1967].

2.4.3. Least squares by Householder QR decomposition. An algorithm for solving the linear least squares problem

(2.4.10) $$\min_x \|Ax - b\|_2$$

based on the Householder QR decomposition was first developed in an important paper by Golub [364, 1965]. This paper, which also discusses column pivoting, iterative refinement, regularization, and updating, started a new epoch in least squares methods.

Golub's method is easily generalized to apply to the more general problems (2.4.1)–(2.4.2).

ALGORITHM 2.4.6. GOLUB'S METHOD. Given $A \in R^{m \times n}$ with rank$(A) = n$, $b \in R^m$, and $c \in R^n$, compute R and $P_1, P_2, \ldots, P_n$ by Algorithm 2.4.1. Then compute z, d, x, and y by

$$(2.4.11) \qquad R^T z = c, \qquad d = P_n \cdots P_2 P_1 b = \begin{pmatrix} d_1 \\ d_2 \end{pmatrix} \begin{matrix} \}n \\ \}m-n \end{matrix},$$

$$(2.4.12) \qquad Rx = (d_1 - z), \qquad y = P_1 \cdots P_{n-1} P_n \begin{pmatrix} z \\ d_2 \end{pmatrix}. \qquad \blacksquare$$

In the standard least squares problem (2.4.10) we have $c = 0$, which implies $z = 0$. Further, the residual vector $r = b - Ax$ equals y, and hence $\|r\|_2 = \|d_2\|_2$. To compute $d = Q^T b$ in (2.4.11) requires $(2mn - n^2)$ flops, and thus x can be computed in $(2mn - n^2/2)$ flops. For the minimum norm problem we have $b = 0$, which implies $d = 0$. In this special case two triangular solves and one multiplication with Q are required, which takes $2mn$ flops.

REMARK 2.4.8. Golub's method for solving the standard least squares problem is normwise backward stable. The computed solution $\hat{x}$ can be shown to be the exact solution of a slightly perturbed least squares problem

$$\min_x \|(A + \delta A)x - (b + \delta b)\|_2,$$

where the perturbations satisfy the bounds

$$(2.4.13) \qquad \|\delta A\|_2 \leq cun^{1/2}\|A\|_2, \quad \|\delta b\|_2 \leq cu\|b\|_2,$$

and $c = (6m - 3n + 41)n$; see Lawson and Hanson [520, 1974, pp. 90 ff]. $\blacksquare$

Golub's method is also normwise backward stable in the special case when $b = 0$; see Higham [467, 1996, Chap. 16]. The stability properties of Golub's method in the general case, $b \neq 0$, $c \neq 0$, are discussed by Björck and Paige [114, 1994].

2.4.4. Least squares problems by MGS. We now consider the use of the MGS algorithm for solving the linear least squares problem. It is important to note that because of the loss of orthogonality in Q_1 that takes place also in MGS, computing $d_1 = Q_1^T b$ and then x from $Rx = d_1$ *will not give an accurate solution*. Fortunately, a backward stable algorithm can be derived by applying the MGS algorithm to the bordered matrix (A, b) to compute the decomposition

$$(2.4.14) \qquad (A, b) = (Q_1, q_{n+1}) \begin{pmatrix} R & z \\ 0 & \rho \end{pmatrix}.$$

Then the solution to the linear least squares problem $\min_x \|Ax - b\|_2$ is given by

$$(2.4.15) \qquad Rx = z, \quad r = \rho q_{n+1}.$$

2.4. Methods Based on the QR Decomposition

To show that (2.4.15) holds, we have from (2.4.14),

$$Ax - b = (A, b)\begin{pmatrix} x \\ -1 \end{pmatrix} = (Q_1, q_{n+1})\begin{pmatrix} R & z \\ 0 & \rho \end{pmatrix}\begin{pmatrix} x \\ -1 \end{pmatrix}$$
$$= Q_1(Rx - z) - \rho q_{n+1}.$$

If q_{n+1} is orthogonal to Q_1, then the minimum of $\|Ax - b\|_2$ occurs when $Rx = z$ and the residual is ρq_{n+1}. Note that it is not necessary to assume that Q_1 is orthogonal for this conclusion to hold. The resulting algorithm can be written as follows.

ALGORITHM 2.4.7. LINEAR LEAST SQUARES SOLUTION BY MGS. Carry out MGS on $A \in R^{m \times n}$, rank$(A) = n$, to give $Q_1 = (q_1, \ldots, q_n)$ and R, and put $b^{(1)} = b$. Compute the vector $z = (z_1, \ldots, z_n)^T$ by

$$\text{for } k = 1, 2, \ldots, n$$
$$z_k = q_k^T b^{(k)}; \quad b^{(k+1)} = b^{(k)} - z_k q_k;$$
$$\text{end}$$
$$\text{solve } Rx = z;$$

The backward stability of Algorithm 2.4.7 can be proved by interpreting the MGS algorithm for QR decomposition as Householder's method applied to the matrix A *augmented with a square matrix of zero elements on top*; see Björck and Paige [113, 1992]. We now outline this relationship. The Householder method computes the decomposition

$$(2.4.16) \qquad \tilde{Q}^T \begin{pmatrix} 0 \\ A \end{pmatrix} = \begin{pmatrix} R \\ 0 \end{pmatrix}, \qquad \tilde{Q}^T = \tilde{P}_n \cdots \tilde{P}_2 \tilde{P}_1,$$

where $\tilde{P}_k = I - 2\hat{v}_k \hat{v}_k^T / \|\hat{v}_k\|_2^2$, $k = 1, 2, \ldots, n$, are Householder transformations. Because of the special structure of the augmented matrix the vectors v_k have the form

$$\hat{v}_k = (-r_{kk} e_k, \hat{q}_k)^T, \qquad r_{kk} = \|\hat{q}_k\|_2,$$

where e_k is the kth unit vector, and we have chosen the sign so that R has a positive diagonal. It follows that

$$(2.4.17) \qquad \tilde{P}_k = I - v_k v_k^T, \qquad v_k = \begin{pmatrix} -e_k \\ q_k \end{pmatrix}, \qquad \|v_k\|_2^2 = 2,$$

where $q_k = \hat{q}_k / r_{kk}$, and since the first n rows are initially zero, the scalar products of the vector v_k with later columns will only involve q_k. Using this observation it is easily verified that the quantities r_{kj} and q_k are *numerically* equivalent to the quantities in the MGS method (2.4.16)–(2.4.17).

Because of the numerical equivalence outlined above the backward error analysis for the Householder QR decomposition can be applied to the MGS algorithm. It follows (see Björck and Paige [113, 1992]) that there exists an *exactly* orthonormal matrix $\hat{Q} \in \mathbf{R}^{m \times n}$ such that for the *computed* matrix $\bar{R}$,

$$(2.4.18) \qquad A + E = \hat{Q}\bar{R}, \qquad \|\tilde{E}\|_2 \leq cu\|A\|_2,$$

where $c = c(m,n)$ are constants depending on the m, n and the details of the arithmetic.

To solve a least squares problem we then apply the orthogonal transformations also to the right-hand side,

$$\begin{pmatrix} d_1 \\ d_2 \end{pmatrix} = \tilde{P}_n \cdots \tilde{P}_2 \tilde{P}_1 \begin{pmatrix} 0 \\ b \end{pmatrix}.$$

Again, using the special form of the Householder matrices $\tilde{P}_k$, it follows that this is numerically equivalent to Algorithm 2.4.7 and gives $d_1 = z$. Hence Algorithm 2.4.7 is a backward stable method for solving the linear least squares problem. Indeed, according to numerical experiments of Jordan [492, 1968] and Wampler [813, 1970] it seems to be slightly more accurate than other orthogonalization methods. In [658, 1970] Peters and Wilkinson write "*Evidence is accumulating that the modified Gram–Schmidt method gives better results than Householder.... The reasons for this phenomenon appear not to have been elucidated yet.*" A possible explanation could be that Householder may be sensitive to row permutations of A (see Section 4.4.3), whereas MGS is (almost) numerically invariant under row permutations.

Using the equivalence, we can also derive a backward stable algorithm for computing the minimum norm solution of an underdetermined linear system,

$$\min \|y\|_2, \quad A^T y = c.$$

ALGORITHM 2.4.8. MINIMUM NORM SOLUTION BY MGS. Carry out MGS on $A^T \in R^{m \times n}$, with rank$(A) = n$ to give $Q_1 = (q_1, \ldots, q_n)$ and R. Then the minimum norm solution $y = y^{(0)}$ is obtained from:

$$R^T (\zeta_1, \ldots, \zeta_n)^T = c;$$
$$y^{(n)} = 0;$$
$$\text{for } k = n, \ldots, 2, 1$$
$$\quad \omega_k = q_k^T y^{(k)}; \quad y^{(k-1)} = y^{(k)} - (\omega_k - \zeta_k) q_k;$$
$$\text{end}$$

REMARK 2.4.9. Note that if the columns of Q_1 are orthogonal, then $\omega = (\omega_1, \ldots, \omega_n) = 0$, but otherwise ω will compensate for the lack of orthogonality.

2.4.5. Gram–Schmidt with reorthogonalization. The Gram–Schmidt algorithm explicitly computes the matrix Q_1, which theoretically provides an orthogonal basis for $R(A)$. This is in contrast to other numerical methods for computing the QR decomposition, in which Q is implicitly defined as a product of Householder or Givens matrices. However, due to roundoff, there will generally be a gradual loss of orthogonality in the computed columns of Q_1, and the computed matrix $\bar{Q}_1$ will not be orthogonal to working accuracy. In the MGS algorithm the loss of orthogonality can be bounded in terms of $\kappa(A)$,

$$\|I - \bar{Q}_1^T \bar{Q}_1\|_2 \leq \frac{c_1}{1 - c_2 \kappa u} \kappa u,$$

2.4. METHODS BASED ON THE QR DECOMPOSITION

where c_1 and c_2 are constants and u is the machine precision; see Björck and Paige [113, 1992]. However, in the CGS algorithm, the computed vectors q_k may depart from orthogonality to an almost arbitrary extent. As pointed out by Gander [316, 1980], even computing Q_1 via the Cholesky decomposition of $A^T A$ seems superior to CGS. A rounding error analysis for the CGS algorithm with reorthogonalization has been given by Abdelmalek [1, 1971].

The more gradual loss of orthogonality in the computed vectors q_i for MGS is illustrated in the example below from Björck [85, 1967].

EXAMPLE 2.4.1. Let A be the matrix in Example 2.2.1 and assume that ϵ is so small that $fl(1 + \epsilon^2) = 1$. If no other rounding errors are made then the orthogonal matrices computed by CGS and MGS, respectively, are

$$\hat{Q}_{CGS} = \begin{pmatrix} 1 & 0 & 0 \\ \epsilon & -\epsilon & -\epsilon \\ 0 & \epsilon & 0 \\ 0 & 0 & \epsilon \end{pmatrix}, \quad \hat{Q}_{MGS} = \begin{pmatrix} 1 & 0 & 0 \\ \epsilon & -\epsilon & -\epsilon/2 \\ 0 & \epsilon & -\epsilon/2 \\ 0 & 0 & \epsilon \end{pmatrix}.$$

For simplicity we have omitted the normalization of $\hat{Q}$. It is easily verified that the maximum deviation from orthogonality of the computed columns are

$$CGS: |q_3^T q_2| = \frac{1}{2}, \quad MGS: |q_3^T q_1| = \left(\frac{2}{3}\right)^{1/2} \epsilon.$$

The condition number of A is $\kappa = \epsilon^{-1}(3 + \epsilon^2)^{1/2} \approx \epsilon^{-1}\sqrt{3}$, and our assumption above implies that $\epsilon^2 \leq u$ (u is the unit roundoff). It follows that for MGS $|q_3^T q_1| \leq \frac{\sqrt{2}}{3}\kappa(A)u$, but for CGS orthogonality has been completely lost. ∎

The reason why the computed columns of Q_1 may depart from orthogonality is that cancellation may take place when the orthogonal projection on q_i is subtracted. In Algorithm 2.4.3 cancellation will occur in the statement $a_k^{(i+1)} = a_k^{(i)} - r_{ik} q_i$ if

(2.4.19) $$\|a_k^{(i+1)}\|_2 < \alpha \|a_k^{(i)}\|_2$$

for some small constant α. To exhibit the loss of orthogonality we consider a case of orthogonalizing two vectors.

EXAMPLE 2.4.2. (See Rutishauser [696, 1976, pp. 96–97].) For the matrix

$$A = \begin{pmatrix} 8 & 21 \\ 13 & 34 \\ 21 & 55 \\ 34 & 89 \end{pmatrix}$$

we get using either of the Gram–Schmidt algorithms and four-digit computation,

$$r_{11} = (a_1^T a_1)^{1/2} = 42.78, \quad q_1 = (0.1870, 0.3039, 0.4909, 0.7948)^T,$$
$$r_{12} = q_1^T a_2 = 112.0, \quad a_2^{(2)} = (0.06, -0.04, 0.02, -0.02)^T.$$

Severe cancellation has taken place since $\|a_2^{(2)}\|_2 = 0.07746 \ll \|a_2\|_2 = 112.0$. This leads to a serious loss of orthogonality between q_1 and q_2,

$$q_1^T q_2 = q_1^T a_2^{(2)}/\|a_2^{(2)}\|_2 = -0.007022/0.07746 = -0.09065.$$

In some applications it is important to compute Q_1 and R such that $Q_1 R$ accurately represents A, and Q_1 is accurately orthogonal. This is the **orthogonal bases problem**. Then it is necessary to reorthogonalize the computed vectors q_k.

EXAMPLE 2.4.3. Continuing the previous example we compute using four-digit computation $\delta r_{12} = q_1^T a_2^{(2)} = -0.007022$ and reorthogonalize by taking

$$\hat{q}_2 := a_2^{(2)} - \delta r_{12} q_1 = (0.06131, -0.03787, 0.02345, -0.01442)^T.$$

Note that the correction δr_{12} is too small to affect $r_{12} = 112.0$. The new vector $\hat{q}_2$ is now accurately orthogonal to q_1,

$$q_1^T \hat{q}_2 / \|\hat{q}_2\|_2 = \frac{6.866 \cdot 10^{-6}}{7.713 \cdot 10^{-2}} = 0.8902 \cdot 10^{-4},$$

and we normalize to get $q_2 = (0.7949, -0.4910, 0.3040, -0.1870)^T$. ∎

In the above example one reorthogonalization was sufficient. It can be shown, in a sense made more precise below, that this is true in general. Thus, reorthogonalization will at most double the cost of the Gram–Schmidt method.

We now describe the **Kahan–Parlett algorithm** (see Parlett [651, 1980, pp. 105–110]) which is based on unpublished notes of Kahan on the fact that "twice is enough." Given vectors q_1, $\|q_1\|_2 = 1$, and a_2, we want to compute

(2.4.20) $$q_2 = a_2 - r_{12} q_1, \quad r_{12} = q_1^T a_2.$$

Assume that we can perform the computation (2.4.20) with an error

$$\|\bar{q}_2 - q_2\|_2 \leq \epsilon \|a_2\|_2,$$

for some small positive ϵ independent of q_1 and a_2, e.g., $\epsilon = O(u)$. Let α be a fixed value in the range $1.2\epsilon \leq \alpha \leq 0.83 - \epsilon$. Then a vector $\tilde{q}_2$ is computed as follows:

$$\bar{q}_2 = fl(\bar{a}_2 - \bar{r}_{12} q_1), \quad \bar{r}_{12} = fl(q_1^T a_2),$$

where fl denotes floating point computation. If $\|\bar{q}_2\|_2 \geq \alpha \|a_2\|_2$, then put $\tilde{q}_2 = \bar{q}_2$; else reorthogonalize $\bar{q}_2$,

$$\check{q}_2 := fl(\bar{q}_2 - \delta r_{12} q_1), \quad \delta r_{12} := fl(q_1^T \bar{q}_2).$$

If $\|\check{q}_2\|_2 \geq \alpha \|\bar{q}_2\|_2$ then accept $\tilde{q}_2 := \check{q}_2$; else accept $\tilde{q}_2 := 0$. The computed $\tilde{q}_2$ then satisfies

(2.4.21) $$\|\tilde{q}_2 - q_2\|_2 \leq (1+\alpha)\epsilon \|a_2\|_2, \quad \|q_1^T \tilde{q}_2\| \leq \epsilon \alpha^{-1} \|\tilde{q}_2\|_2.$$

Note that when α is large, say 0.5, then the bounds (2.4.21) are very good but reorthogonalization will occur more frequently. If α is small reorthogonalization will be rarer, but the bound on orthogonality will be less good. (Rutishauser [694, 1967] used $\alpha = 0.1$.)

Hoffmann [471, 1989] has given an error analysis for both the classical and modified Gram–Schmidt methods with reorthogonalization. He concludes that when Q_1 is required to be orthogonal to full working precision we should choose $\alpha = 0.5$. In this case the iterative classical method is not inferior to the iterative modified method. If less than full precision orthogonality is wanted, then the modified version is better. For the iterated MGS method it has been observed that for all values of α orthogonality is bounded roughly by ϵ/α. The solution of the orthogonal basis problem with Householder's method requires $2(mn^2 - n^3/3)$ flops. Hence, if the average number of reorthogonalizations needed is ν, then the Gram–Schmidt method requires fewer operations when $\nu < 2 - 2n/(3m)$.

Ruhe [690, 1983] analyzes iterated classical and modified Gram–Schmidt methods for orthogonalizing a vector a_j against vectors $Q = (q_1, \ldots, q_{j-1})$, which need not be accurately orthogonal. The resulting vector is $q_j = a_j - Qr_j$, where r_j is given by the solution to the linear system $Q^T Q r_j = Q^T a_j$. The iterated classical algorithm is shown to correspond to the Jacobi iterative method for solving this system, and the MGS method to the Gauss–Seidel iterative method. Ruhe also generalizes the MGS algorithm to oblique projections, which have applications to orthogonalization in elliptic norms and to bi-orthogonalization.

Molinari [581, 1977] points out that there are special situations where even better orthogonality is required than what can be obtained by *one* reorthogonalization. He gives an Algol procedure for "superorthogonalization" which, depending on a parameter, may carry out several reorthogonalizations.

2.4.6. Hybrid algorithms. Sometimes it may be advantageous to carry out a **partial** QR decomposition, where only the first $k < n$ columns are reduced to upper triangular form. If we have computed

$$(2.4.22) \qquad P_k \cdots P_1 A = \begin{pmatrix} R_{11} & R_{12} \\ 0 & A_{22} \end{pmatrix}, \quad P_k \cdots P_1 b = \begin{pmatrix} b_1 \\ b_2 \end{pmatrix},$$

then by orthogonality

$$\|Ax - b\|_2 = \left\| \begin{pmatrix} R_{11} & R_{12} \\ 0 & A_{22} \end{pmatrix} \begin{pmatrix} x_1 \\ x_2 \end{pmatrix} - \begin{pmatrix} b_1 \\ b_2 \end{pmatrix} \right\|_2.$$

Hence, if we solve the reduced least squares problem for x_2,

$$(2.4.23) \qquad \min_{x_2} \|A_{22} x_2 - b_2\|_2,$$

x_1 is obtained by back-substitution from

$$(2.4.24) \qquad R_{11} x_1 = b_1 - R_{12} x_2.$$

Note that any method, e.g., the normal equations, can be used to solve the reduced problem (2.4.23). Such hybrid methods are of interest when the reduced problem is more well-conditioned than the initial least squares problem.

A similar hybrid algorithm based on MGS is easily derived. Assume that after k steps of MGS we have obtained the partial decomposition

$$(A, b) = (Q_k, A^{(k+1)}, b^{(k+1)}) \begin{pmatrix} R_{11} & R_{12} & z_k \\ 0 & I & 0 \\ 0 & 0 & 1 \end{pmatrix}.$$

Then the least squares problem decomposes into (compare (2.4.23))

$$\min_{x_2} \|b^{(k+1)} - A^{(k+1)} x_2\|_2, \qquad R_{11} x_1 = z_k - R_{12} x_2,$$

where $x = (x_1, x_2)^T$. The reduced least squares subproblem may be solved, e.g, by the method of normal equations; see Foster [310, 1991]. A special example occurs in regression analysis, where often the first column of A equals the vector $(1, 1, \ldots, 1)^T$. Taking $k = 1$ in the above partial decomposition is equivalent to "subtracting out the means."

If A is large and sparse, then it is often uneconomical to store and access Q or Q_1. If A is stored, then a possibility for the treatment of additional right-hand sides is to use the seminormal equations (SNE)

(2.4.25) $$R^T R x = A^T b.$$

A similar approach was suggested by Saunders [701, 1972] for use in sparse linear programs to solve the minimum norm problem $\min_x \|x\|_2$, subject to $A^T x = b$. Instead of computing x from $R^T y = b$, $x = Q_1 y$, he used

(2.4.26) $$R^T R w = b, \qquad x = Aw.$$

The method of seminormal equations for the minimum norm problem was analyzed by Paige [623, 1973]. He proved that the method (2.4.26) is numerically quite satisfactory and that "the bound on the error in x is proportional to κu rather than $\kappa^2 u$ as has often been thought." (Here κ denotes the condition number of A, and u the machine precision.)

A similar analysis for the SNE method (2.4.25) given by Björck [91, 1987] shows that unfortunately the satisfactory properties of the SNE method for the minimum norm problem do not carry over. The error in x for the SNE method is of the same size as that for the method of normal equations, even though an R-factor of better quality than that obtained from a Cholesky factorization of $A^T A$ is used. By adding a correction step to the SNE we get a method (CSNE) which, although not backward stable, is much more satisfactory. Here the corrected solution is computed by performing one step of iterative refinement in fixed precision of the solution computed by SNE. The CSNE method is further discussed in Sections 2.9.3 and 6.6.5. An analysis which applies to the more general least squares problems (2.4.1)–(2.4.2) is given in Björck and Paige [114, 1994].

2.4.7. Block algorithms.

Many current computing architectures require code that is dominated by matrix-matrix multiplication in order to attain near-peak performance. This explains the current interest in developing block algorithms, where the main work can be done by so-called Level 3 BLAS; see Dongarra et al. [229, 1990]. Schreiber and Van Loan [713, 1989] describe an efficient block version of the Householder QR decomposition, which is currently implemented in LAPACK; see [16, 1995].

Assume that the matrix $A \in \mathbf{R}^{m \times n}$, is partitioned into p blocks of columns

$$(2.4.27) \qquad A = (A_1, A_2, \ldots, A_N), \qquad A_k \in \mathbf{R}^{m \times p},$$

where for simplicity we assume that $n = Np$. In the first step we compute the Householder QR decomposition

$$A_1 = Q_1 \begin{pmatrix} R_{11} \\ 0 \end{pmatrix}, \qquad Q_1 = H_1 H_2 \cdots H_p,$$

where $H_k = I - u_k u_k^T / \gamma_k$ are Householder reflections. The remaining columns of A are then updated through premultiplication by Q_1^T,

$$(2.4.28) \qquad A_k^{(2)} = Q_1^T A_k, \qquad k = 2, \ldots, N, \qquad Q_1^T = H_p \cdots H_2 H_1.$$

The block algorithm requires that this updating be performed as a matrix multiplication. We now show that Q_1 can be expressed in the form

$$Q_1 = I - P_1 U_1 P_1^T,$$

where $U_1 \in \mathbf{R}^{p \times p}$ is upper triangular. We need the following simple result.

LEMMA 2.4.1. *Given a matrix $Q = (I - PUP^T) \in \mathbf{R}^{m \times m}$, and a vector $u \in \mathbf{R}^m$, we have*

$$(2.4.29) \qquad (I - PUP^T)(I - uu^T) = (I - \hat{P}\hat{U}\hat{P}^T),$$

where

$$(2.4.30) \qquad \hat{P} = (P, u), \quad \hat{U} = \begin{pmatrix} U & v \\ 0 & 1 \end{pmatrix}, \quad v = -U(P^T u).$$

Proof. The result follows by identifying terms in the expression

$$\begin{aligned}(I - PUP^T)(I - uu^T) &= I - PUP^T - uu^T + P(UP^T u)u^T \\ &= I - \hat{P}\hat{U}\hat{P}^T. \quad \blacksquare\end{aligned}$$

The formulas (2.4.29)–(2.4.30) can be used to recursively generate

$$Q_1 = (I - u_1 u_1^T / \gamma_1) \cdots (I - u_n u_n^T / \gamma_n) = I - P_1 U_1 P_1^T$$

by taking $P := u_1/\sqrt{\gamma_1}$, $U := 1$, and for $k = 2, \ldots, p$, computing

$$v_k := -U(P^T u_k)/\sqrt{\gamma_k}, \quad U := \begin{pmatrix} U & v_k \\ 0 & 1 \end{pmatrix}, \quad P := (P, u_k/\sqrt{\gamma_k}).$$

This requires about $\frac{1}{2}r^2(m+\frac{1}{3}r)$ flops. Now (2.4.28) can be performed as

$$A_k^{(2)} = Q_1^T A_k = A_k - P_1(U_1^T(P_1^T A_k)), \quad k = 2,\ldots,N.$$

This requires $2(n-p)(mp+\frac{1}{4}p^2)$ flops, and can be expressed in Level 3 BLAS as two matrix-matrix multiplications and one rank p update. In the next step we compute the QR decomposition of $A_2^{(2)}$, generate $Q_2^T = I - P_2 U_2 P_2^T$, and apply it to the rest of the block columns, etc. All the remaining steps are similar, and the total operation count becomes:

1. Compute the QR decompositions of $A_k^{(k)}$: mnp flops.
2. Generate the matrices U_k: $\frac{1}{2}np(m+\frac{1}{3}p)$ flops.
3. Apply reflections (Level 3 BLAS): $(n^2-np)(m+\frac{1}{4}p)$ flops.

In practice, typically $p = 16$ or $p = 32$ and $p < 0.1n$, and then the overhead $3mnp/2 + n^2p/4$ is small compared to mn^2. The traditional column pivoting strategy cannot be used with the block algorithm since it requires the update of all remaining columns at every step. Bischof [75, 1989] suggests a local pivoting strategy based on an incremental condition estimator. Columns which are found to be nearly linearly dependent on the space spanned by previously chosen columns are rejected and permuted to the end of the matrix. In numerical tests this pivoting strategy correctly identified the rank of A and generated a well-conditioned matrix R. It was also observed that the pivoting did not introduce much extra overhead compared to the block algorithm without pivoting currently implemented in LAPACK.

An analogous block version of MGS can easily be developed. In the first step a factorization $A_1 = Q_1 R_{11}$, $Q_1 = (q_1,\ldots,q_p)$ is computed. In MGS the remaining block columns are updated by

$$A_k^{(2)} = P_1 A_k, \quad k = 2,\ldots,N, \quad P_1 = (I - q_p q_p^T)\cdots(I - q_1 q_1^T).$$

Here P_1^T can again be expressed in the form $P_1^T = I - Q_1 U_1 Q_1^T$ using Lemma 2.4.1.

An even simpler version can be obtained by using a GS decomposition of the column blocks where the matrix Q_k in the decomposition $A_k = Q_k R_{kk}$ is orthogonal to working accuracy. This is achieved by *reorthogonalization* (see Section 2.4.5) and then $U_k = I$ and $P_k = I - Q_k Q_k^T$. Reorthogonalization will increase the operation count in step 1 with mnr flops. On the other hand, we save about $\frac{1}{2}mnr$ flops in step 2 and $n^2r/4$ flops in step 3. Also, the storage space $nr/2$ for the matrices U_k is saved. This method has been analyzed by Jalby and Philippe [484, 1991].

In CGS a new vector is *simultaneously* orthogonalized against previously computed vectors, and the decompositions of A_k can also be performed by Level 3 BLAS. Hence CGS may be preferred to the MGS algorithm in this context.

Malard [559, 1992, Chap. 5] describes block MGS algorithms for local memory multiprocessors. Parallel algorithms for MGS and QR decompositions on message passing systems in which the matrix is distributed by blocks of rows are discussed in O'Leary and Whitman [610, 1990].

2.5. Methods Based on Gaussian Elimination

2.5.1. The Peters–Wilkinson method.
Standard algorithms for solving nonsymmetric linear systems $Ax = b$ are usually based on Gaussian elimination with partial pivoting. Several extensions of this method to solving least squares problems have been suggested; see, e.g., Ben-Israel and Wersan [59, 1963], Tewarson [774, 1968], and the excellent survey by Noble [602, 1976]. Peters and Wilkinson [658, 1970] developed methods based on Gaussian elimination from a uniform standpoint.

A rectangular matrix $A \in \mathbf{R}^{m \times n}$, $m \geq n$, can be reduced by Gaussian elimination with partial pivoting to an upper triangular form U. In general column interchanges are needed to ensure numerical stability. In the full rank case, rank$(A) = n$, the resulting LU factorization becomes

$$(2.5.1) \qquad \Pi_1 A \Pi_2 = \begin{pmatrix} A_1 \\ A_2 \end{pmatrix} = LU = \begin{pmatrix} L_1 \\ L_2 \end{pmatrix} U,$$

where $L_1 \in \mathbf{R}^{n \times n}$ is unit lower triangular and $U \in \mathbf{R}^{n \times n}$ is upper triangular and nonsingular. Thus the matrix L has the same dimensions as A and a lower trapezoidal structure. Computing this factorization requires $\frac{1}{2}n^2(m - \frac{1}{3}n)$ flops.

Using the LU factorization (2.5.1) and setting $\tilde{x} = \Pi_2^T x$, $\tilde{b} = \Pi_1 b$, the least squares problem $\min_x \|Ax - b\|_2$ is reduced to

$$(2.5.2) \qquad \min_y \|Ly - \tilde{b}\|_2, \qquad U\tilde{x} = y.$$

If partial pivoting by rows is used in the factorization (2.5.1), then L is usually a well-conditioned matrix. In this case the solution to the least squares problem (2.5.2) can be computed from the normal equations

$$L^T L y = L^T \tilde{b},$$

without substantial loss of accuracy. This is the approach taken by Peters and Wilkinson [658, 1970]. The following example shows that this is a more stable method than using the normal equation $A^T A x = A^T b$.

EXAMPLE 2.5.1. (Noble [602, 1976]) Consider the matrix A and its generalized inverse

$$A = \begin{pmatrix} 1 & 1 \\ 1 & 1 + \epsilon^{-1} \\ 1 & 1 - \epsilon^{-1} \end{pmatrix}, \qquad A^\dagger = \frac{1}{6} \begin{pmatrix} 2 & 2 - 3\epsilon^{-1} & 2 + 3\epsilon^{-1} \\ 0 & 3\epsilon^{-1} & -3\epsilon^{-1} \end{pmatrix}.$$

The (exact) matrix of normal equations is

$$A^T A = \begin{pmatrix} 3 & 3 \\ 3 & 3+2\epsilon^2 \end{pmatrix}.$$

If $\epsilon \leq \sqrt{u}$, then in floating point computation $fl(3+2\epsilon^2) = 3$, and the computed matrix $fl(A^T A)$ has rank one. However, in the LU factorization of A

$$A = LU = \begin{pmatrix} 1 & 0 \\ 1 & 1 \\ 1 & -1 \end{pmatrix} \begin{pmatrix} 1 & 1 \\ 0 & \epsilon \end{pmatrix}.$$

Here L is well-conditioned, and the pseudoinverse can be stably computed from $A^\dagger = U^{-1}(L^T L)^{-1} L^T$. ∎

Forming the symmetric matrix $L^T L$ requires $\frac{1}{2}n^2(m - \frac{2}{3}n)$ flops, and computing its Cholesky factorization takes $n^3/6$ flops. Hence, neglecting terms of order n^2, the total number of flops to compute the least squares solution by the Peters–Wilkinson method is $n^2(m - \frac{1}{3}n)$. This is always more expensive than the method of normal equations applied to $A^T A$.

Sautter [704, 1978] has shown that when $m-n < n$ an algebraic reformulation is advantageous. If we let $T = L_2 L_1^{-1}$ and $L_1 y = z$, problem (2.5.2) becomes

$$\min_z \left\| \begin{pmatrix} I_n \\ T \end{pmatrix} z - \begin{pmatrix} \tilde{b}_1 \\ \tilde{b}_2 \end{pmatrix} \right\|_2.$$

The solution z can be computed from

$$\begin{aligned} z &= (I_n + T^T T)^{-1}(\tilde{b}_1 + T^T \tilde{b}_2) \\ &= \tilde{b}_1 + (I_n + T^T T)^{-1} T^T (\tilde{b}_2 - T\tilde{b}_1) \\ &= \tilde{b}_1 + T^T (I_{m-n} + TT^T)^{-1}(\tilde{b}_2 - T\tilde{b}_1). \end{aligned}$$ (2.5.3)

The last expression can be evaluated more efficiently if $m-n < n$ and leads to the most efficient method for solving slightly overdetermined least squares problems. (Note that for $m = n+1$ the inversion in (2.5.3) is reduced to a division.)

Methods based on the factorization (2.5.1) for solving the minimum norm problem $\min \|y\|_2$, subject to $A^T y = c$ can be similarly developed. Setting $\tilde{c} = \Pi_2^T c$ and $\tilde{y} = \Pi_1 y$, we have

$$\tilde{y} = (U^T L^T)^\dagger \tilde{c} = L(L^T L)^{-1} U^{-T} \tilde{c}.$$

For the case $m - n < n$ we note that from $U^T L^T \tilde{y} = \tilde{c}$ we have

$$\tilde{y}_1 = L_1^{-T} U^{-T} \tilde{c} - (L_2 L_1^{-1})^T \tilde{y}_2 = e - T^T \tilde{y}_2.$$ (2.5.4)

Hence $\tilde{y}_2$ can be obtained as the solution to the least squares problem

$$\min_{\tilde{y}_2} \left\| \begin{pmatrix} T^T \\ I_{m-n} \end{pmatrix} \tilde{y}_2 - \begin{pmatrix} e \\ 0 \end{pmatrix} \right\|_2,$$

2.5. Methods Based on Gaussian Elimination

or using the normal equations,

(2.5.5) $$\tilde{y}_2 = (I_{m-n} + TT^T)^{-1}Te.$$

The reformulation used above for the almost square case follows from a useful identity, which holds for any matrix S of dimension $r \times (n-r)$ of rank r:

(2.5.6) $$(I_r + S^TS)^{-1}S^T = S^T(I_{n-r} + SS^T)^{-1}.$$

This identity is easily proved using the Woodbury formula (3.1.6). It can be interpreted as an algebraic relation between certain pseudoinverses.

THEOREM 2.5.1. *For any matrix* $S \in \mathbf{R}^{r \times (n-r)}$ *we have* $(\,I_r \quad S\,)^\dagger = C$, *where*

(2.5.7) $$C = \begin{pmatrix} C_1 \\ C_2 \end{pmatrix}, \quad C_1 = I_r - SC_2, \quad C_2 = \begin{pmatrix} S \\ -I_{n-r} \end{pmatrix}^\dagger \begin{pmatrix} I_r \\ 0 \end{pmatrix}.$$

The above theorem reduces the computation of the pseudoinverse of a matrix of rank p to the computation of the pseudoinverse of a matrix of rank $(n-p)$. If $n - p \ll p$, there is a great gain in efficiency.

Several methods combining LU factorization and orthogonalization methods have been given in the literature. Tewarson [774, 1968] suggested that the least squares problem in (2.5.2) is solved by an orthogonal reduction of L to lower triangular form. Thus the solution is obtained by solving $\tilde{L}y = c_1$ by forward substitution, where

$$L = Q\begin{pmatrix} \tilde{L} \\ 0 \end{pmatrix}, \quad Q^T\Pi_1 b = \begin{pmatrix} c_1 \\ c_2 \end{pmatrix}.$$

Cline [172, 1973] developed such an algorithm, which uses Householder transformations to perform this reduction of L. The kth Householder transformation P_k is here chosen to affect only rows $k, n+1, \ldots, m$ and zero elements in column k below row n. These transformations require $n^2(m-n)$ flops. The total number of flops required for computing the least squares solution x by Cline's method is about $n^2(\frac{3}{2}m - \frac{7}{6}n)$ flops. Since the method of normal equations using the Cholesky factorization on A^TA requires $n^2(\frac{1}{2}m + \frac{1}{6}n)$ flops Cline's method uses fewer operations if $m \le \frac{4}{3}n$. (The Golub method requires $n^2(m - \frac{1}{3}n)$ flops, and this is more if $m \le \frac{2}{3}n$.) Hence for slightly overdetermined least squares problems, the elimination method combined with Householder transformations is very efficient.

A version solving (2.5.2) with the MGS method has been analyzed by Plemmons [662, 1974]. If the lower triangular structure of L is taken advantage of then this method requires $n^2(\frac{3}{2}m - \frac{5}{6}n)$ flops, which is slightly more than Cline's variant. Similar methods for the underdetermined case $(m < n)$ based on the LU decomposition of A have been studied by Cline and Plemmons [176, 1976].

Sometimes it may suffice to compute a partial LU factorization. If the Gaussian elimination is stopped after $p < n$ steps, the resulting factorization is

(2.5.8) $$\tilde{A} = \Pi_1 A \Pi_2 = L_p U_p,$$

where the first p rows of the matrix L_p have nonzeros only in a unit lower triangle and the last $n - p$ rows of U_p equal the last $n - p$ rows of the unit matrix I_n,

(2.5.9) $$L_p = \begin{pmatrix} L_{1p} & 0 \\ L_{2p} & \tilde{A}_p \end{pmatrix}, \quad U_p = \begin{pmatrix} U_{1p} & U_{2p} \\ 0 & I \end{pmatrix}.$$

For some problems $\kappa(L_p) \ll \kappa(A)$ already for $p \ll n$. Then p steps of Gaussian elimination suffice, and we can solve the transformed problem

(2.5.10) $$\min_y \|L_p y - \tilde{b}\|_2, \quad \tilde{b} = \Pi_1 b, \quad U_p \Pi_2^T x = D_p^{-1} y,$$

using the method of normal equations. This approach can be useful for weighted least squares problems where the first p equations have a large weight. For such problems it is important to scale the matrix L_p in (2.5.10) so that its columns have equal norm; see Section 4.5.2. Such a method for solving sparse and weighted least squares problems is discussed by Björck and Duff in [104, 1980].

2.5.2. Pseudoinverse solutions from LU decompositions. Peters and Wilkinson [658, 1970] also considered the case when $\mathrm{rank}(A) = r < n$, and showed how pseudoinverse solutions can be computed from LU factorizations. In case the rank is not known beforehand the pivoting strategy is important, and partial pivoting alone will not be sufficient. After p steps the matrix $A = A_0$ has been transformed into a matrix A_p, which is $m \times n$ and of the form

$$A_p = \begin{pmatrix} U_p & V_p \\ 0 & \tilde{A}_p \end{pmatrix}.$$

In complete pivoting if $\tilde{a}_{qs}$ is the element in $\tilde{A}_p$ of largest magnitude, we would interchange columns s and $p+1$ and rows q and $p+1$. Moreover if $|\tilde{a}_{qs}| < tol$, where tol is a tolerance, then $\tilde{A}_p$ is regarded as zero and the elimination stopped. Otherwise $\tilde{a}_{qs}$ is used as a pivot in the next elimination step.

Instead of complete pivoting, it will usually be sufficient to use partial pivoting with a linear independence check. Let $\tilde{a}_{q,p+1}$ be the element of largest magnitude in column $p+1$. If $|\tilde{a}_{q,p+1}| < tol$, column $p+1$ is considered to be linearly dependent and is placed last. We then go on to column $p+2$, etc. Sautter [704, 1978] gives a detailed analysis of stability and rounding errors of the LU algorithm for computing pseudoinverse solutions. Rank revealing LU decompositions have more recently been studied by Hwang, Lin, and Yang in [481, 1992].

Assume that we have computed the LU factorization

(2.5.11) $$\Pi_1 A \Pi_2 = \begin{pmatrix} L_{11} \\ L_{21} \end{pmatrix} \begin{pmatrix} U_{11} & U_{12} \end{pmatrix},$$

where $L_{11}, U_{11} \in \mathbf{R}^{r \times r}$ are triangular and nonsingular. This can be written

$$\Pi_1 A \Pi_2 = \begin{pmatrix} I_r \\ T \end{pmatrix} L_{11} U_{11} \begin{pmatrix} I_r & S \end{pmatrix},$$

2.5. Methods Based on Gaussian Elimination

where T and S can be computed by back-substitution

$$L_{11}T = L_{21}L_{11}^{-1}, \qquad S = U_{11}^{-1}U_{12}.$$

We then get for the pseudoinverse solution

(2.5.12) $$x = A^\dagger \Pi_1^T \tilde{b} = \Pi_2 \begin{pmatrix} I_r & S \end{pmatrix}^\dagger U_{11}^{-1} L_{11}^{-1} \begin{pmatrix} I_r \\ T \end{pmatrix}^\dagger \tilde{b}.$$

Here the pseudoinverse $\begin{pmatrix} I_r & S \end{pmatrix}^\dagger$ of rank r can be reduced using Theorem 2.5.1 to the pseudoinverse

$$\begin{pmatrix} S \\ -I_{n-r} \end{pmatrix}^\dagger$$

of rank $n-r$. If $n-r \ll r$ this is a great savings. If $m-r < r$ a similar reduction can be applied to compute

$$\begin{pmatrix} I_r \\ T \end{pmatrix}^\dagger = \left(\begin{pmatrix} I_r & T^T \end{pmatrix}^\dagger \right)^T.$$

Note the symmetry in the treatment of the L and U factors!

2.5.3. The augmented system method. As remarked in Section 1.1.4, the normal equations (1.2.3) and the defining equations for the residual $r = b - Ax$ combine to form an augmented system of $m+n$ equations

(2.5.13) $$\begin{pmatrix} I & A \\ A^T & 0 \end{pmatrix} \begin{pmatrix} r \\ x \end{pmatrix} = \begin{pmatrix} b \\ 0 \end{pmatrix} \Leftrightarrow Mz = d.$$

The use of this system for solving least squares problems seems first to have been suggested by Siegel [720, 1965]. The system (2.5.13) is square and symmetric but indefinite if $A \neq 0$. It is nonsingular if and only if rank$(A) = n$.

We first note that if m steps of symmetric Gaussian elimination with pivots chosen from the diagonal block I are applied to the system (2.5.13), then the reduced block upper triangular system simply is

$$\begin{pmatrix} I & A \\ 0 & -A^T A \end{pmatrix} \begin{pmatrix} r \\ x \end{pmatrix} = \begin{pmatrix} b \\ -A^T b \end{pmatrix}.$$

Hence, for this choice of pivots we just recover the normal equations. To get a more stable method it is necessary that pivots also be chosen outside the block I.

For a general symmetric indefinite matrix M a factorization LDL^T may not exist and can be unstable, even when symmetric row and column permutations are used at each stage to choose the largest diagonal element in the reduced matrix as a pivot. However, a stable symmetric factorization can always be found if 2×2 symmetric block pivots are also used. We now describe the **Bunch–Kaufman** pivoting scheme. [136, 1977] It suffices to consider the first stage of elimination, since all later stages proceed similarly.

An efficient pivotal strategy is needed that guarantees control of element growth without requiring too much search. The constraint of symmetry allows row and column permutations, which bring any diagonal element $d_1 = b_{rr}$, or any 2×2 submatrix of the form

$$\begin{pmatrix} b_{rr} & b_{rs} \\ b_{sr} & b_{ss} \end{pmatrix} \quad (b_{rs} = b_{sr}),$$

to the pivot position. Using the 2×2 submatrix as a pivot is equivalent to a double step of Gaussian elimination, pivoting first on b_{rs} and then on b_{sr}. It is easily seen that such a double step preserves symmetry, and hence only the elements on and below the main diagonal of the reduced matrix need to be computed. Ultimately, a factorization $A = LDL^T$ is obtained in which D is block diagonal with, in general, a mixture of 1×1 and 2×2 blocks, and L is unit lower triangular with $l_{k+1,k} = 0$ when $B^{(k)}$ is reduced by a 2×2 pivot.

A possible strategy would be to search until two columns r and s have been found for which the common element b_{rs} bounds in modulus the other off-diagonal elements in the r and s columns. Then a 2×2 pivot on these two columns or the largest in modulus of the two diagonal elements as a 1×1 pivot is taken, according to the test

$$\max(|b_{rr}|, |b_{ss}|) \geq \rho |b_{rs}|, \quad \rho = (\sqrt{17} + 1)/8 \approx 0.6404.$$

The number ρ has here been chosen so as to minimize the bound on the growth per stage of elements of B, allowing for the fact that two stages are taken by a 2×2 pivot. With this choice the element growth is bounded by $g_n \leq (1 + 1/\rho)^{n-1} < (2.57)^{n-1}$. No example is known where significant element growth occurs at every step. This bound can be compared to the bound 2^{n-1}, which holds for Gaussian elimination with partial pivoting.

The above bound for element growth can be achieved with fewer comparisons, using the following strategy due to Bunch and Kaufman. First determine the off-diagonal element of largest magnitude in the first column,

$$\lambda = |b_{r1}| = \max_{2 \leq i \leq n} |b_{i1}|.$$

If $|b_{11}| \geq \rho\lambda$, then take b_{11} as a pivot. Else, determine the largest off-diagonal element in column r,

$$\sigma = \max_{1 \leq i \leq n} |b_{ir}|, \quad i \neq r.$$

If $|b_{11}| \geq \rho\lambda^2/\sigma$, then again take b_{11} as a pivot; else if $|b_{rr}| \geq \rho\sigma$, take b_{rr} as a pivot. Otherwise take the 2×2 pivot corresponding to the off-diagonal element b_{1r}. Note that at most two columns need to be searched in each step, and at most n^2 comparisons are needed in all.

When the factorization $M = LDL^T$ has been obtained the solution of $Mz = d$ is obtained in the three steps

$$Lv = d, \quad Dw = v, \quad L^T z = w.$$

2.5. Methods Based on Gaussian Elimination

It has been shown by Higham [468, 1995] that for stability it is necessary to solve the 2×2 systems arising in $Dw = v$ using partial pivoting or the explicit 2×2 inverse. The proof of this is nontrivial and makes use of the special relations satisfied by the elements of the 2×2 pivots in the Bunch–Kaufman pivoting scheme.

Whenever a 2×2 pivot is used, it holds that $b_{11}b_{rr} \leq \rho^2 |b_{1r}|^2 < |b_{1r}|^2$. It follows that the corresponding 2×2 block in D has a negative determinant $\delta_{1r} = b_{11}b_{rr} - b_{1r}^2 < 0$. Since δ_{1r} equals the product of the eigenvalues, a 2×2 block in D corresponds to a positive and a negative eigenvalue. It follows that when B is positive definite, then all pivots chosen by the Bunch–Kaufman strategy will be 1×1.

Unfortunately, the Bunch–Kaufman pivoting scheme does not generally give a stable method for the least squares problem, since the perturbations introduced by roundoff do not respect the structure of the augmented system. This can be made clear by introducing the scaled vector $s = \alpha^{-1}r$; the augmented system can then be written

$$(2.5.14) \qquad \begin{pmatrix} \alpha I & A \\ A^T & 0 \end{pmatrix} \begin{pmatrix} \alpha^{-1}r \\ x \end{pmatrix} = \begin{pmatrix} b \\ 0 \end{pmatrix} \quad \Longleftrightarrow \quad M_\alpha z_\alpha = d_\alpha,$$

where we assume that

$$0 \leq \alpha \leq \|A\|_2 = \sigma_1(A).$$

Using the pivoting scheme described above, the choice of pivots will depend on the value of α. Note that the scaling parameter α only affects the accuracy by influencing the choice of pivots.

For sufficiently large values of α the Bunch–Kaufman scheme will choose the first m pivots from the diagonal $(1,1)$ block. The resulting reduced system equals the normal equations, which is not a backward stable method. Using smaller values of α will introduce 2×2 pivots of the form

$$\begin{pmatrix} \alpha & a_{1r} \\ a_{1r} & 0 \end{pmatrix},$$

which may improve the stability. This raises the question of the *optimal choice of α for stability*.

The eigenvalues λ of M_α can be expressed in terms of the singular values σ_i, $i = 1, \ldots, n$ of A; see Björck [84, 1967]. If $M_\alpha z = \lambda z$, $z = (s, x)^T \neq 0$, then

$$\alpha s + Ax = \lambda s, \qquad A^T s = \lambda x,$$

or eliminating s, $\alpha \lambda x + A^T A x = \lambda^2 x$. Hence, if $x \neq 0$ then x is an eigenvector and $(\lambda^2 - \alpha \lambda)$ an eigenvalue of $A^T A$. On the other hand, $x = 0$ implies that

$$A^T s = 0, \qquad \alpha s = \lambda s, \qquad s \neq 0.$$

It follows that the $m+n$ eigenvalues of M_α are

$$\lambda = \begin{cases} \frac{\alpha}{2} \pm \sqrt{\frac{\alpha^2}{4} + \sigma_i^2}, & i = 1, 2, \ldots, n, \\ \alpha. \end{cases}$$

If $\text{rank}(A) = r \leq n$, then the eigenvalue α has multiplicity $(m-r)$, and 0 is an eigenvalue of multiplicity $(n-r)$. From this it is easily deduced that if $\sigma_n > 0$ then

$$\min_\alpha \kappa_2(M_\alpha) \approx \sqrt{2}\kappa_2(A)$$

is attained for

$$\alpha = \tilde{\alpha} = 2^{-1/2}\sigma_n(A).$$

Because of the above result $\tilde{\alpha}$ (or σ_n) has been suggested as the optimal scaling factor in the augmented system method. Minimizing $\kappa_2(M_\alpha)$ will minimize the forward bound for the error in z_α,

$$\|\bar{z}_\alpha - z_\alpha\|_2 \leq \frac{\epsilon\kappa(M_\alpha)}{1 - \epsilon\kappa(M_\alpha)}\|z_\alpha\|_2, \quad z_\alpha = \begin{pmatrix} \alpha^{-1}r \\ x \end{pmatrix}.$$

However, α also influences the norm in which the error is measured.

A more refined error analysis which instead minimizes a bound for the errors in $\bar{x}$ and $\bar{y}$ separately has been given by Björck [99, 1992]. It is shown that upper bounds for roundoff errors in the computed solution satisfy

(2.5.15)
$$\begin{pmatrix} \|\bar{r} - r\|_2 \\ \|\bar{x} - x\|_2 \end{pmatrix} \leq cguf(\alpha)\begin{pmatrix} \sigma_1(A) \\ \kappa_2(A) \end{pmatrix},$$

where c is a low degree polynomial, g the growth factor, and

$$f(\alpha) = \left(1 + \frac{\alpha}{\sigma_n}\right)\left(\frac{1}{\alpha}\|r\|_2 + \|x\|_2\right).$$

Here if $x \neq 0$, then $f(\alpha)$ is minimized for

(2.5.16)
$$\alpha = \alpha_{\text{opt}} = \left(\frac{\sigma_n\|r\|_2}{\|x\|_2}\right)^{1/2}.$$

The corresponding minimum value of $f(\alpha)$ is

(2.5.17)
$$f_{\min} = \left(1 + \frac{\alpha_{\text{opt}}}{\sigma_n}\right)^2\|x\|_2 = \left(1 + \frac{\sigma_n}{\alpha_{\text{opt}}}\right)^2\sigma_n^{-1}\|r\|_2.$$

Taking $\alpha = \sigma_n$ we find

$$f(\sigma_n) = 2\left(\frac{1}{\sigma_n}\|r\|_2 + \|x\|_2\right) \leq 2f_{\min},$$

i.e., using $\alpha = \sigma_n$ will at most double the error bound.

We recall that an *acceptable-error stable* algorithm is defined as one which gives a solution whose error size is never significantly worse than the error bound obtained from a tight perturbation analysis. It can be shown that the augmented system method is acceptable-error stable both with $\alpha = \sigma_n$ and $\alpha = \alpha_{\text{opt}}$.

2.6. Computing the SVD

2.6.1. SVD and least squares problems.
If the singular value decomposition (SVD) $A = U\Sigma V^T \in \mathbf{R}^{m \times n}$ is known, then by Theorem 1.2.10 the minimum norm least squares solution of $\min_x \|Ax - b\|_2$ is given by

$$(2.6.1) \qquad x = A^{\dagger}b = V \begin{pmatrix} \Sigma_r^{-1} & 0 \\ 0 & 0 \end{pmatrix} c, \qquad c = U^T b,$$

where $\text{rank}(A) = r \leq n$. Here $\Sigma_r = \text{diag}(\sigma_1, \ldots, \sigma_r)$, V, and $U^T b$ are required, but U need not be explicitly formed.

The lack of an efficient and stable numerical method for computing the SVD until the late sixties is the main reason why the SVD was not widely used as a computational tool much earlier. The singular values of A are equal to the positive square roots of the eigenvalues of the symmetric matrix $A^T A$ and $A A^T$, and the matrices U and V of left and right singular vectors are the corresponding eigenvectors. However, forming $A^T A$ or $A A^T$ explicitly will lead to a severe loss of accuracy in the smaller singular values. Therefore, this approach will not directly lead to stable numerical methods for computing the SVD.

A stable algorithm for the SVD was first outlined by Golub and Kahan [370, 1965]. They suggested that the matrix A first should be reduced to bidiagonal form by Householder transformation of a Lanczos process. The singular values and vectors can then be computed as eigenvalues and eigenvectors of a special tridiagonal matrix, by a method based on Sturm sequences. Later, Golub [365, 1968] gave an adaptation of the QR algorithm for computing the SVD of the bidiagonal matrix, and described a simplified interpretation of this process due to Wilkinson. The final form of the QR algorithm for computing the SVD was given by Golub and Reinsch [382, 1970]. Basically this is still the method of choice for dense matrices A, and is described in detail below. Later in this section we describe Jacobi-type methods for computing the SVD and some newer methods for computing singular values of bidiagonal matrices to high relative accuracy.

2.6.2. Transformation to bidiagonal form.
In the Golub–Reinsch method the first step in computing the SVD is to reduce A to upper bidiagonal form. This reduction can be performed using a finite sequence of Householder transformations from left and right as follows.

It is no restriction to assume that $m \geq n$, since otherwise we consider A^T. We compute a sequence of matrices, $A = A^{(1)}, A^{(2)}, \ldots, A^{(n-1)}$, where $A^{(k+1)} = Q_k A^{(k)} P_k$, and Q_k and P_k are Householder matrices. After the first step we have

$$A^{(2)} = Q_1 A P_1 = \begin{pmatrix} q_1 & e_2 & 0 & \cdots & 0 \\ 0 & \tilde{a}_{22} & \tilde{a}_{23} & \cdots & \tilde{a}_{2n} \\ 0 & \tilde{a}_{32} & \tilde{a}_{33} & \cdots & \tilde{a}_{3n} \\ \vdots & \vdots & \vdots & & \vdots \\ 0 & \tilde{a}_{m2} & \tilde{a}_{m3} & \cdots & \tilde{a}_{mn} \end{pmatrix}.$$

Here Q_1 is chosen so that $Q_1 A$ has zeros in the first column under the main diagonal. P_1 is then chosen to zero the last $n-2$ elements in the first row. The first column is obviously not touched by P_1, which only affects the last $n-1$ columns. All later steps are similar. In the kth step we take $A^{(k+1)} = Q_k A^{(k)} P_k$, where Q_k is chosen to zero the last $m-k$ elements in the kth column and P_k the last $n-k-1$ elements in the kth row of the matrix $\tilde{A}^{(k)}$. After n steps we have the required form

$$(2.6.2) \qquad Q_B^T A P_B = \begin{pmatrix} B \\ 0 \end{pmatrix}, \quad B = \begin{pmatrix} q_1 & e_2 & & & \\ & q_2 & e_3 & & \\ & & \ddots & \ddots & \\ & & & q_{n-1} & e_n \\ & & & & q_n \end{pmatrix},$$

where
$$Q_B = Q_1 \cdots Q_n \in \mathbf{R}^{m \times m}, \qquad P_B = P_1 \cdots P_{n-2} \in \mathbf{R}^{n \times n}.$$

Since Q_B and P_B are orthogonal the singular values of B equal those of A.

The above reduction requires $2n^2(m - \frac{1}{3}n)$ flops and was first described by Golub and Kahan [370, 1965]. The Householder vectors associated with Q_B can be stored in the lower triangular part of A, and those associated with P_B, in the upper triangular part of A. If Q_B and P_B are explicitly required they can be accumulated in $2n(m(m-n) + \frac{1}{3}n^2)$ and $\frac{2}{3}n^3$ flops, respectively.

A similar algorithm will reduce a complex matrix $A \in \mathbf{C}^{m \times n}$ to *real* bidiagonal form using a sequence of complex Householder transformations; see (2.3.8)–(2.3.9). An algorithm for the singular value decomposition of a complex matrix is given by Businger and Golub in [143, 1969].

A modified algorithm for computing the bidiagonal decomposition (2.6.2) of A, which is more efficient when $m \geq \frac{5}{3}n$, has been developed by Chan [151, 152, 1982]. The idea, also mentioned in Lawson and Hanson [520, 1974, pp. 119, 122], is to begin by computing the QR decomposition of A:

$$(2.6.3) \qquad Q^T A = \begin{pmatrix} R \\ 0 \end{pmatrix}, \qquad R \in \mathbf{R}^{n \times n}.$$

If the SVD algorithm is then applied to R to get $R = U_R \Sigma V^T$, then the SVD of A is given by

$$(2.6.4) \qquad A = Q \begin{pmatrix} U_R \\ 0 \end{pmatrix} \Sigma V^T.$$

The QR decomposition and Householder reduction to bidiagonal form here take $n^2(m - \frac{1}{3}n)$ and $\frac{4}{3}n^3$ operations, respectively. Hence the modified reduction to bidiagonal form requires $n^2(m+n)$ flops.

Using Householder transformations it is not possible to take advantage of the triangular structure when reducing R to bidiagonal form. Using instead Givens transformations, this can be done as follows. In the first step the elements $r_{1n}, \ldots, r_{13}$ in the first row are annihilated in this order. To zero out the element

2.6. COMPUTING THE SVD

r_{1j} a Givens rotation $G_{j-1,j}$ is applied from the right. This introduces a new nonzero element $r_{j,j-1}$, which is annihilated by a rotation $\tilde{G}_{j-1,j}$ from the left. The first few rotations in the process are pictured below:

$$\begin{pmatrix} \times & \times & \times & \times & \oplus \\ & \times & \times & \times & \times \\ & & \times & \times & \times \\ & & & \times & \times \\ & & & + & \times \end{pmatrix} \Rightarrow \begin{pmatrix} \times & \times & \times & \times & 0 \\ & \times & \times & \times & \times \\ & & \times & \times & \times \\ & & & \times & \times \\ & & & \oplus & \times \end{pmatrix} \Rightarrow \begin{pmatrix} \times & \times & \times & \oplus & 0 \\ & \times & \times & \times & \times \\ & & \times & \times & \times \\ & & + & \times & \times \\ & & & 0 & \times \end{pmatrix}.$$

(Here $\oplus$ denotes the element to be zeroed, and fill-in elements are denoted $+$.)

Since *two* Givens rotations are needed to zero each element, the operation count is the same as that for the Householder reduction. If the product of the transformations needs to be accumulated, the Givens reduction will require more work, unless fast Givens transformations are used. A thorough analysis of the computational details is given in Chan [151, 1982].

The reduction by Householder transformations is stable in the following sense. The computed $\bar{B}$ can be shown to be the exact result of an orthogonal transformation from left and right of a matrix $A + E$, where

$$\|E\|_F \leq cn^2 u \|A\|_F$$

and c is a constant of order unity. Moreover, if we use the information stored to generate the products $Q_B = Q_1 \cdots Q_n$ and $P_B = P_1 \cdots P_{n-2}$ then the computed result is close to the matrices Q and P which reduce $A + E$. This will guarantee that the singular values and transformed singular vectors of $\bar{B}$ are accurate approximations to those of a matrix close to A.

2.6.3. The QR algorithm for real symmetric matrices. We first briefly review the implicit QR algorithm for a real symmetric matrix $A \in \mathbf{R}^{n \times n}$, on which the SVD algorithm is based. In this a sequence of matrices $A = A_0, A_1, A_2, \ldots$ is computed by

$$(2.6.5) \quad A_k - \tau_k I = Q_k R_k, \quad R_k Q_k + \tau_k I = A_{k+1}, \quad k = 0, 1, 2, \ldots,$$

where Q_k is orthogonal, R_k is upper triangular, and τ_k is a shift. Hence in each step we first compute the QR decomposition of the matrix $A_k - \tau_k I$. We then multiply the factors in reverse order and add back the shift. From (2.6.5) it follows that

$$A_{k+1} = Q_k^T (A_k - \tau_k I) Q_k + \tau_k I = Q_k^T A_k Q_k.$$

Hence each QR iteration is an orthogonal similarity transformation, and thus the QR iteration preserves the eigenvalues. Under rather general conditions A_k will tend to a diagonal matrix Λ whose elements are the eigenvalues of A. For an account of the convergence theory of the symmetric QR algorithm see

Parlett [651, 1980]. The eigenvectors of A are found by accumulating the product $P_k = Q_0 Q_1 \cdots Q_k$, $k = 1, 2, \ldots$, of the transformations.

The cost for one QR iteration for a full symmetric matrix $A \in \mathbf{R}^{n \times n}$ is about $4n^3/3$ flops, which is too much to make it a practical algorithm. However, a real symmetric matrix A can be reduced to symmetric tridiagonal form by an orthogonal similarity transformation

$$(2.6.6) \qquad Q_T^T A Q_T = T = \begin{pmatrix} \delta_1 & \alpha_2 & & & \\ \alpha_2 & \delta_2 & \alpha_3 & & \\ & \ddots & \ddots & \ddots & \\ & & \alpha_{n-1} & \delta_{n-1} & \alpha_n \\ & & & \alpha_n & \delta_n \end{pmatrix}.$$

The QR algorithm can now be applied to the reduced matrix. It is easily verified that the QR algorithm preserves symmetric tridiagonal form; that is, if T is symmetric tridiagonal, then so is T', where

$$(2.6.7) \qquad T - \tau I = QR, \qquad T' = RQ + \tau I.$$

This reduces the cost of each QR step to only $O(n)$ flops per iteration.

If τ approximates an eigenvalue of A, then it follows from the convergence theory that the elements α_n in the last row and last column of T in (2.6.6) will approach zero very quickly. When $\alpha_n = 0$, δ_n must be an eigenvalue, and the QR iterations can be continued on the deflated matrix obtained by deleting the last row and column in T. The shift τ is usually determined as the eigenvalue λ closest to δ_n of the submatrix

$$\begin{pmatrix} \delta_{n-1} & \alpha_n \\ \alpha_n & \delta_n \end{pmatrix},$$

which is called the **Wilkinson shift**. This shift gives guaranteed *global* convergence and almost always *local cubic convergence*, although quadratic convergence might be possible. The Wilkinson shift may not give the eigenvalues in monotonic order. There are many other possible schemes for choosing the shift; see Parlett [651, 1980, Chap. 8].

We say that T is **unreduced** if all off-diagonal elements α_k, $k = 2, \ldots, n$, are nonzero. Let T be unreduced and λ an eigenvalue of T. Then $\text{rank}(T - \lambda I) = n - 1$ (the submatrix obtained by crossing out the first row and last column of $T - \lambda I$ has nonzero determinant). Hence there is only one eigenvector corresponding to λ and so λ must have multiplicity one. Thus all eigenvalues of an unreduced symmetric tridiagonal matrix are distinct. In the following we assume that T is unreduced, since otherwise it can be split up into smaller unreduced tridiagonal matrices. For such matrices the following important uniqueness theorem holds, which is the basis for the implicit QR algorithm.

THEOREM 2.6.1. *Let $Q = (q_1, \ldots, q_n)$ and $V = (v_1, \ldots, v_n)$ be orthogonal matrices such that both $Q^T A Q = T$ and $V^T A V = S$ are real, symmetric tridiagonal matrices. If $v_1 = q_1$ and T is unreduced then $v_i = \pm q_i$, $i = 2, \ldots, n$.*

2.6. Computing the SVD

In the QR step (2.6.7) the first transformation $P_1 = R_{12}$ is chosen so that $P_1 t_1 = \pm \|t_1\|_2 e_1$, where t_1 is the first column in $T - \tau I$,

$$t_1 = (\delta_1 - \tau, \alpha_2, 0, \ldots, 0)^T.$$

The transformation is determined by a call to Algorithm 2.4.2 to set up a Givens rotation. The result of applying this transformation is pictured below (for $n = 5$)

$$P_1^T T = \begin{pmatrix} \rightarrow \\ \rightarrow \\ \\ \\ \end{pmatrix} \begin{pmatrix} \times & \times & + & & \\ \times & \times & \times & & \\ & \times & \times & \times & \\ & & \times & \times & \times \\ & & & \times & \times \end{pmatrix}, \quad P_1^T T P_1 = \begin{pmatrix} \downarrow & \downarrow & & & \\ \times & \times & + & & \\ \times & \times & \times & & \\ + & \times & \times & \times & \\ & & \times & \times & \times \\ & & & \times & \times \end{pmatrix}.$$

To preserve the tridiagonal form we now choose the transformation $P_2 = R_{23}$ to zero out the element $+$, and postmultiply to get

$$P_2^T P_1^T T P_1 P_2 = \begin{pmatrix} \times & \times & 0 & & \\ \times & \times & \times & + & \\ 0 & \times & \times & \times & \\ & + & \times & \times & \times \\ & & & \times & \times \end{pmatrix}.$$

We continue to chase the element $+$ down, with transformations $P_3 = R_{34}$, and after $P_{n-1} = R_{n-1,n}$ it disappears. We have then obtained a symmetric tridiagonal matrix $Q^T T Q$, where the first column in Q is $P_1 P_2 \cdots P_{n-1} e_1 = P_1 e_1$, which by Theorem 2.6.1 must be the matrix T'.

The simplest way to check when convergence has taken place is to use the criterion

$$|\alpha_i| \leq u(|\delta_{i-1}| + |\delta_i|),$$

where the elements refer to the current values. If this is satisfied we set α_i to zero. The tridiagonal matrix is then split into two smaller tridiagonal matrices, and the QR algorithm applied to each of them. This criterion is sufficient to ensure backward stability of the algorithm in the normwise sense.

2.6.4. The QR algorithm for the SVD. We start with describing a QR algorithm without shift for computing the SVD of a triangular matrix R, which has been discussed by Chandrasekaran and Ipsen [157, 1992]. Although this algorithm is not advocated as a practical algorithm for computing the SVD, it gives insight into the relationship between the QR algorithm for the SVD and that for the symmetric eigenvalue problem. The algorithm starts with $R_0 = R$, and for $k = 0, 1, 2, \ldots$ computes the R_{k+1} from the QR decomposition:

(2.6.8) $$R_k^T = Q_{k+1} R_{k+1}.$$

Hence, in the first iteration, the QR decomposition of the lower triangular matrix $R_0^T = Q_1 R_1$ is computed. In the second iteration, the QR decomposition

$R_1^T = Q_2 R_2$ of the transpose of the resulting upper triangular matrix R_1 is determined. Then R_2 is related to the original matrix by

$$R_2 = Q_2^T R_1^T = Q_2^T R_0 Q_1,$$

and hence R_2 has the same singular values as R_0. These two transformations are mathematically equivalent to one step of the unshifted QR algorithm (see Section 2.6.4) applied to $R_0^T R_0 = A^T A$, since

(2.6.9) $$R_2^T R_2 = R_1 R_1^T = Q_1^T (R_0^T R_0) Q_1.$$

The transformations simultaneously amount to one step of the unshifted QR algorithm to $R_0 R_0^T = AA^T$, since

$$R_2 R_2^T = R_1^T R_1 = Q_2^T (R_0 R_0^T) Q_2.$$

This has been achieved without explicitly forming $A^T A$ or AA^T. The algorithm is of interest in refining complete orthogonal decompositions; see Section 2.7.

If we start by reducing A to a bidiagonal matrix B, the algorithm described above can then be applied to B. We now describe a practical algorithm, which works on B using an implicit shift technique. We first notice that if in (2.6.2), $e_i = 0$, then the matrix B breaks into two upper bidiagonal matrices, for which the singular values can be computed independently. If $q_i = 0$, then B has a singular value equal to zero. Applying a sequence $G_{i,i+1}, G_{i,i+2}, \ldots, G_{i,n}$ of Givens rotations from the left the ith row is zeroed out, and again the matrix breaks up into two parts. Hence we may, without loss of generality, assume that none of the elements $q_1, q_i, e_i,\ i = 2, \ldots, n$ are zero. This assumption implies that the matrix $B^T B$ has nondiagonal elements $\alpha_{i+1} = q_i e_{i+1} \neq 0$, and hence is unreduced. It follows (see Parlett [651, 1981, Sec. 7.7]) that all eigenvalues of $B^T B$ are positive and distinct, and hence for the singular values of B we have

$$\sigma_1 > \cdots > \sigma_n > 0.$$

When the bidiagonal matrix B has been computed we could proceed by forming the symmetric matrix

$$C = \begin{pmatrix} 0 & B^T \\ B & 0 \end{pmatrix} \in \mathbf{R}^{2n \times 2n},$$

whose eigenvalues are $\pm \sigma_i$, $i = 1, \ldots, n$. After reordering rows and columns by an odd/even permutation C becomes a symmetric tridiagonal matrix with zeros on the main diagonal,

(2.6.10) $$T = P^T C P = \begin{pmatrix} 0 & q_1 & & & & & \\ q_1 & 0 & e_2 & & & & \\ & e_2 & 0 & q_2 & & & \\ & & q_2 & 0 & \ddots & & \\ & & & \ddots & \ddots & q_n & \\ & & & & q_n & 0 & \end{pmatrix},$$

2.6. COMPUTING THE SVD

where P is the permutation matrix whose columns are those of the identity in the order $(1, n+1, 2, n+2, \ldots, n, 2n)$. The implicit QR algorithm could then be applied to this special tridiagonal matrix.

A disadvantage of applying the QR algorithm to T is that the dimension is essentially doubled. However, a closer inspection reveals that this algorithm is equivalent to an algorithm where the iterations are carried out directly on B. It is also equivalent to an *implicit* version of the QR algorithm applied to the symmetric tridiagonal matrix $T = B^T B$, which we now describe.

Forming $B^T B$ would lead to a severe loss of accuracy in the small singular values. It is therefore essential to work directly with the matrix B. To determine the Wilkinson shift Golub and Reinsch [382, 1970] used the lower right 2×2 submatrix in $B^T B$. A more robust way to compute the shift is to $\tau = \sigma^2$, where σ is the smallest singular value of the 2×2 upper triangular submatrix in B,

$$\begin{pmatrix} q_{n-1} & e_n \\ 0 & q_n \end{pmatrix}.$$

An algorithm for computing the singular values of such a matrix to full relative accuracy has been given by Demmel and Kahan [218, 1990]; see (2.6.20).

In the implicit shift QR algorithm for $B^T B$ we determine a Givens rotation $T_1 = R_{12}$ so that

$$(2.6.11) \qquad T_1^T t_1 = \pm \|t_1\|_2 e_1, \qquad t_1 = (q_1^2 - \tau, q_1 e_2, 0, \ldots, 0)^T,$$

where t_1 is the first column in $B^T B$ and τ is the computed shift. This rotation can be determined by a call to the subroutine named *givrot* in Algorithm 2.4.2. Next we should apply a sequence of Givens transformations such that

$$T_{n-1}^T \cdots T_2^T T_1^T B^T B T_1 T_2 \cdots T_{n-1}$$

is tridiagonal, but we wish to avoid doing this explicitly. Let us start by applying the transformation T_1 to B. Then we get (take $n = 5$)

$$BT_1 = \begin{pmatrix} \times & \times & & & \\ + & \times & \times & & \\ & & \times & \times & \\ & & & \times & \times \\ & & & & \times \end{pmatrix}.$$

If we now premultiply by a Givens rotation $S_1^T = R_{12}$ to zero out the $+$ element, this creates a new nonzero element in the $(1,3)$ position. To preserve the bidiagonal form we then choose the transformation $T_2 = R_{23}$ to zero out the element $+$:

$$Q_1^T B T_1 = \begin{pmatrix} \times & \times & + & & \\ 0 & \times & \times & & \\ & & \times & \times & \\ & & & \times & \times \\ & & & & \times \end{pmatrix}, \qquad Q_1^T B T_1 T_2 = \begin{pmatrix} \times & \times & 0 & & \\ & \times & \times & & \\ & + & \times & \times & \\ & & & \times & \times \\ & & & & \times \end{pmatrix}.$$

We can now continue to chase the element + down, with transformations alternately from the right and left until we get a new bidiagonal matrix

$$\hat{B} = (S_{n-1}^T \cdots S_1^T) B (T_1 \cdots T_{n-1}) = U^T B P.$$

But then the matrix

$$\hat{T} = \hat{B}^T \hat{B} = P^T B^T U U^T B P = P^T T P$$

is tridiagonal, where the first column of P equals the first column of T_1. If T is unreduced $\hat{T}$ must be the result of one QR iteration on T with shift equal to σ.

The subdiagonal entries of T equal $q_i e_{i+1}$, $i = 1, \ldots, n-1$. If some element e_{i+1} is zero, then the bidiagonal matrix splits into two smaller bidiagonal matrices

$$B = \begin{pmatrix} B_1 & 0 \\ 0 & B_2 \end{pmatrix}.$$

If $q_i = 0$, then we can zero the ith row by premultiplication by a sequence of Givens transformations $R_{i,i+1}, \ldots, R_{i,n}$, and the matrix then splits as above. In practice, two convergence criteria are used. After each QR step if

(2.6.12) $$|e_{i+1}| \leq 0.5 u (|q_i| + |q_{i+1}|),$$

where u is the machine unit, we set $e_{i+1} = 0$. We then find the smallest p and the largest q such that B splits into quadratic subblocks

$$\begin{pmatrix} B_1 & 0 & 0 \\ 0 & B_2 & 0 \\ 0 & 0 & B_3 \end{pmatrix},$$

of dimensions p, $n - p - q$, and q, where B_3 is diagonal and B_2 has a nonzero subdiagonal.

Second, if diagonal elements in B_2 satisfy

(2.6.13) $$|q_i| \leq 0.5 u (|e_i| + |e_{i+1}|),$$

set $q_i = 0$, zero the superdiagonal element in the same row, and repartition B. Otherwise continue the QR algorithm on B_2. A justification for these tests is that roundoff in a rotation could make the matrix indistinguishable from one with a q_i or e_{i+1} equal to zero. Also, the error introduced in the singular values by the tests is not larger than some constant times $u \|B\|_2$.

When all the superdiagonal elements in B have converged to zero we have

(2.6.14) $$Q_S^T B T_S = \Sigma = \operatorname{diag}(\sigma_1, \ldots, \sigma_n).$$

Hence the SVD of A is

(2.6.15) $$A = U \begin{pmatrix} \Sigma \\ 0 \end{pmatrix} V^T, \quad U = Q_B \begin{pmatrix} Q_S & 0 \\ 0 & I_{m-n} \end{pmatrix}, \quad V = T_B T_S.$$

2.6. COMPUTING THE SVD

Usually less than $2n$ iterations are needed in the second phase. One QR iteration requires $14n$ multiplications and $2n$ calls to *givrot*. Accumulating the rotations into U requires $6mn$ flops. Accumulating the rotations into V requires $6n^2$ flops.

An important implementation issue is that the bidiagonal matrix is often graded, i.e., the elements may be large at one end and small at the other. For example, if in the Chan SVD, column pivoting is used in the initial QR decomposition, then the matrix is usually graded from large at upper left to small at lower right, as illustrated below:

$$\begin{pmatrix} 1 & 10^{-1} & & & \\ & 10^{-2} & 10^{-3} & & \\ & & 10^{-4} & 10^{-5} & \\ & & & & 10^{-6} \end{pmatrix}.$$

The QR algorithm tries to converge to the singular values from smallest to largest, and "chases the bulge" from top to bottom. Convergence will then be fast. However, if B is graded the opposite way then the QR algorithm may require many more steps. To avoid this the rows and columns of B should in this case be reversed before the QR algorithm is applied. Many implementations of the algorithm check for the direction of grading. If the matrix breaks up into diagonal blocks which are graded in different ways, the bulge is chased in the appropriate direction.

If singular vectors are desired, the cost of a QR iteration goes up to $4n^2$ flops, and the overall cost, to $O(mn^2)$. To reduce the number of QR iterations where transformations are accumulated, we can first compute the singular values without accumulating singular vectors. If we then run the QR algorithm a second time with shifts based on the computed singular values (*perfect shifts*) convergence occurs in one iteration. This may reduce the cost of the overall computations by about 40%. However, if fewer than 25% of the singular vectors are wanted, then inverse iteration should be used instead. The drawback of this approach is the difficulty of getting orthogonal singular vectors to clustered singular values.

The QR-SVD algorithm can be shown to be normwise backward stable, i.e., the computed singular values $\bar{\Sigma} = \text{diag}(\bar{\sigma}_k)$ are the exact singular values of a nearby matrix $A + \delta A$, where $\|\delta A\|_2 \leq c(m,n) \cdot u\sigma_1$. Here $c(m,n)$ is a constant depending on m, n, and the machine unit u. From Theorem 1.2.7 it follows that

$$|\bar{\sigma}_k - \sigma_k| \leq c(m,n) \cdot u\sigma_1.$$

Thus, if A is nearly rank deficient this will always be revealed by the computed singular values. Note, however, that the smaller singular values may not be computed with high relative accuracy.

For a detailed description of the SVD algorithm we refer to Golub and Reinsch [382, 1970] and Dongarra et al. [228, 1979, Chap. 11]. Chan in [152, 1982] compares the operation count for the two variants of SVD algorithms in four different cases depending on whether U and V are explicitly required, where $U = (U_1, U_2)$. Only the highest order terms in m and n are kept, and so the results are correct for relatively large dimensions. It is assumed that the iterative phase of the SVD takes on average two complete QR iterations per singular value. In Table 2.6.1 $C = 4$ if standard Givens transformations are used, and $C = 2$ if "fast" Givens transformations are used.

TABLE 2.6.1

Comparison of multiplications for SVD algorithms.

Case	Required	Golub–Reinsch SVD	Chan SVD
a	Σ, U_1, V	$(3+C)mn^2 + \frac{11}{3}n^3$	$3mn^2 + 2(C+1)n^3$
b	Σ, U_1	$(3+C)mn^2 - n^3$	$3mn^2 + (C+4/3)n^3$
c	Σ, V	$2mn^2 + Cn^3$	$mn^2 + (C+5/3)n^3$
d	Σ	$2mn^2 - 2n^3/3$	$mn^2 + n^3$

Here case a arises in the computation of the pseudoinverse, case c in least squares applications, and case d in the estimation of condition numbers and rank determination.

2.6.5. Zero shift QR algorithm. Recently Demmel and Kahan [218, 1990] have shown how all singular values of bidiagonal matrices may be computed with high relative accuracy. This is theoretically possible because of the following perturbation result.

THEOREM 2.6.2. *Let $B \in \mathbf{R}^{n \times n}$ be a bidiagonal matrix and suppose $|\delta B| \le \omega |B|$. Let $\sigma_1 \ge \cdots \ge \sigma_n$ be the singular values of B, and let $\bar{\sigma}_1 \ge \cdots \ge \bar{\sigma}_n$ be the singular values of $\bar{B} = B + \delta B$. Then if $\eta = (2n-1)\omega < 1$,*

$$(2.6.16) \qquad |\bar{\sigma}_i - \sigma_i| \le \frac{\eta}{1-\eta} |\sigma_i|, \quad i = 1, \ldots, n,$$

and

$$(2.6.17) \qquad \max\left(\sin \theta(u_i, \tilde{u}_i), \sin \theta(v_i, \tilde{v}_i)\right) \le \frac{\sqrt{2}\eta(1+\eta)}{\text{relgap}_i - \eta},$$

where the **relative gap** *between singular values is*

$$(2.6.18) \qquad \text{relgap}_i = \min_{j \ne i} \frac{|\sigma_i - \sigma_j|}{\sigma_i + \sigma_j}.$$

Proof. See Demmel and Kahan [218, 1990]. ∎

The Demmel–Kahan algorithm is a hybrid of the conventional shifted QR algorithm and a new, stable implementation of QR with a zero shift. As in the standard QR algorithm the singular vectors are also available.

A zero shift algorithm introduces simplifications in the QR step as follows. By (2.6.11), if $\tau = 0$, the first transformation is determined so that

$$T_1^T t_1 = \pm \|t_1\|_2 e_1, \qquad t_1 = (q_1, e_2, 0, \ldots, 0)^T.$$

2.6. Computing the SVD

Hence, after the first rotation we have an extra zero in the $(1,2)$ position:

$$BT_1 = \begin{pmatrix} \downarrow & \downarrow & & & \\ \times & 0 & & & \\ + & \times & \times & & \\ & & \times & \times & \\ & & & \times & \times \\ & & & & \times \end{pmatrix}.$$

This zero will propagate through the algorithm, so that after two more rotations we have an extra zero in the $(2,3)$ position,

$$Q_1^T BT_1 = \begin{pmatrix} \rightarrow & \times & \times & + & \\ \rightarrow & 0 & \times & \times & \\ & & & \times & \times \\ & & & & \times & \times \\ & & & & & \times \end{pmatrix}, \quad Q_1^T BT_1 T_2 = \begin{pmatrix} & & \downarrow & \downarrow & \\ \times & \times & 0 & & \\ & \times & 0 & & \\ & + & \times & \times & \\ & & & \times & \times \\ & & & & \times \end{pmatrix}.$$

The work in one QR iteration is thereby reduced to only $4n$ multiplications and $2n$ calls to *givrot*. Remarkably this algorithm uses no addition or subtractions, and hence the zero shift QR algorithm has the property that it computes each entry of the transformed matrix to high *relative* accuracy.

ALGORITHM 2.6.1. THE DEMMEL–KAHAN ZERO SHIFT QR ALGORITHM. The algorithm performs one step of the implicit zero shift QR algorithm on the bidiagonal matrix B in (2.6.2):

$$\begin{aligned}
&oldc = 1; \\
&f = q_1; \quad g = e_2; \\
&\text{for } i = 1, n-1 \\
&\qquad \text{call givrot}(f, g, c, s, r); \\
&\qquad \text{if } (i \neq 1) \; e_i = olds * r; \\
&\qquad f = oldc * r; \\
&\qquad g = q_{i+1} * s; \quad h = q_{i+1} * c; \\
&\qquad \text{call givrot}(f, g, c, s, r); \\
&\qquad q_i = r; f = h; \\
&\qquad \text{if } (i \neq n-1) \; g = e_{i+2}; \\
&\qquad oldc = c; \quad olds = s; \\
&\text{end} \\
&e_n = h * s; \quad q_n = h * c;
\end{aligned}$$

The zero shift QR algorithm is only used on deflated submatrices of B whose condition number $\kappa = \sigma_{\max}/\sigma_{\min}$ is so large that the standard shifted QR algorithm would make unacceptably large changes in the computed $\sigma_{\min}$.

Although the zero shift algorithm only has linear convergence, it will converge quickly since $\sigma_{\min}/\sigma_{\max}$ is very small.

The Demmel–Kahan algorithm will compute the smallest singular values to maximal relative accuracy and the others to maximal absolute accuracy. Their algorithm represents a major improvement over the original SVD algorithm as implemented in the LINPACK library. The new algorithm is generally faster, and occasionally much faster than the standard algorithm. The zero shift step has several remarkable features. It uses only about a third as many operations as the standard shift version. Further, since no additions or subtractions occur, possible errors from cancellation are avoided. Other important features of the Demmel–Kahan algorithm are stricter convergence criteria and a very sophisticated algorithm for computing accurate singular values and vectors of an upper triangular 2×2 matrix; see Section 2.6.6.

Fernando and Parlett [291, 1994] have given an algorithm based on a variant of the qd-algorithm of Rutishauser [692, 1954]. This algorithm can exploit shift strategies that are at least quadratically convergent, while preserving maximal relative accuracy for all the singular values of the bidiagonal matrix. The error bounds for their algorithm are significantly smaller than those for the Demmel–Kahan approach. However, there are some difficulties when the singular vectors also are desired. A related algorithm using an orthogonal qd-algorithm has been given by von Matt [809, 1995]. This algorithm achieves the same high accuracy for the singular values and has no difficulties with computing both the left and right singular vectors.

2.6.6. Jacobi methods for the SVD. Before the advent of the QR algorithm two different Jacobi-type methods for computing the SVD were developed. In **Kogbetliantz's method** [506, 1955] the "norm" of the off-diagonal elements in A is successively reduced by a sequence of two-sided Givens transformations. Hestenes [449, 1958] developed an algorithm using one-sided Givens transformations. These methods have several features which have made them very popular algorithms for implementation on machines with parallel architectures. These Jacobi-type algorithms are slower than the QR algorithm, but can compute singular values of a general matrix more accurately than any algorithm based on the bidiagonal reduction in Section 2.6.2; see Drmač [233].

In **Hestenes' method** one-sided Givens transformations are used to find an orthogonal matrix V such that the matrix AV has orthogonal columns. Then $AV = U\Sigma$ and the SVD of A is readily obtained. In Hestenes' original algorithm the columns are explicitly interchanged so that the final columns of AV appear in order of decreasing norm. The basic step rotates two columns:

$$(2.6.19) \qquad (\hat{a}_p, \hat{a}_q) = (a_p, a_q) \begin{pmatrix} c & s \\ -s & c \end{pmatrix}, \qquad p < q.$$

The parameters c, s are determined so that the rotated columns are orthogonal, or equivalently so that

2.6. Computing the SVD

$$\begin{pmatrix} c & s \\ -s & c \end{pmatrix}^T \begin{pmatrix} \|a_p\|_2^2 & a_p^T a_q \\ a_q^T a_p & \|a_q\|_2^2 \end{pmatrix} \begin{pmatrix} c & s \\ -s & c \end{pmatrix}$$

is diagonal. This is a 2×2 symmetric eigenproblem and can be solved by a Jacobi transformation. However, this approach may lead to numerical problems since we are squaring part of the matrix. To determine the rotation we instead first compute the QR factorization

$$(a_p, a_q) = (q_1, q_2) \begin{pmatrix} r_{pp} & r_{pq} \\ 0 & r_{qq} \end{pmatrix},$$

and then the 2×2 SVD $R = U \Sigma V^T$ using one of the algorithms given below. Since $(a_p, a_q) V = (q_1, q_2) U \Sigma$ will then have orthogonal columns, V is the desired rotation in (2.6.19).

A normwise backward stable algorithm for computing the SVD of an upper triangular 2×2 matrix

$$\begin{pmatrix} c_l & s_l \\ -s_l & c_l \end{pmatrix} \begin{pmatrix} r_{11} & r_{12} \\ 0 & r_{22} \end{pmatrix} \begin{pmatrix} c_r & -s_r \\ s_r & c_r \end{pmatrix} = \begin{pmatrix} \sigma_1 & 0 \\ 0 & \sigma_2 \end{pmatrix}$$

is outlined in Golub and Van Loan [389, 1989, Problem 8.5.1]. In the first step a Givens rotation is determined to symmetrize the matrix R. In the second step the symmetric matrix B is diagonalized by a Jacobi transformation.

A method which always gives high *relative accuracy* in the singular values and vectors has been developed by Demmel and Kahan. They note that the singular values are given by the explicit expression

$$(2.6.20) \qquad \frac{1}{2} \left| \sqrt{(r_{11} + r_{22})^2 + r_{12}^2} \pm \sqrt{(r_{11} - r_{22})^2 + r_{12}^2} \right|,$$

of which the larger is σ_1 and the smaller $\sigma_2 = |r_{11} r_{22}|/\sigma_1$. The right singular vector $(-s_r, c_r)$ is parallel to $(r_{11}^2 - \sigma_1^2, r_{11} r_{12})$ and the left singular vectors are determined by $(c_l, s_l) = (r_{11} c_r - r_{12} s_r, r_{22} s_r)/\sigma_1$. These expressions should not be used directly, since they suffer from possible over/underflow in the squared subexpressions. However, the computation can be reorganized to provide results with nearly full machine precision. A Fortran code and a sketch of its error analysis are given in the appendix of Bai and Demmel [32, 1993].

Note that the SVD produced by Hestenes' method will by construction have U orthogonal to working accuracy. However, loss of orthogonality in V may occur, and the columns of V should be reorthogonalized using a Gram–Schmidt process at the end.

Clearly, Hestenes' algorithm is mathematically equivalent to applying Jacobi's method to diagonalize $C = A^T A$, and hence its convergence properties are the same. Convergence of Jacobi's method is related to the fact that in each step the sum of squares of the off-diagonal elements

$$S(C) = \sum_{i \ne j} c_{ij}^2, \qquad C = A^T A$$

is reduced. There are various strategies for choosing the order in which the off-diagonal elements are annihilated. In a cyclic Jacobi method, the off-diagonal elements are annihilated in some predetermined order, each element being rotated exactly once in any sequence of $N = n(n-1)/2$ rotations called a **sweep**. For sequential computers the most popular cyclic ordering is the row-wise scheme, i.e., the rotations are performed in the order

$$(2.6.21) \quad \begin{array}{cccc} (1,2) & (1,3) & \ldots & (1,n) \\ & (2,3) & \ldots & (2,n) \\ & & \ldots & \ldots \\ & & & (n-1,n), \end{array}$$

which is cyclically repeated. In practice, with the cyclic Jacobi method not more than about five sweeps are needed to obtain singular values of more than single precision accuracy even when n is large. The number of sweeps grows approximately as $O(\log n)$.

Convergence of any cyclic Jacobi method can be guaranteed if any rotation is omitted for which the off-diagonal element is smaller in magnitude than some threshold. To ensure a good rate of convergence this threshold tolerance should be successively decreased after each sweep. It has been shown that the rate of convergence is ultimately quadratic, so that for k large enough, we have $S(C_{k+1}) < cS(C_k)^2$ for some constant c. The iterations are repeated until $S(C_k) < \delta \|C\|_F$, where δ is a tolerance, which can be chosen equal to the machine unit u. The Bauer–Fike theorem (see Golub and Van Loan [389, 1989, Thm. 7.2.2]) then shows that the diagonal elements of C_k then approximate the eigenvalues of $A^T A$ with an error less than $\delta \|C\|_F$.

Parallel implementation can take advantage of the fact that noninteracting rotations, (p_i, q_i) and (p_j, q_j), where p_i, q_i and p_j, q_j are distinct, can be performed simultaneously. If n is even $n/2$ transformations can be performed simultaneously, and a sweep needs at least $n-1$ such parallel steps. Several parallel schemes which use this minimum number of steps have been constructed. These can be illustrated in the $n = 8$ case by

$$(p,q) = \begin{array}{cccc} (1,2), & (3,4), & (5,6), & (7,8), \\ (1,4), & (2,6), & (3,8), & (5,7), \\ (1,6), & (4,8), & (2,7), & (3,5), \\ (1,8), & (6,7), & (4,5), & (2,3), \\ (1,7), & (8,5), & (6,3), & (4,2), \\ (1,5), & (7,3), & (8,2), & (6,4), \\ (1,3), & (5,2), & (7,4), & (8,6). \end{array}$$

The rotations associated with *each row* of the above can be calculated simultaneously.

Hestenes' method works on general real (or complex) matrices $A \in \mathbf{R}^{m \times n}$, $m \geq n$. In the following we can assume without restriction that $m = n$. If $m > n$ we first compute the QR decomposition of A and apply the algorithm to the upper triangular matrix $R \in \mathbf{R}^{n \times n}$. Indeed, an initial QR decomposition of A can be recommended also when $m = n$, since it tends to speed up convergence and simplifies the transformations.

2.6. Computing the SVD

In Kogbetliantz's method applied to a square matrix A the elementary step consists of two-sided Givens transformations

$$A' = J_{pq}(\phi) A J_{pq}^T(\psi), \tag{2.6.22}$$

where $J_{pq}(\phi)$ and $J_{pq}(\psi)$ are determined so that $a'_{pq} = a'_{qp} = 0$. Note that only rows and columns p and q in A are affected by the transformation. The rotations $J_{pq}(\phi)$ and $J_{pq}(\psi)$ are determined by computing the SVD of a 2×2 submatrix

$$A = \begin{pmatrix} a_{pp} & a_{pq} \\ a_{qp} & a_{qq} \end{pmatrix}, \quad a_{pp} \geq 0, \quad a_{qq} \geq 0.$$

The assumption of nonnegative diagonal elements is no restriction, since we can change the sign of these by premultiplication with an orthogonal matrix diag $(\pm 1, \pm 1)$.

Since the Frobenius norm is invariant under orthogonal transformations it follows that

$$S(A') = S(A) - (a_{pq}^2 + a_{qp}^2), \quad S(A) = \|A - D\|_F^2.$$

This relation is the basis for a proof that the matrices generated by Kogbetliantz's method converge to a diagonal matrix containing the singular values of A. Orthogonal systems of left and right singular vectors can be obtained by accumulating the product of all the transformations. Convergence is analyzed in Paige and Van Dooren [640, 1986] and Fernando [290, 1989].

It can be shown that if Kogbetliantz's method is applied to a triangular matrix then after one sweep of the row cyclic algorithm (2.6.21) an upper (lower) triangular matrix becomes lower (upper) triangular. Below we picture the annihilation of the elements in the first row for $n = 4$, using Jacobi rotations in the order $(1,2), (1,3), (1,4)$:

$$\begin{pmatrix} x & a_0 & b_0 & c_0 \\ 0 & x & d_0 & e_0 \\ 0 & 0 & x & f_0 \\ 0 & 0 & 0 & x \end{pmatrix} \Rightarrow \begin{pmatrix} x & 0 & b_1 & c_1 \\ 0 & x & d_1 & e_1 \\ 0 & 0 & x & f_0 \\ 0 & 0 & 0 & x \end{pmatrix} \Rightarrow \begin{pmatrix} x & 0 & 0 & c_2 \\ g_0 & x & d_2 & e_1 \\ 0 & 0 & x & f_1 \\ 0 & 0 & 0 & x \end{pmatrix} \Rightarrow \begin{pmatrix} x & 0 & 0 & 0 \\ g_1 & x & d_2 & e_2 \\ h_0 & 0 & x & f_2 \\ 0 & 0 & 0 & x \end{pmatrix}.$$

The switching between upper and lower triangular format can be avoided by a simple permutation scheme; see Fernando [290, 1989]. This makes it possible to reorganize the algorithm so that at each stage of the recursion one needs only to store and process a triangular matrix. The reorganization of the row cyclic scheme is achieved by the following algorithm (see also Luk [551, 1986] and Charlier, Vanbegin, and Van Dooren [161, 1988]):

$$\text{for } i = 1 : n - 1$$
$$\quad \text{for } i_k = 1 : n - i$$
$$\quad\quad A = P_{i_k} J_{i_k, i_k+1}(\phi_k) A J_{i_k, i_k+1}^T(\psi_k) P_{i_k}^T$$
$$\quad \text{end}$$
$$\text{end}$$

Here P_i denotes a permutation matrix that interchanges column (or rows) i and $i+1$. The permutations will shuffle the rows and columns of A_k so that each index pair (i_k, j_k) in the row cyclic scheme becomes an adjacent pair of the type $(i_k, i_k + 1)$ when it is its turn to be processed. The permutations involved are performed simultaneously with the rotations at no extra cost. Note also that in this scheme only rotations on *adjacent* rows and columns occur. Below we picture the annihilation of the elements in the first row for $n = 4$ for the reorganized scheme. After the elimination of a_0 the first and second rows and columns are interchanged. Element b_1 is now in the first superdiagonal and can be annihilated. Again, by interchanging rows and columns 3 and 4, c_2 is brought to the superdiagonal and can be eliminated. The resulting matrix is still upper triangular!

$$\begin{pmatrix} x & a_0 & b_0 & c_0 \\ 0 & x & d_0 & e_0 \\ 0 & 0 & x & f_0 \\ 0 & 0 & 0 & x \end{pmatrix} \Rightarrow \begin{pmatrix} x & 0 & d_1 & e_1 \\ 0 & x & b_1 & c_1 \\ 0 & 0 & x & f_0 \\ 0 & 0 & 0 & x \end{pmatrix} \Rightarrow \begin{pmatrix} x & d_2 & g_0 & e_1 \\ 0 & x & 0 & f_1 \\ 0 & 0 & x & c_2 \\ 0 & 0 & 0 & x \end{pmatrix} \Rightarrow \begin{pmatrix} x & d_2 & e_2 & g_1 \\ 0 & x & f_2 & h_0 \\ 0 & 0 & x & 0 \\ 0 & 0 & 0 & x \end{pmatrix}.$$

In the triangular version of Kogbetliantz's method the main transformation is reduced to finding the SVD of an upper triangular 2×2 matrix. This can be solved using the method by Demmel and Kahan described above. Because of its simplicity Kogbetliantz's algorithm has been adapted for the computation of generalized singular value decompositions; see Paige [633, 1986], Heath et al. [444, 1986], and Bai and Demmel [32, 1993]. Kogbetliantz's algorithm has been implemented on systolic arrays used for real-time signal processing; see Luk [551, 1986].

2.6.7. Singular values by spectrum slicing. Let A be a real symmetric matrix. The transformation $\hat{A} = T^T A T$ is called a **congruence transformation** of A. Congruence transformations do not, in general, preserve the eigenvalues of A. However, Sylvester's famous law of inertia says that the *signs* of eigenvalues are preserved by congruence transformations

THEOREM 2.6.3. Sylvester's Law of Inertia. *Let A be Hermitian and define the inertia of A to be the number triple $in(A) = (\pi, \nu, \delta)$ of positive, negative, and zero eigenvalues of A. Then if T is nonsingular A and $\hat{A} = T^H A T$ have the same inertia.*

Proof. See Golub and Van Loan [389, 1989, Thm. 8.1.12]. ∎

Sylvester's law of inertia leads to a simple and important method called **spectrum slicing** for counting the eigenvalues greater than a given real number σ of a real symmetric matrix A. Assume that symmetric Gaussian elimination can be carried through for $A - \sigma I$, yielding the factorization

$$A - \sigma I = LDL^T, \quad D = \text{diag}(d_1, \ldots, d_n),$$

where L is a unit lower triangular matrix. Then $A - \sigma I$ is congruent to D, and hence by Sylvester's Law of Inertia the number of eigenvalues of A greater than

2.6. Computing the SVD

σ equals the number of positive elements $\pi(D)$ in the sequence $d_1, \ldots, d_n$.

EXAMPLE 2.6.1. The LDL^T factorization

$$A - I = \begin{pmatrix} 1 & 2 & \\ 2 & 2 & -4 \\ & -4 & -6 \end{pmatrix} = \begin{pmatrix} 1 & & \\ 2 & 1 & \\ & 2 & 1 \end{pmatrix} \begin{pmatrix} 1 & & \\ & -2 & \\ & & 2 \end{pmatrix} \begin{pmatrix} 1 & 2 & \\ & 1 & 2 \\ & & 1 \end{pmatrix}$$

shows that the matrix A has two eigenvalues greater than 1. ∎

The LDL^T factorization may fail to exist if $A - \sigma I$ is not positive definite. This will happen, for example, if we choose the shift $\sigma = 2$ for the matrix in the example above. Then $a_{11} - \sigma = 0$, and the first step in the factorization cannot be carried out. A closer analysis shows that the factorization will fail if and only if σ equals an eigenvalue to one or more of the $n-1$ leading principal submatrices of A. If σ is chosen in a small interval around each of these values, big growth of elements occurs and the factorization may give the wrong count. In such cases one should perturb σ by a small amount and restart the factorization from the beginning.

When A is a symmetric tridiagonal matrix the procedure outlined above becomes particularly efficient and reliable. Here the factorization is $T - \sigma I = LDL^T$, where L is unit lower bidiagonal and $D = \mathrm{diag}\,(d_1, \ldots, d_n)$. The remarkable fact is that if we only take care to avoid over/underflow, then *element growth will not affect the accuracy of the slice*. We can apply this procedure to the special symmetric tridiagonal matrix T in (2.6.10). The algorithm can be simplified by taking advantage of the zero diagonal. Also, we need only consider the nonnegative part of the spectrum.

ALGORITHM 2.6.2. SPECTRUM SLICING FOR SINGULAR VALUES. Let $q_1, \ldots, q_n$ and $e_2, \ldots, e_n$ be the elements of the special tridiagonal matrix T in (2.6.10). The algorithm generates the number π of singular values of T greater than a given value $\sigma > 0$.

$$d_1 = -\sigma; \quad \pi = 0;$$
for $k = 2 : 2n$
$\quad \alpha_k = $ if k even then $q_{k/2}$ else $e_{(k+1)/2}$;
$\quad d_k = -\sigma - \alpha_k(\alpha_k/d_{k-1})$;
$\quad$ if $|d_k| < \sqrt{\omega}$ then $d_k = \sqrt{\omega}$;
$\quad$ if $d_k > 0$ then $\pi = \pi + 1$;
end

REMARK 2.6.1. To prevent breakdown of the recursion the algorithm should be modified so that a small $|d_k|$ is replaced by $\sqrt{\omega}$, where ω is the underflow threshold.

One slice in using Algorithm 2.6.2 requires only $2n$ flops, and it is not necessary to store the elements d_k. The number of multiplications can be halved by computing α_k^2 initially, but this may cause unnecessary over/underflow. ∎

A roundoff error analysis shows that, assuming that no over/underflow occurs, the computed values $\bar{d}_k$ satisfy exactly (let $\alpha_1 = 0$)

$$\begin{aligned}
\bar{d}_k &\equiv fl((-\sigma - \alpha_k(\alpha_k/\bar{d}_{k-1}))) \\
&= \left(-\sigma - \frac{\alpha_k^2}{\bar{d}_{k-1}}(1+\epsilon_{1k})(1+\epsilon_{2k})\right)(1+\epsilon_{3k}) \\
&\equiv -\sigma' - (\alpha'_k)^2/\bar{d}_{k-1}, \quad k = 1,\ldots,2n,
\end{aligned}$$

where $|\epsilon_{ik}| \leq u$. Hence, the computed number $\bar{\pi}$ is the exact number of singular values greater than σ of a tridiagonal matrix T', where T' has elements satisfying

$$|\sigma' - \sigma| \leq u\sigma, \quad |\alpha'_k - \alpha_k| \leq 2u|\alpha_k|,$$

which is a very satisfactory backward error bound. Combined with Theorem 2.6.2 it shows that the bisection algorithm computes singular values of a bidiagonal matrix B with small *relative* errors.

The above technique can be used to locate any individual singular value σ_k of B; cf. the bisection algorithm by Barth, Martin, and Wilkinson [56, 1967]. Assume we have two values σ_l and σ_u such that for the corresponding diagonal factors we have

$$\pi(D_l) \geq k, \quad \pi(D_u) < k.$$

Then σ_k lies in the interval $[\sigma_l, \sigma_u)$, and using p steps of the bisection method σ_k can be located in an interval of length $(\sigma_u - \sigma_l)/2^p$. The initial interval can be determined from Gershgorin's theorem. For the tridiagonal matrix (2.6.10) all the singular values are contained in the interval

$$\left[0, \max_i(|q_i|) + \max(|e_i|)\right].$$

An alternative suggested by Golub and Kahan [370, 1965] is to use, instead of the sequence of pivot elements $d_1, \ldots, d_n$, the Sturm sequence of characteristic polynomials p_j of leading principal minors of T. If $T - \sigma I = LDL^T$ then

$$(-1)^n p_j(\sigma) = d_1(\sigma) \cdots d_j(\sigma), \quad j = 1,\ldots,n,$$

and we have the relation $d_j(\sigma) = -p_j(\sigma)/p_{j-1}(\sigma)$. The principal minors can be computed directly from a linear three-term recurrence but this may suffer severe under/overflow problems.

An algorithm based on similar recursions as in spectrum slicing for computing the singular values of a bidiagonal matrix has been given by Li, Rhee, and Zeng [528, 1995]. This uses Laguerre's iteration instead of bisection and a divide-and-conquer technique, which makes it more efficient and inherently parallel.

We finally mention that so-called divide-and-conquer methods also show great promise for computing singular values in parallel; see Jessup and Sorenson [489, 1992] and Gu and Eisenstat [403, 1992].

2.7. Rank Deficient and Ill-Conditioned Problems

Because of roundoff errors, any numerical scheme for computing the pseudoinverse of a matrix A will, at best, produce the pseudoinverse of a perturbed matrix $A + E$. However, Theorem 1.4.1 states that if a matrix $A + E$ is close to A, but has rank different than A, then its pseudoinverse $(A + E)^\dagger$ will be different from $A^\dagger$, and the smaller E is, the greater the difference will be. Therefore, the rank of A must be explicitly determined before we can compute its pseudoinverse numerically. A similar conclusion holds for solving least squares problems which are rank deficient or ill-conditioned.

The discontinuity of the pseudoinverse also means that the mathematical notion of rank is not appropriate in numerical computations. For example, if the exact matrix $\tilde{A}$ has (mathematical) rank $k < n$, then the perturbed matrix $A = \tilde{A} + E$ most likely has full rank n. However, if E is small then A is very close to a rank deficient matrix and should be considered as *numerically* rank deficient. It is important that this is recognized, since overestimating the rank of A can lead to a computed solution of very large norm, since then components corresponding to small (numerically zero) singular values are included. A good discussion of numerical rank deficiency is given by Stewart [744, 1984].

2.7.1. SVD and numerical rank. From the above considerations it follows that the **numerical rank** assigned to matrix A should depend on a tolerance which reflects the error level. We make the following definition in terms of the singular values of A.

DEFINITION 2.7.1. *A matrix A is said to have numerical δ-rank equal to k if*

$$(2.7.1) \qquad k = \min\{\mathrm{rank}\,(B) \mid \|A - B\|_2 \leq \delta\}.$$

It follows from this definition, using Theorem 1.2.3, that if $k < n$ then

$$(2.7.2) \qquad \inf_{\mathrm{rank}\,(B) \leq k} \|A - B\|_2 = \sigma_{k+1},$$

where $\sigma_i, i = 1, 2, \ldots, \min(m, n)$ are the singular values of A, and this infimum is attained for the matrix

$$B = \sum_{i=1}^{k} \sigma_i u_i v_i^T.$$

Hence a matrix A has numerical δ-rank k if and only if

$$(2.7.3) \qquad \sigma_1 \geq \cdots \geq \sigma_k > \delta \geq \sigma_{k+1}, \geq \cdots \geq \sigma_n.$$

The definition (2.7.3) is satisfactory only when there is a well-defined gap between σ_{k+1} and σ_k. If the exact matrix $\tilde{A}$ is rank deficient but well-conditioned, then this should be the case.

The choice of the parameter δ in (2.7.3) is not always an easy matter. Let the error in the matrix A be $E = (e_{ij})$. Assume that the absolute size of the elements e_{ij} in the error matrix are all about the same size, and that $|e_{ij}| \leq \epsilon$, for all i,j. Then, since

$$\|E\|_2 \leq \|E\|_F \leq (mn)^{1/2}\epsilon,$$

a reasonable choice in (2.7.3) is to take $\delta = (mn)^{1/2}\epsilon$. If the absolute size of the elements e_{ij} are not about the same, one could try to scale rows and columns of A so that they become nearly equal; see the discussion in Dongarra et al. [228, 1979, pp. I.10–12]. Note that any diagonal scaling $D_1 A D_2$ will induce the same scaling $D_1 E D_2$ of the error matrix. However, in the least squares setting scaling the rows of A is normally not allowed, since it would change the solution to the least squares problem. For an interval analysis approach to rank determination, see Manteuffel [564, 1981].

There are matrices for which no gap exists for any k. Suppose, e.g., that $\sigma_{i+1} = 0.9\sigma_i$, $i = 1, 2, \ldots, n-1$. In this case the rank of the matrix obviously is not well determined. Problems which yield matrices that lack a well-determined numerical rank often are discretizations of continuous ill-posed problems. Examples of such problems are Laplace transformation, inverse heat equation, and digital image restoration. For such difficult problems additional information is usually needed to get a meaningful solution. Often a quadratic constraint on the solution is added in order to get a more well-conditioned problem; see Section 5.3.

2.7.2. Truncated SVD solutions and regularization. We now consider solving the linear least squares problem

(2.7.4) $$\min_x \|Ax - b\|_2,$$

where the matrix A is ill-conditioned and possibly rank deficient. The Gauss–Markoff theorem (Theorem 1.1.1) states that the least squares solution is the best unbiased linear estimator of x, in the sense that it has minimum variance. If A is ill-conditioned this minimum variance is still large. If the estimator is allowed to be biased, the variance can be drastically reduced. One way to achieve this is to compute the **truncated SVD solution**.

Assume that the singular value decomposition of A is

$$A = U \Sigma V^T = \sum_{i=1}^{n} u_i \sigma_i v_i^T.$$

Given δ and using Definition 2.7.1 we assign a numerical rank k to A. Setting to zero all singular values σ_i, $i > k$, the corresponding solution can then be written as an expansion in the right singular vectors

(2.7.5) $$x = \sum_{i=1}^{k} \frac{c_i}{\sigma_i} v_i, \quad c = U^T b,$$

2.7. Rank Deficient Problems

which is the truncated SVD (TSVD) solution. The TSVD solution solves the related least squares problem

$$\min_x \|A_k x - b\|_2, \quad A_k = \sum_{i=1}^{k} u_i \sigma_i v_i^T,$$

where A_k is the best rank k approximation of A. From Definition 2.7.1 it follows that

$$\|A - A_k\|_2 = \|AV_2\|_2 \leq \delta, \quad V_2 = (v_{k+1}, \ldots, v_n),$$

and

(2.7.6) $$\mathcal{R}(V_2) = \text{span}\{v_{k+1}, \ldots, v_n\}$$

is called the **numerical nullspace** of A.

An alternative to TSVD is **Tikhonov regularization**, introduced by Tikhonov [775, 1963]. Here one considers the regularized problem

(2.7.7) $$\min_x \|Ax - b\|_2^2 + \tau^2 \|Dx\|_2^2,$$

where $D = \text{diag}(d_1, \ldots, d_n)$ is a positive diagonal matrix. The problem (2.7.7), also called a **damped least squares**, is equivalent to the least squares problem

(2.7.8) $$\min_x \left\| \begin{pmatrix} \tau D \\ A \end{pmatrix} x - \begin{pmatrix} 0 \\ b \end{pmatrix} \right\|_2,$$

where the matrix A has been modified by appending the matrix τD. When $\tau > 0$ this problem is always of full column rank and has a unique solution.

If $D = I$ the singular values of the modified matrix in (2.7.8) are equal to $\tilde{\sigma}_i = (\sigma_i^2 + \tau^2)^{1/2}$, $i = 1, \ldots, n$. In this case the solution can be expressed in terms of the SVD as

(2.7.9) $$x(\tau) = \sum_{i=1}^{n} \frac{c_i f_i}{\sigma_i} v_i, \quad f_i = \frac{\sigma_i^2}{\sigma_i^2 + \tau^2}.$$

The quantities f_i are often called **filter factors**. Notice that as long as $\tau \ll \sigma_i$ we have $f_i \approx 1$, and if $\tau \gg \sigma_i$ then $f_i \ll 1$. This establishes a relation to the truncated SVD solution (2.7.5), and $x(\tau)$ will approximately equal the truncated SVD solution for $\delta = \tau$.

An advantage of the regularized problem (2.7.8) is that its solution can be computed from the QR decomposition

(2.7.10) $$\begin{pmatrix} \tau D \\ A \end{pmatrix} = Q \begin{pmatrix} R \\ 0 \end{pmatrix}.$$

This can be obtained by a small modification of any of the methods described in Section 2.4. As an example, consider the Householder QR factorization. The shape of the transformed matrix after $k = 2$ steps is shown ($m = n = 4$):

$$\begin{pmatrix} \times & + & + & + \\ & \times & + & + \\ & & \times & \\ & & & \times \\ 0 & 0 & \times & \times \\ 0 & 0 & \times & \times \\ 0 & 0 & \times & \times \\ 0 & 0 & \times & \times \end{pmatrix}$$

Notice that the first two rows of D have filled in, but the remaining rows of D are still not touched. For each step $k = 1, \ldots, n$ there are m elements in the current column to be annihilated. Therefore the operation count for the Householder QR decomposition will increase with $n^3/3$ to mn^2 flops. A similar increase in operations occurs in Givens or MGS QR decompositions.

If there is no a priori information which can be used to establish the numerical rank, or in the case of regularization the value of the parameter τ, then the so-called L-curve method can be useful. It derives its name from a plot of the curve $(\|b - Ax_\tau\|, \|x_\tau\|)$ in a doubly logarithmic scale, which typically forms an L-shaped curve. The idea is to choose the regularization parameter τ near the "corner" of this L-curve, since this represents a compromise between a small residual and a small solution; see Hansen [432, 1992]. Using this curve for determining the optimal regularization parameter seems to have been suggested first by Lawson and Hanson [520, 1974, Chap. 26].

Even with regularization we may not be able to compute the solution of an ill-conditioned problem with the accuracy that the data allows. In those cases it is possible to improve the solution by the following **iterated regularization** scheme due to Riley [687, 1956] and analyzed by Golub [364, 1965]. Take $x^{(0)} = 0$, and compute a sequence of approximate solutions by

$$x^{(q+1)} = x^{(q)} + \delta x^{(q)},$$

where $\delta x^{(q)}$ solves the least squares problem

(2.7.11) $$\min_{\delta x} \left\| \begin{pmatrix} \tau I \\ A \end{pmatrix} \delta x - \begin{pmatrix} 0 \\ r^{(q)} \end{pmatrix} \right\|_2, \qquad r^{(q)} = b - Ax^{(q)}.$$

This iteration may be implemented very effectively since only the QR decomposition (2.7.10) (with $D = I$) is needed. The convergence of iterated regularization can be expressed in terms of the SVD of A.

(2.7.12) $$x^{(q)}(\tau) = \sum_{i=1}^n \frac{c_i f_i^{(q)}}{\sigma_i} v_i, \qquad f_i^{(q)} = 1 - \left(\frac{\tau^2}{\sigma_i^2 + \tau^2}\right)^q.$$

Thus for $q = 1$ we have the standard regularized solution and as $q \to \infty$ $x^{(q)} \to A^\dagger b$. A related scheme has been suggested by Rutishauser [695, 1968].

2.7.3. QR decompositions with column pivoting.

The SVD is in general the most reliable method for determining the numerical rank of a matrix. However, in practice, the QR decomposition often works as well and requires less work. Further, if a basic solution is wanted rather than the minimal norm solution (see Section 1.3.3) then the QR decomposition is the relevant tool.

To compute a QR decomposition which reveals the rank of A, we first consider modifying the algorithms given in Section 2.4 by introducing column pivoting. A QR decomposition with column pivoting was first introduced by Golub [364, 1965]. We now show how to incorporate this **standard column pivoting** scheme into the QR decomposition by Householder transformations.

Suppose that after $(k-1)$ steps in Algorithm 2.4.1 we have computed

$$(2.7.13) \qquad A^{(k)} = Q^T A \Pi = \begin{pmatrix} R_{11} & R_{12} \\ 0 & A_{22}^{(k)} \end{pmatrix} \begin{matrix} \}k-1 \\ \}m-k+1 \end{matrix},$$

where R_{11} is upper triangular, $Q = P_1 \cdots P_{k-1}$ is a product of Householder matrices, and $\Pi = \Pi_1 \cdots \Pi_{k-1}$ is a product of elementary permutation matrices. We let

$$A_{22}^{(k)} = (\tilde{a}_k^{(k)}, \ldots, \tilde{a}_n^{(k)})$$

and choose the permutation matrix Π_k to interchange columns p and k, where p is the smallest index that satisfies

$$s_p^{(k)} \geq s_j^{(k)}, \quad s_j^{(k)} = \|\tilde{a}_j^{(k)}\|_2^2, \quad j = k, \ldots, n.$$

Hence the pivoting scheme is equivalent to searching for the column of largest norm in the submatrix $A_{22}^{(k)}$. In particular, in the first step the column of largest norm in A is selected as the pivot column. Obviously, for this column pivoting to work it is essential that the columns of A are well scaled.

With the column pivoting scheme described above, the diagonal elements in R will form a nonincreasing sequence $r_{11} \geq r_{22} \geq \cdots \geq r_{nn}$. It is easily shown that in fact the elements in R will satisfy the stronger inequalities

$$(2.7.14) \qquad r_{kk}^2 \geq \sum_{i=k}^{j} r_{ij}^2, \quad j = k+1, \ldots, n.$$

In particular, this implies that if $r_{kk} = 0$, then $r_{ij} = 0$, $i, j \geq k$.

REMARK 2.7.1. If the column norms in $A_{22}^{(k)}$ are recomputed at each stage, then column pivoting will increase the operation count of the Algorithm 2.4.1 by one-half. An alternative is to compute the norms of the columns of A initially:

$$(2.7.15) \qquad s_j^{(1)} = \|a_j\|_2^2, \quad j = 1, \ldots, n,$$

and then update these values for $k = 1, 2, \ldots, r+1$ using the recursion

$$(2.7.16) \qquad s_j^{(k+1)} = s_j^{(k)} - r_{kj}^2, \quad j = k+1, \ldots, n.$$

(Naturally, the $s_j^{(k)}$'s must be interchanged if the columns of $A_{22}^{(k)}$ are interchanged.) Using (2.7.16) will reduce the overhead of column pivoting to $O(mn)$ operations. However, some care must be taken to avoid numerical problems; see Dongarra et al. [228, 1979, pp. 9.16–9.18].

REMARK 2.7.2. The pivoting scheme will in the kth step select a pivot column which maximizes the diagonal element r_{kk}. This is equivalent to choosing at the kth step a pivot column with largest distance to the subspace

$$\mathcal{R}(A_{k-1}) = \text{span}\,(a_{c_1},\ldots,a_{c_{k-1}}),$$

where A_{k-1} is the submatrix of A formed by the columns corresponding to the first $k-1$ selected pivots. In other words the pivot column maximizes

(2.7.17) $$s_j^{(k)} = \min_y \|A_{k-1}y - a_{c_j}\|_2^2, \qquad j = k,\ldots,n.$$

Note that this column pivoting scheme applies to pivoting in the QR decomposition of A as well as in the Cholesky factorization of $A^T A$.

REMARK 2.7.3. Column pivoting can similarly be implemented in the rowwise version of the MGS method, Algorithm 2.4.5. Here, after $(k-1)$ steps, we have transformed the nonpivotal columns according to

$$a_j^{(k)} = a_j - \sum_{i=1}^{k-1} r_{ij} q_i, \qquad j = k,\ldots,n.$$

It follows that $a_j^{(k)}$ is just the orthogonal projection of a_j onto the orthogonal complement of $\mathcal{R}(A_{k-1}) = \text{span}\{q_1,\ldots,q_{k-1}\}$. Hence, in the kth step we should maximize

$$s_j^{(k)} = \|a_j^{(k)}\|_2^2, \qquad j = k,\ldots,n.$$

These quantities can be updated by the same formulas (2.7.16) as for the Householder and Cholesky algorithms, but again some care is necessary to avoid numerical cancellation. ∎

Suppose that in the decomposition (2.7.13) we have

(2.7.18) $$\|A_{22}^{(k)}\|_2 \le \epsilon \sigma_1(A) = \delta,$$

for some small tolerance $\epsilon > 0$. It follows that by putting $A_{22}^{(k)} = 0$, we introduce a perturbation E_k in the original matrix A such that $A + E_k$ has rank $k-1$, and $\|E_k\|_2 \le \delta$. Hence A has numerical δ-rank equal to at most $k-1$. From (2.7.14) it follows that

$$\|A_{22}^{(k)}\|_2 \le \|A_{22}^{(k)}\|_F \le (n-k+1)^{1/2}|r_{kk}|, \qquad k = 1,\ldots,n.$$

In particular we have
(2.7.19) $$|r_{11}| \le \sigma_1(A) \le n^{1/2}|r_{11}|.$$

2.7. Rank Deficient Problems

Hence, if
(2.7.20) $$|r_{kk}| \leq \bar{\epsilon}|r_{11}|,$$
then (2.7.18) holds with $\epsilon = (n-k+1)^{1/2}\bar{\epsilon}$. Hence the criterion (2.7.20) can be used to determine *an upper bound* for the numerical rank.

REMARK 2.7.4. Above it was assumed that rounding errors in the QR decompositions can be neglected. From the rounding error analysis of the Householder algorithm (cf. Remark 2.4.2) we know that $A^{(k)}$ is exactly orthogonally equivalent to $A + E_k$, where

(2.7.21) $$\|E_k\|_2 \leq c_1 u \|A\|_F \leq c_1 u n^{1/2} \sigma_1(A).$$

Using Theorem 1.2.7 we get from (2.7.18)–(2.7.21)

$$\sigma_k(A) \leq \sigma_k(A^{(k)}) + c_1 u n^{1/2} \sigma_1(A) \leq (\epsilon + c_1 u n^{1/2}) \sigma_1(A).$$

Hence if (2.7.18) is satisfied, then A has numerical δ-rank at most equal to $r = k - 1$, for

$$\delta = (\epsilon + c_1 u n^{1/2}) \sigma_1(A). \blacksquare$$

Although the pivoting strategy described has been found to seldom fail in practice, *it cannot be guaranteed to reveal the rank of A*. Although it is true that if the algorithm with column pivoting for the QR decomposition terminates with a small diagonal element r_{kk}, then A is close to a matrix of rank $k-1$, the converse does not hold. As the following example shows, a triangular matrix R can be nearly rank deficient *without any diagonal element r_{kk} being small*.

EXAMPLE 2.7.1. (See Kahan [495, 1966, pp. 791–792].) Consider the upper triangular matrix

$$R_n = \mathrm{diag}(1, s, \ldots, s^{n-1}) \begin{pmatrix} 1 & -c & -c & \cdots & -c \\ & 1 & -c & \cdots & -c \\ & & 1 & & \vdots \\ & & & \ddots & \vdots \\ & & & & 1 \end{pmatrix}.$$

The matrix R_n is upper triangular, and it can be verified that it satisfies the inequalities (2.7.14). Therefore R_n is invariant under the algorithm for QR decomposition with column pivoting. For $n = 100, c = 0.2$ the smallest singular value is $\sigma_n = 0.368 \cdot 10^{-8}$, but $r_{nn} = s^{n-1} = 0.133$. Hence, the near singularity of R_n is not revealed! $\blacksquare$

The inequalities (2.7.19) give upper and lower bounds for $\sigma_1(R)$ in terms of r_{11}. For the smallest singular value $\sigma_n(R)$ we have, assuming that $r_{nn} \neq 0$,

(2.7.22) $$\sigma_n^{-1} = \|R^{-1}\|_2 \geq |r_{nn}^{-1}|,$$

since the diagonal elements of R^{-1} equal r_{ii}^{-1}, $i = 1, \ldots, n$. Hence $|r_{nn}|$ is an *upper bound* for σ_n. If the elements of R satisfy (2.7.14), then we also have the lower bound

(2.7.23) $$|r_{nn}| \geq \sigma_n \geq \frac{3|r_{nn}|}{\sqrt{4^n + 6n - 1}} \geq 2^{1-n}|r_{nn}|.$$

For a discussion of this lower bound, which was first given in Faddeev, Kublanovskaya, and Faddeeva [284, 1968], we refer to the excellent survey by Higham [455, 1987]. The matrices R_n in Example 2.7.1 show that this lower bound can almost be attained.

2.7.4. Pseudoinverse solutions from QR decompositions. Suppose we have obtained a QR decomposition of the form

$$A^{(r+1)} = Q^T A\Pi = \begin{pmatrix} R_{11} & R_{12} \\ 0 & 0 \end{pmatrix} \begin{matrix} \}r \\ \}m-r \end{matrix}, \quad (2.7.24)$$

with R_{11} nonsingular. Then it is easily verified that an explicit basis for the nullspace of $A\Pi$ is given by the columns of the matrix

$$W = (w_k, \ldots, w_n) = \begin{pmatrix} R_{11}^{-1} R_{12} \\ -I \end{pmatrix}. \quad (2.7.25)$$

Using the decomposition (2.7.24) the least squares problem (2.7.4) becomes

$$\min_x \left\| \begin{pmatrix} R_{11} & R_{12} \\ 0 & 0 \end{pmatrix} \begin{pmatrix} \tilde{x}_1 \\ \tilde{x}_2 \end{pmatrix} - \begin{pmatrix} d_1 \\ d_2 \end{pmatrix} \right\|_2, \quad (2.7.26)$$

where $d = Q^T b$, and $\tilde{x} = \Pi^T x$ have been partitioned conformably. It is easily seen that for any given $\tilde{x}_2$ we can always determine $\tilde{x}_1$ so that the first r components of the residual vector in (2.7.26) is zero. It follows that $\tilde{x}_2$ can be chosen arbitrarily, and that the general solution of (2.7.26) can be written

$$x = \Pi \begin{pmatrix} R_{11}^{-1}(d_1 - R_{12}\tilde{x}_2) \\ \tilde{x}_2 \end{pmatrix}. \quad (2.7.27)$$

In particular, if we take $\tilde{x}_2 = 0$, we obtain the solution

$$x = \Pi \begin{pmatrix} x_b \\ 0 \end{pmatrix}, \quad x_b = R_{11}^{-1} d_1 = R_{11}^{-1} Q_1^T b, \quad (2.7.28)$$

with at most $r = \text{rank}(A)$ nonzero components corresponding to the r columns in A_1, where $A\Pi = (A_1, A_2)$. Any such solution, where Ax only involves at most r columns of A, is called a **basic solution**. In several applications a basic solution is desired, e.g., when the columns of A represent factors in a linear model, and the vector of observations b should be fitted using *as few variables as possible*.

In order to simplify notations we assume in the following that in (2.7.13) we have $\Pi = I$. This is no restriction, since the permutation of the columns of A can always be assumed to have been carried out in advance. The QR decomposition (2.7.13) can also be used to compute the pseudoinverse solution to (2.7.4). This solution minimizes $\|x\|_2$, which by (2.7.27) is equivalent to the linear least squares problem

$$\min_{x_2} \left\| \begin{pmatrix} S \\ -I_p \end{pmatrix} x_2 - \begin{pmatrix} x_b \\ 0 \end{pmatrix} \right\|_2, \quad (2.7.29)$$

2.7. Rank Deficient Problems

where we have put $p = n - r$. Note that this has the form of a regularized least squares problem; see Section 2.7.2. S and x_b can be computed in about $\frac{1}{2}r^2(p+1)$ flops by back-substitution in

$$R_{11}S = R_{12}, \qquad R_{11}x_b = Q_1^T b.$$

Note that S and x_b can overwrite R_{12} and d_1.

Since the matrix in (2.7.29) always has full column rank, x_2 can be computed from the normal equations

$$x_2 = (S^T S + I_{n-r})^{-1} S^T x_b,$$

using a Cholesky factorization of $(S^T S + I_{n-r})$. This method, which is further studied by Deuflhard and Sautter [227, 1980], takes about $\frac{1}{2}rp(p+1) + \frac{1}{6}p^3$ flops. When x_2 has been determined, then we have $x_1 = x_b - Sx_2$. Alternatively, (2.7.29) can be solved using the QR decomposition

$$(2.7.30) \qquad \begin{pmatrix} S \\ -I_{n-r} \end{pmatrix} = U^T \begin{pmatrix} R_S \\ 0 \end{pmatrix}, \qquad f = \begin{pmatrix} f_1 \\ f_2 \end{pmatrix} = U^T \begin{pmatrix} x_b \\ 0 \end{pmatrix}.$$

Here R_S is nonsingular, and it follows that $x_2 = R_S^{-1} f_1$. This algorithm is only slightly less efficient than using the normal equations and takes rp^2 flops. Both methods have good stability provided that column pivoting has chosen R_{11} so that $\kappa(R_{11})$ is not much larger than $\kappa(A)$. This is usually the case if standard column pivoting is used, although counterexamples can be constructed where $\kappa(R_{11}) \approx 2^n \kappa(A)$. An alternative method which is backward stable can be obtained by using a complete orthogonal decomposition of A; see Section 2.7.6.

The pseudoinverse solution will minimize $\|x\|_2$. It should be stressed that this is not always a good way to resolve rank deficiency, and the following more general approach is often to be preferred. For a given matrix $B \in R^{p \times n}$ with linearly independent rows, consider the problem

$$(2.7.31) \qquad \min_{x \in S} \|Bx\|_2, \quad S = \{x \mid \min_x \|Ax - b\|_2\}.$$

Substituting the general solution (2.7.27) we find that (2.7.31) is equivalent to

$$(2.7.32) \qquad \min_{x_2} \left\| B \begin{pmatrix} S \\ -I_{n-r} \end{pmatrix} x_2 - B \begin{pmatrix} x_b \\ 0 \end{pmatrix} \right\|_2,$$

which is a linear least squares problem of dimension $p \times (n - r)$. If this problem is not rank deficient, then (2.7.31) has a unique solution which can be computed by solving (2.7.32) for x_2 and then substituting the solution in (2.7.27).

EXAMPLE 2.7.2. The special case $B = I$ gives the minimal norm solution. Often one wants to choose B so that $\|Bx\|_2$ is a measure of the smoothness of the solution x. For example, we can let B be a discrete approximation to the second derivative operator,

$$(2.7.33) \qquad B = \begin{pmatrix} 1 & -2 & 1 & & & \\ & 1 & -2 & 1 & & \\ & & \ddots & \ddots & \ddots & \\ & & & 1 & -2 & 1 \end{pmatrix} \in R^{(n-2) \times n}.$$

2.7.5. Rank revealing QR decompositions. According to Theorem 1.3.5, if $A \in R^{m \times n}$, $m \geq n$, then there exists for any $0 < k < n$ a rank revealing column permutation Π such that the QR decomposition of $A\Pi$ has the form

$$(2.7.34) \qquad A\Pi = Q \begin{pmatrix} R_{11} & R_{12} \\ 0 & R_{22} \end{pmatrix} \begin{matrix} \}k \\ \}m-k \end{matrix},$$

where $\sigma_k(R_{11}) \geq \frac{1}{c}\sigma_k$, $\|R_{22}\|_2 \leq c\sigma_{k+1}$, and $c < \sqrt{r(n-r) + \min(r, n-r)}$, where $m \geq n$ and $0 < r < n$. Thus $c < (n+1)/2$. In particular, if A has numerical δ-rank equal to k, then there is a column permutation such that $\|R_{22}\|_2 \leq c\delta$. If $\|R_{22}\|_2$ is small then the matrix W in (2.7.25) gives a basis for the numerical nullspace, since

$$A\Pi W = Q \begin{pmatrix} R_{11} & R_{12} \\ 0 & R_{22} \end{pmatrix} W = Q \begin{pmatrix} 0 \\ -R_{22} \end{pmatrix}.$$

A systematic study of algorithms for determining rank revealing QR (RRQR) decompositions is found in Chandrasekaran and Ipsen [159, 1994]. We here describe an algorithm developed by Chan [153, 1987] for computing a rank revealing QR decomposition. (A similar algorithm was proposed independently by Foster [308, 1986].) The algorithm proceeds in three steps. First we compute any QR decomposition of A. Then an approximate right singular vector corresponding to the smallest singular value of R is computed, and a column permutation Π is determined by inspecting its elements. The initial QR factorization is then updated to give the QR decomposition for $A\Pi$. This process is repeated until a rank revealing decomposition is obtained.

The idea of finding column permutations from singular vectors was first used in a procedure for subset selection by Golub, Klema, and Stewart [371, 1976]; see Section 2.7.6. Here we need the following simple result.

LEMMA 2.7.1. (See Chan [153, 1987].) *Let the vector* v, $\|v\|_2 = 1$, *satisfy* $\|Av\|_2 = \epsilon$, *and let* Π *be a permutation such that if* $\Pi^T v = w$, *then* $|w_n| = \|w\|_\infty$. *If*

$$A\Pi = Q \begin{pmatrix} R \\ 0 \end{pmatrix}$$

is the QR decomposition of $A\Pi$, *then* $|r_{nn}| \leq n^{1/2}\epsilon$.

Proof. Since $|w_n| = \|w\|_\infty$ and $\|v\|_2 = \|w\|_2 = 1$, it follows that $|w_n| \geq n^{-1/2}$. Further,

$$Q^T Av = Q^T A\Pi(\Pi^T v) = \begin{pmatrix} Rw \\ 0 \end{pmatrix},$$

and since the last component of the vector Rw is $r_{nn}w_n$ we have

$$\epsilon = \|Av\|_2 = \|Q^T Av\|_2 = \|Rw\|_2 \geq |r_{nn}w_n|,$$

from which the lemma follows. ∎

In particular, if we take $v = v_n$, where v_n is the right singular vector corresponding to the singular value $\sigma_n(A)$, we have $Av = \sigma_n u_n$, and hence from Lemma 2.7.1
$$(2.7.35) \qquad \sigma_n \geq n^{-1/2}|r_{nn}|.$$

2.7. RANK DEFICIENT PROBLEMS

Chan suggests that an approximation w to v_n is computed by one or two steps of inverse iteration applied implicitly to $R^T R$: for $i = 1, 2, \ldots$,

(2.7.36) $\qquad R^T y = w^{(i)}, \qquad Rz = y, \qquad w^{(i+1)} = z/\|z\|_2.$

Usually two steps suffice if an efficient condition estimator (see Section 2.8.1) is used to select the initial vector $w^{(1)}$. Since only two triangular solves are needed in each iteration, this step only costs $O(n^2)$ flops.

From w an elementary permutation matrix Π is determined as above. Then the QR decomposition $R\Pi = \bar{Q}\bar{R}$ is computed, which gives

$$A\Pi = Q \begin{pmatrix} R\Pi \\ 0 \end{pmatrix} = Q \begin{pmatrix} \bar{Q} & 0 \\ 0 & I_{m-n} \end{pmatrix} \begin{pmatrix} \bar{R} \\ 0 \end{pmatrix},$$

an RRQR decomposition of $A\Pi$. The QR decomposition of $R\Pi$ can be computed in less than $2n^2$ flops using updating techniques to be described in Section 3.2. Hence the computational complexity of the RRQR decomposition is only slightly larger than that of the standard QR decomposition.

Chan [153, 1987] has extended the above procedure to the case when $\text{rank}(A) < n - 1$, by repeatedly applying the one-dimensional algorithm to smaller and smaller leading blocks of R. He shows that the resulting algorithm is guaranteed to work for matrices of small rank deficiency, and that it is very likely to work also for large rank deficiency.

ALGORITHM 2.7.1. RRQR. Compute any QR decomposition $A\Pi = Q \begin{pmatrix} R \\ 0 \end{pmatrix}$. Let τ be a user tolerance, and for $k = n, n-1, \ldots$ do:

1. Partition
$$R = \begin{pmatrix} R_{11} & R_{12} \\ 0 & R_{22} \end{pmatrix} \begin{matrix} \}k \\ \}n-k \end{matrix},$$
and determine $\delta_k = \sigma_{\min}(R_{11})$ and the corresponding right singular vector w_k.

2. If $\delta_k > \tau$ then set $\text{rank}(A) = k$ and finish.

3. Determine a permutation matrix P such that $|(P^T w_k)_k| = \|P^T w_k\|_\infty$.

4. Compute the QR decomposition $R_{11} P = \tilde{Q}\tilde{R}_{11}$, and update
$$\Pi := \Pi \begin{pmatrix} P & 0 \\ 0 & I_{n-k} \end{pmatrix}, \quad Q := Q \begin{pmatrix} \tilde{Q} & 0 \\ 0 & I_{n-k} \end{pmatrix}, \quad R := \begin{pmatrix} \tilde{R}_{11} & \tilde{Q}^T R_{12} \\ 0 & R_{22} \end{pmatrix}.$$

5. Let $w_k = \begin{pmatrix} w_k \\ 0 \end{pmatrix}$, assign to the kth column of W, and update
$$W = \begin{pmatrix} W_1 \\ W_2 \end{pmatrix} := \begin{pmatrix} P^T & 0 \\ 0 & I_{n-k} \end{pmatrix} W,$$

where W_2 is upper triangular and nonsingular. ∎

By the interlacing property of singular values, Theorem 1.2.9, it follows that the δ_i are nonincreasing and that the singular values σ_i of A satisfy $\delta_i \leq \sigma_i$, $k+1 \leq i \leq n$. Chan [153, 1987] proves the following upper and lower bounds.

THEOREM 2.7.1. *Let $R_{22}^{(i)}$ and $W_2^{(i)}$ denote the lower right submatrices of dimension $(n-i+1) \times (n-i+1)$ of R_{22} and W_2, respectively. Let δ_i denote the smallest singular value of the leading principal $i \times i$ submatrices of R. Then for $i = k+1, \ldots, n$,*

$$\frac{\sigma_i}{\sqrt{n-i+1}\|(W_2^{(i)})^{-1}\|_2} \leq \delta_i \leq \sigma_i \leq \|R_{22}^{(i)}\|_2 \leq \sigma_i \sqrt{n-i+1}\|(W_2^{(i)})^{-1}\|_2.$$

Hence $\|R_{22}^{(i)}\|_2$ are easily computable upper bounds for σ_i. Further, the outermost bounds in the theorem show that if $\|(W_2^{(i)})^{-1}\|_2$ is not large then δ_i and $\|R_{22}^{(i)}\|_2$ are guaranteed to be tight bounds, and hence the decomposition will have revealed the rank.

The matrix W determined by the RRQR algorithm satisfies

$$(2.7.37) \qquad \|A\Pi W\|_2^2 \leq \|A\Pi W\|_F^2 = \sum_{i=k+1}^n \delta_i^2.$$

Therefore $\mathcal{R}(\Pi W)$ in the RRQR algorithm is a good approximation to the numerical nullspace $\mathcal{N}_k(A)$. A more accurate and orthogonal basis for $\mathcal{N}_k(A)$ can be determined by simultaneous inverse iteration with $R^T R$ starting with W. If R has zero or nearly zero diagonal elements a small multiple of the machine unit is substituted. The use of RRQR decompositions for computing truncated SVD solutions is discussed by Chan and Hansen [154, 1990].

If the matrix A has low rank rather than low rank deficiency, it is more efficient to build up the rank revealing QR decomposition from estimates of singular vectors corresponding to the large singular values. Such algorithms are described by Chan and Hansen in [156, 1994].

2.7.6. Complete orthogonal decompositions. The RRQR decomposition has the advantage over the SVD that it is cheaper to compute and can be cheaply updated; see Section 3.2. However, it is less suitable in applications where a basis for the approximate nullspace is needed since the matrix W in (2.7.25) cannot easily be updated.

The decomposition (2.7.34) can be carried one step further to obtain a **complete orthogonal decomposition** of A. Assume first that $R_{22} = 0$, and postmultiply by a sequence of Householder transformations such that

$$(R_{11} \; R_{12}) P_k \cdots P_1 = (R \; 0).$$

Here $P_j = I - \gamma_j^{-1} u_j u_j^T$, $j = k, k-1, \ldots, 1$, where u_j has nonzero elements only in positions $j, k+1, \ldots, n$ and chosen so that the elements in row j of R_{12} are annihilated. These transformations require $k^2(n-k)$ flops. Then we obtain

$$(2.7.38) \qquad A = Q \begin{pmatrix} R & 0 \\ 0 & 0 \end{pmatrix} V^T, \qquad V = \Pi P_1 \cdots P_k.$$

2.7. Rank Deficient Problems

Hanson and Lawson [437, 1969] first suggested the use of a complete orthogonal decomposition to solve rank deficient least squares problems. It has the advantage that it explicitly provides an orthogonal basis for the nullspace of A, and a representation for the pseudoinverse.

THEOREM 2.7.2. *Assume that we have a complete orthogonal decomposition (2.7.38) of A. Then if $V = (V_1, V_2)$, V_2 gives an orthogonal basis for the nullspace of A, and the pseudoinverse of A is*

$$A^\dagger = V \begin{pmatrix} R^{-1} & 0 \\ 0 & 0 \end{pmatrix} Q^T. \tag{2.7.39}$$

Proof. It is immediately verified that $AV_2 = 0$, and a dimensional argument shows that V_2 spans the nullspace of A. Using the orthogonal invariance of the l_2-norm it follows that $V_1 R^{-1} Q_1^T b$ gives the minimum norm solution of the least squares problem $\min_x \|Ax - b\|_2$. Since the pseudoinverse is uniquely defined by this property, the theorem follows. ∎

We define a rank revealing complete orthogonal decomposition to have the form

$$A = U \begin{pmatrix} R_{11} & R_{12} \\ 0 & R_{22} \end{pmatrix} V^T, \tag{2.7.40}$$

where U and V are orthogonal matrices, $R_{11} \in \mathbf{R}^{k \times k}$, and

$$\sigma_k(R_{11}) \geq \frac{1}{c} \sigma_k, \qquad (\|R_{12}\|_F^2 + \|R_{22}\|_F^2)^{1/2} \leq c \sigma_{k+1}. \tag{2.7.41}$$

This is also often called a rank revealing URV decomposition. From (2.7.40) we have

$$\|AV_2\|_2 = \left\| \begin{pmatrix} R_{12} \\ R_{22} \end{pmatrix} \right\|_F \leq c \sigma_{k+1}.$$

Hence the orthogonal matrix V_2 can be taken as an approximation to the numerical nullspace $\mathcal{N}_k$.

Stewart [749, 1992] has shown how to compute a rank revealing complete orthogonal decomposition from an RRQR decomposition (2.7.38). Let w_n be the last column in the matrix W giving the numerical nullspace. Determine a sequence of Givens rotations that eliminates the first $n - 1$ components of w_n,

$$Q^T w_n = G_{n-1,n}^T \cdots G_{12}^T w_n = \|w_n\|_2 e_n.$$

Next an orthogonal matrix P such that $P^T R Q$ is upper triangular is determined. This is done by applying the rotations $G_{12}, \ldots, G_{n-1,n}$ from the right to R as shown below. When a rotation $G_{j-1,j}$ is applied to two columns of R a nonzero element is introduced just below the diagonal of R. A left rotation can be used to eliminate this nonzero and restore the triangular form.

$$\begin{pmatrix} \downarrow & \downarrow & & \\ r & r & r & r \\ + & r & r & r \\ 0 & 0 & r & r \\ 0 & 0 & 0 & r \end{pmatrix} \overset{\rightarrow}{\underset{\rightarrow}{\Rightarrow}} \begin{pmatrix} r & r & r & r \\ \oplus & r & r & r \\ 0 & 0 & r & r \\ 0 & 0 & 0 & r \end{pmatrix} \Rightarrow \begin{pmatrix} \downarrow & \downarrow & & \\ r & r & r & r \\ 0 & r & r & r \\ 0 & + & r & r \\ 0 & 0 & 0 & r \end{pmatrix}$$

$$\begin{pmatrix} r & r & r & r \\ 0 & r & r & r \\ \to \\ \to \end{pmatrix} \begin{pmatrix} r & r & r & r \\ 0 & r & r & r \\ 0 & \oplus & r & r \\ 0 & 0 & 0 & r \end{pmatrix} \Rightarrow \begin{pmatrix} \downarrow & \downarrow & & \\ r & r & r & r \\ 0 & r & r & r \\ 0 & 0 & r & r \\ 0 & 0 & + & r \end{pmatrix} \Rightarrow \begin{pmatrix} r & r & r & e \\ 0 & r & r & e \\ 0 & 0 & r & e \\ 0 & 0 & \oplus & e \end{pmatrix}.$$

This process requires $O(n^2)$ multiplications. We now have

$$P^T R w_n = (P^T R Q)(Q^T w_n) = \|w_n\|_2 \tilde{R} e_n.$$

If $\|Rw_n\|_2 < |r_{nn}|$, then since P is orthogonal it follows that

$$\|\tilde{R} e_n\|_2 < \delta / \|w_n\|_2.$$

This bounds the norm for the last column of the transformed matrix $\tilde{R}$. If $|r_{n-1,n-1}|$ is small we can continue this deflation on the principal submatrix of order $n-1$ of $\tilde{R}$.

Stewart [748, 1991] has suggested a refinement process for the decomposition (2.7.40), which reduces the size of the block R_{12}, and increases the accuracy in the nullspace approximation. The refinement step can be viewed as one step of the unshifted QR algorithm, (2.6.9), and can be iterated. It will converge quickly if there is a large relative gap between the singular values σ_k and σ_{k+1}. Alternatively one can work with the corresponding decomposition of lower triangular form, the rank revealing ULV decomposition

(2.7.42) $$A = U \begin{pmatrix} L_{11} & 0 \\ L_{21} & L_{22} \end{pmatrix} V^T.$$

For this decomposition with the partitioning $V = (V_1, V_2)$,

$$\|AV_2\|_2 = \|L_{22}\|_F.$$

Hence the size of $\|L_{21}\|_F$ does not adversely affect the nullspace approximation.

Suppose we have a rank revealing decomposition

$$A\Pi = Q \begin{pmatrix} L_{11} & 0 \\ L_{21} & L_{22} \end{pmatrix},$$

where L_{11} and L_{22} are lower triangular and

$$\sigma_k(L_{11}) \geq \frac{1}{c}\sigma_k, \qquad \|L_{22}\|_2 \leq c\sigma_{k+1}.$$

(Note that such a decomposition can be obtained from a rank revealing QR decomposition by reversing the rows and columns of the R-factor.) Then a rank revealing ULV decomposition can be obtained by a similar procedure as shown above for the URV decomposition. Suppose we have a vector w such that $\|w^T L\|_2$ is small. Then, as before, w is first reduced to the unit vector e_n,

$$Q^T w_n = G_{n-1,n}^T \cdots G_{12}^T w_n = \|w_n\|_2 e_n.$$

The sequence of Givens rotations are then applied to L from the left, and extra rotations from the right used to preserve the lower triangular form; see Stewart [752, 1993].

2.7. Rank Deficient Problems

2.7.7. Subset selection by SVD and RRQR. Given a matrix $A \in \mathbf{R}^{m \times n}$ the **subset selection** problem is the problem of determining the $k < n$ most linearly independent columns of A. In other words we want to find a permutation Π such that the k first columns of $A\Pi$ are as well-conditioned as possible. The RRQR algorithm can be used to compute a solution to this problem. A basic solution to the least squares problem $\min_x \|Ax - b\|_2$ can then be computed from the QR decomposition of the k columns selected from A.

We now consider an SVD-based algorithm for solving the subset selection problem given by Golub, Klema, and Stewart [371, 1976]; see also Golub and Van Loan [389, 1989, Sec. 12.2].

ALGORITHM 2.7.2. SUBSET SELECTION BY SVD. Given $A \in \mathbf{R}^{m \times n}$, $b \in \mathbf{R}^m$, the following algorithm computes a permutation matrix Π such that the first k columns of $A\Pi$ are sufficiently independent and a vector $z \in \mathbf{R}^k$ which solves

$$\min_z \left\| A\Pi \begin{pmatrix} z \\ 0 \end{pmatrix} - b \right\|_2.$$

1. Compute Σ and V in the SVD of A,

$$A = U\Sigma V^T, \quad \Sigma = \operatorname{diag}(\sigma_1, \ldots, \sigma_n),$$

and use it to determine the numerical rank k of A.

2. Partition the matrix of right singular vectors according to

$$V = \begin{pmatrix} V_{11} & V_{12} \\ V_{21} & V_{22} \end{pmatrix} \begin{matrix} \}k \\ \}n-k \end{matrix},$$

and use QR with column pivoting to compute

$$Q^T (V_{22}^T, V_{12}^T) \bar{\Pi} = (R_{11}, R_{12}).$$

3. Let Π_I be the permutation matrix which performs the interchange

$$(V_{12}^T, V_{22}^T) \Pi_I = (V_{22}^T, V_{12}^T)^T.$$

Set $A\Pi = (B_1, B_2)$ where $\Pi = \Pi_I \bar{\Pi} \Pi_I$, $\bar{\Pi}$ is the permutation matrix from step 2, and $B_1 \in \mathbf{R}^{m \times r}$. Compute the QR decomposition of B_1 and solve

$$\min_z \|B_1 z - b\|_2.$$

If the Chan SVD algorithm, which computes the QR decomposition of A as an intermediate step, is used in step 1, Householder QR in step 2 and updating techniques in step 3, then this algorithm requires a total of $mn^2 + 19n^3/3 - n^2 k + 4k^3/3$ flops. The key step is step 2, where the permutation $\bar{\Pi}$ will tend to make $\tilde{V}_{22}$ well-conditioned, where

$$(\tilde{V}_{22}^T, \tilde{V}_{12}^T) = (V_{22}^T, V_{12}^T) \bar{\Pi}.$$

It can be shown (Golub and Van Loan [389, 1989, Thm. 12.2.1]) that the singular value $\sigma_k(B_1)$ is bounded by

$$\sigma_k(A)/\|\tilde{V}_{22}^{-1}\|_2 \leq \sigma_k(B_1) \leq \sigma_k(A), \tag{2.7.43}$$

which is the theoretical basis for this selection strategy.

A comparison between the RRQR and the above SVD-based algorithms is given by Chan and Hansen [155, 1992]. Although in general the methods will not necessarily compute equivalent solutions, the subspaces spanned by the two sets of selected columns are still almost identical whenever the ratio σ_{k+1}/σ_k is small.

2.8. Estimating Condition Numbers and Errors

In Sections 1.4.2 and 1.4.4 we gave normwise and componentwise perturbation bounds for the solution to least squares problems. The bounds for the normwise perturbation in x depend critically on the condition number $\kappa(A) = \sigma_1(A)/\sigma_n(A)$. However, unless the SVD of A is available, the computation of $\sigma_n(A)$ is more demanding than the computation of the least squares solution x. Hence there is a need for efficient and reliable methods to estimate $\kappa(A)$. Since the condition number is invariant under an orthogonal transformation we have

$$\kappa(A) = \kappa(R) \leq \|R\|_F \|R^{-1}\|_F,$$

where R is the upper triangular factor in the QR decomposition of A. Hence a bound for $\kappa(A)$ can be obtained from R^{-1}. However, to compute R^{-1} requires $n^3/6$ flops, and would significantly increase the work. In this section we discuss methods to estimate the condition number of a triangular matrix R, which only require $O(n^2)$ flops. An excellent survey of such condition estimators is given by Higham [455, 1987].

2.8.1. The LINPACK condition estimator.
Combining the estimates (2.7.19) and (2.7.22) we obtain the lower bound

$$\kappa(A) = \sigma_1/\sigma_n \geq |r_{11}/r_{nn}|. \tag{2.8.1}$$

As shown by Example 2.7.1, this is not entirely reliable as a condition estimate since $|r_{nn}|$ may considerably overestimate the smallest singular value σ_n even when column pivoting is used in the QR decomposition. However, there is much empirical evidence to suggest that it is very rare for the lower bound in (2.8.1) to considerably differ from $\kappa(A)$. In extensive numerical testing by Stewart [736, 1980] on randomly generated test matrices the bound usually underestimated the condition number only by a factor of 2–3 and never by more than 10.

The number $\rho = |r_{nn}|$ has a nice interpretation: it is the norm of the smallest perturbation of the last column in A that will make A exactly rank deficient. Stewart [744, 1984] has suggested computing the set of numbers $\{\rho_1, \rho_2, \ldots, \rho_n\}$ for a sequence of permutations, which moves each column of A to the last position, using the updating techniques referred to above.

2.8. Estimating Condition Numbers and Errors

We now describe a condition estimator given by Cline et al. [175, 1979]. It is included in the LINPACK collection as the Fortran subroutine STRCO (Dongarra et al. [228, 1979]), and therefore often referred to as the LINPACK condition estimator. The basic algorithm is as follows:

1. Choose a vector d such that $\|y\|/\|d\|$ is large where $R^T y = d$.
2. Solve $Rz = y$, and estimate

$$\|R^{-1}\| \approx \|z\|/\|y\| \leq \|R^{-1}\|.$$

In STRCO the norm is the 1-norm, but the algorithm can be used also for the 2-norm or ∞-norm.

In the LINPACK estimator we have $R^T R z = d$, and hence

$$(2.8.2) \qquad z = (R^T R)^{-1} d = (A^T A)^{-1} d.$$

Hence the LINPACK estimator is equivalent to performing one step of inverse iteration with $A^T A$. If $R = U \Sigma V^T$ is the SVD of R, expanding d in terms of the right singular vectors V, we have

$$d = \sum_{i=1}^{n} \alpha_i v_i, \quad y = \sum_{i=1}^{n} (\alpha_i/\sigma_i) u_i, \quad z = \sum_{i=1}^{n} (\alpha_i/\sigma_i^2) v_i.$$

Hence, provided α_n, the component of d along v_n, is not very small, the vector z is likely to be dominated by its component of v_n, and

$$(2.8.3) \qquad \sigma_n^{-1} \approx \|z\|_2/\|y\|_2$$

will usually be a good estimate of σ_n^{-1}.

In the LINPACK algorithm the vector d is chosen as

$$d = (\pm 1, \pm 1, \ldots, \pm 1)^T,$$

where the sign of d_j is determined adaptively at the stage when y_j is computed. We note that the equation $Ry = d$ can be solved by a column-oriented version of back-substitution:

$$\begin{aligned}
&\text{for } i = 1:n, \quad s_i = 0; \text{ end} \\
&\text{for } j = n:-1:1 \\
&\qquad y_j = (d_j - s_j)/r_{jj}; \\
&\qquad \text{for } i = 1:j-1 \\
&\qquad\qquad s_i = s_i + r_{ij} y_j; \\
&\qquad \text{end} \\
&\text{end}
\end{aligned}$$

At the jth step the element $d_j = \pm 1$ is chosen so as to maximize a weighted sum of $d_j - s_j$ and the partial sums $s_{j-1}, \ldots, s_1$, which are to be computed in this step.

For details of this strategy see Cline et al. [175, 1979]. This estimator requires only about $2n^2$ flops. Examples of parametrized matrices have been constructed for which the LINPACK estimate can underestimate the true condition number by an arbitrarily large factor. However, in practice it performs very reliably and produces good order of magnitude estimates; see Higham [455, 1987].

O'Leary [605, 1980] has suggested a modification to the LINPACK condition estimator, which can improve the estimate. She estimates $\|T^{-1}\|_1$ by

$$\max\{\|y\|_\infty/\|d\|_\infty, \|z\|_1/\|y\|_1\},$$

which also makes use of information from the first step. Cline, Conn, and Van Loan [174, 1982] have described a generalization of the LINPACK algorithm which incorporates a "look-behind" technique, which allows for the possibility of modifying previously chosen d_j's, and gives an algorithm for the 2-norm which requires $5n^2$ flops.

Another possibility (already mentioned in Cline et al. [175, 1979]) is to choose the vector d as a random vector of unit length and perform several steps of inverse iteration,

$$x^{(0)} = d, \quad x^{(j)} = (R^T R)^{-1} x^{j-1}, \quad j = 1, \ldots, p.$$

As an estimate of the condition,

$$\gamma = \max_{j \le p} \gamma_j, \quad \gamma_j = \|x_j\|^{1/(2j)}$$

is used. Higham [455, 1987] reports that there is often a significant improvement in γ_2 over γ_1, and suggests that the number of steps p is chosen adaptively.

The condition estimator will detect near rank deficiency of the matrix A even in the (unusual) case when this is not revealed by a small diagonal element in R. This is important, since failure to detect near rank deficiency can lead to meaningless solutions of very large norm, or even to failure of the algorithm.

2.8.2. Hager's condition estimator. Hager [420, 1984] has given an algorithm for estimating the 1-norm (or ∞-norm) of a matrix $B \in \mathbf{R}^{n \times n}$. Hager's algorithm is based on convex optimization and uses the observation that $\|B\|_1$ is the maximal value of the convex function

$$f(x) = \|Bx\|_1 = \sum_{i=1}^{n} |y_i|, \quad y = Bx,$$

over the convex set $S = \{x \in \mathbf{R}^n \mid \|x\|_1 \le 1\}$. From convexity results it follows that the maximum is attained at one of the vertices e_j, $j = 1, \ldots, n$, of S. From this observation Hager derives the following algorithm for finding a local maximum that with high probability is also the global maximum.

2.8. Estimating Condition Numbers and Errors

ALGORITHM 2.8.1. HAGER'S NORM ESTIMATOR. Given a matrix $B \in \mathbf{R}^{n \times n}$ this algorithm computes $y = Bx$ such that $\gamma = \|y\|_1/\|x\|_1 \leq \|B\|_1$. Let $e = (1, 1, \ldots, 1)^T$, e_j be the jth unit vector, and $\xi = \operatorname{sign}(y)$ where $\xi_i = \pm 1$ according to whether $y_i \geq 0$ or $y_i < 0$. Set $x = e/n$ and repeat:

$$y = Bx;$$
$$\xi = \operatorname{sign}(y)$$
$$z = B^T \xi;$$
$$\text{if } \|z\|_\infty \leq z^T x$$
$$\quad \gamma = \|y\|_1; \text{quit}$$
$$\text{end}$$
$$x = e_j; \quad \text{where } |z_j| = \|z\|_\infty;$$

The algorithm tries to maximize the function $f(x) = \|Bx\|_1$ subject to $\|x\|_1 = 1$. The vector z computed at each step can be shown to be a subgradient of f at x. From convexity properties,

$$f(\pm e_j) \geq f(x) + z^T(\pm e_j - x), \quad j = 1, \ldots, n.$$

Hence if $|z_j| > z^T x$ for some j, then f can be increased by moving from x to the vertex e_j of S. If, however, $\|z\|_\infty \leq z^T x$, and if $y_j \neq 0$ for all j, then x can be shown to be a local maximum point for f over S.

The estimates produced by Hager's algorithm are generally sharper than those produced by the LINPACK estimator. Its results are frequently exact, usually good ($\gamma \geq 0.1 \|B\|_1$), but sometimes poor. The algorithm almost always converges after at most four iterations, and Higham [461, 1990] recommends that between two and five iterations be used. The cost for estimating $\|R\|_1$ for a triangular matrix R is then approximately $3n^2$ flops in practice. Note that if Hager's algorithm is applied to B^T we obtain an estimate of $\|B\|_\infty$.

Higham has made several modifications to Hager's norm estimator, which improves its performance. In [458, 1988] he gives two Fortran 77 codes implementing the modified algorithm for the 1-norm of a real or complex matrix. This algorithm is used for all the condition number estimations in LAPACK; see [16, 1995]. For further details and comments on the algorithm, see Hager [420, 1984] and Higham [462, 1990].

An important feature of Hager's norm estimator is that to estimate $\|B^{-1}\|_1$ we only need to be able to solve linear systems $By = x$ and $B^T z = \xi$. This feature makes it useful for estimating the componentwise error bounds given in Section 1.4.4. For the least squares problem the bound (1.4.25) can be written in the form $\|\delta x\|_\infty \leq \omega \operatorname{cond}(A, b) \|x\|_\infty$, where

(2.8.4) $$\operatorname{cond}(A, b) \leq \left(\||B_1|g_1\|_\infty + \||B_2|g_2\|_\infty \right)/\|x\|_\infty,$$

and

(2.8.5) $$B_1 = A^\dagger, \quad g_1 = |A||x| + b, \quad B_2 = (A^T A)^{-1}, \quad g_2 = |A|^T |r|.$$

It is possible to get an inexpensive and reliable estimate of $\kappa_r(A, b)$ using Hager's algorithm. The key idea is to note that the terms in (2.8.4) are all of the form $\||B|g\|_\infty$, where $g > 0$. Following Arioli, Demmel, and Duff [20, 1989] we let $G = \operatorname{diag}(g)$. Then $g = Ge$ where e is a column vector of all ones, and using the properties of the l_∞-norm we have

$$\||B|g\|_\infty = \||B|Ge\|_\infty = \||BG|\|_\infty = \||BG|e\|_\infty = \|BG\|_\infty.$$

Hence Hager's algorithm can be applied to estimate $\||B|g\|_\infty$ provided that matrix-vector products BGx and $G^T B^T y$ can be computed efficiently. Hence to estimate $\operatorname{cond}(A, b)$ we need to be able to compute matrix-vector products of the forms $A^\dagger x$, $(A^\dagger)^T y$, and $(A^T A)^{-1} x$. This can be done efficiently if, for example, the QR decomposition of A is known.

2.8.3. Computing the variance-covariance matrix. Consider the linear statistical model

(2.8.6) $$Ax = b + \epsilon, \qquad A \in \mathbf{R}^{m \times n}.$$

Assume that $\operatorname{rank}(A) = n$, and that the random vector ϵ has zero mean and variance-covariance matrix $\sigma^2 I$, i.e., $\mathcal{E}(\epsilon) = 0$ and $\mathcal{V}(\epsilon) = \sigma^2 I$. Then by Theorem 1.1.1 (the Gauss–Markoff theorem) the least squares estimate $\hat{x}$ is the linear unbiased estimator of x with minimum variance equal to

(2.8.7) $$V_x = \sigma^2 C_x, \quad C_x = (A^T A)^{-1} = R^{-1} R^{-T},$$

where R is the Cholesky factor of $A^T A$. An unbiased estimate of σ^2 is given by

(2.8.8) $$s^2 = \|\hat{r}\|_2^2/(m-n), \quad \hat{r} = b - A\hat{x}.$$

In order to assess the accuracy of the computed estimate of x it is often required to compute the matrix V_x or part of it. In particular the variance of the component $\hat{x}_i$ is given by the diagonal element v_{ii} in V_x.

The residual vector $\hat{r} = b - A\hat{x}$ has variance-covariance matrix equal to

(2.8.9) $$V_r = \sigma^2 A(A^T A)^{-1} A^T = \sigma^2 Q_1 Q_1^T = \sigma^2 P_{\mathcal{R}(A)},$$

where $Q = (Q_1, Q_2)$ is the orthogonal matrix in the QR decomposition of A. The **normalized residuals**

$$\tilde{r} = (\operatorname{diag} V_r)^{-1/2} \hat{r}$$

are often used to detect and identify single or multiple bad data, which is assumed to correspond to large components in $\tilde{r}$.

In many situations the matrix V_x only occurs as an intermediate quantity in a formula. For example, the variance of a linear functional $\varphi = f^T \hat{x}$ is equal to

(2.8.10) $$\mathcal{V}(\varphi) = f^T V_x f = \sigma_2 f^T R^{-1} R^{-T} f = \sigma^2 z^T z,$$

where $z = R^{-T} f$. Thus the variance may be computed by solving the triangular system $R^T z = f$ and forming $z^T z$. This is a more stable and efficient approach than using the expression involving V_x.

2.8. ESTIMATING CONDITION NUMBERS AND ERRORS

To compute $V_x = \sigma^2 C_x$ in the nonsingular case we can first compute the inverse $S = R^{-1}$ and then form $C_x = SS^T$. The matrix S satisfies the triangular system $RS = I$. It follows that S is also upper triangular, and its elements can be computed by the following algorithm:

$$
\begin{aligned}
&\text{for } j = n, \ldots, 1 \\
&\quad s_{jj} = 1/r_{jj}; \\
&\quad \text{for } i = j-1, \ldots, 1 \\
&\qquad s_{ij} = -\left(\sum_{k=i+1}^{j} r_{ik} s_{kj}\right)/r_{ii}; \\
&\quad \text{end} \\
&\text{end}
\end{aligned}
$$

Here the elements of S can overwrite the corresponding elements of R in storage. The algorithm requires $n^3/6$ flops.

The elements of $\operatorname{diag}(C_x) = \operatorname{diag}(SS^T)$ are just the 2-norms squared of the rows of S and can be computed in a further $n^2/2$ flops by $c_{ii} = \sum_{j=i}^{n} s_{ij}^2$, $i = 1, 2, \ldots, n$. The matrix C_x is symmetric, and therefore we need only compute its upper triangular part. This takes $n^3/6$ flops and can be sequenced so that the elements of C_x overwrite those of S.

In case C_x is needed there is an alternative way of computing C_x without inverting R. We have from (2.8.10), multiplying by R from the left,

$$(2.8.11) \qquad RC_x = R^{-T}.$$

The diagonal elements of R^{-T} are simply r_{kk}^{-1}, $k = n, \ldots, 1$, and since R^{-T} is lower triangular it has $\frac{1}{2}n(n-1)$ zero elements. Hence, $\frac{1}{2}n(n+1)$ elements of R^{-T} are known and the corresponding equations in (2.8.11) suffice to determine the elements in the upper triangular part of the symmetric matrix C_x.

To compute the elements in the last column c_n of C_x we solve the system

$$Rc_n = r_{nn}^{-1} e_n, \quad e_n = (0, \ldots, 0, 1)^T$$

by back-substitution. This gives

$$(2.8.12) \qquad c_{nn} = r_{nn}^{-2}, \quad c_{in} = -r_{ii}^{-1} \sum_{j=i+1}^{n} r_{ij} c_{jn}, \quad i = n-1, \ldots, 1.$$

By symmetry $c_{ni} = c_{in}$, $i = n-1, \ldots, 1$, so we also know the last row of C_x. Now assume that we have computed the elements $c_{ij} = c_{ji}$, $j = n, \ldots, k+1$, $i \leq j$. We next determine the elements c_{ik}, $i \leq k$. We have

$$c_{kk} r_{kk} + \sum_{j=k+1}^{n} r_{kj} c_{jk} = r_{kk}^{-1},$$

and since the elements $c_{kj} = c_{jk}$, $j = k+1, \ldots, n$, have already been computed,

$$(2.8.13) \qquad c_{kk} = r_{kk}^{-1}\left(r_{kk}^{-1} - \sum_{j=k+1}^{n} r_{kj} c_{kj}\right).$$

Similarly, for $i = k-1, \ldots, 1$,

$$(2.8.14) \qquad c_{ik} = -r_{ii}^{-1}\left(\sum_{j=i+1}^{k} r_{ij} c_{jk} + \sum_{j=k+1}^{n} r_{ij} c_{kj}\right).$$

Using the formulas (2.8.12)–(2.8.14) all elements of C_x can be computed in about $n^3/3$ flops. For the case when the Cholesky factor R is sparse it is possible to use the above algorithm to compute *elements of C_x in positions where R has nonzero elements very efficiently*; see Section 6.7.4. Note that since R is nonsingular these include the diagonal elements of C_x.

2.9. Iterative Refinement

2.9.1. Iterative refinement for linear systems. In iterative refinement a computed solution to a linear problem is regarded as an initial approximation to the true solution and is corrected in an iterative process. Let $\bar{x}$ be any approximate solution to the linear system of equations $Ax = b$. Then the process of **iterative refinement** can be described as follows: put $x^{(1)} = \bar{x}$, and

$$\begin{aligned}
&\text{for } s = 1, 2, \ldots; \\
&\quad r^{(s)} = b - Ax^{(s)}; &&\text{(in precision } \epsilon_2\text{)} \\
&\quad \text{solve } A\delta^{(s)} = r^{(s)}; &&\text{(in precision } \epsilon_1\text{)} \\
&\quad x^{(s+1)} = x^{(s)} + \delta^{(s)}; \\
&\text{end};
\end{aligned}$$

When $\bar{x}$ has been computed by a direct method the same factorization can be used to solve the sequence of systems $A\delta^{(s)} = r^{(s)}$, $s = 1, 2, \ldots$. Hence, the arithmetic cost of the refinement is quite small compared to the cost of computing the initial approximation. However, the refinement requires that the data A and b be saved. This extra storage requirement can make the technique fairly costly for large matrices. Note the possibility of using *extended precision* $\epsilon_2 \ll \epsilon_1$ for accumulating the inner products in computing the residuals $r^{(s)}$; these are then rounded to precision ϵ_1 before solving for $\delta^{(s)}$.

The behavior of iterative refinement using double precision residuals has been summed up in popular terms by Wilkinson [840, 1977] as follows:

- If A is almost singular to the working precision the first solution has no correct significant digits, and the same is true of all subsequent refined solutions. The matrix A is too ill-conditioned (possibly singular) for solution to be possible without working to higher precision in the factoring.

- If A is not too ill-conditioned the first solution has some accuracy. Let us assume that $\|x - \bar{x}\|/\|x\| \approx \beta^{-k}$ in some natural norm, where we are concerned with a t-digit floating point computation with base β. Then the relative error diminishes by a factor of roughly β^{-k} with each step of refinement until we reach a stage at which $\|\delta_c\|/\|x_c\| < \beta^{-t}$, when we may say that the solution is correct to working precision.

Note that although the computed solution improves progressively with each iteration, this is *not* reflected in a corresponding decrease in the norm of the residual, which stays about the same when the solution is based on the LU factorization.

There are at least two distinct uses of iterative refinement with double precision residuals:

1. To compute a more accurate solution, e.g., in the case where A and b are exactly known and A is ill-conditioned.

2. The difference $x^{(2)} - x^{(1)}$ usually gives a very good estimate of the effect of relative changes in the data A, b of the order of the machine precision.

The first use may not be very important in practice if a backward stable method has been used. Since most problems involve inexact data, obtaining highly accurate solutions may not be justified. The second property offers a useful alternative to condition estimators for estimating the accuracy and reliability of computed solutions.

2.9.2. Extended precision iterative refinement. A straightforward generalization of iterative refinement to improve a computed solution $\bar{x}$ to the linear least squares problem (1.1) is as follows: put $x^{(1)} = \bar{x}$, and

$$
\begin{aligned}
&\text{for } s = 1, 2, \ldots \\
&\quad r^{(s)} = b - Ax^{(s)}; \qquad \text{(in precision } \epsilon_2\text{)} \\
&\quad \text{solve } \min_{\delta^{(s)}} \|A\delta^{(s)} - r^{(s)}\|_2; \quad \text{(in precision } \epsilon_1\text{)} \\
&\quad x^{(s+1)} = x^{(s)} + \delta^{(s)}; \\
&\text{end}
\end{aligned}
$$

The correction vector is itself the solution to a linear least squares problem, and if $\bar{x}$ has been computed from the QR decomposition of A, the same factorization can be used to solve for $\delta^{(s)}$.

This scheme is implemented in the procedure by Businger and Golub [142, 1965] and analyzed by Golub and Wilkinson [393, 1966]. They found that it performed well only when the true residual vector $r = b - Ax$ was sufficiently small, and noted in particular "whatever precision of computation is used there will be right-hand sides for which iterative refinement will never give solutions which are correct to working accuracy. This is in striking contrast with the linear equation case."

The key to a more successful algorithm for iterative refinement of least squares solutions is to use the augmented system of $m + n$ equations

$$(2.9.1) \quad \begin{pmatrix} I & A \\ A^T & 0 \end{pmatrix} \begin{pmatrix} y \\ x \end{pmatrix} = \begin{pmatrix} b \\ c \end{pmatrix},$$

and to simultaneously refine both y and x. For the least squares problem we have $c = 0$ and the residual $r = y = b - Ax$. We now describe this scheme, which was developed and analyzed in Björck [84, 110, 1967]. To make the description more compact we take as initial approximations

$$(2.9.2) \quad y^{(0)} = 0, \quad x^{(0)} = 0.$$

ALGORITHM 2.9.1. EXTENDED PRECISION ITERATIVE REFINEMENT. Assume that we have computed the QR decomposition of A by Householder. Take $y^{(0)} = 0$, $x^{(0)} = 0$, and compute a sequence of fixed precision approximations $y^{(s+1)}, x^{(s+1)}$, $s = 0, 1, \ldots$, where the sth iteration consists of three steps:

1. Compute the residual vectors to the system (2.9.1):

$$(2.9.3) \quad \begin{aligned} f^{(s)} &= fl_2(b - y^{(s)} - Ax^{(s)}), \\ g^{(s)} &= fl_2(c - A^T y^{(s)}), \end{aligned}$$

where $fl_2(E)$ indicates that the expression E is computed using extended precision accumulation of the inner products defining E.

2. Using fixed precision, solve for the corrections $\delta r^{(s)}$ and $\delta x^{(s)}$ from

$$(2.9.4) \quad \begin{pmatrix} I & A \\ A^T & 0 \end{pmatrix} \begin{pmatrix} \delta y^{(s)} \\ \delta x^{(s)} \end{pmatrix} = \begin{pmatrix} f^{(s)} \\ g^{(s)} \end{pmatrix},$$

using Golub's method, Algorithm 2.4.6.

3. Compute the new approximations

$$(2.9.5) \quad \begin{aligned} y^{(s+1)} &= y^{(s)} + \delta y^{(s)}, \\ x^{(s+1)} &= x^{(s)} + \delta x^{(s)}. \end{aligned}$$

REMARK 2.9.1. Because of the special initial values (2.9.2) we have $f^{(0)} = b$ and $g^{(0)} = 0$. It follows that $h = 0$ and $y^{(1)}$ and $x^{(1)}$ will just be the solution computed by QRD. If for convenience we drop the superscript, Algorithm 2.4.6 for the system (2.9.4) becomes

$$(2.9.6) \quad R^T h = g, \quad \begin{pmatrix} c \\ d \end{pmatrix} = Q^T f, \quad \delta y = Q \begin{pmatrix} h \\ d \end{pmatrix}, \quad R \delta x = c - h.$$

Assuming that Q is stored as a product of Householder transformations, for $s \geq 1$ the solution in step 2 takes $4mn - n^2$ flops.

2.9. ITERATIVE REFINEMENT

REMARK 2.9.2. The system (2.9.4) can also be solved by the MGS decomposition. Using the interpretation of MGS as a Householder method, one obtains the following algorithm. Assume that we have computed R and $Q_1 = (q_1, q_2, \ldots, q_n)$ by MGS. The first and last steps are the same as in (2.9.6). We compute $c = (c_1, \ldots, c_n)^T$ and $d = d^{(n+1)}$ by orthogonalizing $d^{(1)} = f$ against q_k, for $k = 1, 2, \ldots, n$,

$$(2.9.7) \qquad c_k = q_k^T d^{(k)}, \quad d^{(k+1)} = d^{(k)} - c_k q_k.$$

Then $\delta y = d^{(1)}$ is obtained by backward recursion,

$$(2.9.8) \qquad d^{(k)} = d^{(k+1)} - z_k q_k, \quad z_k = q_k^T d^{(k+1)} - h_k,$$

for $k = n, n-1, \ldots, 1$. This algorithm improves that given by Björck [86, 1968]. It requires $4mn + n^2$ flops, and can be proved to give accuracy similar to that of Golub's method; see Björck and Paige [113, 1992]. ∎

An error analysis of the iterative refinement method (2.9.3)–(2.9.5) has been given by Björck [84, 1967], [86, 1968]. It is shown that the initial rate of improvement of the solution is linear with rate

$$(2.9.9) \qquad \rho_s = \|x^{(s)} - x\|_2 / \|x^{(s-1)} - x\|_2 < cu\kappa(A),$$

$s = 2, 3, \ldots$, where $c = c(m, n)$ is an error constant. This convergence rate is similar to that for the linear system case, even though the condition of the least squares problem involves a term proportional to $\kappa^2(A)$! Hence, in a sense, iterative refinement is *even more satisfactory* for large residual least squares problems, and may give solutions to full single precision accuracy even when the initial solution may have *no correct significant figures*!

REMARK 2.9.3. The estimate in (2.9.9) can be improved by substituting for $\kappa(A)$,

$$\kappa'(A) = \min_{D>0} \kappa(AD),$$

where D is a positive diagonal matrix. This rate of convergence is achieved even without actually carrying out the scaling of A by the optimal D.

EXAMPLE 2.9.1. (See Björck and Golub [110, 1967].) To illustrate the method of iterative refinement we consider the linear least squares problem where A is the last six columns of the inverse of the Hilbert matrix $H_8 \in R^{8 \times 8}$, which has elements

$$h_{ij} = 1/(i+j-1), \quad 1 \le i, j \le 8.$$

Two right-hand sides b_1 and b_2 are chosen so that the exact solution equals

$$x = (1/3, 1/4, 1/5, 1/6, 1/7, 1/8)^T.$$

For $b = b_1$ the system $Ax = b$ is compatible; for $b = b_2$ the norm of the residual $r = b - Ax$ equals $1.04 \cdot 10^7$. Hence for b_2 the term proportional to $\kappa^2(A)$ in the perturbation bound (1.4.25) dominates.

The refinement algorithm (2.9.2)–(2.9.5) was run on a computer with unit roundoff $u = 1.46 \cdot 10^{-11}$. The systems (2.9.4) were solved by the method (2.9.6) using a QR decomposition computed using Householder transformations. We stress that it is essential that double precision accumulation of inner products be used in step (i), but otherwise all computations can be performed in fixed precision. We give below the first component of the successive approximations $x^{(s)}, r^{(s)} s = 1, 2, 3 \ldots$ for the right-hand sides b_1 and b_2.

rhs b_1		rhs b_2	
$x_1^{(s)} =$	$3.33323\ 25269 \cdot 10^{-1}$	$x_1^{(s)} =$	$5.56239\ 01547 \cdot 10^{+1}$
	$3.33333\ 35247 \cdot 10^{-1}$		$3.37777\ 18060 \cdot 10^{-1}$
	$3.33333\ 33334 \cdot 10^{-1}$		$3.33311\ 57908 \cdot 10^{-1}$
	$3.33333\ 33334 \cdot 10^{-1}$		$3.33333\ 33117 \cdot 10^{-1}$
$r_1^{(s)} =$	$9.32626\ 24303 \cdot 10^{-5}$	$r_1^{(s)} =$	$2.80130\ 68864 \cdot 10^{6}$
	$5.05114\ 03416 \cdot 10^{-7}$		$2.79999\ 98248 \cdot 10^{6}$
	$3.65217\ 71718 \cdot 10^{-11}$		$2.79999\ 99995 \cdot 10^{6}$
	$-1.95300\ 70174 \cdot 10^{-13}$		$2.80000\ 00000 \cdot 10^{6}$

We observe a gain of almost three digits accuracy per step in the approximations to x_1 and r_1 for both right-hand sides b_1 and b_2. This is consistent with the estimate (2.9.9) since

$$\kappa(A) = 5.03 \cdot 10^8, \quad u\kappa(A) = 5.84 \cdot 10^{-3}.$$

For the right-hand side b_1 the approximation $x_1^{(4)}$ is correct to full fixed precision accuracy. It is interesting to note that for the right-hand side b_2 the effect of the error term proportional to $u\kappa^2(A)$ is evident in that the computed solution $x_1^{(1)}$ is in error by a factor of 10^3. However, $x_1^{(4)}$ has eight correct digits and $r_1^{(4)}$ is close to the true value $2.8 \cdot 10^6$.

2.9.3. Fixed precision iterative refinement. In most early descriptions of iterative refinement it is stressed that it is absolutely essential that the residuals be computed with a higher precision than that of the rest of the computation. However, more recently, Skeel [722, 1980] proved that for linear systems solved by Gaussian elimination and partial pivoting, iterative refinement with fixed precision residual will give improved stability. More precisely, he showed that if A is not too ill-conditioned, and if the components of the vector $|A||x|$ do not vary too much, then the solution computed by Gaussian elimination with one step of iterative refinement will have a *componentwise relative backward error* on the order of machine precision. We here consider similar refinement processes for linear least squares problems and for minimum norm solutions of underdetermined systems.

When fixed precision is used to compute the residuals in (2.9.3) the roundoff errors in the computation of the vectors f and g become more important. We now show that these are equivalent to small componentwise perturbations in the

2.9. ITERATIVE REFINEMENT

nonzero blocks of the augmented matrix. By a standard backward error analysis it follows that

$$\bar{f} = b - \delta b - (A + \delta A_1)\bar{x} - (I + \delta I)\bar{r},$$
$$\bar{g} = -(A + \delta A_2)^T \bar{r},$$

where δI is diagonal and hence

(2.9.10) $\qquad |\delta A_1| \leq 1.06(n+3)u|A|, \quad |\delta b| \leq 1.06u|b|,$
(2.9.11) $\qquad |\delta A_2| \leq 1.06(m+2)u|A|, \quad |\delta I| \leq uI,$

and the inequalities are to be interpreted componentwise. It follows that the computed residuals $\bar{f}$ and $\bar{g}$ are the *exact* residuals corresponding to the perturbed system

$$\begin{pmatrix} \bar{f} \\ \bar{g} \end{pmatrix} = \begin{pmatrix} b + \delta b \\ 0 \end{pmatrix} - \begin{pmatrix} I + \delta I & A + \delta A_1 \\ (A + \delta A_2)^T & 0 \end{pmatrix} \begin{pmatrix} \bar{r} \\ \bar{x} \end{pmatrix},$$

where the perturbations satisfy the componentwise bounds derived above. A perturbation $|\delta I|$ can be considered as a small perturbation in the weights of the rows of $Ax - b$.

Roundoff errors also occur in solving the equations (2.9.8), and we will get a small componentwise relative backward error only if the iterative refinement converges and the roundoff errors in the solution of the final corrections are negligible. Higham [464, 1991] has shown that although results are weaker for least squares problems than for square linear systems, iterative refinement improves a componentwise measure of backward stability.

In fixed precision the general algorithm for iterative refinement can be simplified for the least squares problem. The resulting algorithm only requires one matrix-vector multiplication each with A and A^T, and the solution of two triangular systems.

ALGORITHM 2.9.2. FIXED PRECISION ITERATIVE REFINEMENT $(c = 0)$. Set $x_0 = 0$, $r_0 = 0$. For $s = 0, 1, 2, \ldots$ until convergence do

$$r_s = b - Ax_s,$$
$$R^T(R\delta x_s) = A^T r_s,$$
$$x_{s+1} = x_s + \delta x_s.$$

Fixed precision iterative refinement is of most interest when a method which is not backward stable has been used. For example, it can be very efficient for improving a solution obtained from the normal equations. Then R is computed by Cholesky factorization of the matrix $A^T A$, and the first step, $i = 0$, is identical to the method of normal equations. However, with R from the Cholesky factorization the favorable rate of convergence in (2.9.9) is not obtained; see Björck [88, 1978]. Instead we have

$$\bar{\rho} = cu\kappa'(A)^2 = \kappa'(A)\rho.$$

However, if several steps of refinement are carried out, this will also lead to good accuracy for a large class of problems; see the discussion in Section 6.6.5.

Another application of fixed precision iterative refinement is to sparse problems, where R has been computed by a sparse QR decomposition, but the matrix Q has not been saved. Then the system of **seminormal equations**

$$(2.9.12) \qquad R^T(Rx) = A^T b$$

is often used to solve a least squares problem. It was shown in Björck [91, 1987] that the solution computed by (2.9.12) is in general no more accurate than a solution obtained from the normal equations using the Cholesky factorization. However, one refinement step is usually sufficient to get the same error level as for a backward stable method. We refer to the algorithm with *one step* of refinement as the *corrected seminormal equations* (CSNE). A more detailed discussion of this algorithm is given in Section 6.6.5.

Iterative refinement can also be used to improve computed solutions to the minimum norm problem

$$\min \|y\|_2, \text{ subject to } A^T y = c,$$

which corresponds to the special case $b = 0$ in the augmented system (2.9.1). For fixed precision refinement the general algorithm can be simplified for this special case as follows.

ALGORITHM 2.9.3. FIXED PRECISION ITERATIVE REFINEMENT $(b=0)$. Set $y_0 = 0$, $x_0 = 0$. For $s = 0, 1, 2, \ldots$ until convergence do

$$\begin{aligned} g_s &= c - A^T y_s, \\ R^T(R\delta x_s) &= -g_s, \\ x_{s+1} &= x_s + \delta x_s, \\ y_{s+1} &= y_s - A\delta x_s. \end{aligned}$$

For sparse least squares a method based on the factorization of the augmented system has been considered by Arioli, Duff, and de Rijk [21, 1989]. As described in Section 2.5.3, 1×1 and 2×2 diagonal pivots are used. Here the optimal scaling cannot be obtained a priori. The corresponding loss of accuracy can often be compensated for by the use of fixed precision iterative refinement.

Chapter 3
Modified Least Squares Problems

3.1. Introduction

3.1.1. Updating problems. It is often desired to solve a sequence of modified least squares problems

$$(3.1.1) \qquad \min_x \|Ax - b\|_2, \qquad A \in \mathbf{R}^{m \times n},$$

where in each step rows of data in (A, b) are added, deleted, or both. This need arises, e.g., when data are arriving sequentially. In various time-series problems a window moving over the data is used; when a new observation is added, an old one is deleted as the window moves to the next step in the sample. In other applications columns of the matrix A may be added or deleted. Such modifications are usually referred to as updating (downdating) of least squares solutions.

Important applications where modified least squares problems arise include statistics, optimization, and signal processing. In statistics an efficient and stable procedure for adding and deleting rows to a regression model is needed; see Chambers [147, 1971]. In regression one also wants to examine different regression models, which can be achieved by adding and deleting columns (or permuting columns) in the model.

Applications in signal processing often require near real-time solutions, and the following requirements are critical; see Alexander, Pan, and Plemmons [9, 1988].

1. The modification should be performed with *as few operations and as little storage requirement as possible*. Recomputing the QR decomposition is too costly since it requires $O(mn^2)$ operations.

2. To make it possible to use a computer with *short word-length*, the solution should be accurate up to the limitations of data and conditioning of the problem; i.e., a *stable numerical method* must be used.

However, methods based on the normal equations and/or updating of the Cholesky factorization are still used in statistics and signal processing, although they do not fulfill the second requirement. An example is the recursive least squares algorithm given in Section 3.1.4.

3.1.2. Modified linear systems. We first review some classical results for solving linear systems, where the matrix has been modified by a correction of low rank. Let $A \in \mathbf{R}^{n \times n}$, and consider a system where A has been modified by a correction of rank p,

$$(3.1.2) \qquad (A + UDV^T)\hat{x} = b, \quad U, V \in \mathbf{R}^{n \times p},$$

where D is a nonsingular diagonal matrix. The modified linear system (3.1.2) is closely related to the bordered system

$$(3.1.3) \qquad \begin{pmatrix} A & U \\ V^T & -D^{-1} \end{pmatrix} \begin{pmatrix} \hat{x} \\ \theta \end{pmatrix} = \begin{pmatrix} b \\ 0 \end{pmatrix}.$$

Indeed, by eliminating θ it can be verified that $\hat{x}$ satisfies the linear system (3.1.2).

We now give some formulas for the inverse of a bordered matrix

$$M = \begin{pmatrix} A & B \\ C^T & D \end{pmatrix},$$

where $A \in \mathbf{R}^{n \times n}$ and $D \in \mathbf{R}^{p \times p}$ are square matrices. If A is nonsingular we can factor M into the product of a block lower and a block upper triangular matrix

$$M = \begin{pmatrix} I & 0 \\ C^T A^{-1} & I \end{pmatrix} \begin{pmatrix} A & B \\ 0 & S \end{pmatrix}, \quad S = D - C^T A^{-1} B.$$

This is equivalent to block Gaussian elimination, and the matrix S is called the **Schur complement** of A in M. From $M^{-1} = (LU)^{-1} = U^{-1}L^{-1}$, using formulas for the inverses of 2×2 block triangular matrices, we get the Schur-Banachiewicz inverse formula

$$M^{-1} = \begin{pmatrix} A^{-1} + A^{-1}BS^{-1}C^T A^{-1} & -A^{-1}BS^{-1} \\ -S^{-1}C^T A^{-1} & S^{-1} \end{pmatrix}.$$

Similarly, if D is assumed to be nonsingular we can factor M into a product of a block upper and a block lower triangular matrix

$$M = \begin{pmatrix} I & BD^{-1} \\ 0 & I \end{pmatrix} \begin{pmatrix} T & 0 \\ C^T & D \end{pmatrix}, \quad T = A - BD^{-1}C^T,$$

where T is the Schur complement of D in M. This is equivalent to block Gaussian elimination in reverse order. From this an alternative expression for M^{-1} can be derived,

$$M^{-1} = \begin{pmatrix} T^{-1} & -T^{-1}BD^{-1} \\ -D^{-1}C^T T^{-1} & D^{-1} + D^{-1}C^T T^{-1}BD^{-1} \end{pmatrix}.$$

By equating the $(1,1)$ block in the inverse M^{-1} we get the useful relation

$$(3.1.4) \qquad (A - BD^{-1}C^T)^{-1} = A^{-1} + A^{-1}B(D - C^T A^{-1}B)^{-1}C^T A^{-1},$$

which is often called the **Woodbury formula**. This formula, which assumes that both A and D are nonsingular, provides the inverse of A after it is modified

3.1. INTRODUCTION

by a matrix of rank p, is very useful when $p \ll n$. If we apply this formula to (3.1.3) we obtain

(3.1.5) $\quad (A + UDV^T)^{-1} = A^{-1} - A^{-1}U(D^{-1} + V^T A^{-1} U)^{-1} V^T A^{-1}.$

Hence the solution to the modified system (3.1.2) can be obtained from

$$\hat{x} = (A + UDV^T)^{-1} b = x - W(D^{-1} + V^T W)^{-1} V^T x, \quad W = A^{-1} U.$$

If an LU factorization of A is available, computing $W = A^{-1}U$ requires $n^2 p$ flops, $(D^{-1} + V^T W)$ and its LU factorization takes $np^2 + p^3/3$ flops, and finally $2np + p^2$ flops for computing $V^T x$ and the correction to x. If $p \ll n$ this is computationally advantageous compared to solving the system from scratch. In case A is symmetric and $U = V$ the formula simplifies further.

If we specialize the relation (3.1.5) to the case $p = 1$, when

(3.1.6) $\quad\quad\quad (A + \rho uv^T)\hat{x} = b, \quad u, v \in \mathbf{R}^n, \quad \rho \neq 0,$

we obtain the **Sherman–Morrison formula**.

(3.1.7) $\quad (A + \rho uv^T)^{-1} = A^{-1} - \alpha A^{-1} uv^T A^{-1}, \quad \alpha = 1/(\rho^{-1} + v^T A^{-1} u).$

Hence the solution to the modified system (3.1.6) can be obtained from

(3.1.8) $\quad\quad\quad \hat{x} = (A + \rho uv^T)^{-1} b = x - \alpha(v^T x) w, \quad w = A^{-1} u.$

For the history of these and similar updating formulas and a survey of various applications, see Hager [421, 1989].

If A is either a rectangular or square singular matrix, and we want the least squares solution, then it is not always possible to obtain similar formulas with A^{-1} replaced by $A^\dagger$, where $A^\dagger$ denotes the pseudoinverse of A. For the case $p = 1$ Meyer [577] gives explicit expressions for $(A + uv^H)^\dagger$. He shows that there are six different cases depending on which of the three conditions $u \in \mathcal{R}(A)$, $v \in \mathcal{R}(A^H)$, and $1 + v^H A^\dagger u \neq 0$ are satisfied. Generalizations of Meyer's results to $p > 1$ are not known

3.1.3. Modifying matrix factorizations. Solving a modified system by the Woodbury or Sherman–Morrison formula will not always lead to stable methods. Stability will be a problem whenever the unmodified problem is worse-conditioned than the modified problem. In general, methods which instead rely on modifying a matrix factorization of A are to be preferred. The first systematic use of such algorithms seems to have been in optimization. Numerous aspects of updating various matrix factorizations are discussed in Gill et al. [354, 1974].

There is a simple relationship between the problem of updating matrix factorizations and that of updating the least squares solutions. Recall that, by Theorem 2.2.1, if A has full column rank and the R-factor of the matrix (A, b) is

(3.1.9) $\quad\quad\quad\quad\quad\quad \begin{pmatrix} R & z \\ 0 & \rho \end{pmatrix},$

then the solution to the least squares problem (3.1.1) is given by

(3.1.10) $$Rx = z, \quad \|Ax - b\|_2 = \rho.$$

The upper triangular matrix (3.1.9) can be computed either from the QR decomposition of (A, b) or as the Cholesky factor of $(A, b)^T(A, b)$. Hence, updating algorithms for matrix factorizations applied to (A, b) give updating algorithms for least squares solutions.

The updating of the Householder QR decomposition of A, where Q is stored as a product of Householder transformations, is not feasible. This is because there seems to be no efficient way to update a product of Householder transformations, e.g., when a row is added. In Section 3.2 we give algorithms for updating the factorization

(3.1.11) $$A = Q \begin{pmatrix} R \\ 0 \end{pmatrix},$$

where $Q \in \mathbf{R}^{m \times m}$ is stored explicitly as an $m \times m$ matrix. These updating algorithms require $O(m^2)$ multiplications, and are almost normwise backward stable; see the error analysis given by Paige [629, 1980]. In many applications it suffices to update the "Gram–Schmidt" QR decomposition

(3.1.12) $$A = Q_1 R, \quad Q_1 \in \mathbf{R}^{m \times n},$$

where $Q_1 \in \mathbf{R}^{m \times n}$ consists of the first n columns of Q. In Section 3.2.7 we describe stable updating algorithms for this decomposition, developed by Daniel et al. [202, 1976]. These only require $O(mn)$ storage and operations. A set of FORTRAN routines implementing these algorithms is given by Reichel and Gragg [678, 1990].

In some applications, e.g., when A is large and/or sparse, one would like to update only the R-factor without the extra cost of storing and updating the matrix Q. Such algorithms, as shown in Section 3.5, are inherently less stable than methods that update both Q and R. However, they can be stabilized by a refinement technique using the original data matrix A.

All known updating algorithms for the singular value decomposition (SVD)

$$A = U \begin{pmatrix} \Sigma \\ 0 \end{pmatrix} V^T,$$

where $U \in \mathbf{R}^{m \times m}$ and $V \in \mathbf{R}^{n \times n}$, require $O(mn^2)$ flops, which is the same order as recomputing the SVD from scratch. However, since the order constant is less, there is still a gain. Bunch and Nielsen [138, 1978] have developed methods for updating the SVD of A, when A is modified by adding or deleting a row or column. Also, algorithms for solving the correspondingly modified least squares problems are developed. These updating methods, however, all require $O(n^3)$ operations when $A \in \mathbf{R}^{m \times n}$, $m \geq n$.

For applications where the matrix A may not have full numerical column rank, a rank revealing QR decomposition, or a complete QR decomposition can be updated. This is a more difficult problem, but intensive work on updating algorithms for such decompositions has recently begun.

3.1.4. Recursive least squares.

The solution to the least squares problem (3.1.1) satisfies the normal equations

$$A^T A x = A^T b, \tag{3.1.13}$$

where we assume that $A \in \mathbf{R}^{m \times n}$ and rank$(A) = n$. If the observation

$$w^T x = \beta \tag{3.1.14}$$

is added, then the updated solution $\tilde{x}$ satisfies the modified normal equations

$$(A^T A + w w^T) \tilde{x} = A^T b + \beta w. \tag{3.1.15}$$

A straightforward method for computing $\tilde{x}$ is based on updating the matrix

$$C = (A^T A)^{-1}. \tag{3.1.16}$$

Since the matrix C is the scaled covariance matrix such a method is called a **covariance matrix method**. Since $\tilde{C}^{-1} = C^{-1} + w w^T$ we have by the Sherman–Morrison formula (3.1.7)

$$\tilde{C} = C - \frac{1}{1 + w^T u} u u^T, \qquad u = C w. \tag{3.1.17}$$

Adding a term $w w^T x$ to both sides of (3.1.13) and subtracting from (3.1.15) gives

$$(A^T A + w w^T)(\tilde{x} - x) = (\beta - w^T x) w.$$

Solving for the updated solution gives the following basic formula:

$$\tilde{x} = x + (\beta - w^T x) \tilde{u}, \qquad \tilde{u} = \tilde{C} w. \tag{3.1.18}$$

Equations (3.1.17)–(3.1.18) define a **recursive least squares** (RLS) algorithm associated with the Kalman filter. The vector $\tilde{u} = \tilde{C} w$, which weights the predicted residual $\beta - w^T x$ of the new observation, is called the **Kalman gain vector**.

As pointed out by Pan and Plemmons [645, 1989] the equations (3.1.17)–(3.1.18) can with slight modifications be used also for *deleting* the observation (3.1.14). We have

$$\tilde{C} = C + \frac{1}{1 - w^T u} u u^T, \qquad \tilde{x} = x - (\beta - w^T x) \tilde{u}, \tag{3.1.19}$$

provided that $1 - w^T u \neq 0$.

The simplicity and recursive nature of this updating algorithm has made it very popular for many applications. The main disadvantage of the RLS algorithm is its serious sensitivity to roundoff errors. The updating algorithms based on orthogonal transformations developed in the following sections are generally to be preferred.

3.2. Modifying the Full QR Decomposition

3.2.1. Introduction. In this section algorithms are given for updating the QR decomposition (3.1.11) for three important kinds of modifications.

1. General rank one change of A.

2. Deleting (adding) a column of A.

3. Adding (deleting) a row of A.

We will see that deleting a column and adding a row are "easy" operations, whereas adding a column and deleting a row are more delicate operations. This is because in the latter cases the modified matrix may become rank deficient, and if we are close to this case the problem becomes ill-conditioned.

We stress again that the algorithms for updating and downdating the QR decomposition can be used to add and remove equations from a least squares problem by applying them to the augmented matrix (A, b). More details on this use can be found in Dongarra et al. [228, 1979, Chap. 10].

The algorithms given in this section can be modified in a straightforward fashion to treat cases where a block of rows/columns are added or deleted. Such block algorithms are more amenable to efficient implementation on vector and parallel computers. Block methods for downdating are studied in Eldén and Park [274, 1994] and Olszanskyj, Lebak, and Bojanczyk [611, 1994].

3.2.2. General rank one change. Assume that we know the complete QR decomposition (3.1.11) of the matrix $A \in \mathbf{R}^{m \times n}$. We want to compute the decomposition

$$(3.2.1) \qquad \tilde{A} = A + uv^T = \tilde{Q} \begin{pmatrix} \tilde{R} \\ 0 \end{pmatrix},$$

where $u \in \mathbf{R}^m$ and $v \in \mathbf{R}^n$ are given. For simplicity we assume that $\text{rank}(A) = \text{rank}(\tilde{A}) = n$, so that R and $\tilde{R}$ are uniquely determined.

We first compute $w = Q^T u \in \mathbf{R}^m$, so that

$$(3.2.2) \qquad A + uv^T = Q\left[\begin{pmatrix} R \\ 0 \end{pmatrix} + wv^T\right].$$

Next we determine a sequence of Givens rotations $J_k = G_{k,k+1}(\theta_k)$, $k = m-1, \ldots, 1$ such that

$$J_1^T \cdots J_{m-1}^T w = \alpha e_1, \qquad \alpha = \pm \|w\|_2.$$

Note that these transformations zero the last $m - 1$ components of w from bottom up. (For details on how to compute J_k see Algorithm 2.3.1.) If these transformations are applied to the R-factor in (3.2.2) we obtain

$$(3.2.3) \qquad \tilde{H} = J_1^T \cdots J_{m-1}^T \left[\begin{pmatrix} R \\ 0 \end{pmatrix} + wv^T\right] = H + \alpha e_1 v^T.$$

3.2. Modifying the Full QR Decomposition

(Note that $J_{n+1}, \ldots, J_{m-1}$ have no effect on R.) Because of the structure of the Givens rotations the matrix H will be an upper Hessenberg matrix, i.e., H is triangular except for extra nonzero elements $h_{k+1,k}, k = 1, 2, \ldots, n$ (e.g., $m = 6$, $n = 4$),

$$H = \begin{pmatrix} \times & \times & \times & \times \\ \times & \times & \times & \times \\ 0 & \times & \times & \times \\ 0 & 0 & \times & \times \\ 0 & 0 & 0 & \times \\ 0 & 0 & 0 & 0 \end{pmatrix}.$$

Since only the first row of H is modified by the term $\alpha e_1 v^T$, $\tilde{H}$ is also upper Hessenberg. Then we can determine Givens rotations $\tilde{J}_k = G_{k,k+1}(\phi_k)$, $k = 1, \ldots, n$, which zero the element in position $(k+1, k)$, so that

$$\tilde{J}_n^T \cdots \tilde{J}_1^T \tilde{H} = \begin{pmatrix} \tilde{R} \\ 0 \end{pmatrix} \tag{3.2.4}$$

is upper triangular. Finally the transformations are accumulated into Q to get

$$\tilde{Q} = QU, \qquad U = J_{m-1} \cdots J_1 \tilde{J}_1 \cdots \tilde{J}_n.$$

$\tilde{Q}$ and $\tilde{R}$ now give the desired decomposition (3.2.1). The work needed for this update is as follows: computing $w = Q^T u$ takes m^2 flops. Computing H and $\tilde{R}$ takes $4n^2$ flops and accumulating the transformations J_k and $\tilde{J}_k$ into Q takes $4(m^2 + mn)$ flops for a total of $5m^2 + 4mn + 4n^2$ flops. Hence the work has been decreased from $O(mn^2)$ to $O(m^2)$. However, if n is small updating may still be more expensive than computing the decomposition from scratch.

To update a least squares solution we apply the procedure above to compute the QR decomposition of

$$(A + uv^T, b) = (A, b) + u(v^T, 0).$$

Given the QR decomposition of $A = (a_1, \ldots, a_n) \in \mathbf{R}^{m \times n}$, it is often required to compute the QR decomposition of a matrix $\tilde{A}$ obtained from A by adding or deleting a column. An important application occurs in active set methods for solving least squares problems with inequality constraints; see Section 3.5. The algorithms developed below apply with trivial modifications to computing the QR decomposition of (A, b) when a column is deleted/added.

3.2.3. Deleting a column. We first observe that the QR decomposition behaves nicely under partitioning. Assume that

$$A = (A_1, A_2) = Q \begin{pmatrix} R_{11} & R_{12} \\ 0 & R_{22} \\ 0 & 0 \end{pmatrix},$$

where

$$A_1 = (a_1, \ldots, a_p), \qquad A_2 = (a_{p+1}, \ldots, a_n).$$

From this the QR decomposition of A_1 is trivially obtained by deleting the $n - p$ trailing columns from the decomposition.

Suppose now that we want to compute the QR decomposition of the matrix resulting from deleting the kth column in A, $k < n$,

$$\tilde{A} = (a_1, \ldots, a_{k-1}, a_{k+1}, \ldots, a_n).$$

From the above observation it follows that this decomposition can readily be obtained from the QR decomposition of the permuted matrix

$$(3.2.5) \qquad AP_L = (a_1, \ldots, a_{k-1}, a_{k+1}, \ldots, a_n, a_k) = Q \begin{pmatrix} RP_L \\ 0 \end{pmatrix},$$

where P_L is a permutation matrix which performs a *left circular shift* of the columns $a_k, \ldots, a_n$. The matrix RP_L will have the structure

$$RP_L = (r_1, \ldots, r_{k-1}, r_{k+1}, \ldots, r_n, r_k) = \begin{pmatrix} R_{11} & R_{13} & v \\ 0 & w^T & r_{kk} \\ 0 & R_{33} & 0 \end{pmatrix},$$

where $R_{11} \in \mathbf{R}^{(k-1) \times (k-1)}$ and $R_{33} \in \mathbf{R}^{(n-k) \times (n-k)}$ are upper triangular ($k = 3$, $n = 5$),

$$RP_L = \begin{pmatrix} \times & \times & \times & \times & \times \\ 0 & \times & \times & \times & \times \\ 0 & 0 & \times & \times & \times \\ 0 & 0 & \times & \times & 0 \\ 0 & 0 & 0 & \times & 0 \end{pmatrix}.$$

Hence it is possible to determine a sequence of Givens rotations $G_k = R_{k,k+1}$ so that

$$G_{n-1}^T \cdots G_k^T \begin{pmatrix} w^T & r_{kk} \\ R_{33} & 0 \end{pmatrix} = \tilde{R}_{22}$$

is upper triangular (cf. (3.2.4)). Note that the last column will fill in. With

$$\tilde{R} = \begin{pmatrix} R_{11} & \tilde{R}_{12} \\ 0 & \tilde{R}_{22} \end{pmatrix}, \qquad \tilde{R}_{12} = (R_{13}, v),$$

we have the updated decomposition

$$\tilde{A} = AP_L = \tilde{Q} \begin{pmatrix} \tilde{R} \\ 0 \end{pmatrix}, \qquad \tilde{Q} = QG_k \cdots G_{n-1}.$$

The required QR decomposition of $\tilde{A}$ is then obtained by deleting the last column in $\tilde{A}$ and $\tilde{R}$.

We remark that more generally one may want to compute the QR decomposition of

$$(a_1, \ldots, a_{k-1}, a_{k+1}, \ldots, a_p, a_k, a_{p+1}, \ldots a_n),$$

i.e., of the matrix resulting from a left circular shift applied to the columns $a_k, \ldots, a_p$; cf. Dongarra et al. [228, 1979, p. 10.2]. This can be done by an obvious extension of the above algorithm.

3.2. Modifying the Full QR Decomposition

3.2.4. Appending a column. We now consider the problem of computing the QR decomposition of a matrix where the column a_{n+1} has been appended in the kth position,

$$(3.2.6) \qquad (a_1, \ldots, a_{k-1}, a_{n+1}, a_k, \ldots, a_n) = (A, a_{n+1})P_R = \tilde{A}P_R,$$

where P_R is a permutation matrix which performs a *right circular shift* on the columns $a_k, \ldots, a_{n+1}$.

We first compute the QR decomposition of $\tilde{A} = (A, a_{n+1})$ from that of A. This is straightforward using, e.g., Householder's method, since this algorithm can be used to process A a column at a time. We form the vector w,

$$w = Q^T a_{n+1} = \begin{pmatrix} u \\ v \end{pmatrix} \begin{matrix} \} \; n \\ \} \; m-n \end{matrix},$$

and construct a Householder transformation P_n such that

$$(3.2.7) \qquad P_n^T \begin{pmatrix} u \\ v \end{pmatrix} = \begin{pmatrix} u \\ \gamma \\ 0 \end{pmatrix}, \qquad \gamma = \|v\|_2.$$

This gives the decomposition

$$(A, a_{n+1}) = \tilde{Q} \begin{pmatrix} \tilde{R} \\ 0 \end{pmatrix}, \quad \tilde{Q} = QP_n, \quad \tilde{R} = \begin{pmatrix} R & u \\ 0 & \gamma \end{pmatrix}.$$

The matrix $\tilde{R}P_R$ now has the structure

$$\tilde{R}P_R = (r_1, \ldots, r_{k-1}, r_{n+1}, r_k, \ldots, r_n) = \begin{pmatrix} R_{11} & u_1 & R_{12} \\ 0 & u_2 & R_{22} \\ 0 & \gamma & 0 \end{pmatrix},$$

where $R_{11} \in \mathbf{R}^{(k-1)\times(k-1)}$ and $R_{22} \in \mathbf{R}^{(n-k)\times(n-k)}$ are upper triangular, e.g.,

$$\tilde{R}P_R = \begin{pmatrix} \times & \times & \times & \times & \times \\ 0 & \times & \times & \times & \times \\ 0 & 0 & \times & \times & \times \\ 0 & 0 & \times & 0 & \times \\ 0 & 0 & \times & 0 & 0 \end{pmatrix},$$

($k=3$, $n=4$). We determine Givens rotations $J_i = G_{i-1,i}$, $i = n, \ldots, k$, to zero the last $n-k+1$ elements in the kth column of $\tilde{R}P_R$. Then

$$J_k^T \cdots J_{n-1}^T \begin{pmatrix} u_2 & R_{22} \\ \gamma & 0 \end{pmatrix} = \tilde{R}_{22}$$

is upper triangular and the updated R-factor is given by

$$(3.2.8) \qquad \bar{R} = \begin{pmatrix} R_{11} & \tilde{R}_{12} \\ 0 & \tilde{R}_{22} \end{pmatrix}, \qquad \tilde{R}_{12} = (u_1, R_{12}).$$

The QR decomposition of $\tilde{A}$ in (3.2.6) becomes

$$\tilde{A} = \tilde{Q}\begin{pmatrix} \tilde{R} \\ 0 \end{pmatrix}, \qquad \tilde{Q} = \bar{Q}J_{n-1}\cdots J_k.$$

Again, the above method easily generalizes to compute the QR decomposition of

$$(a_1,\ldots,a_{k-1},a_p,a_k,\ldots,a_{p-1},a_{p+1},\ldots,a_{n+1}),$$

i.e., of the matrix resulting from a right circular shift of the columns $a_k,\ldots,a_p$.

It is important to note that when a column is *deleted* the new R-factor can be computed without Q being available. However, when a column is *added* it is essential that Q be known in the above algorithm. When a column is deleted, then from the interlacing property (Theorem 1.4.2) it follows that the smallest singular value will not decrease. On the other hand, when a column is added, the smallest singular value will not increase, and indeed the new triangular factor may become singular.

3.2.5. Appending a row. Given the QR decomposition (3.1.11) of a matrix $A \in \mathbf{R}^{m \times n}$ we first consider the problem of computing the QR decomposition of

(3.2.9) $$\tilde{A} = \begin{pmatrix} A \\ w^T \end{pmatrix}, \qquad w \in \mathbf{R}^n$$

where a row w^T has been appended. From (3.1.11) we have

$$\begin{pmatrix} Q^T & 0 \\ 0 & 1 \end{pmatrix}\begin{pmatrix} A \\ w^T \end{pmatrix} = \begin{pmatrix} R \\ 0 \\ w^T \end{pmatrix} = \Pi_{n+1,m+1}\begin{pmatrix} R \\ w^T \\ 0 \end{pmatrix},$$

where $\Pi_{n+1,m+1}$ denotes a permutation matrix interchanging the rows $n+1$ and $m+1$. The updating consists of n Givens rotations $J_k = G_{k,n+1}$, $k = 1,\ldots,n$, the kth of which annihilates the kth element in the last row to get

$$J_n \cdots J_1 \begin{pmatrix} R \\ w^T \end{pmatrix} = \begin{pmatrix} \tilde{R} \\ 0 \end{pmatrix}.$$

The updated Q-factor becomes

$$\tilde{Q} = \text{diag}(Q,1)\Pi_{n+1,m+1}J_1^T \cdots J_n^T.$$

Note that R can be updated without Q being available. Also from the interlacing property (Theorem 1.4.2) it follows that the smallest singular value of R will increase, and hence this procedure is stable. This scheme requires $2n^2 + O(n)$ multiplications if standard Givens rotations are used.

3.2.6. Deleting a row.

One way of incorporating changes in data is the sliding window method. Here, when a new data row has been added, an old data row is deleted. Another instance when a data row has to be removed is when it has somehow been identified as faulty.

We now consider modifying the QR decomposition when a row is *deleted*, which is the **downdating** problem. This corresponds to the problem of deleting the effects of an observation in a least squares problem. There is no loss of generality in assuming that the *first* row of A is to be deleted. To delete the kth row we merely apply the algorithm below with A and Q replaced by $\Pi_{1,k}A$ and $\Pi_{1,k}Q$, where $\Pi_{1,k}$ is a permutation matrix interchanging rows 1 and k.

We wish to obtain the QR decomposition of the matrix $\tilde{A} \in \mathbf{R}^{(m-1)\times n}$ when

(3.2.10) $$A = \begin{pmatrix} a_1^T \\ \tilde{A} \end{pmatrix} = Q \begin{pmatrix} R \\ 0 \end{pmatrix}$$

is known. We now show that this is equivalent to finding the QR decomposition of (e_1, A), where a dummy column $e_1 = (1, 0, \ldots, 0)^T$ has been added. We have

$$Q^T(e_1, A) = \begin{pmatrix} q_1 & R \\ q_2 & 0 \end{pmatrix},$$

where $q^T = (q_1^T, q_2^T) \in \mathbf{R}^m$ is the *first* row of Q. We now determine Givens rotations $J_k = G_{k,k+1}$, $k = m-1, \ldots, 1$, so that

(3.2.11) $$J_1^T \cdots J_{m-1}^T q = \alpha e_1, \quad \alpha = \pm 1.$$

Then we have

(3.2.12) $$J_1^T \cdots J_{m-1}^T \begin{pmatrix} q_1 & R \\ q_2 & 0 \end{pmatrix} = \begin{pmatrix} \alpha & v^T \\ 0 & \tilde{R} \\ 0 & 0 \end{pmatrix},$$

where the matrix $\tilde{R}$ is upper triangular. Note that the transformations $J_{n+1}, \ldots, J_{m-1}$ will not affect R. Further, if we compute

$$\bar{Q} = Q J_{m-1} \cdots J_1,$$

it follows from (3.2.11) that the first row of $\bar{Q}$ equals αe_1^T. Since $\bar{Q}$ is orthogonal it must have the form

$$\bar{Q} = \begin{pmatrix} \alpha & 0 \\ 0 & \tilde{Q} \end{pmatrix},$$

with $|\alpha| = 1$ and $\tilde{Q} \in \mathbf{R}^{(m-1)\times(m-1)}$ orthogonal. Hence, from (3.2.10),

$$\begin{pmatrix} a_1^T \\ \tilde{A} \end{pmatrix} = \begin{pmatrix} \alpha & 0 \\ 0 & \tilde{Q} \end{pmatrix} \begin{pmatrix} v^T \\ \tilde{R} \\ 0 \end{pmatrix},$$

and hence the desired decomposition is

$$\tilde{A} = \tilde{Q} \begin{pmatrix} \tilde{R} \\ 0 \end{pmatrix}.$$

This algorithm for downdating is a special case of the rank one change algorithm, and is obtained by taking $u = -e_1$, $v^T = a_1^T$ in (3.2.1).

3.2.7. Modifying the Gram–Schmidt decomposition. In many applications, especially if $m \gg n$, it is too costly to save and modify the full QR decomposition. When we use the Gram–Schmidt (GS) QR decomposition,

$$(3.2.13) \qquad A = Q_1 R, \quad Q_1 \in \mathbf{R}^{m \times n},$$

then the storage requirement is reduced to mn for Q_1 from m^2, which is required for the full Q-factor.

Daniel et al. [202, 1976] develop stable algorithms for modifying the GS decomposition of a matrix A when A is changed by a matrix of rank one, or when a row or column is added or deleted. A principal tool of the algorithms is the GS process used with *reorthogonalization* to ensure orthogonality to working precision. A slightly simplified algorithm given in Reichel and Gragg [678, 1990] relies on the fact that in the full rank case one reorthogonalization is always enough; see also Parlett [651, 1980].

The algorithms given in Section 3.2.3 (and 3.2.5) for updating the full QR decomposition when deleting a column (adding a row) apply with trivial modifications to the GS decomposition as well. Adding a column is also straightforward, using the columnwise GS algorithm with reorthogonalization; see Algorithms 2.4.3 and 2.4.4. Here we consider the more delicate problems of deleting a row.

Assume that we have the QR factorization

$$(3.2.14) \qquad A = \begin{pmatrix} a_1^T \\ \tilde{A} \end{pmatrix} = Q_1 R = \begin{pmatrix} q_1^T \\ \hat{Q}_1 \end{pmatrix} R,$$

and want to delete the first row $z^T = a_1^T$. Note that (3.2.14) can be written as

$$(3.2.15) \qquad \begin{pmatrix} z^T \\ \tilde{A} \end{pmatrix} = \begin{pmatrix} q_1^T & 1 \\ \hat{Q}_1 & 0 \end{pmatrix} \begin{pmatrix} R \\ 0 \end{pmatrix}.$$

Following Daniel et al. [202, 1976] we first apply the GS algorithm (with reorthogonalization) so that the appended column $e_1 = (1, 0, \ldots, 0)^T$ is orthogonalized to

$$Q_1 = \begin{pmatrix} q_1^T \\ \hat{Q}_1 \end{pmatrix} \in \mathbf{R}^{m \times n}.$$

Because of the special form of the appended column, the result has the form

$$(3.2.16) \qquad \begin{pmatrix} q_1^T & 1 \\ \hat{Q}_1 & 0 \end{pmatrix} = \begin{pmatrix} q_1^T & \gamma \\ \hat{Q}_1 & h \end{pmatrix} \begin{pmatrix} I & q_1 \\ 0 & \gamma \end{pmatrix}$$

where $1 = q_1^T q_1 + (\gamma)^2$. Using (3.2.15)

$$\begin{pmatrix} z^T \\ \tilde{A} \end{pmatrix} = \begin{pmatrix} q_1^T & \gamma \\ \hat{Q}_1 & h \end{pmatrix} \begin{pmatrix} R \\ 0 \end{pmatrix},$$

and we now determine a sequence of plane rotations J_k, $k = n, n-1, \ldots, 1$, in the plane $(k, n+1)$ such that

$$(3.2.17) \qquad \begin{pmatrix} q_1^T & \gamma \\ \hat{Q}_1 & h \end{pmatrix} U = \begin{pmatrix} 0 & \tau \\ \tilde{Q}_1 & \tilde{h} \end{pmatrix}, \qquad U = J_n \cdots J_1,$$

3.2. MODIFYING THE FULL QR DECOMPOSITION

where J_k is chosen to annihilate the kth component in $(q_1^T \; \gamma)$. Since orthogonal transformations preserve length we can make $\tau = 1$, and because the transformed matrix has orthonormal columns, $\tilde{h} = 0$. Thus we have

$$\begin{pmatrix} z^T \\ \tilde{A} \end{pmatrix} = \begin{pmatrix} 0 & 1 \\ \tilde{Q}_1 & 0 \end{pmatrix} \begin{pmatrix} \tilde{R} \\ z^T \end{pmatrix},$$

where

(3.2.18) $$U^T \begin{pmatrix} R \\ 0 \end{pmatrix} = J_1^T \cdots J_n^T \begin{pmatrix} R \\ 0 \end{pmatrix} = \begin{pmatrix} \tilde{R} \\ 0 \end{pmatrix}$$

with $\tilde{R}$ upper triangular. Hence the downdated QR decomposition becomes

(3.2.19) $$\tilde{A} = \tilde{Q}_1 \tilde{R}.$$

As mentioned, it is necessary to perform reorthogonalization when the Gram–Schmidt algorithm is used to orthogonalize e_1 to Q_1 in (3.2.15). Let $v = e_1 - Q_1 q_1$. If $\|v\|_2 \geq 1/\sqrt{2}$ then take $v := v/\|v\|_2$; else reorthogonalize

$$s := Q_1^T v, \qquad v' := v - Q_1 s.$$

If $\|v'\|_2 \geq \|v\|_2/\sqrt{2}$, then take

$$v := v'/\|v'\|_2, \qquad \begin{pmatrix} \gamma \\ h \end{pmatrix} := v.$$

Otherwise e_1 is (numerically) linearly dependent on the columns of Q_1. Then we take $\gamma := 0$ and determine $h \in \mathbf{R}^{(n-1)}$ orthogonal to $\hat{Q}_1$; see Daniel et al. [202, 1976].

With one reorthogonalization, the GS downdating algorithm requires about $7mn + 2.5n^2$ flops. This can be reduced to $5mn + 1.5n^2$ flops when fast scaled rotations are used in (3.2.17) and (3.2.18). Note that the data matrix A is never needed; to delete the first row of A, only the R-factor and the corresponding row in Q_1 are needed. Thus, the storage requirement is about $mn + 0.5n^2$ for Q_1 and R.

The reason for the reorthogonalization step in the GS downdating algorithm is that in the modified Gram–Schmidt (MGS) QR decomposition the computed Q_1 will not be accurately orthogonal when the condition number of A is large. Yoo and Park [846, 1996] have devised an accurate MGS downdating algorithm that does not use reorthogonalization. This is based on a relation between MGS applied to $A \in \mathbf{R}^{m \times n}$ and Householder QR on A augmented with an $n \times n$ zero matrix on top shown by Björck and Paige [113, 1992].

Let MGS applied to A gives the decomposition

$$A = Q_1 R, \qquad Q_1 = (q_1 \; q_2 \; \ldots \; q_n).$$

Suppose first that the columns in Q_1 are exactly orthonormal. Then this decomposition is equivalent to the Householder QR decomposition

$$\begin{pmatrix} 0_{n \times n} \\ A \end{pmatrix} = P \begin{pmatrix} R \\ 0 \end{pmatrix},$$

where
$$(3.2.20) \quad P = P_1 \cdots P_n = I - \sum_{k=1}^n u_k u_k^T = \begin{pmatrix} 0_{n \times n} & Q_1^T \\ Q_1 & I - Q_1 Q_1^T \end{pmatrix},$$

and $P_k = I - u_k u_k^T$, $u_k^T = (-e_k^T \; q_k^T)$.

The outlined equivalence also holds *numerically*. However, if $\bar{Q}_1 = (\bar{q}_1 \; \ldots \; \bar{q}_n)$ is the computed factor Q_1 in MGS, the departure of $\bar{Q}_1^T \bar{Q}_1$ can be significant. Then, instead of (3.2.20) in the equivalent Householder decomposition, we have

$$(3.2.21) \quad \bar{P} = \bar{P}_1 \cdots \bar{P}_n = I - \sum_{k=1}^n \bar{u}_k \bar{u}_k^T = \begin{pmatrix} \bar{P}_{11} & \bar{P}_{12} \\ \bar{P}_{21} & \bar{P}_{22} \end{pmatrix}$$
$$= \begin{pmatrix} \bar{P}_{11} & (I - \bar{P}_{11}) \bar{Q}_1^T \\ \bar{Q}_1 (I - \bar{P}_{11}) & I - \bar{Q}_1 (I - \bar{P}_{11}) \bar{Q}_1^T \end{pmatrix},$$

where $\bar{P}_k = I - \bar{u}_k \bar{u}_k^T$, $\bar{u}_k^T = (-e_k^T \; \bar{q}_k^T)$. The key idea in the downdating algorithm is that downdating the GS decomposition when the first row in A is deleted can be done by downdating the equivalent Householder decomposition, deleting the $(n+1)$st row in
$$\begin{pmatrix} 0_{n \times n} \\ A \end{pmatrix}.$$

However, this can be done stably if the $(n+1)$st row in $\bar{P}$ is available. It can be shown (see Björck and Paige [113, 1992]) that

$$\bar{P}_{21} = (\bar{q}_1, \; \bar{M}_1 \bar{q}_2, \; \ldots, \; \bar{M}_1 \bar{M}_2 \cdots \bar{M}_{n-1} \bar{q}_n) \in \mathbf{R}^{m \times n},$$
$$\bar{P}_{22} = \bar{M}_1 \bar{M}_2 \cdots \bar{M}_n \in \mathbf{R}^{m \times m},$$

where $\bar{M}_i = I - \bar{q}_i \bar{q}_i^T$, $i = 1, \ldots, n$. Using these expressions, it is possible to recursively evaluate $\bar{P}_{21} e_1$ and $\bar{P}_{22} e_1$ in $O(mn)$ operations. Yoo and Park [846, 1996] give a detailed description of such a Householder GS downdating algorithm which uses $10mn + 2n^2$ flops. They present numerical results which show that this algorithm can be much more accurate than previous GS downdating algorithms.

3.3. Downdating the Cholesky Factorization

3.3.1. Introduction. Suppose that the factor $R \in \mathbf{R}^{n \times n}$ is given from a Cholesky factorization of $A^T A$ (or QR decomposition of A), where

$$A = \begin{pmatrix} z^T \\ \tilde{A} \end{pmatrix}.$$

In the Cholesky downdating problem we want to find the Cholesky factor of $\tilde{R}$ such that
$$\tilde{R}^T \tilde{R} = \tilde{A}^T \tilde{A} = R^T R - z z^T.$$

This problem is analytically the same as that considered in Section 3.2.6, except that Q is not used.

3.3. Downdating the Cholesky Factorization

In this formulation the downdating problem is inherently more ill-conditioned than if Q is also available. The problem is that if the elements in the row z^T dominate those of $\tilde{A}$, then there is not sufficient information stored in R about $\tilde{R}$. The best we can hope for is that the downdating method is backward stable in the sense that it computes the *exact* Cholesky factor of $(R+E)^T(R+E)-(z+e)(z+e)^T$ with $\|E\|$ and $\|e\|$ small. It is important to note that *this does not guarantee that we obtain an $\tilde{R}$ that is the exact Cholesky factor of $(\tilde{A}+E)^T(\tilde{A}+E)$ for some small $\|E\|$*!

3.3.2. The Saunders algorithm. In the downdating algorithm the first row of Q played an essential role. If Q is not available the following algorithm due to Saunders [701, 1972] can be used. It is implemented in LINPACK as subroutine SCHDD (see Dongarra et al. [228, 1979, Chap. 10]), and often referred to as the **LINPACK** algorithm. A rounding error analysis of this method is given in Stewart [734, 1979].

The first row of the QR decomposition (3.2.10) can be written

$$a_1^T = q^T \begin{pmatrix} R \\ 0 \end{pmatrix} = (q_1^T, q_2^T) \begin{pmatrix} R \\ 0 \end{pmatrix},$$

where q has been partitioned conformably with the R-factor. Hence we can obtain q_1 by solving

$$R^T q_1 = a_1,$$

and then, since q is of unit length,

$$(3.3.1) \qquad \gamma = \|q_2\|_2 = (1 - \|q_1\|_2^2)^{1/2}.$$

The transformations $J_{n+1}, \ldots, J_{m-1}$ in (3.2.11) will only have the effect of computing

$$J_{n+1}^T \cdots J_{m-1}^T \begin{pmatrix} q_1 \\ q_2 \end{pmatrix} = \begin{pmatrix} q_1 \\ \pm\gamma \\ 0 \end{pmatrix}$$

and, as remarked above, will not affect R. Thus, we may determine the Givens transformations J_k, $k=1,\ldots,n$, by

$$J_1^T \cdots J_n^T \begin{pmatrix} q_1 \\ \gamma \end{pmatrix} = \alpha e_1, \quad \alpha = \pm 1,$$

and obtain the updated factor $\tilde{R}$ as in (3.2.12).

We note that if $\gamma \approx u^{1/2}$, where u is the unit roundoff, then γ cannot be computed stably from (3.3.1) because of severe cancellation in the subtraction $1 - \|q_1\|_2^2 \approx u$. Therefore, this algorithm will not be as stable as that using information from Q. The possible failure of the Saunders algorithm can be illustrated by the following simple example from Björck, Eldén, and Park [106, 1994].

EXAMPLE 3.3.1. Consider the least squares problem $\min \|Ax - b\|_2$, where

$$A = \begin{pmatrix} \tau \\ 1 \end{pmatrix}, \quad b = \begin{pmatrix} 1 \\ 1 \end{pmatrix}, \quad \tau = 1/\sqrt{u},$$

and u is the unit roundoff. We may think of the first row of A as an outlier. The QR decomposition of A, correctly rounded to single precision, is

$$A = \begin{pmatrix} \tau \\ 1 \end{pmatrix} = \begin{pmatrix} 1 & -\epsilon \\ \epsilon & 1 \end{pmatrix} \begin{pmatrix} \tau \\ 0 \end{pmatrix},$$

where $\epsilon = 1/\tau$. The Saunders algorithm will compute

$$q_1 = \tau/\tau = 1, \quad \gamma^2 = 1 - 1 = 0, \quad J_1 = I.$$

Hence it gives the downdated factor $\tilde{R} = 0$, and the downdated least squares solution is not defined. It is easily verified that if we downdate using Q we get the correct result $\tilde{R} = 1$ and the downdated solution $x = 1$. ∎

Sun [766, 1995] gave a first order perturbation analysis of the block downdating problem for the Cholesky factor

$$R^T R - ZZ^T = \tilde{R}^T \tilde{R}, \quad Z \in \mathbf{R}^{n \times r}.$$

This bound was later improved by Chang and Paige [160, 1996].

THEOREM 3.3.1. *Let $R, E \in \mathbf{R}^{n \times n}$ be upper triangular and $Z, F \in \mathbf{R}^{n \times r}$. Assume that the Cholesky factorization $R^T R - ZZ^T = \tilde{R}^T \tilde{R}$ exists and define $V = R^{-T} Z$. Let $\alpha > 0$ be small enough so that the factorization*

$$(R + \epsilon E)^T (R + \epsilon E) - (Z + \epsilon F)(Z + \epsilon F)^T = (\tilde{R} + T(\epsilon))^T (\tilde{R} + T(\epsilon))$$

exists for all $\epsilon \in (-\alpha, \alpha)$. Then we have the bound

$$(3.3.2) \qquad \frac{\|T(\epsilon)\|_F}{\|\tilde{R}\|_p} \leq \frac{\sqrt{2}\|\tilde{R}^{-1}\|_2}{\sqrt{1 - \|V\|_2^2}} (\|E\|_F + \|V\|_2 \|F\|_2) |\epsilon| + O(\epsilon^2).$$

The above perturbation analysis shows that using R to form the downdating transformations may be a much more ill-conditioned problem than downdating the original matrix A. This is because the original row in A is not perturbed in the same way as the vector $(q_1^T \ \gamma)$, which is computed by solving a triangular system to determine the downdating transformation in the Saunders algorithm. Hence, any method that uses R alone to recover the necessary elements of Q cannot be backward stable in the same sense as the downdating algorithms that use Q directly.

3.3.3. The corrected seminormal equations. In Example 3.3.1 the information from the second row in A is not present in R, only in Q, and therefore *no method working only from R can hope to do better*. However, for the case when A

is available a more accurate method for adding a column (or deleting a row) can be developed based on the corrected seminormal equations (CSNE); see Björck [91, 1987].

Assume first that we want to compute the R-factor in the QR decomposition of (A, a_{n+1}); cf Section 3.2.4. Then we compute the vector u and scalar γ in (3.2.7) as follows. A vector z is computed as the solution to the least squares problem

$$\min_z \|Az - a_{n+1}\|_2,$$

using CSNE. That is, we solve $R^T R z = A^T a_{n+1}$, compute a correction δz from

$$r = a_{n+1} - Az, \quad R^T R \delta z = r,$$

and take $z = z + \delta z$. Then u and γ are obtained from

$$u = Rz, \quad \gamma = \|Az - a_{n+1}\|_2,$$

and the updating of R proceeds as in Section 3.2.4.

Since deleting a row is equivalent to adding the special column e_1, the method of CSNE can also be used for the downdating problem. This leads to a modification of the Saunders algorithm, where the vector q_1 is constructed as follows. Let v be the solution to

$$\min_v \|Av - e_1\|_2,$$

and R the R-factor of A. Then the R-factor of (A, e_1) is

$$(3.3.3) \qquad \begin{pmatrix} R & q_1 \\ 0 & \gamma \end{pmatrix}, \quad q_1 = Rv, \quad \gamma = \|e_1 - Av\|_2.$$

The downdated R-factor is then obtained by applying orthogonal transformations to transform the last column into the vector e_1.

A similar procedure can be used to downdate the augmented R-factor (3.1.9) by solving the least squares problem

$$\min_{x,\phi} \left\| (A,\ b) \begin{pmatrix} x \\ \phi \end{pmatrix} - e_1 \right\|_2$$

using the CSNE. This leads to a downdating algorithm for least squares problems. However, the modifications are not trivial partly because the condition number of the augmented R-factor is large when ρ is small. For details we refer to the description in Björck, Eldén, and Park [106, 1994].

3.3.4. Hyperbolic rotations. We describe here another downdating algorithm using only R, which is of interest because it is recursive, requires fewer operations than the Saunders algorithm, and is better suited to parallel implementation. It is algorithmically very similar to the method for adding a row, and was originally suggested by Golub [366, 1969]. This is based on the observation that *deleting* the row w^T is formally equivalent to *adding* the row iw^T, where $i^2 = -1$. Hence the algorithm from Section 3.2.2, which uses a sequence of Givens rotations for adding a row, can be applied. The resulting algorithm can be expressed entirely in real arithmetic; see Lawson and Hanson [520, 1974, pp. 229–231].

A more modern description of this algorithm (see, e.g., [9, 1988]) is in terms of so-called **hyperbolic rotations**, which are of the form

$$(3.3.4) \qquad H = \begin{pmatrix} c & -s \\ -s & c \end{pmatrix}, \quad c = \cosh\theta, \quad s = \sinh\theta,$$

where $c^2 - s^2 = 1$. (Note that the condition number of H in (3.3.4) is not bounded.) A hyperbolic rotation in the plane (i,j) can be used to zero the component x_j of a vector x provided that $|x_i| > |x_j|$. The jth component of $H_{i,j}x$ is zero if

$$\sigma = \sqrt{x_i^2 - x_j^2}, \quad c = x_i/\sigma, \quad s = x_j/\sigma.$$

Hyperbolic rotations have been generalized in Rader and Steinhardt [673, 1988] to hyperbolic Householder transformations, designed to zero several components at a time in a vector x.

An error analysis given by Bojanczyk et al. [123, 1987] indicates that the stability properties of the algorithm using hyperbolic rotations is inferior to that of the following variant suggested by Chambers [147, 1971]. Chambers noted that it is possible to recreate the steps that would have updated $\tilde{R}$ with the row w^T to obtain R. Let the last row after $k-1$ steps be $w^{(k-1)}$. Then the kth step operates on rows k and $n+1$ as follows, for $j = k, \ldots, n$:

$$(3.3.5) \qquad r_{kj} = \cos\theta_k \tilde{r}_{kj} + \sin\theta_k w_j^{(k-1)},$$

$$(3.3.6) \qquad w_j^{(k)} = -\sin\theta_k \tilde{r}_{kj} + \cos\theta_k w_j^{(k-1)}.$$

The angle θ_k is chosen to make $w_k^{(k)} = 0$, i.e.,

$$\cos\theta_k = r_{kk}/\sigma, \quad \sin\theta_k = w_k^{(k-1)}/\sigma, \quad \sigma = r_{kk} = \sqrt{\tilde{r}_{kk}^2 + (w_k^{(k-1)})^2}$$

(see Algorithm 2.3.1 for a more precise description).

The downdating uses the same equations. Considering R and w^T as known and $\tilde{R}$ to be determined, and assuming $r_{kk} > w_k^{(k-1)}$, we can compute

$$\tilde{r}_{kk} = \sqrt{r_{kk}^2 - (w_k^{(k-1)})^2}, \quad \sin\theta_k = w_k^{(k-1)}/r_{kk}, \quad \cos\theta_k = \tilde{r}_{kk}/r_{kk}.$$

Rearranging the equation (3.3.6) gives for $j = k+1, \ldots, n$

$$(3.3.7) \qquad \tilde{r}_{kj} = (r_{kj} - \sin\theta_k w_j^{(k-1)})/\cos\theta_k,$$

$$(3.3.8) \qquad w_j^{(k)} = -\sin\theta_k \tilde{r}_{kj} + \cos\theta_k w_j^{(k-1)}.$$

This combines a hyperbolic rotation in the first equation with a Givens rotation in the second, and allows us to compute $\tilde{r}_{kj}$ and $w_j^{(k)}$ given r_{kj} and $w_j^{(k-1)}$. In Stewart [753, 1994] it is shown that the downdating via hyperbolic transformation is neither forward nor backward stable, whereas downdating using (3.3.7)–(3.3.8) is relationally stable in a certain sense.

3.4. Modifying the Singular Value Decomposition

3.4.1. Introduction. The singular value decomposition (SVD) of a matrix $A \in \mathbf{R}^{m \times n}$, $m \geq n$, of rank $r \leq n$ can be written

$$(3.4.1) \qquad A = U \begin{pmatrix} \Sigma \\ 0 \end{pmatrix} V^T,$$

where $U \in \mathbf{R}^{m \times n}$ and $V \in \mathbf{R}^{n \times n}$ are orthogonal, $\Sigma = \mathrm{diag}\,(\sigma_1, \sigma_2, \ldots, \sigma_n) \in \mathbf{R}^{m \times n}$, and $\sigma_1 \geq \sigma_2 \geq \cdots \geq \sigma_n \geq 0$. The SVD is of great importance, e.g., for determining the numerical rank of A and for solving rank deficient least squares problems involving A.

Assume that the SVD of A has been computed. When A is modified by appending or deleting a row or a column we would like to take advantage of this to reduce the amount of work necessary to compute the SVD of the modified matrix. Alternatively, if the SVD of A has been used to solve the least squares problem (3.1.1),

$$x = V \Sigma^\dagger c, \qquad c = U^T b,$$

we would like to update V, Σ, and c, which allows us to update x.

We first note that in the SVD rows and columns are treated the same. Hence appending or deleting a column in A can be treated by appending or deleting a row in A^T. This simplifies the updating problem in that only modifications of rows need be considered. (Of course, in the least squares updating, there is a lack of symmetry.) The general rank one change is more difficult.

The problem of updating the SVD was first considered by Businger [141, 1970], and is related to the technique used above for updating the QR decomposition; see also Barlow, Zha, and Yoon [46, 1993]. Bunch and Nielsen [138, 1978] develop updating methods related to updating symmetric eigendecompositions. Related schemes are given by Gu and Eisenstat [404, 1993], [407, 1995]. Unfortunately, all known exact updating schemes require on the order of mn^2 operations, which is the same order as recomputing the SVD from scratch. However, since the order constant is smaller there is still a gain. Moonen and Van Dooren [582, 1993] and Moonen, Van Dooren, and Vandewalle [583, 1992], have considered updating algorithms that approximate the SVD. The idea is to append (delete) a row, reduce to triangular form, and then use a Jacobi-type sweep to restore approximate diagonality. These algorithms are suitable for the implementation on systolic arrays.

3.4.2. Appending a row. Given the SVD of A we first consider the problem of computing the SVD when a row w^T is appended:

$$(3.4.2) \qquad \tilde{A} = \begin{pmatrix} A \\ w^T \end{pmatrix}, \qquad w \in \mathbf{R}^n.$$

One approach to this updating problem is to use the relationship between the SVD of A and the symmetric eigenvalue problem for $A^T A$. From

$$\tilde{A}^T \tilde{A} = A^T A + w w^T = V \Sigma^2 V^T + w w^T = V(\Sigma^2 + \rho^2 z z^T) V^T = \tilde{V} \tilde{\Sigma}^2 \tilde{V}^T,$$

where $\rho = \|w\|_2$, $z = V^T w/\rho$, it follows that $\tilde{\Sigma}^2$ and $\tilde{V}$ are the solution to a symmetric eigenvalue problem modified by a perturbation of rank one. Efficient algorithms for solving such problems have been given by Golub [367, 1973]. These are based on the observation that the eigenvalues of the matrix

$$(3.4.3) \qquad C = D + \rho^2 zz^T, \quad D = \mathrm{diag}\,(d_1, d_2, \ldots, d_n), \quad \|z\|_2 = 1,$$

$d_1 \geq d_2 \geq \cdots \geq d_n$, are the zeros of $g(\lambda)$, where

$$(3.4.4) \qquad g(\lambda) = 1 + \rho^2 \sum_{j=1}^{n} \frac{\zeta_j^2}{(d_j - \lambda)}.$$

Since by the interlacing property (Theorem 1.2.9)

$$\tilde{d}_1 \geq d_1 \geq \tilde{d}_2 \geq \cdots \geq d_{n-1} \geq \tilde{d}_n \geq d_n,$$

good initial approximations to the roots are known, and (3.4.4) can be solved by a method based on rational approximation safeguarded with bisection. The details of this algorithm are quite subtle. A stopping criterion together with a reformulation of (3.4.4) which allows an iterative method to satisfy it are given by Gu and Eisenstat [404, 1993].

When the modified eigenvalues $\tilde{d}_i = \tilde{\sigma}_i^2$ have been calculated, the corresponding eigenvectors of the modified eigenproblem (3.4.3) can be found by solving

$$(D_i + \rho^2 zz^T)x_i = 0, \qquad D_i = D - \tilde{d}_i I.$$

Provided that D_i is nonsingular (this can be assured by an initial deflation) we have

$$x_i = D_i^{-1} z / \|D_i^{-1} z\|_2.$$

(Note that forming $D_i^{-1} z$ explicitly should be avoided in practice, and a different scheme is used by Bunch and Nielsen [138].) The updated right singular vectors are $\tilde{V} = VX$, where $X = (x_1, \ldots, x_n)$. If the matrix A (or $\tilde{A}$) is still available the updated left singular vectors $\tilde{U}$ can be found from

$$\tilde{U} = \tilde{A} \tilde{V} \tilde{\Sigma}^{-1}.$$

An alternative approach for appending a row is given by Businger [141, 1970]. We have

$$\begin{pmatrix} U^T & 0 \\ 0 & 1 \end{pmatrix} \begin{pmatrix} A \\ w^T \end{pmatrix} V = \Pi_{n+1,m+1} \begin{pmatrix} L \\ 0 \end{pmatrix}, \qquad L = \begin{pmatrix} \Sigma \\ w^T V \end{pmatrix},$$

where $\Pi_{n+1,m+1}$ denotes a permutation matrix interchanging the rows $n+1$ and $m+1$, and L is a special lower triangular matrix. Businger's updating algorithm consists of two major phases. The first phase is a finite process that transforms $L \in \mathbf{R}^{(n+1) \times n}$ into upper bidiagonal form using Givens transformations from left and right,

$$G_1 L G_2 = \begin{pmatrix} \tilde{B} \\ 0 \end{pmatrix}, \qquad \tilde{B} \in \mathbf{R}^{n \times n}.$$

3.4. Modifying the Singular Value Decomposition

The second phase is an implicit QR diagonalization of $\tilde{B}$ (see Section 2.6.3), which reduces $\tilde{B}$ to diagonal form $\tilde{\Sigma}$.

In phase 1 there are $n-1$ major steps, of which the kth eliminates the kth element of $w^T V$ using a chasing scheme of Givens rotations on rows and columns. The elimination in phase 1 is pictured below for $n=5$ and $k=3$.

$$\begin{pmatrix} \times & \times & & & \\ & \times & \times & + & \\ & & \times & + & \\ & & + & \times & \\ 0 & 0 & \oplus & \times & \times \end{pmatrix} \Rightarrow \begin{pmatrix} \times & \times & & & \\ & \times & \times & \times & \\ & & \times & \times & \\ & & \oplus & \times & \\ 0 & 0 & 0 & \times & \times \end{pmatrix} \Rightarrow \begin{pmatrix} \times & \times & & & \\ & \times & \times & \oplus & \\ & & + & \times & \times \\ & & & & \times \\ 0 & 0 & 0 & \times & \times \end{pmatrix}$$

$$\begin{pmatrix} \times & \times & + & & \\ & \times & \times & & \\ & \oplus & \times & \times & \\ & & & \times & \\ 0 & 0 & 0 & \times & \times \end{pmatrix} \Rightarrow \begin{pmatrix} \times & \times & \oplus & & \\ + & \times & \times & & \\ & & \times & \times & \\ & & & \times & \\ 0 & 0 & 0 & \times & \times \end{pmatrix} \Rightarrow \begin{pmatrix} \times & \times & & & \\ \oplus & \times & \times & & \\ & & \times & \times & \\ & & & \times & \\ 0 & 0 & 0 & \times & \times \end{pmatrix}.$$

Phase 1 uses $1 + 2 + \cdots + (n-1) = n^2/2$ rows and the same number of column rotations. Most of the computing work goes into applying these rotations to U and V, which requires $2n^2(m+n)$ flops if standard Givens rotations are used. For the least squares updating we only need to update V, Σ, and $c = U^T b$, and the dominating term is then reduced to $2n^3$ flops. Zha [853, 1992] has shown that the work can be halved by using a two-way chasing scheme in the reduction to bidiagonal form. Phase 2 typically requires about n^3 additions and $2n^3$ multiplications.

Note that Σ and V can be updated without U being available. From the interlacing property (Theorem 1.4.2) it follows that the smallest singular value will increase, and hence the rank cannot decrease.

3.4.3. Deleting a row.

We now consider modifying the SVD when a row is *deleted*. There is no loss of generality in assuming that the *first* row of A is to be deleted, i.e., we wish to determine the SVD of the matrix $\tilde{A} \in \mathbf{R}^{(m-1)\times n}$ when we know the SVD

$$(3.4.5) \qquad A = \begin{pmatrix} z^T \\ \tilde{A} \end{pmatrix} = U \begin{pmatrix} \Sigma \\ 0 \end{pmatrix} V^T.$$

We first note that the problem can be reduced to a modified eigenvalue problem of the form

$$(3.4.6) \qquad C = D - \rho^2 z z^T, \quad D = \mathrm{diag}\,(d_1, d_2, \ldots, d_n), \quad \|z\|_2 = 1.$$

The interlacing property now gives $d_1 \geq \tilde{d}_1 \geq d_2 \geq \cdots \geq \tilde{d}_{n-1} \geq d_n \geq \tilde{d}_n \geq 0$. Hence the Bunch–Nielsen scheme is readily adapted to solving this problem.

Businger reduces the downdating problem to that of adding a row $\sqrt{-1}\,z^T$. This approach has the advantage that it does not require U, but as shown in Section 3.3.4, it is unstable. Park and Van Huffel [650] have given a backward stable algorithm based on finding the SVD of (e_1, A), where e_1 is an added dummy column. Then

$$U^T(e_1, A) \begin{pmatrix} 1 & 0 \\ 0 & V \end{pmatrix} = \begin{pmatrix} u_1 & \Sigma \\ u_2 & 0 \end{pmatrix},$$

where (u_1^T, u_2^T) is the first row of U. We first determine row and column Givens rotations so that

(3.4.7)
$$G_1 \begin{pmatrix} u_1 & \Sigma \\ u_2 & 0 \end{pmatrix} G_2 = \begin{pmatrix} 1 & w^T \\ 0 & \tilde{B} \\ 0 & 0 \end{pmatrix},$$

where the matrix $\tilde{B}$ is upper bidiagonal. This is achieved by a chasing scheme similar to that used when adding a row. The desired bidiagonal form is built from bottom to top while nonzeros are chased away to the lower-right corner. We picture the reduction below for $k = 3$, $n = 4$:

$$\begin{matrix} \rightarrow \\ \rightarrow \end{matrix} \begin{pmatrix} \times & \times & & \\ & \times & \times & + \\ \oplus & & + & \times \\ 0 & & \times & \times \\ 0 & & & \times \end{pmatrix} \Rightarrow \begin{pmatrix} \times & \times & & \\ & \times & \times & \oplus \\ 0 & & \times & \times \\ 0 & & + & \times & \times \\ 0 & & & & \times \end{pmatrix} \Rightarrow \begin{matrix} \rightarrow \\ \rightarrow \end{matrix} \begin{pmatrix} \times & \times & & & \\ & \times & \times & & \\ 0 & & \times & \times & + \\ 0 & & \oplus & \times & \times \\ 0 & & & & \times \end{pmatrix}$$

$$\begin{pmatrix} \times & \times & & \\ & \times & \times & \\ 0 & & \times & \times & \oplus \\ 0 & & & \times & \times \\ 0 & & + & & \times \end{pmatrix} \Rightarrow \begin{matrix} \rightarrow \\ \rightarrow \end{matrix} \begin{pmatrix} \times & \times & & \\ & \times & \times & \\ 0 & & \times & \times & \\ 0 & & & \times & \times \\ 0 & & & \oplus & \times \end{pmatrix}.$$

A total of $(n-1)^2 + 1$ plane rotations are needed.

By construction the first column of $G_1 U^T$ equals e_1^T, and by orthogonality this matrix must have the form

$$G_1 U^T = \begin{pmatrix} 1 & 0 \\ 0 & \bar{U}^T \end{pmatrix},$$

with $\bar{U}$ orthogonal. Further, since no rotation from the right involves the first column the transformed matrix has the form

$$\begin{pmatrix} \alpha & 0 \\ 0 & V \end{pmatrix} G_2 = \begin{pmatrix} 1 & 0 \\ 0 & \bar{V} \end{pmatrix}.$$

It now follows from (3.2.10) that

$$\begin{pmatrix} \alpha & 0 \\ 0 & \bar{U}^T \end{pmatrix} \begin{pmatrix} 1 & z^T \\ 0 & A \end{pmatrix} \begin{pmatrix} 1 & 0 \\ 0 & \bar{V} \end{pmatrix} = \begin{pmatrix} \alpha & w^T \\ 0 & \tilde{B} \\ 0 & 0 \end{pmatrix}.$$

This gives the bidiagonal reduction of the downdated A,

$$\bar{U}^T \tilde{A} \bar{V} = \begin{pmatrix} \tilde{B} \\ 0 \end{pmatrix}.$$

In the second phase the implicit QR diagonalization is used to reduce $\tilde{B}$ to diagonal form $\tilde{\Sigma}$ and update $\bar{U}$ and $\bar{V}$.

3.5. Modifying Rank Revealing QR Decompositions

3.5.1. Appending a row.
The difficulties of updating the SVD have led to a renewed interest in rank revealing QR decompositions and complete orthogonal decompositions, which can be updated more cheaply. These are effective in exhibiting the rank and nullspace of A, and can be thought of as compromises between the SVD and QR decompositions.

In Section 2.7.4 we defined a rank revealing complete orthogonal decomposition (CQRD) as a decomposition of the form

$$(3.5.1) \qquad A = U \begin{pmatrix} R & F \\ 0 & G \end{pmatrix} V^T,$$

where U and V are orthogonal matrices, and $R \in \mathbf{R}^{k \times k}$ and $G \in \mathbf{R}^{(m-k) \times (n-k)}$ are upper triangular. Further, for some constant c,

$$\sigma_k(R) \geq \frac{1}{c}\sigma_k, \qquad (\|F\|_F^2 + \|G\|_F^2)^{1/2} \leq c\sigma_{k+1},$$

where $\sigma_1 \geq \sigma_2 \geq \cdots \geq \sigma_n$ are the singular values of A. This is also often called a rank revealing URV decomposition; see Stewart [749, 1992]. For the factorization to be useful it should hold that $\sigma_k \gg \sigma_{k+1}$, and the numerical rank of A should equal k.

The CQRD has the advantage that it can be updated when a row is added to A, by an algorithm which only requires $O(n^2)$ operations. Assume that we are given a tolerance δ so that we can accept the decomposition (3.5.1) as rank revealing provided that

$$(3.5.2) \qquad (\|F\|_F^2 + \|G\|_F^2)^{1/2} = \nu \leq \delta.$$

Given the CQRD (3.5.1) of A we want to compute this decomposition for

$$\tilde{A} = \begin{pmatrix} A \\ w^T \end{pmatrix}, \quad w \in \mathbf{R}^n.$$

Following the algorithm given by Stewart [749, 1992], we write

$$(3.5.3) \quad \begin{pmatrix} U^T & 0 \\ 0 & 1 \end{pmatrix} \begin{pmatrix} A \\ w^T \end{pmatrix} V = \begin{pmatrix} R & F \\ 0 & G \\ x^T & y^T \end{pmatrix},$$

where $w^T V = (x^T y^T)$. In the simplest case the relation

$$(3.5.4) \quad \sqrt{\nu^2 + \|y\|_2^2} \leq \delta$$

is satisfied. Then it suffices to reduce the matrix in (3.5.3) to upper triangular form by a sequence of left Givens rotations, as described in Section 3.2.5. Note that the updated matrix R cannot become effectively rank deficient, since all its singular values will increase.

If (3.5.4) is not satisfied we first reduce y^T in (3.5.3) so that it becomes proportional to e_1^T, while keeping the upper triangular form of G. This can be done by a sequence of right and left Givens rotations as illustrated below. (Note that here the f's represent entire columns of F.)

$$\begin{pmatrix} f & f & f & f \\ g & g & g & g \\ 0 & g & g & g \\ 0 & 0 & g & g \\ 0 & 0 & + & g \\ y & y & y & 0 \end{pmatrix} \Rightarrow \begin{pmatrix} f & f & f & f \\ g & g & g & g \\ 0 & g & g & g \\ 0 & 0 & g & g \\ 0 & 0 & \oplus & g \\ y & y & y & 0 \end{pmatrix} \Rightarrow \begin{pmatrix} f & f & f & f \\ g & g & g & g \\ 0 & g & g & g \\ 0 & + & g & g \\ 0 & 0 & 0 & g \\ y & y & 0 & 0 \end{pmatrix} \Rightarrow$$

$$\begin{pmatrix} f & f & f & f \\ g & g & g & g \\ 0 & g & g & g \\ 0 & \oplus & g & g \\ 0 & 0 & 0 & g \\ y & y & 0 & 0 \end{pmatrix} \Rightarrow \begin{pmatrix} f & f & f & f \\ g & g & g & g \\ + & g & g & g \\ 0 & \oplus & g & g \\ 0 & 0 & 0 & g \\ \sigma & 0 & 0 & 0 \end{pmatrix} \Rightarrow \begin{pmatrix} f & f & f & f \\ g & g & g & g \\ \oplus & g & g & g \\ 0 & 0 & 0 & g \\ 0 & 0 & g & g \\ \sigma & 0 & 0 & 0 \end{pmatrix}.$$

Here $\sigma = \|y\|_2$, and R and x^T are not involved in this part of the reduction.

After this step the matrix has been reduced to the form

$$\begin{pmatrix} \tilde{R} & f & \tilde{F} \\ 0 & g & \tilde{G} \\ x^T & \sigma & 0 \end{pmatrix}.$$

This matrix is then reduced to triangular form using left Givens rotations as in Section 3.2.5, and k is increased by 1. Finally, the new R is checked for degeneracy and possibly reduced by deflation, as described in Section 2.7.5. This completes the updating of the decomposition. The complete update takes $O(n^2)$ operations.

Stewart [752, 1993] has pointed out that although the decomposition (3.5.1) is very satisfactory for recursive least squares problems, it is less suited for

3.5. MODIFYING RANK REVEALING QR DECOMPOSITIONS

applications where an approximate nullspace is to be recursively updated. Let (U_1, U_2) and (V_1, V_2) be partitionings of U and V conformally with (3.5.1). Then we have

$$\|AV_2\|_2 = \left\|\begin{pmatrix} F \\ G \end{pmatrix}\right\|_2. \tag{3.5.5}$$

Hence the orthogonal matrix V_2 can be taken as an approximation to the numerical nullspace $\mathcal{N}_k$. On the other hand we have that $\|U_2^T A\|_2 = \|G\|_2$, and therefore the last $n - k$ singular values of A are less than or equal to $\|G\|_2$. Thus, V_2 is not the best available approximate nullspace, since F is also involved in the bound (3.5.5).

As mentioned in Section 2.7.5 this problem can be resolved by working instead with the corresponding rank revealing ULV decomposition

$$A = U \begin{pmatrix} L & 0 \\ H & E \end{pmatrix} V^T, \tag{3.5.6}$$

where L and E have *lower* triangular form, and

$$\sigma_k(L) \geq \frac{1}{c}\sigma_k, \qquad (\|H\|_F^2 + \|E\|_F^2)^{1/2} = \nu \leq \delta.$$

For this decomposition, $\|AV_2\|_2 = \|E\|_F$, where $V = (V_1, V_2)$ is a conformal partitioning of V. Hence the size of $\|H\|_2$ does not affect the nullspace approximation.

Stewart [752, 1993] has presented an updating scheme for the decomposition (3.5.6). With $w^T V = (x^T, y^T)$, the problem reduces to updating

$$\begin{pmatrix} L & 0 \\ H & E \\ x^T & y^T \end{pmatrix}.$$

The main difference compared to the scheme for updating the URV decomposition is that there is now not the same simplification when $\sqrt{\nu^2 + \|y\|_2^2} \leq \delta$. We first reduce y^T to $\|y\|_2 e_1^T$ by right rotations while keeping the triangular form of E. At the end of this reduction the matrix has the form

$$\begin{pmatrix} l & 0 & 0 & 0 & 0 \\ l & l & 0 & 0 & 0 \\ l & l & l & 0 & 0 \\ h & h & h & e & 0 \\ h & h & h & e & e \\ x & x & x & y & 0 \end{pmatrix}.$$

This matrix can be reduced to lower triangular form by a sequence of left rotations, and k is increased by 1. (For the case above we would use $Q = G_{16}G_{26}G_{36}G_{46}$.) In case there has been no effective increase in rank, a deflation process has to be applied.

If

$$\sqrt{\nu^2 + \|y\|_2^2} \leq \delta$$

then the rank cannot increase. In that case we perform the reduction, but skip the first rotation (G_{46}). This gives us a matrix of the form

$$\begin{pmatrix} l & 0 & 0 & y & 0 \\ l & l & 0 & y & 0 \\ l & l & l & y & 0 \\ h & h & h & e & 0 \\ h & h & h & e & e \\ 0 & 0 & 0 & y & 0 \end{pmatrix}.$$

The y elements above the main diagonal can be eliminated using right rotations. This fills out the last row again, but now with elements the same size as y. Now the last row can be reduced by the procedure described above without destroying the rank revealing structure; see again Stewart [752, 1993].

We have seen that the updating of the rank revealing ULV decomposition is more involved than that for the URV decomposition. Hence, even if the refinement steps in the decomposition URV are an added complication it is not clear under what circumstances one of these two decompositions is to be preferred to the other.

3.5.2. Deleting a row. In some applications one wants to delete old data from the problem, and hence to downdate the rank revealing orthogonal decomposition. This problem is closely related to the problem of downdating the usual QR decomposition, which has been discussed at length in Sections 3.2.6–3.2.7 and 3.3.

When downdating the rank revealing URV decomposition (3.5.1) a complication is that when $m \gg n$, the extra storage for $U \in \mathbf{R}^{m \times m}$ may be prohibitive, and only V and the triangular factor are stored. Then we must use methods related to those described in Section 3.3 like the Saunders (LINPACK) algorithm, possibly stabilized with the CSNE algorithm (Section 3.3.3). Other possible algorithms are based on the use of hyperbolic rotations as modified by Chambers; see Section 3.3.4. Neither of these methods will be as satisfactory as methods using U (cf. Section 3.2.6) or Gram–Schmidt-based methods using U_1 described in Section 3.2.7. Because there are so many possible variants we will not describe methods in detail here, but refer to work by Park and Eldén [649, 1995] and Barlow, Yoon, and Zha [45, 1996].

Chapter 4
Generalized Least Squares Problems

4.1. Generalized QR Decompositions

4.1.1. Introduction. In Section 1.1.3 we introduced the general univariate linear model, where the covariance matrix was proportional to a semidefinite symmetric matrix $W \in \mathbf{R}^{m \times m}$. When W is positive definite we can write $W = BB^T$, and if A has full column rank the best linear unbiased estimate (BLUE) for x is the solution of the generalized least squares problem

$$(4.1.1) \qquad \min_x \|B^{-1}(Ax - b)\|_2.$$

This problem now involves two matrices A and B, and in general models, both can be rank deficient. In this section we consider generalized QR decompositions, which can be used to solve such problems in a stable manner.

4.1.2. Computing the GQR and PQR. In many applications an effective computational tool is the QR decomposition of a matrix

$$(4.1.2) \qquad C = B^T A, \qquad A \in \mathbf{R}^{m \times n}, \qquad B \in \mathbf{R}^{m \times p},$$

for given A and B. We refer to this as a **product QR decomposition** (PQR). In principle the PQR could be computed by first explicitly forming the matrix C. However, in order to obtain a computed result which is backward stable in the sense that it corresponds to the exact result to nearby data $A + \delta A$ and $B + \delta B$ this must be avoided.

Similar considerations apply to computing the QR decompositions for a matrix of the form

$$(4.1.3) \qquad C = B^{-1} A, \qquad A \in \mathbf{R}^{m \times n}, \qquad B \in \mathbf{R}^{m \times m},$$

for given A and nonsingular B. Such a QR decomposition will be called a **generalized QR decomposition** (GQR), as suggested by Hammarling [423, 1985]. (It is also called the quotient QR decomposition.) When B is singular or not square then C does not exist. We would like the GQR to be defined for *general* matrices $A \in \mathbf{R}^{m \times n}$ and $B \in \mathbf{R}^{m \times p}$ with the same number of rows.

To construct the GQR in the general case we proceed in two steps. First we compute a rank revealing QR decomposition of A; see Section 2.7.4. If the numerical rank $(A) = r$, then we obtain

$$(4.1.4) \qquad Q^T A \Pi = \begin{pmatrix} U_{11} & U_{12} \\ 0 & 0 \end{pmatrix} \begin{matrix} \}r \\ \}m-r \end{matrix},$$

where Q is orthogonal, Π is a permutation matrix, and $U_{11} \in \mathbf{R}^{r \times r}$ is upper triangular and nonsingular. The orthogonal transformation Q_A^T is then applied also to B:

$$Q^T B = \begin{pmatrix} B_1 \\ B_2 \end{pmatrix} \begin{matrix} \}r \\ \}m-r \end{matrix}.$$

Next an orthogonal matrix $\tilde{Q}$ is constructed so that

$$(4.1.5) \quad Q^T B \tilde{Q} = R = \begin{pmatrix} R_{11} & R_{12} & 0 \\ 0 & R_{22} & 0 \\ 0 & 0 & 0 \end{pmatrix} \begin{matrix} \}r \\ \}q \\ \}m-t \end{matrix}, \quad \tilde{Q} = (Q_1 \ Q_2 \ Q_3),$$

where rank $(B_2) = q$, $t = r + q$, $R_{11} \in \mathbf{R}^{r \times k_1}$, $R_{22} \in \mathbf{R}^{(n-q) \times k_2}$ are upper trapezoidal, and rank $(R_{11}) = k_1$, rank $(R_{22}) = k_2$. If rank $(B) = p$ there will be no zero columns. Note that row interchanges can be performed on the block B_2 if Q is modified accordingly.

If B is square ($p = m$) and nonsingular then so is R, and from (4.1.4)–(4.1.5) we have

$$(4.1.6) \qquad \tilde{Q}^T (B^{-1} A) \Pi = \begin{pmatrix} R_{11}^{-1} U \\ 0 \end{pmatrix} = S, \quad U = (U_{11} \ U_{12}),$$

which is the QR decomposition of $B^{-1} A \Pi$. Even in this case one should avoid computing R, since in most applications it is not needed and it is usually more effective to use R_{11} and U separately. Another advantage of keeping R_{11} and U is that the corresponding decompositions (4.1.4)–(4.1.5) can be updated by the standard methods described in Chapter 3 when columns or rows are added or deleted from A and B. Even when S is defined by (4.1.6) it cannot generally be updated in a stable way.

Another approach to handling the case when B is not square or singular would be to define the GQR as the QR decomposition of $B^\dagger A$, where $B^\dagger$ denotes the pseudoinverse of B. It has been pointed out by Paige [634, 1990] that this will not be very useful since it does not produce the correct solution for many of the applications of the GQR.

The PQR decomposition of $A \in \mathbf{R}^{m \times n}$ and $B \in \mathbf{R}^{m \times p}$ can be computed in a similar manner. We again use (4.1.4) as the first step, and then replace (4.1.5) by

$$(4.1.7) \quad Q^T B \tilde{Q} = \begin{pmatrix} L_{11} & 0 & 0 \\ L_{21} & L_{22} & 0 \\ 0 & 0 & 0 \end{pmatrix} \begin{matrix} \}r \\ \}q \\ \}m-t \end{matrix}, \quad \tilde{Q} = (Q_1 \ Q_2 \ Q_3),$$

where $L_{11} \in \mathbf{R}^{q \times r_1}$, $L_{22} \in \mathbf{R}^{(n-q) \times r_2}$, and $\operatorname{rank}(L_{11}) = r_1$, $\operatorname{rank}(L_{22}) = r_2$. This gives the product PQR since

$$(4.1.8) \qquad \tilde{Q}^T B^T A = \begin{pmatrix} L_{11}^T U \\ 0 \end{pmatrix} = \begin{pmatrix} L^T \\ 0 \end{pmatrix},$$

with $L^T \in \mathbf{R}^{r_1 \times n}$ upper trapezoidal. Again one should avoid computing L^T, since it is not needed in most applications and it usually leads to more accurate methods if L_{11} and U are kept separate. (A trivial example of this is the case when $B = A$.)

An excellent survey of the use of the GQR decomposition for solving different generalized least squares problems is given by Paige in [634, 1990]. The applications include the preprocessing stage in computing the generalized SVD of A and B, generalized least squares problems (4.1.1), and equality-constrained linear least squares problems. Implementation aspects are considered in Anderson, Bai, and Dongarra [18, 1991], and some generalizations discussed by De Moor and Van Dooren [212, 1992]. Luk and Qiao [553, 1994] have developed rank revealing GQR decompositions.

4.2. The Generalized SVD

We shall introduce a **generalized singular value decomposition** (GSVD) for two matrices $A \in \mathbf{R}^{m \times n}$ and $B \in \mathbf{R}^{p \times n}$ with the same number of columns. The GSVD and its application to certain constrained least squares problems was first studied by Van Loan [796, 1976]. Paige and Saunders [637, 1981] extended the GSVD to handle all possible cases, and gave a computationally more amenable form.

4.2.1. The CS decomposition.
In the special case when A and B are blocks of a partitioned matrix having orthonormal columns the GSVD simplifies to the CS decomposition, which is of interest in its own right.

THEOREM 4.2.1. CS Decomposition. *Let $Q \in \mathbf{R}^{(m+p) \times n}$ have orthonormal columns, and be partitioned as*

$$(4.2.1) \qquad Q = \begin{pmatrix} Q_1 \\ Q_2 \end{pmatrix} \begin{matrix} \}m \\ \}p \end{matrix} \in \mathbf{R}^{(m+p) \times n}, \qquad m \geq n,$$

i.e., $Q^T Q = Q_1^T Q_1 + Q_2^T Q_2 = I_n$. Then there are orthogonal matrices $U_1 \in \mathbf{R}^{m \times m}$, $U_2 \in \mathbf{R}^{p \times p}$, and $V \in \mathbf{R}^{n \times n}$, and square nonnegative diagonal matrices

$$(4.2.2) \quad C = \operatorname{diag}(c_1, \ldots, c_q), \quad S = \operatorname{diag}(s_1, \ldots, s_q), \quad q = \min(n, p),$$

satisfying $C^2 + S^2 = I_q$ such that

$$(4.2.3) \qquad \begin{pmatrix} U_1^T & 0 \\ 0 & U_2^T \end{pmatrix} \begin{pmatrix} Q_1 \\ Q_2 \end{pmatrix} V = \begin{pmatrix} U_1^T Q_1 V \\ U_2^T Q_2 V \end{pmatrix} = \begin{pmatrix} \Sigma_1 \\ \Sigma_2 \end{pmatrix} \begin{matrix} \}m \\ \}p \end{matrix}$$

has one of the following forms:

$$p \geq n: \begin{pmatrix} C \\ 0 \\ S \\ 0 \end{pmatrix} \begin{matrix} \}n \\ \}m-n \\ \}n \\ \}p-n \end{matrix}, \qquad p < n: \begin{pmatrix} C & 0 \\ 0 & I \\ 0 & 0 \\ \underbrace{S}_{p} & \underbrace{0}_{n-p} \end{pmatrix} \begin{matrix} \}p \\ \}n-p \\ \}m-n \\ \}p \end{matrix}.$$

The diagonal elements c_i and s_i are

$$c_i = \cos(\theta_i), \quad s_i = \sin(\theta_i), \quad i = 1, \ldots, q,$$

where without loss of generality, we may assume that

$$0 \leq \theta_1 \leq \theta_2 \leq \cdots \leq \theta_q \leq \pi/2.$$

Proof. (See Stewart [738, 1982].) To construct U_1, V, and C, note that since U_1 and V are orthogonal and C is a nonnegative diagonal matrix, (4.2.3) is the SVD of Q_1. Hence the elements c_i are the singular values of Q_1. If we put $\tilde{Q}_2 = Q_2 V$, then the matrix

$$\begin{pmatrix} C \\ 0 \\ \tilde{Q}_2 \end{pmatrix} = \begin{pmatrix} U_1^T & 0 \\ 0 & I_p \end{pmatrix} \begin{pmatrix} Q_1 \\ Q_2 \end{pmatrix} V$$

has orthonormal columns. Thus

$$C^2 + \tilde{Q}_2^T \tilde{Q}_2 = I_n,$$

which implies that $\tilde{Q}_2^T \tilde{Q}_2 = I_n - C^2$ is diagonal and hence the matrix $\tilde{Q}_2 = (\tilde{q}_1^{(2)}, \ldots, \tilde{q}_n^{(2)})$ has orthogonal columns.

We assume that the singular values $c_i = \cos(\theta_i)$ of Q_1 have been ordered according to (4.2.1) and that $c_r < c_{r+1} = 1$. Then the matrix $U_2 = (u_1^{(2)}, \ldots, u_p^{(2)})$ is constructed as follows. Since $\|\tilde{q}_j^{(2)}\|_2^2 = 1 - c_j^2 \neq 0$, $j \leq r$ we take

$$u_j^{(2)} = \tilde{q}_j^{(2)} / \|\tilde{q}_j^{(2)}\|_2, \qquad j = 1, \ldots, r,$$

and fill the possibly remaining columns of U_2 with orthonormal vectors in the orthogonal complement of $\mathcal{R}(\tilde{Q}_2)$. From the construction it follows that $U_2 \in \mathbf{R}^{p \times p}$ is orthogonal and that

$$U_2^T \tilde{Q}_2 = U_2 Q_2 V = \begin{pmatrix} S & 0 \\ 0 & 0 \end{pmatrix}, \qquad S = \mathrm{diag}(s_1, \ldots, s_q)$$

with

$$s_j = \begin{cases} (1 - c_j^2)^{1/2} > 0, & j = 1, \ldots, r, \\ 0, & j = r+1, \ldots, q. \end{cases} \qquad \blacksquare$$

The assumption $m \geq n$ in the theorem is made for notational convenience only. Stewart [738, 1982] treats the general case, which gives rise to four different forms corresponding to cases where Q_1 and/or Q_2 have too few rows to accommodate a full diagonal matrix of order n.

The proof of the CS decomposition is constructive. In particular U_1, V, and C can be computed by a standard SVD algorithm. However, the algorithm for computing U_2 is unstable when some singular values c_i are close to 1. Stewart [738, 1982] and Van Loan [798, 1985] describe two modified stable algorithms for computing the CS decomposition.

The CS decomposition can be stated in a related form which is often useful. Let $Q \in \mathbf{R}^{m \times m}$ be orthogonal and consider the partitioning

$$Q = \begin{pmatrix} Q_{11} & Q_{12} \\ Q_{21} & Q_{22} \end{pmatrix} \begin{matrix} \}j \\ \}k \end{matrix}, \qquad j \geq k,$$
$$\underbrace{\phantom{Q_{11}}}_{j} \underbrace{\phantom{Q_{12}}}_{k}$$

Then there exist orthogonal matrices $U_1, V_1 \in \mathbf{R}^{j \times j}$ and $U_2, V_2 \in \mathbf{R}^{k \times k}$ such that

$$(4.2.4) \quad \begin{pmatrix} U_1 & 0 \\ 0 & U_2 \end{pmatrix} \begin{pmatrix} Q_{11} & Q_{12} \\ Q_{21} & Q_{22} \end{pmatrix} \begin{pmatrix} V_1 & 0 \\ 0 & V_2 \end{pmatrix} = \begin{pmatrix} C & 0 & -S \\ 0 & I & 0 \\ S & 0 & C \end{pmatrix} \begin{matrix} \}k \\ \}j-k \\ \}k \end{matrix},$$

where

$$C = \mathrm{diag}(c_1, \ldots, c_k), \qquad S = \mathrm{diag}(s_1, \ldots, s_k),$$
$$c_i = \cos(\theta_i), \quad s_i = \sin(\theta_i), \quad 0 \leq \theta_j \leq \pi/2, \quad i = 1, \ldots, k.$$

The decomposition (4.2.4) can be proved in a way similar to the proof of Theorem 4.2.1. A proof of a slightly more general decomposition, where Q_{11} and Q_{22} are not required to be square matrices, is given in Paige and Saunders [637, 1981]. Stewart [732, 1977] first put forward an explicit form of the CS decomposition, although it is implicit in the works by Davis and Kahan [203, 1970] and Björck and Golub [111, 1973]. For an account of the history and development of the CS decomposition, see Paige [635, 1994].

Note that the decomposition (4.2.4) treats rows and columns of Q in a symmetric way. The matrix on the right-hand side of (4.2.4) is a generalization of a Givens rotations matrix (see (2.3.9)) and its transpose is its inverse. As remarked by Stewart [732, 1977], the decomposition (4.2.4) "often enables one to obtain routine computational proofs of geometric theorems that would otherwise require considerable ingenuity to establish."

4.2.2. The generalized SVD. The CS decomposition enables us to give a constructive development of the GSVD of two matrices A and B with the same number of columns. The assumption $m > n$ in the theorem is made for notational convenience only. For a general formulation and proof, see Paige and Saunders [637, 1981].

THEOREM 4.2.2. *The Generalized Singular Value Decomposition (GSVD).* Let $A \in \mathbf{R}^{m \times n}$, $m \geq n$, and $B \in \mathbf{R}^{p \times n}$ *be given matrices. Assume that*

$$\mathrm{rank}\,(M) = k \leq n, \qquad M = \begin{pmatrix} A \\ B \end{pmatrix}.$$

Then there exist orthogonal matrices $U_A \in \mathbf{R}^{m \times m}$ and $U_B \in \mathbf{R}^{p \times p}$ and a matrix $Z \in \mathbf{R}^{k \times n}$ of rank k such that

(4.2.5) $$U_A^T A = \begin{pmatrix} D_A \\ 0 \end{pmatrix} Z, \qquad U_B^T B = \begin{pmatrix} D_B & 0 \\ 0 & 0 \end{pmatrix} Z,$$

where

$$D_A = \mathrm{diag}(\alpha_1, \ldots, \alpha_k), \qquad D_B = \mathrm{diag}(\beta_1, \ldots, \beta_q), \qquad q = \min(p, k).$$

Further, we have

$$0 \leq \alpha_1 \leq \cdots \leq \alpha_k \leq 1, \qquad 1 \geq \beta_1 \geq \cdots \geq \beta_q \geq 0,$$
$$\alpha_i^2 + \beta_i^2 = 1, \quad i = 1, \ldots, q, \qquad \alpha_i = 1, \quad i = q+1, \ldots, k,$$

and the singular values of Z equal the nonzero singular values of M.

Proof. Let the SVD of M be

$$M = \begin{pmatrix} A \\ B \end{pmatrix} = Q \begin{pmatrix} \Sigma_1 & 0 \\ 0 & 0 \end{pmatrix} P^T,$$

where Q and P are orthogonal matrices of order $(m+p)$ and n, respectively, and

$$\Sigma_1 = \mathrm{diag}(\sigma_1, \ldots, \sigma_k), \qquad \sigma_1 \geq \cdots \geq \sigma_k > 0.$$

Set $t = m + p - k$ and partition Q and P as follows:

$$Q = \begin{pmatrix} Q_{11} & Q_{12} \\ Q_{21} & Q_{22} \end{pmatrix} \begin{matrix} \}m \\ \}p \end{matrix}, \qquad P = (\underbrace{P_1}_{k}, \underbrace{P_2}_{n-k}).$$
$$\underbrace{\phantom{Q_{11}}}_{k} \underbrace{\phantom{Q_{12}}}_{t}$$

Then the SVD of M can be written

(4.2.6) $$\begin{pmatrix} A \\ B \end{pmatrix} P = \begin{pmatrix} AP_1 & 0 \\ BP_1 & 0 \end{pmatrix} = \begin{pmatrix} Q_{11} \\ Q_{21} \end{pmatrix} (\Sigma_1 \quad 0).$$

Now let

$$Q_{11} = U_A \begin{pmatrix} C \\ 0 \end{pmatrix} V^T, \qquad Q_{21} = U_B \begin{pmatrix} S & 0 \\ 0 & 0 \end{pmatrix} V^T$$

be the CS decomposition of Q_{11} and Q_{21}. Substituting this into (4.2.6) we obtain

$$AP = U_A \begin{pmatrix} C \\ 0 \end{pmatrix} V^T (\Sigma_1 \quad 0),$$
$$BP = U_B \begin{pmatrix} S & 0 \\ 0 & 0 \end{pmatrix} V^T (\Sigma_1 \quad 0),$$

and (4.2.5) follows with

$$D_A = C, \quad D_B = S, \quad Z = V^T (\Sigma_1 \quad 0) P^T.$$

Here $\sigma_1 \geq \cdots \geq \sigma_k > 0$ are the singular values of Z. ∎

When $B \in \mathbf{R}^{n \times n}$ is square and nonsingular the GSVD of A and B corresponds to the SVD of AB^{-1}. However, when A or B is ill-conditioned, then computing AB^{-1} would usually lead to unnecessarily large errors, so this approach is to be avoided. It is important to note that when B is not square, or is singular, then the SVD of $AB^\dagger$ does not always correspond to the GSVD.

4.2.3. Computing the GSVD.

Two stable algorithms for computing the GSVD were developed by Stewart [738, 1982], [740, 1983] and Van Loan [798, 1985]. Their algorithms resemble the proof of Theorem 4.2.2, but in the first phase a QR decomposition is used instead of the SVD. The second phase is to compute the CS decomposition.

A different algorithm was proposed by Paige [633, 1986]. In the first phase of the algorithm A and B are reduced to the following generalized triangular form:

$$U^T A P = \begin{pmatrix} U_{11} & U_{12} & U_{13} \\ 0 & 0 & 0 \\ 0 & 0 & 0 \end{pmatrix} \begin{matrix} \}r \\ \}q \\ \}m-t \end{matrix},$$

$$\underbrace{\phantom{U_{11}}}_{r} \underbrace{\phantom{U_{12}}}_{q} \underbrace{\phantom{U_{13}}}_{n-t}$$

$$V^T B P = \begin{pmatrix} R_{11} & R_{12} & R_{13} \\ 0 & R_{22} & R_{23} \\ 0 & 0 & 0 \end{pmatrix} \begin{matrix} \}r \\ \}q \\ \}p-t \end{matrix},$$

$$\underbrace{\phantom{R_{11}}}_{r} \underbrace{\phantom{R_{12}}}_{q} \underbrace{\phantom{R_{13}}}_{n-t}$$

where U_{11} is nonsingular upper triangular, R_{11} is upper triangular, $r = \text{rank}(A)$, $t = r + q$, and if $q > 0$ then R_{22} is nonsingular upper triangular. This transformation can be done first by a rank revealing QR decomposition with column pivoting of the matrix A. Second, the columns of B are permuted in the same way, and then a QR decomposition is performed on B, where column pivoting is used on the block of the last $p - r$ rows and $n - r$ columns of B, and q is the rank of this block. Another approach begins by applying a reduction by Bai and Zha [33, 1993], which extracts a regular pair (A, B), with A and B upper triangular and B nonsingular, from a general matrix pair. In the second phase the GSVD of two $n \times n$ upper triangular matrices is computed by a Kogbetliantz-type method; see Section 2.6.6.

The GSVD is also called the quotient SVD or QSVD. Heath et al. [444, 1986] give a generalized Kogbetliantz algorithm for computing the SVD of the product $B^T A$. It accurately computes very small singular values of the product $B^T A$. Another variation of Paige's algorithm has been given by Bai and Demmel [32, 1993]. This contains a new preprocessing step to reduce A and B to the above upper triangular form. The second innovation is a new 2×2 triangular GSVD algorithm, which is proved to be stable and accurate.

The Kogbetliantz-type algorithms for the GSVD extend fairly easily to products and quotients of several matrices. Since the method of choice for computing the standard SVD is the implicit QR algorithm, extensions of this method should also be promising for the GSVD. Such a QR-like algorithm for computing the SVD of a product or quotient of two or more matrices has been outlined by Golub, Solna, and Van Dooren [384, 1995]. We discuss here the problem of computing the SVD of a product of two matrices

$$A = A_2^{s_2} A_1^{s_1}, \qquad s_i = \pm 1,$$

i.e., a product or quotient of two matrices. Consider the expression

(4.2.7) $$Q_2^T A Q_0 = (Q_2^T A_2^{s_2} Q_1)(Q_1^T A_1^{s_1} Q_0),$$

where U, V, and Q_1 are orthogonal matrices. We can here choose Q_1 and Q_2 so that $Q_i^T A_i Q_{i-1} = T_i$, $i = 1, 2$, are upper triangular. But then it holds that $Q_i^T A_i^{s_i} Q_{i-1} = T_i^{s_i}$ is also upper triangular for $s_i = \pm 1$. It can be shown that the remaining orthogonal matrix Q_0 leaves enough freedom to diagonalize the product A. It is shown in [384, 1995] how to implicitly construct the matrices Q_i, $i = 0, 1, 2$, so that the product (4.2.7) is the bidiagonalization of A. Typically the bidiagonal matrix will be graded, and will allow small singular values to be computed with high relative precision by the QR algorithm. The method generalizes to the product and/or quotient of an arbitrary number of matrices.

4.3. General Linear Models and Generalized Least Squares

4.3.1. Gauss–Markoff linear models.
Let $A \in \mathbf{R}^{m \times n}, m \geq n$, be a known matrix, $b \in \mathbf{R}^m$ a known vector, and $x \in \mathbf{R}^n$ an unknown parameter vector which is to be estimated. The **general Gauss–Markoff linear model** has the form

$$(4.3.1) \qquad Ax + \epsilon = b, \qquad V(\epsilon) = \sigma^2 W,$$

where ϵ is a random vector with zero mean and covariance matrix $\sigma^2 W$, and W is a symmetric nonnegative definite matrix. For $W = I$ we get the special linear model discussed in Section 1.1.3. In this section we will treat the general case, when both matrices A and W may be rank deficient. The special case when W is a positive diagonal matrix, **weighted linear models**, will be discussed in Section 4.5.

We will assume that W is given in factored form

$$(4.3.2) \qquad W = BB^T, \qquad B \in \mathbf{R}^{m \times p}, \qquad p \leq m.$$

If W is initially given, then B can be computed as the Cholesky factor of W. We replace (4.3.1) by the equivalent model

$$(4.3.3) \qquad Ax + Bu = b, \qquad V(u) = \sigma^2 I,$$

where the random vector $u \in \mathbf{R}^p$ has covariance matrix $\sigma^2 I$. We now show how the generalized singular value decomposition can be used to analyze the model (4.3.3). The following analysis is based on the work by Paige [632, 1985].

Since the matrices A and B have the same number of rows a slightly modified version of the generalized singular value decomposition of Section 4.2.2 can be applied to $A^T \in \mathbf{R}^{n \times m}$ and $B^T \in \mathbf{R}^{p \times m}$. We state the resulting decomposition using slightly different notations.

THEOREM 4.3.1. *Assume that*

$$r = \operatorname{rank}(A), \quad s = \operatorname{rank}(B), \quad k = \operatorname{rank}(A, B),$$

where it follows from the assumptions that $r \leq n$, $s \leq p$, $k \leq r + s$. Then there exist orthogonal matrices $U \in \mathbf{R}^{n \times n}$ and $V \in \mathbf{R}^{p \times p}$ and a matrix $Z \in \mathbf{R}^{m \times k}$ of rank k such that

$$(4.3.4) \qquad AU = Z \begin{pmatrix} 0 & 0 & 0 \\ 0 & D_A & 0 \\ 0 & 0 & I_{k-s} \end{pmatrix} \begin{matrix} \}k-r \\ \}q \\ \}k-s \end{matrix},$$

$$\underbrace{}_{n-r} \underbrace{}_{r}$$

4.3. General Linear Models and Generalized Least Squares

$$(4.3.5) \qquad BV = Z \begin{pmatrix} I_{k-r} & 0 & 0 \\ 0 & D_B & 0 \\ 0 & \underbrace{0}_{s} & \underbrace{0}_{p-s} \end{pmatrix} \begin{matrix} \}k-r \\ \}q \\ \}k-s \end{matrix},$$

where $q = r + s - k$,

$$(4.3.6) \qquad \begin{aligned} D_A &= \mathrm{diag}(\alpha_1, \ldots, \alpha_q), & 0 < \alpha_1 \leq \cdots \leq \alpha_q < 1, \\ D_B &= \mathrm{diag}(\beta_1, \ldots, \beta_q), & 1 > \beta_1 \geq \cdots \geq \beta_q > 0, \end{aligned}$$

and $D_A^2 + D_B^2 = I_q$.

Note that the row partitionings in (4.3.4) are the same. If we partition the orthogonal matrices U and V conformably with the column blocks on the right-hand sides in (4.3.4),

$$U = (U_1, U_2, U_3), \qquad V = (V_1, V_2, V_3),$$

then we note that $AU_1 = 0$, $BV_3 = 0$, and hence U_1 and V_3 span the nullspaces of A and B, respectively. The decomposition (4.3.4) separates out the common column space of A and B. We have $AU_2 = ZD_A$ and $BV_2 = ZD_B$, and thus $AU_2 D_B = BV_2 D_A$. Since $D_A > 0$ and $D_B > 0$ it follows that

$$\mathcal{R}(AU_2) = \mathcal{R}(BV_2) = \mathcal{R}(A) \cap \mathcal{R}(B),$$

and has dimension q. For the special case $B = I$ we have $s = k = m$ and then $q = \mathrm{rank}(A)$.

Now let the QR decomposition of the matrix Z in (4.3.4) be

$$(4.3.7) \qquad Q^T Z = \begin{pmatrix} R \\ 0 \end{pmatrix}, \qquad Q = (Q_1, Q_2),$$

where $R \in \mathbf{R}^{k \times k}$ is upper triangular and nonsingular. In the model (4.3.3) we make the orthogonal transformations of the variables

$$(4.3.8) \qquad \tilde{x} = U^T x, \qquad \tilde{u} = V^T u.$$

Then, using (4.3.4) and (4.3.7) the model (4.3.3) becomes

$$(4.3.9) \qquad \begin{pmatrix} R \\ 0 \end{pmatrix} \left\{ \begin{pmatrix} 0 & 0 & 0 \\ 0 & D_A & 0 \\ 0 & 0 & I_{k-s} \end{pmatrix} \begin{pmatrix} \tilde{x}_1 \\ \tilde{x}_2 \\ \tilde{x}_3 \end{pmatrix} + \begin{pmatrix} I_{k-r} & 0 & 0 \\ 0 & D_B & 0 \\ 0 & 0 & 0 \end{pmatrix} \begin{pmatrix} \tilde{u}_1 \\ \tilde{u}_2 \\ \tilde{u}_3 \end{pmatrix} \right\} = \begin{pmatrix} Q_1^T b \\ Q_2^T b \end{pmatrix},$$

where $\tilde{x}_i = U_i^T x_i$, $\tilde{u}_i = V_i^T u_i$, $i = 1, 2, 3$.

It immediately follows that the model is correct only if $Q_2^T b = 0$, which is equivalent to the condition $b \in \mathcal{R}(A, B)$. If this condition is not satisfied, then b could not have come from the model.

The remaining part of (4.3.9) can now be written

$$(4.3.10) \qquad \begin{pmatrix} R_{11} & R_{12} & R_{13} \\ 0 & R_{22} & R_{23} \\ 0 & 0 & R_{33} \end{pmatrix} \begin{pmatrix} \tilde{u}_1 \\ D_A \tilde{x}_2 + D_B \tilde{u}_2 \\ \tilde{x}_3 \end{pmatrix} = \begin{pmatrix} c_1 \\ c_2 \\ c_3 \end{pmatrix} \begin{matrix} \}k-r \\ \}q \\ \}k-s \end{matrix},$$

where we have partitioned R and $c = Q_1^T b$ conformably with the block rows of the two block diagonal matrices in (4.3.9).

We first note that $\tilde{x}_1$ has no effect on b, and therefore cannot be estimated. The decomposition
$$x = x^n + x^e, \qquad x^n = U_1 \tilde{x}_1, \qquad x^e = U_2 \tilde{x}_2 + U_3 \tilde{x}_3$$
splits x into a nonestimable part x^n and an estimable part x^e. Further, $\tilde{x}_3$ can be determined exactly from
$$R_{33} \tilde{x}_3 = c_3.$$
Note that $\tilde{x}_3$ has dimension $k - s = \text{rank}(A, B) - \text{rank}(B)$, so that this can only occur when $\text{rank}(B) < m$.

The second block row in (4.3.10) gives the linear model
$$D_A \tilde{x}_2 + D_B \tilde{u}_2 = R_{22}^{-1}(c_2 - R_{23} \tilde{x}_3),$$
where from (4.3.8) we have that $V(\tilde{u}_2) = \sigma^2 I$. Here the right-hand side is known and the best linear unbiased estimate of $\tilde{x}_2$ is

(4.3.11) $$\hat{x}_2 = D_A^{-1} R_{22}^{-1}(c_2 - R_{23} \tilde{x}_3).$$

The error satisfies $D_A(\hat{x}_2 - \tilde{x}_2) = D_B \tilde{u}_2$, and hence the error covariance is
$$V(\hat{x}_2 - \tilde{x}_2) = \sigma^2 (D_A^{-1} D_B)^2,$$
and so has uncorrelated components.

The random vector $\tilde{u}_3$ has no effect on b. The dimension of $\tilde{u}_3$ is $p - s = p - \text{rank}(B)$, and so is zero if B has independent columns. Finally the vector $\tilde{u}_1$ can be solved exactly from the system (4.3.10). Since $\tilde{u}_1$ has zero mean and covariance matrix $\sigma^2 I$ it can be used in estimating σ^2. Note that $\tilde{u}_1$ has dimension $k - r = \text{rank}(A, B) - \text{rank}(A)$.

REMARK 4.3.1. The GSVD could be used to compute the estimate (4.3.11) and its error covariance matrix. However, a reliable and more efficient method has been given by Paige [628, 1979]. This method will be described for the case of a positive definite W in Section 4.3.3.

4.3.2. Generalized linear least squares problems. The best linear unbiased estimate (BLUE) of any estimable function of x in the general linear model (4.3.3) is given by the solution to

(4.3.12) $$\min_{v,x} \|v\|_2^2 \quad \text{subject to} \quad Ax + Bv = b,$$

which is a **generalized least squares problem**. In the case when $W = BB^T \in R^{m \times m}$ is positive definite this problem reduces to that of finding a vector $x \in R^n$ that solves

(4.3.13) $$\min_x (Ax - b)^T W^{-1}(Ax - b).$$

4.3. GENERAL LINEAR MODELS AND GENERALIZED LEAST SQUARES

The solution to this problem gives the least squares estimate of the vector x in the linear model

$$(4.3.14) \qquad Ax = b + \epsilon, \qquad \mathcal{E}(\epsilon) = 0, \qquad \mathcal{V}(\epsilon) = \sigma^2 W.$$

Problems where W is singular correspond to linear least squares problems with linear constraints, and will be treated in Section 5.1.

The solution of problem (4.3.13) is given by the normal equations

$$(4.3.15) \qquad A^T W^{-1} A x = A^T W^{-1} b.$$

If $A^T W^{-1} A = R^T R$ is the Cholesky factorization of the matrix of normal equations then the variance-covariance matrix of the least squares solution is $\sigma^2 C_x$, where

$$C_x = (A^T W^{-1} A)^{-1} = R^{-1} R^{-T}.$$

Introducing the vector $y = W^{-1}(b - Ax)$ the normal equations are equivalent to the augmented linear system

$$(4.3.16) \qquad \begin{pmatrix} W & A \\ A^T & 0 \end{pmatrix} \begin{pmatrix} y \\ x \end{pmatrix} = \begin{pmatrix} b \\ 0 \end{pmatrix}.$$

Rao [675, 1973] called the matrix in (4.3.16) the **fundamental matrix**, and showed in theory how to obtain the solution from a generalized inverse of the fundamental matrix. Wedin [828, 1985] gives a unified treatment and a perturbation analysis of generalized and constrained least squares problems based on the system (4.3.15).

If a given W is symmetric positive definite, it has a factorization

$$(4.3.17) \qquad W = BB^T, \qquad B \in R^{m \times m},$$

where B is nonsingular. Then (4.3.13) is equivalent to

$$(4.3.18) \qquad \min_x \| B^{-1}(Ax - b) \|_2.$$

Often B itself is given rather than W. In such cases forming W can lose important information, and could cause the computed W to be singular even when B has full rank. In general it is preferable to work with B instead of W. Even if B is not given, we can compute $B = L$ from the Cholesky factorization $W = LL^T$; see Section 2.2.2.

The problem (4.3.18) can be transformed into a standard least squares problem by computing $\bar{A} = B^{-1} A$, and $\bar{b} = B^{-1} b$. When B is at all ill-conditioned this is not a stable computational approach, and instead the GQR decomposition of A and B should be used.

4.3.3. Paige's method.

A stable algorithm for the generalized linear least squares problem (4.3.12) has been given by Paige [627, 1979], [628, 1979]. Although this algorithm allows for rank deficiency in A and B we consider here only the case when B is nonsingular and A has full rank n. For a treatment of the general case we refer to Paige [627, 1979] and Section 4.3.1.

Paige's method begins with computing the QR decomposition of A,

$$\text{(4.3.19)} \qquad Q^T A = \begin{pmatrix} R \\ 0 \end{pmatrix}, \quad Q = (Q_1, Q_2),$$

and then applies the orthogonal transformation Q^T also to b and B

$$Q^T b = \begin{pmatrix} c_1 \\ c_2 \end{pmatrix} \begin{matrix} \} \ n \\ \} \ m-n \end{matrix}, \quad Q^T B = \begin{pmatrix} C_1 \\ C_2 \end{pmatrix} \begin{matrix} \} \ n \\ \} \ m-n \end{matrix}.$$

The constraints in (4.3.12) can now be written in partitioned form

$$\text{(4.3.20)} \qquad \begin{pmatrix} C_1 \\ C_2 \end{pmatrix} v + \begin{pmatrix} R \\ 0 \end{pmatrix} x = \begin{pmatrix} c_1 \\ c_2 \end{pmatrix}.$$

For any vector $v \in R^m$ we can always determine x so that the first block of these equations is satisfied. An orthogonal matrix $P \in R^{m \times m}$ is then determined such that

$$\text{(4.3.21)} \qquad P^T C_2^T = \begin{pmatrix} 0 \\ S^T \end{pmatrix} \begin{matrix} \} \ n \\ \} \ m-n \end{matrix},$$

where the matrix S is upper triangular. By the nonsingularity of B it follows that C_2 will have linearly independent rows, and hence the matrix S will be nonsingular. Note that (4.3.21) after reversing the rows and columns is just the QR decomposition of C_2^T. Now the second set of constraints in (4.3.20) becomes

$$\text{(4.3.22)} \qquad S u_2 = c_2, \quad \text{where} \quad P^T v = u = \begin{pmatrix} u_1 \\ u_2 \end{pmatrix} \begin{matrix} \} \ n \\ \} \ m-n \end{matrix}.$$

Since P is orthogonal we have $\|v\|_2 = \|u\|_2$ and so the minimum in (4.3.12) is found by taking

$$u_1 = 0, \quad u_2 = S^{-1} c_2, \quad v = P_2 u_2,$$

where $P = (P_1, P_2)$. Finally x is obtained by solving the triangular system $Rx = c_1 - C_1 v$ in (4.3.20).

Paige's algorithm (4.3.19)–(4.3.22) requires a total of about $2m^3/3 + m^2 n$ flops. If $m \gg n$ the work in the QR decomposition of C_2 dominates. It can be shown that the computed solution is an unbiased estimate of x for the model (4.3.12) with covariance matrix $\sigma^2 C$, where

$$\text{(4.3.23)} \qquad C = R^{-1} L^T L R^{-T}, \quad L^T = C_1^T P_1.$$

Paige [627, 1979] obtains a perturbation analysis for the problem (4.3.18) by using the formulation (4.3.12), and gives a rounding error analysis to show that the above algorithm is numerically stable. The algorithm can be generalized in a straightforward way to rank deficient A and B. For details see Paige [627, 1979].

The algorithm above does not take advantage of any special structure the matrix B may have. If B has been obtained from the Cholesky factorization $W = BB^T$ it is of lower triangular form. In this case, and also when W is diagonal, it is advantageous to carry out the two QR decompositions in (4.3.19) and (4.3.21) together, maintaining the lower triangular form throughout. Paige [628, 1979] has given such a variation of the algorithm using a "zero chasing technique," with a careful sequencing of Givens transformations. With fast Givens rotations this requires a total of about $m^2n + 2mn^2 - 4n^3/3$ flops.

REMARK 4.3.2. In some applications, notably from interior point methods, one needs to solve a sequence of problems of the form (4.3.12), with A constant but $B = B_k$, $k = 1, \ldots, p$. The QR decomposition (4.3.19) can then be computed once and for all. In case $m = n$ this reduces the work for solving an additional problem from $5n^3/3$ to n^3.

4.4. Weighted Least Squares Problems

4.4.1. Introduction.
In this section we consider the special linear model (4.3.1) where the components in the random error vector ϵ are uncorrelated. In this case the covariance matrix W is a positive diagonal matrix

$$W = \operatorname{diag}(w_1, w_2, \ldots, w_m) > 0.$$

The corresponding least squares problem, $\min_x (Ax - b)^T W^{-1}(Ax - b)$, can be written as a **weighted linear least squares problem**

$$(4.4.1) \qquad \min_x \|D(Ax - b)\|_2,$$

where we have introduced the diagonal **weight matrix**

$$D = W^{-1/2} = \operatorname{diag}(d_1, d_2, \ldots, d_m).$$

In many cases it is possible to solve (4.4.1) as a standard linear least squares problem

$$\min_x \|\tilde{A}x - \tilde{b}\|_2, \quad \tilde{A} = DA, \quad \tilde{b} = Db.$$

However, in applications where the weights $d_1, \ldots, d_m$ vary widely in size this is not generally a numerically stable approach.

Note that the weight matrix in (4.4.1) is not unique. Therefore we will in the following assume that the matrix A has been row equilibrated, that is,

$$\max_{1 \leq j \leq n} |a_{ij}| = 1, \quad i = 1, \ldots, m.$$

We also assume here and in the following that the rows of A are ordered so that the weights satisfy

$$(4.4.2) \qquad \infty > d_1 \geq d_2 \geq \cdots \geq d_m > 0.$$

Then $d_1/d_m = \gamma \gg 1$ corresponds to the case when some components of the error vector in the linear model have much smaller variance than the rest, and we call such weighted problems **stiff**. Note that in the limit when some d_i tend to infinity, the corresponding ith equation becomes a linear constraint.

For stiff problems the condition number $\kappa(DA)$ will be large. An upper bound is given by
$$\kappa(DA) \le \kappa(D)\kappa(A) = \gamma\kappa(A).$$
It is important to note that this does *not* mean that the problem of computing x from given data $\{D, A, b\}$ is ill-conditioned. For the weighted problem (4.4.1) the perturbations in DA and Db will have a special form, and the normwise perturbation analysis given in Section 1.4.2 is not relevant; see Remark 1.4.3. However, that $\kappa(DA) \gg 1$ correctly warns us that special care may be needed in solving stiff weighted linear least squares problems.

REMARK 4.4.1. Problems with extremely ill-conditioned weight matrices arise, e.g., in electrical networks, certain classes of finite element problems, and interior point methods for constrained optimization. Vavasis [806, 1994] and Hough and Vavasis [474, 1994] have developed special methods for such applications, which satisfy a strong type of stability. ∎

It is easily seen that in general the method of normal equations is not well suited for solving stiff problems. To illustrate this, we consider the important special case where only the first p equations are weighted:

$$\text{(4.4.3)} \qquad \min_x \left\| \begin{pmatrix} \gamma A_1 \\ A_2 \end{pmatrix} x - \begin{pmatrix} \gamma b_1 \\ b_2 \end{pmatrix} \right\|_2^2,$$

$A_1 \in \mathbf{R}^{p \times n}$ and $A_2 \in \mathbf{R}^{(m-p) \times n}$. Such problems occur, for example, when the method of weighting is used to solve least squares problems with the linear equality constraints $A_1 x = b_1$; see Section 5.1.4. For this problem the matrix of normal equations becomes

$$B = \begin{pmatrix} \gamma A_1^T & A_2^T \end{pmatrix} \begin{pmatrix} \gamma A_1 \\ A_2 \end{pmatrix} = \gamma^2 A_1^T A_1 + A_2^T A_2.$$

If $\gamma > u^{-1/2}$ (u is the unit roundoff) and $A_1^T A_1$ is dense, then $B = A^T A$ will be completely dominated by the first term and the data contained in A_2 may be lost. However, if the number p of very accurate observations is less than n, then the solution depends critically on the less precise data in A_2. (The matrix in Example 2.2.1 is of this type.) We conclude that for weighted least squares problems with $\gamma \gg 1$ the method of normal equations generally is not well behaved.

4.4.2. Methods based on Gaussian elimination. In Section 2.5 several methods based on a preliminary factorization by Gaussian elimination were discussed. In the Peters–Wilkinson method (see Section 2.5.1) A is first reduced by Gaussian elimination to upper triangular form. It was pointed out by Björck and Duff [104, 1980] that this method is suitable for weighted problems.

4.4. Weighted Least Squares Problems

Assume that $\text{rank}(A_1) = p$, and that p steps of Guassian elimination are performed on the weighted matrix $\tilde{A} = DA$ using row and column pivoting. Then the resulting factorization can be written

(4.4.4) $$\Pi_1 \tilde{A} \Pi_2 = L_p D U_p,$$

where Π_1 and Π_2 are permutation matrices,

$$L_p = \begin{pmatrix} L_{11} & \\ L_{21} & L_{22} \end{pmatrix} \in \mathbf{R}^{m \times n}, \quad U_p = \begin{pmatrix} U_{11} & U_{12} \\ & I \end{pmatrix} \in \mathbf{R}^{n \times n},$$

$L_{11} \in \mathbf{R}^{p \times p}$ is unit lower triangular, and $U_{11} \in \mathbf{R}^{p \times p}$ unit upper triangular. Assuming that $\tilde{A}$ has full rank, D is nonsingular. Then (4.4.1) is equivalent to

$$\min_y \|L_p y - \Pi_1 \tilde{b}\|_2, \quad U_p \Pi_2^T x = D^{-1} y.$$

This least squares problem is usually well-conditioned, since any ill-conditioning in $\tilde{A}$ is usually reflected in U. We illustrate the method in a simple example.

EXAMPLE 4.4.1. In Example 2.2.1 it was shown that the method of normal equations failed for the problem of Läuchli [517, 1961]. After multiplication with $\gamma = \epsilon^{-1}$ this becomes

$$A = \begin{pmatrix} \gamma & \gamma & \gamma \\ 1 & & \\ & 1 & \\ & & 1 \end{pmatrix}, \quad b = \begin{pmatrix} \gamma \\ 0 \\ 0 \\ 0 \end{pmatrix},$$

which is of the form (4.4.3) with $p = 1$. After one step of Gaussian elimination we obtain the factorization $A = L_1 D_1 U_1$, where

$$L_1 = \begin{pmatrix} 1 & & \\ \gamma^{-1} & -1 & -1 \\ & 1 & \\ & & 1 \end{pmatrix}, \quad D_1 U_1 = \begin{pmatrix} \gamma & \gamma & \gamma \\ & 1 & \\ & & 1 \end{pmatrix}.$$

It is easily verified that L_1 is well-conditioned, and the solution can be accurately obtained by solving $L_1^T L_1 y = L_1^T b$, and back-substitution $D_1 U_1 x = y$. ∎

In general, for a problem of the form (4.4.3) the LU factorization (4.4.4) will have the form

(4.4.5) $$\begin{pmatrix} \gamma A_1 \\ A_2 \end{pmatrix} = \begin{pmatrix} L_{11} & \\ \frac{1}{\gamma} L_{21} & L_{22} \end{pmatrix} \begin{pmatrix} \gamma U_{11} & \gamma U_{12} \\ & I \end{pmatrix} \equiv L(DU),$$

where the blocks L_{ij} and U_{ij} are $O(1)$, and $L_{22} \in \mathbf{R}^{(m-p) \times (n-p)}$ is the reduced matrix. The normal equations for $y = (DU)x$ then equal $L^T L y = L^T b$, where

$$L^T L = \begin{pmatrix} L_{11}^T L_{11} + \frac{1}{\gamma^2} L_{21}^T L_{21} & \frac{1}{\gamma} L_{21}^T L_{22} \\ \frac{1}{\gamma} L_{22}^T L_{21} & L_{22}^T L_{22} \end{pmatrix},$$

$$L^T b = \begin{pmatrix} \gamma L_{11}^T b_1 + \frac{1}{\gamma} L_{21}^T b_2 \\ L_{22}^T b_1 \end{pmatrix}.$$

For $\gamma \gg 1$ the matrix $L^T L$ is almost block diagonal and its condition number is to first approximation independent of γ. If we let R_{11} and R_{22} be the Cholesky factors of $L_{11}^T L_{11}$ and $L_{22}^T L_{22}$, respectively, then the Cholesky factor of $L^T L$ will have the form

$$R = (1 + O(\gamma^{-2})) \begin{pmatrix} R_{11} & \frac{1}{\gamma}(L_{21}R_{11}^{-1})^T L_{22} \\ & R_{22} \end{pmatrix};$$

cf. Stewart [742, 1984]. After solving $RR^T y = L^T b$ the least squares solution is obtained from $DUx = y$, giving

$$x_2 = y_2, \qquad U_{11} x_1 = \frac{1}{\gamma} y_1 - U_{12} y_2.$$

For the weighted least squares problem the augmented system (4.3.16) has the form

(4.4.6) $$\begin{pmatrix} \alpha W & A \\ A^T & 0 \end{pmatrix} \begin{pmatrix} \alpha^{-1} r \\ x \end{pmatrix} = \begin{pmatrix} b \\ 0 \end{pmatrix},$$

where $W = D^{-2}$. The scaling factor α has been introduced for stability reasons; see Section 2.5.2. As before we assume that D has been chosen so that A is *row equilibrated*, which will tend to lower the condition of A. Further results on the prescaling of A before using the augmented system method are given in Duff [239, 1994]. The system can be solved by using the Bunch–Kaufman factorization described in Section 2.5.2. An advantage with this formulation is that linear constraints can be treated by letting $w_i = 0$ in (4.4.6).

A problem with this approach is that it is not easy to get an a priori estimate of the optimal value of α for stability. A second drawback with the method outlined in this section is that it works with a system of order $m + n$, which may be much larger than n. Therefore, the main use of this method seems to be for sparse problems, where the sparsity of the block I can be taken into account; see Arioli, Duff, and de Rijk [20, 1989].

4.4.3. QR decompositions for weighted problems. We now consider the use of methods based on the QR decomposition of A for solving weighted problems. We first examine the Householder QR method, and show by an example that this method can give poor accuracy for stiff problems unless the algorithm is extended to include *row interchanges*.

EXAMPLE 4.4.2. (See Powell and Reid [670, 1969].) Consider the problem $\min_x \|Ax - b\|_2$, where

$$A = \begin{pmatrix} 0 & 2 & 1 \\ \gamma & \gamma & 0 \\ \gamma & 0 & \gamma \\ 0 & 1 & 1 \end{pmatrix}, \qquad b = \begin{pmatrix} 2 \\ 2\gamma \\ 2\gamma \\ 2 \end{pmatrix},$$

with exact solution equal to $x = (1, 1, 1)$. Using exact arithmetic we obtain after the first step of QR decomposition of A by Householder transformations

4.4. Weighted Least Squares Problems

(Algorithm 2.4.1) the reduced matrix

$$\tilde{A}^{(2)} = \begin{pmatrix} \frac{1}{2}\gamma - 2^{1/2} & -\frac{1}{2}\gamma - 2^{-1/2} \\ -\frac{1}{2}\gamma - 2^{1/2} & \frac{1}{2}\gamma - 2^{-1/2} \\ 1 & 1 \end{pmatrix}.$$

If $\gamma > u^{-1}$ the terms $-2^{1/2}$ and $-2^{-1/2}$ in the first and second rows are lost. However, this is equivalent to the loss of all information present in the first row of A. This loss is disastrous because the number of rows containing large elements is less than the number of components in x, so there is a substantial dependence of the solution x on the first row of A. (However, compared to the method of normal equations, which fails already when $\gamma > u^{-1/2}$, this is an improvement!) ∎

Van Loan [799, 1985] has given several examples illustrating that solving

$$(4.4.7) \qquad \min_x \left\| \begin{pmatrix} A_2 \\ \gamma A_1 \end{pmatrix} x - \begin{pmatrix} b_2 \\ \gamma b_1 \end{pmatrix} \right\|_2^2$$

instead of (4.4.3) with Householder will give bad accuracy for large values of γ.

It is also essential that *column pivoting* is performed when QR decomposition is used for weighted problems. Van Loan [799, 1985] gives an example of the form (4.4.3), where

$$A_1 = \begin{pmatrix} 1 & 1 & 1 \\ 1 & 1 & -1 \end{pmatrix},$$

to illustrate the need for column pivoting. Stability is lost here without column pivoting because the first two columns of the matrix A_1 are linearly dependent. When column pivoting is introduced this difficulty disappears.

Powell and Reid [670, 1969] extended the Householder algorithm to include *row interchanges*. In each step a pivot column is first selected in the reduced matrix, and then the element of largest absolute value in the pivot column is permuted to the top. Powell and Reid give an error analysis for this algorithm which shows that it has good stability properties for stiff problems as well.

It seems that there is no need to perform row pivoting in Householder QR, provided that the rows are sorted after decreasing row norm before the factorization, so that the weights satisfy (4.4.2). For example, if in Example 4.4.2 the two large rows are permuted to the top of the matrix A, then the Householder algorithm works well.

An approach related to that of Powell and Reid is taken by Gulliksson and Wedin [413, 1992]. They use scaled Householder transformations $\tilde{P}$ which are W invariant, i.e., satisfy

$$(4.4.8) \qquad \tilde{P} W \tilde{P}^T = W = \mathrm{diag}\,(w_1, \ldots, w_m).$$

It is easy to verify that P must have the form

$$P = I - 2Wvv^T/(v^T W v), \qquad P^2 = I,$$

i.e., P is a reflector. Note that $W^{-1/2}PW^{1/2}$ is an orthogonal reflector.

A sequence of W invariant reflectors is used to transform $A\Pi$, where Π is a permutation matrix, to upper triangular form,

$$Q^T A\Pi = \begin{pmatrix} R \\ 0 \end{pmatrix}, \qquad Q^T = P_n \cdots P_2 P_1.$$

This is equivalent to the ordinary QR factorization

$$W^{-1/2} A\Pi = (W^{-1/2} Q W^{1/2}) \begin{pmatrix} W^{-1/2} R \\ 0 \end{pmatrix}.$$

When $W > 0$ this method is equivalent to the algorithm of Powell and Reid. However, this approach generalizes simply to the case when W has the form $W = \text{diag}\,(0, W_2)$, which corresponds to a constrained least squares problem. A backward error analysis of this method has been given by Gulliksson [410, 1995].

In contrast to the Householder QR method, the modified Gram–Schmidt (MGS) method is numerically invariant under row interchanges (except for effects deriving from different summation orders in the computed inner products). In particular, for problems of the special form (4.4.3) MGS will give accurate solutions independent of row ordering if γ is chosen optimally. However, as illustrated by the numerical results by Anda and Park [15, 1996], MGS will lose accuracy for very large values of γ. Gulliksson [411, 1995] has made a detailed study of the numerical stability of MGS for weighted problems.

Anda and Park [15, 1995] have studied the use of Givens QR algorithms for stiff least squares problems, and developed self-scaling fast plane rotations for such problems. They show that both fast and standard Givens rotations produce accurate results regardless of row sorting.

The following example from [15] illustrates the effect of row sorting in Givens rotation. Let $\gamma \gg 1$, and

$$A = \begin{pmatrix} a_{pp} & a_{pq} \\ \gamma a_{qp} & \gamma a_{qq} \end{pmatrix}, \qquad \bar{A} = \begin{pmatrix} \gamma \bar{a}_{pp} & \gamma \bar{a}_{pq} \\ \bar{a}_{qp} & \bar{a}_{qq} \end{pmatrix}.$$

The Givens transformations that zero the elements a'_{qp} and $\bar{a}'_{qp}$ in $A' = GA$, and $\bar{A}' = \bar{G}\bar{A}$, respectively, are (see (2.3.13))

$$G = \frac{1}{\sigma}\begin{pmatrix} a_{pp} & \gamma a_{qp} \\ -\gamma a_{qp} & a_{pp} \end{pmatrix}, \qquad \bar{G} = \frac{1}{\bar{\sigma}}\begin{pmatrix} \gamma \bar{a}_{pp} & \bar{a}_{qp} \\ -\bar{a}_{qp} & \gamma \bar{a}_{pp} \end{pmatrix},$$

where $\sigma = \sqrt{a_{pp}^2 + \gamma^2 a_{qp}^2}$ and $\bar{\sigma} = \sqrt{\gamma^2 \bar{a}_{pp}^2 + \bar{a}_{qp}^2}$. In each case the more heavily weighted row of the resulting matrix GA and $\bar{G}\bar{A}$ is in top position regardless of its initial position. Hence a sequence of rotations will move rows of large norms to the top of the matrix. The numerical results of Anda and Park also showed that the self-scaling rotations maintained high accuracy for extremely large values of γ. Their tests also showed no significant difference in accuracy between different rotation orderings.

4.4.4. Weighted problems by updating.

A stable method for solving the stiff least squares problem of the special form (4.4.3) can be based on **updating** the solution to the unweighted problem. Let $x(\gamma)$ be the solution to problem (4.4.3) with weight $\gamma > 1$, and let $r(\gamma) = b - Ax(\gamma)$ be the corresponding residual vector, where

$$A = \begin{pmatrix} A_1 \\ A_2 \end{pmatrix}, \qquad b = \begin{pmatrix} b_1 \\ b_2 \end{pmatrix}.$$

We now relate $x(\gamma)$ and $r_1(\gamma) = b_1 - A_1 x(\gamma)$ to the corresponding quantities $x(1)$ and $r_1(1)$ for the unweighted problem. The normal equations for the weighted problem can be written

$$\left(\hat{\gamma}^2 A_1^T A_1 + A^T A\right) x(\gamma) = \hat{\gamma}^2 A_1^T b_1 + A^T b,$$

where $\hat{\gamma}^2 = \gamma^2 - 1$. Subtracting the normal equations for $\gamma = 1$ and solving for $\delta x = x(\gamma) - x(1)$ gives

$$\delta x = \hat{\gamma}^2 \left(A^T A + \hat{\gamma}^2 A_1^T A_1\right)^{-1} A_1^T r_1(1).$$

Using the Woodbury formula (3.1.6) we obtain after some calculation

(4.4.9) $$r_1(\gamma) = \hat{\gamma}^{-2} \left(A_1 (A^T A)^{-1} A_1^T + \hat{\gamma}^{-2} I_p\right)^{-1} r_1(1).$$

When $r_1(\gamma)$ is known we can obtain $x(\gamma)$ from

(4.4.10) $$x(\gamma) = x(1) + \hat{\gamma}^2 (A^T A)^{-1} A_1^T r_1(\gamma).$$

If $p \ll m$, then the extra work in the updating steps (4.4.9) and (4.4.10) is small. If the factorization $A^T A = R^T R$ is known (from Cholesky or QR factorization) it follows that

$$A_1 (A^T A)^{-1} A_1^T = S_1 S_1^T, \qquad R^T S_1^T = A_1^T,$$

where $S_1 \in \mathbf{R}^{p \times n}$ can be computed by forward substitution. We can then compute $r_1(\gamma)$ and the correction $x(\gamma) - x(1)$ by solving

(4.4.11) $$(S_1 S_1^T + \hat{\gamma}^{-2} I_p) r_1(\gamma) = \hat{\gamma}^{-2} r_1(1),$$
(4.4.12) $$R(x(\gamma) - x(1)) = \hat{\gamma}^2 S_1^T r_1(\gamma).$$

Thus we only need to solve a (small) $p \times p$ system, form $S_1^T r_1(\gamma)$, and perform a back-substitution. If the matrix A is not too ill-conditioned the method of normal equations might be used to compute the solution to the unweighted problem and determine R. On the other hand, if the matrix B is ill-conditioned, it is better to solve the $p \times p$ system in (4.4.11) by QR factorization, noting that

$$(\hat{\gamma}^{-2} I_p + S_1 S_1^T) = R_S^T R_S, \qquad \begin{pmatrix} S_1^T \\ \hat{\gamma}^{-1} I_p \end{pmatrix} = Q_S R_S.$$

It follows from (4.4.11) that if $\text{rank}(A_1) = p$, then for large values of γ the residual $r_1(\gamma)$ will be proportional to γ^{-2}. If we let $\gamma \to \infty$ we obtain a solution of a constrained problem which satisfies $A_1 x = b_1$ exactly. In this case the above algorithm simplifies; see Section 5.1.5.

4.5. Minimizing the l_p Norm

4.5.1. Introduction. In some applications it might be more adequate to minimize some other norm than the l_2 norm of the residual $r = b - Ax$. We consider the Hölder vector l_p-norms $\|\cdot\|_p$, which are defined by

$$(4.5.1) \qquad \|r\|_p = \left(\sum_{i=1}^m |r_i|^p\right)^{1/p}, \qquad 1 \le p < \infty,$$

where the Euclidian norm corresponds to $p = 2$. In the limiting case $p \to \infty$ we take

$$(4.5.2) \qquad \|r\|_\infty = \max_{1 \le i \le n} |r_i|.$$

Using these norms leads to the more general minimization problems

$$(4.5.3) \qquad \min_x \|Ax - b\|_p^p \qquad 1 \le p < \infty,$$

and for $p = \infty$

$$(4.5.4) \qquad \min_x \max_i |a_i^T x - b_i|,$$

where a_i^T is the ith row of the matrix A.

The effect of using a Hölder norm with $p \ne 2$ was illustrated in Example 1.1.2 by considering the problem of estimating the scalar γ from m observations $y \in \mathbf{R}^m$. The estimates were shown to correspond to the median, mean, and midrange, respectively. In particular, the estimate γ_1 does not depend on extreme values of y_i. This property carries over to more general problems, and a small number of isolated large errors will generally not change the l_1 solution. However, for $p = 1$ the solution may not be unique, while for $1 < p < \infty$ there exists exactly one l_p solution. Thus it may be desirable to obtain the l_p solution for a noninteger value of $p > 1$.

In Section 1.1.1 it was mentioned that Laplace in 1799 used the principle of l_1 approximation. It can be shown that the solution then must satisfy at least n equations exactly. Likewise, the l_∞ solution has the property that the absolute error in at least n equations equals the maximum error. The advantage of the l_1 solution is that it is **robust** in that a small number of isolated large errors do not usually change the solution. However, a similar effect is also achieved with p greater than but close to 1.

Often it is preferable to solve the approximation problem

$$(4.5.5) \qquad \psi(x) = \min_x \|Ax - b\|_p^p, \qquad 1 < p < 2,$$

for noninteger values of p. Here $p = 1$ is excluded because for this case the function $\psi(x)$ in (4.5.5) is only piecewise differentiable. Further, the l_1 solution may not be unique since for $1 < p < 2$ the problem is strictly convex and hence has exactly one solution.

4.5.2. Iteratively reweighted least squares.

For solving the l_p norm problem when $1 < p < 2$, the **iteratively reweighted least squares** (IRLS) method (see Osborne [618, 1985]) is widely used. This approach reduces the problem to the solution of a sequence of weighted least squares problems, which is attractive since methods and software for weighted least squares are generally available; see Section 4.4.

We start by noting that, provided that $|r_i(x)| = |b - Ax|_i > 0$, $i = 1, \ldots, m$, the problem (4.5.3) can be restated in the form

$$(4.5.6) \quad \min_x \psi(x), \quad \psi(x) = \sum_{i=1}^m |r_i(x)|^p = \sum_{i=1}^m |r_i(x)|^{p-2} r_i(x)^2.$$

This can be interpreted as a weighted least squares problem

$$(4.5.7) \quad \min_x \|D(r)^{(p-2)/2}(b - Ax)\|_2, \quad D(r) = \text{diag}(|r|),$$

where here and in the following the notation $\text{diag}(|r|)$, $r \in \mathbf{R}^m$, denotes the diagonal matrix with ith component $|r_i|$.

The diagonal weight matrix $D(r)^{(p-2)/2}$ in (4.5.7) depends on the unknown solution x, but we can attempt to use the following iterative method.

ALGORITHM 4.5.1. IRLS FOR l_p APPROXIMATION $1 < p < 2$. Let $x^{(0)}$ be an initial approximation such that $r_i^{(0)} = (b - Ax^{(0)})_i \neq 0$, $i = 1, \ldots, m$.

for $k = 0, 1, 2, \ldots$
$\quad r_i^{(k)} = (b - Ax^{(k)})_i;$
$\quad D_k = \text{diag}\,((|r_i^{(k)}|)^{(p-2)/2});$
$\quad$ solve $\delta x^{(k)}$ from
$$\min_{\delta x} \|D_k(r^{(k)} - A\delta x)\|_2;$$
$\quad x^{(k+1)} = x^{(k)} + \delta x^{(k)};$
end

Since
$$D_k b = D_k(r^{(k)} - Ax^{(k)}),$$
$x^{(k+1)}$ in IRLS solves $\min_x \|D_k(b - Ax)\|_2$ but the implementation above is to be preferred. It should be pointed out that the iterations can be carried out entirely in the r space without the x variables. Upon convergence to a residual vector r_{opt} the corresponding solution can be found by solving the consistent linear system $Ax = b - r_{\text{opt}}$.

It has been assumed that in the IRLS algorithm, at each iteration $r_i^{(k)} \neq 0$, $i = 1, \ldots, n$. In practice this cannot be guaranteed, and it is customary (see Li [529, 1993]) to modify the algorithm so that

$$D_k = \text{diag}\,((100ue + |r^{(k)}|)^{(p-2)/2}),$$

where u is the machine precision and $e^T = (1, \ldots, 1)$ is the vector of all ones.

Methods for solving weighted problems have been described in Section 4.5. Because the weight matrix D_k is not constant, the simplest implementations of IRLS recompute, e.g., the QR factorization of $D_k A$ in each step. It is also possible to use updating techniques; see O'Leary [606, 1990].

We now compare IRLS with the Newton method for minimizing $\psi(x) = \phi(r)$, $r = b - Ax$, where $\psi(x)$ is the function in (4.5.6). The gradient of $\phi(r)$ is a vector y with elements

$$y_i = p \,\text{sign}\,(r_i)(|r_i|)^{p-1}, \qquad i = 1, \ldots, m,$$

and the Hessian is a diagonal matrix $D = \text{diag}\,(d_i)$ with entries

$$d_i = p(p-1)(|r_i|)^{p-2}, \qquad i = 1, \ldots, m.$$

It follows that the gradient g and Hessian matrix H of second derivatives of $\psi(x)$ are given by

(4.5.8) $$g = -A^T y, \qquad H = A^T D A.$$

Hence the step s for Newton's method satisfies $Hs = -g$. Apart from the factor $(p-1)$ this is just the normal equations for the weighted least squares problem (4.5.7), and the Newton step for minimizing $\psi(x)$ is related to the IRLS step by

$$s_k = \frac{1}{p-1} \delta x^{(k)}.$$

Since the Newton step is always a descent direction for the objective function $\psi(x)$ it follows that the same is true for the step obtained from the IRLS method. However, because IRLS for $p < 2$ does not take a full Newton step, it is at best only linearly convergent.

Conditions for convergence of IRLS are discussed by Osborne [618, 1985]. It is shown that in the l_p case any fixed point of the IRLS iteration satisfies the necessary conditions for a minimum of $\psi(x)$. It is shown that the IRLS method is convergent for $1 < p < 3$, and also for $p = 1$ provided that the l_1 approximation problem has a unique nondegenerate solution.

Taking a full Newton step may lead to divergence, and hence the Newton method must be combined with a line search to be globally convergent and achieve local quadratic convergence. Such a line search procedure has been developed by Li [529, 1993], who calls the corresponding method IRLSL. This method will also converge quickly for p in the range $2 < p < \infty$. However, for p close to 1 even the IRLSL method converges slowly, which is related to the fact that for $p = 1$ the objective function $\psi(x)$ is not differentiable.

The IRLS method can be extremely slow when p is close to unity. Recently Li [529, 1993], [530, 1993] has developed a method called GNCS based on a globalized Newton method using the complementary slackness condition for the l_1 problem. Far from the solution this algorithm behaves like the IRLSL method. Close to the solution it behaves like a Newton method for an extended nonlinear

4.5. Minimizing the l_p Norm

system of equations capturing the necessary conditions both when $p = 1$ and when $p > 1$. The problem of unbounded second derivatives is handled by a simple technique connected to the line search. The method is globally convergent and the local convergence is superlinear if there are no zero residuals at the solution. Li reports that GNCS is better than IRLSL when $p \leq 1.5$.

4.5.3. Robust linear regression. In robust linear regression possible "outsiders" among the data points are identified and given less weight. For a general treatment of robust statistical procedures, see Huber [478, 1981]. The most popular among the robust estimators is Huber's M-estimator [478, 1981], which uses the least squares estimator for "normal" data but the l_1 norm estimator for data points that disagree too much with the normal picture. More precisely, the Huber M-estimate minimizes the objective function

$$(4.5.9) \qquad \psi(x) = \sum_{i=1}^{m} \rho(r_j(x)/\sigma),$$

where

$$(4.5.10) \qquad \rho(t) = \begin{cases} \frac{1}{2}t^2, & \text{if } |t| \leq \gamma; \\ \gamma|t| - \frac{1}{2}\gamma^2, & \text{if } |t| > \gamma, \end{cases}$$

γ is a tuning parameter, and σ a scaling factor which depends on the data to be estimated. In the following we assume that σ is a constant, and then it is no restriction to take $\sigma = 1$.

Like the l_p estimator for $1 < p < 2$, the Huber estimator can be viewed as a compromise between l_2 and l_1 approximation. For large values of γ it will be close to the least squares estimator; for small values of γ it is close to the l_1 estimator. A great deal of work has been devoted to developing methods for computing the Huber M-estimator; see, e.g., Clark and Osborne [170, 1986], Ekblom [266, 1988], and Madsen and Nielsen [555, 1990]

O'Leary [606, 1990] has studied different implementations of Newton-like methods. The Newton step s for minimizing $\psi(x)$ in (4.5.9) ($\sigma = 1$) is given by the solution of

$$A^T D A s = A^T y,$$

where

$$y_i = \rho'(r_i), \qquad D = \text{diag}\,(\rho''(r_i)), \quad i = 1, \ldots, m.$$

This is similar to (4.5.8) for l_p approximation. O'Leary recommends that at the initial iterations the cutoff value γ in the Huber function (4.5.10) is decreased from a very large number to the desired value. This has the effect of starting the iteration from the least squares solution and helps prevent the occurrence of rank deficient Hessian matrices H.

4.5.4. Algorithms for l_1 and l_∞ approximation. Minimization in the l_1 or l_∞ norm is complicated by the fact that for these values of p the function

$\psi(x) = \|Ax - b\|_p^p$ is not differentiable. In spite of this there are several good, but more costly, computational methods available for these problems.

For the l_1 problem methods which use linear programming techniques have been given by Barrodale and Roberts [50, 1973], [51, 1978]. Their method has been implemented in the Harwell Subroutine Library. Other methods are based on projected gradient techniques; see Bartels, Conn, and Sinclair [54, 1978] for the l_1 problem, and Bartels and Conn [52, 1980] for the l_∞ problem. The general idea in these methods is to use descent methods that explicitly find the correct subset of zero residuals, $p = 1$, and maximum residuals, $p = \infty$.

A globally convergent Newton algorithm for the l_1 problem has been proposed in Coleman and Li [180, 1992]. This algorithm is superlinearly convergent for $p = 1$ when there are no zero residuals at the solution. A similar method for the l_∞ problem is given in Coleman and Li [181, 1992].

Madsen and Nielsen [556, 1993] develop a finite algorithm for l_1 estimation based on smoothing. At each iteration the nondifferentiable function $\psi(x)$, $p = 1$, in (4.5.6) is replaced by the smooth Huber function (4.5.10) for some threshold parameter γ. The parameter γ is successively reduced until, when it is small enough, the l_1 solution can be detected. In comparison with the simplex-type method of Barrodale and Roberts [50, 1973], their implementation is significantly faster.

4.6. Total Least Squares

4.6.1. Errors-in-variables models.

In the standard linear model (see Section 1.1.3) one assumes that the vector $b \in \mathbf{R}^m$ is related to the unknown parameter vector $x \in \mathbf{R}^n$ by a linear relation

$$(4.6.1) \qquad Ax = b + r,$$

where $A \in \mathbf{R}^{m \times n}$ is an exactly known matrix and r a vector of random errors which are uncorrelated and have zero means and the same variance, i.e., $\mathcal{E}(r) = 0$, $\mathcal{V}(r) = \sigma^2 I$. These assumptions are frequently unrealistic since sampling or modeling errors often also affect the matrix A. We now consider a more general linear model, where random errors also occur in the data matrix A. In the **errors-in-variables model** one assumes a linear relation

$$(4.6.2) \qquad (A + E)x = b + r,$$

where the rows of the error matrix (E, r) are independently and identically distributed with zero mean and the same variance. If this assumption is not satisfied, the data (A, b) can be premultiplied in advance by an appropriate matrix $D \in \mathbf{R}^{m \times m}$, such that the data (DA, Db) satisfy the assumptions.

The errors-in-variable model, also known as latent root regression, has been used in statistics for a long time. Orthogonal distance regression with $n = 1$ was already studied in 1878 by Adcock [2, 1878]. (This special case is treated in Section 4.6.6.) Much later, multivariate problems were treated in the statistical

4.6. TOTAL LEAST SQUARES

literature. For more on the history, see [791, 1989]. The first numerically stable algorithm for the model (4.6.2) was given by Golub and Van Loan [388, 1980], who also coined the name **total least squares** (TLS) problem. They analyzed the problem in terms of the SVD of the augmented matrix (A, b). A very complete survey of the theoretical and computational aspects of the TLS problem is given in the excellent monograph by Van Huffel and Vandewalle [792, 1991]. They claim that in typical applications, gains of 10–15% in accuracy can be obtained by using TLS instead of standard least squares methods.

To compute estimates of the true but unknown parameters x in the model (4.6.2) one solves the TLS problem

$$(4.6.3) \qquad \min_{E,r} \|(E, \ r)\|_F, \qquad (A + E)x = b + r,$$

where $\| \cdot \|_F$ denotes the Frobenius matrix norm. We note that the constraint in (4.6.3) implies that $b + r \in \mathcal{R}(A + E)$. If a minimizing $(E, \ r)$ has been found for the problem (4.6.3) then any x satisfying $(A + E)x = b + r$ is said to solve the TLS problem.

In [791, 1989] the more general model is also considered, where the covariance matrix C of the rows are known, up to a factor of proportionality. A simple special case is when the rows of A have a common covariance matrix $\sigma^2 I$, but the variance of the errors in the components of b is $(\sigma/\theta)^2$. This leads to the TLS problem

$$\min_{E,r} \|(E, \ \theta r)\|_F, \qquad (A + E)x = b + r.$$

When θ is small larger perturbations in b will be allowed, and in the limit $\theta \to 0$ we get the ordinary least squares solution.

4.6.2. Total least squares problem by SVD. We first note that the constraint in (4.6.3) can be written

$$(4.6.4) \qquad (\tilde{A}, \ \tilde{b}) \begin{pmatrix} x \\ -1 \end{pmatrix} = 0, \qquad \tilde{A} = A + E, \qquad \tilde{b} = b + r.$$

This shows that the TLS solution has the property that the matrix $(A+E, \ b+r)$ is rank deficient and that $(x, -1)^T$ is a right singular vector corresponding to a zero singular value. Hence the TLS problem involves finding a perturbation matrix $(E, \ r)$ having minimal Frobenius norm, which lowers the rank of the matrix (A, b).

We now show that the TLS problem can be analyzed in terms of the SVD of $(A, \ b)$. Let

$$(A, \ b) = U\Sigma V^T, \qquad \Sigma = \text{diag}(\sigma_1, \ldots, \sigma_{n+1}),$$

where $U^T U = I_m$, $V^T V = I_{n+1}$, and

$$\sigma_1 \geq \sigma_2 \geq \cdots \geq \sigma_{n+1} \geq 0$$

are the singular values of (A, b). By Theorem 1.2.9 the singular values of A,
$$\hat{\sigma}_1 \geq \hat{\sigma}_2 \geq \cdots \geq \hat{\sigma}_n \geq 0,$$
interlace those of (A, b), i.e.,

(4.6.5) $$\sigma_1 \geq \hat{\sigma}_1 \geq \sigma_2 \geq \cdots \geq \sigma_n \geq \hat{\sigma}_n \geq \sigma_{n+1}.$$

Assume that $\text{rank}(A) = n$, or equivalently that $\hat{\sigma}_n > 0$. If $\sigma_{n+1} = 0$, then it follows that $b \in R(A)$. In this case the original system of equations $Ax = b$ is compatible, and we can take $(E, r) = 0$. If $\sigma_{n+1} > 0$, then $b \notin R(A)$ and from the Eckhart and Young theorem (see Remark 1.2.1 to Theorem 1.2.3) we have

(4.6.6) $$\min_{\text{rank}(A+E,\ b+r)<n+1} \|(E,\ r)\|_F = \sigma_{n+1} > 0.$$

If it holds that
$$\sigma_k > \sigma_{k+1} = \cdots = \sigma_{n+1}, \qquad k \leq n,$$
then the minimum is attained for any rank one perturbation of the form

(4.6.7) $$(E,\ r) = -(A,\ b)vv^T, \qquad v \in S = \text{span}[v_{k+1}, \ldots, v_{n+1}],$$

where $v_{k+1}, \ldots, v_{n+1}$ are right singular vectors corresponding to $\sigma_{k+1}, \ldots, \sigma_{n+1}$.

If $e_{n+1} = (0, \ldots, 0, 1)^T$ is orthogonal to S then the TLS problem has no solution. Assume that we can find a unit vector v in S whose $(n+1)$st component γ is nonzero. Then with

(4.6.8) $$v = \begin{pmatrix} z \\ \gamma \end{pmatrix} = -\gamma \begin{pmatrix} x \\ -1 \end{pmatrix}, \qquad x = -\gamma^{-1}z,$$

we have
$$-\gamma(A+E,\ b+r)\begin{pmatrix} x \\ -1 \end{pmatrix} = (A+E,\ b+r)v = (A,\ b)(I - vv^T)v = 0.$$

Hence $(A+E)x = b+r$, which shows that x solves the TLS problem. Using (4.6.8) the minimizing perturbation can be written
$$(E,\ r) = -\gamma^2 (A,\ b)\begin{pmatrix} x \\ -1 \end{pmatrix}(x^T, -1) = \gamma^2(\hat{r}x^T, -\hat{r}),$$

where
(4.6.9) $$\hat{r} = b - Ax, \qquad \gamma^2 = (1 + \|x\|_2^2)^{-1}.$$

When σ_{n+1} is a repeated singular value, i.e., $k < n$ in (4.6.6), the TLS problem may have many solutions. In this case, a unique **minimum norm** TLS solution can be determined as follows. Let Q be an orthogonal matrix of order $n - k + 1$ such that

(4.6.10) $$[v_{k+1}, \ldots, v_{n+1}]Q = \begin{pmatrix} Y & z \\ 0 & \gamma \end{pmatrix} \begin{matrix} \}n \\ \}1 \end{matrix},$$

$\gamma \neq 0$. If we set $x = -\gamma^{-1}z$ then it is easy to show that all other solutions to the TLS problem have larger norms. Here Q can be taken as the Householder transformation which zeros all leading elements in the last row.

We now introduce a sufficient condition for the TLS problem (4.6.3) to have a unique solution.

4.6. TOTAL LEAST SQUARES

THEOREM 4.6.1. *Let the singular values of A be $\hat{\sigma}_1 \geq \hat{\sigma}_2 \geq \cdots \geq \hat{\sigma}_n > 0$. If $\hat{\sigma}_n > \sigma_{n+1}$ then the TLS problem has a unique solution.*

Proof. The interlacing property (4.6.5) implies that if $\hat{\sigma}_n > \sigma_{n+1}$ then σ_{n+1} is not a repeated singular value of (A, b) and v_{n+1} is unique up to a factor ± 1. Assume now that $v_{n+1} = (y\ 0)^T$. Then $(A, b)v_{n+1} = Ay = \sigma_{n+1}u_{n+1}$, where $\sigma_{n+1} = \|Ay\|_2 \geq \hat{\sigma}_n$. But by the interlacing property this implies that $\hat{\sigma}_n = \sigma_{n+1}$, which is a contradiction. Hence v_{n+1} must have a nonzero last component and a TLS solution exists. ∎

When all vectors in $S = \mathrm{span}[v_{k+1}, \ldots, v_{n+1}]$ have a zero last component, $v_{n+1,j} = 0$, $j = k+1, \ldots, n+1$, the problem is called **nongeneric** and the TLS problem fails to have a solution. Two different such cases may occur.

If $\mathrm{rank}(A) < n$ then $\hat{\sigma}_n = \sigma_{n+1} = 0$, and the TLS problem has a solution only in the trivial case when $b \in \mathcal{R}(A)$. Consider the example

$$A = \begin{pmatrix} 1 & 1 \\ 0 & 0 \end{pmatrix}, \qquad b = \begin{pmatrix} 1 \\ 1 \end{pmatrix}.$$

Here the singular vector corresponding to $\sigma_3 = 0$, which is $v_3 = (1, -1, 0)/\sqrt{2}$, only gives information about the rank deficiency in A. Note also that since $b \in \mathcal{R}(A + E)$ for $E = \mathrm{diag}(0, \epsilon)$ and any $\epsilon \neq 0$, there does not exist a smallest value of $\|(E, r)\|_F$. The case when $\mathrm{rank}(A) < n$ can be treated by adding the extra condition

$$\mathrm{rank}(A + E) \leq \mathrm{rank}(A)$$

to the problem (4.6.3). Alternatively, we can select a subset of linearly independent columns A_1 of A, and solve a reduced TLS problem with data (A_1, b). In the example above we can take $A_1 = a_1$, the first column in A. More generally, we can use the technique of subset selection described in Section 2.7.3.

Another nongeneric case can occur when the set of data is highly conflicting. Let, for example, $n = 1$ and

$$(A,\ b) = \begin{pmatrix} 1 & 0 \\ 0 & 2 \end{pmatrix}.$$

Here $(A,\ b)$ is already in diagonal form and has a unique smallest singular value equal to $\sigma_2 = 1$. Clearly $v_{22} = 0$ since $v_2 = e_1$ is orthogonal to e_2. The unique minimizing $(E,\ r)$ gives

$$(A + E, b + r) = \begin{pmatrix} 0 & 0 \\ 0 & 2 \end{pmatrix},$$

so the perturbed system is not compatible. However, an arbitrary small perturbation ϵ in the (2,1) element will give a compatible system with large solution $x = 2/\epsilon$. Hence in this case there is no solution to the TLS problem.

4.6.3. Relationship to the least squares solution. We now consider the conditioning of the TLS problem and its relationship to the least squares problem. To ensure unique solutions to both the TLS and the least squares problems we assume that $\hat\sigma_n > \sigma_{n+1}$ and we denote those solutions by x_{TLS} and x_{LS}, respectively. The vector $(x_{TLS}, -1)^T$ is an eigenvector of $(A, b)^T (A, b)$ with the associated eigenvalue σ_{n+1}^2, i.e.,

$$\begin{pmatrix} A^T A & A^T b \\ b^T A & b^T b \end{pmatrix} \begin{pmatrix} x_{TLS} \\ -1 \end{pmatrix} = \sigma_{n+1}^2 \begin{pmatrix} x_{TLS} \\ -1 \end{pmatrix}.$$

As pointed out by Golub and Van Loan [388, 1980], the first block row of this equation can be written

$$(4.6.11) \qquad (A^T A - \sigma_{n+1}^2 I) x_{TLS} = A^T b.$$

Since here a positive multiple of the unit matrix is *subtracted* from $A^T A$ the TLS problem is a *deregularization* of the least squares problem; cf. Section 2.7.2. Since

$$(4.6.12) \qquad \kappa(A^T A - \sigma_{n+1}^2 I) = \frac{\hat\sigma_1^2 - \sigma_{n+1}^2}{\hat\sigma_n^2 - \sigma_{n+1}^2} > \frac{\hat\sigma_1^2}{\hat\sigma_n^2} = \kappa(A^T A)$$

the TLS problem is always *worse conditioned* than the LS problem.

The least squares solution satisfies the normal equations $A^T A x_{LS} = A^T b$, and thus

$$(4.6.13) \qquad (A^T A - \sigma_{n+1}^2 I) x_{LS} = A^T b - \sigma_{n+1}^2 x_{LS}.$$

Subtracting this from equation (4.6.11) and solving for the difference we obtain

$$x_{TLS} - x_{LS} = \sigma_{n+1}^2 (A^T A - \sigma_{n+1}^2 I)^{-1} x_{LS}.$$

Taking norms and using (4.6.9) we obtain

$$\frac{\|x_{TLS} - x_{LS}\|_2}{\|x_{LS}\|_2} \leq \frac{\sigma_{n+1}^2}{\hat\sigma_n^2 - \sigma_{n+1}^2} = O(\|(E,\, r)\|_F^2),$$

i.e., x_{TLS} and x_{LS} agree up to second-order terms in the error. This was first shown by van der Sluis and Veltkamp [784, 1979]; see also Stewart [743, 1984]. Stewart also proves that up to terms of second order in the error (E, r) the estimate x_{TLS} is insensitive to column scalings of A, or more generally to linear transformations of the variables,

$$x := T^{-1} x, \qquad A := AT.$$

4.6. TOTAL LEAST SQUARES

4.6.4. Multiple right-hand sides. There are several generalizations of the TLS problem. Golub and Van Loan [389, 1989, Sec. 12.3] consider the TLS problem with multiple right-hand sides

$$(4.6.14) \qquad \min_{E,\, F} \|(E,\, F)\|_F, \qquad (A + E)X = B + F,$$

where $B \in \mathbf{R}^{m \times d}$. Writing this,

$$(A + E,\ B + F) \begin{pmatrix} X \\ -I_d \end{pmatrix} = 0,$$

it follows that we now seek a perturbation $(E,\ F)$ that reduces the rank of the matrix $(A,\ B)$ by d. We call this a multidimensional TLS problem. As remarked before, for this problem to be meaningful the rows of the error matrix $(E,\ F)$ should be independently and identically distributed with zero mean and the same variance.

We remark that the multidimensional problem is different from solving d one-dimensional TLS problems with right-hand sides $b_1, \ldots, b_d$. This is because in the multidimensional problem we require that the matrix A be similarly perturbed for all right-hand sides. This is in contrast to the usual least squares solution and may lead to improved predicted power of the TLS solution.

The solution to the TLS problem with multiple right-hand sides can be expressed in terms of the SVD

$$(4.6.15) \qquad (A,\ B) = (U_1,\ U_2)\operatorname{diag}(\Sigma_1,\ \Sigma_2) \begin{pmatrix} V_1^T \\ V_2^T \end{pmatrix},$$

where U, Σ, and V are partitioned conformally with $(A,\ B)$. The minimizing perturbation is given by

$$(E,\ F) = -U_2 \Sigma_2 V_2^T = -(A,\ B) V_2 V_2^T,$$

for which

$$(A + E,\ B + F) V_2 = 0, \qquad V_2 = \begin{pmatrix} V_{12} \\ V_{22} \end{pmatrix}.$$

Let the singular values of A be $\hat{\sigma}_i$, $i = 1, \ldots, n$. If $\hat{\sigma}_n > \sigma_{n+1}$ then it can be shown that the matrix $V_{22} \in \mathbf{R}^{d \times d}$ is nonsingular, and hence the unique solution to the TLS problem is given by

$$X = -V_{12} V_{22}^{-1};$$

see Golub and Van Loan [389, 1989, pp. 577–578]. This condition also ensures that $\sigma_n > \sigma_{n+1}$.

Otherwise assume that $\sigma_k > \sigma_{k+1} = \cdots = \sigma_{n+1}$, $k \leq n$, and let Q be a product of Householder transformations such that

$$(v_{k+1}, \ldots, v_{n+d}) Q = \begin{pmatrix} Y & Z \\ 0 & \Gamma \end{pmatrix},$$

where $\Gamma \in \mathbf{R}^{d \times d}$ is upper triangular. If Γ is nonsingular, then the TLS solution of minimum norm is given by

$$X = -Z\Gamma^{-1}.$$

We remark that if $m > 5(n + d)/3$ then in general it saves operations to transform the TLS problem into upper triangular form. Let Q be an orthogonal matrix such that

$$Q^T(A, B) = \begin{pmatrix} R \\ 0 \end{pmatrix}, \quad R = \begin{pmatrix} R_{11} & C_{12} \\ 0 & C_{22} \end{pmatrix} \in \mathbf{R}^{(n+d) \times (n+d)},$$

where $R_{11} \in \mathbf{R}^{n \times n}$ and $C_{22} \in \mathbf{R}^{d \times d}$ are upper triangular. Since R has the same singular values and right singular vectors as (A, B) we can solve the TLS problem by computing the SVD of R.

In the TLS algorithm one requires only the computation of a basis of the right singular subspace of (A, B) corresponding to its smallest singular values. For this case a **partial SVD** (PSVD) algorithm is given by Van Huffel, Vandewalle, and Haegemans [793, 1987], which can be up to three times faster than the standard SVD algorithm. This algorithm can also be used for solving homogeneous linear systems $Ax = 0$. Fortran codes for the TLS problem are available via netlib; see Dongarra and Grosse [231, 1987].

The SVD is the classical tool for solving TLS problems. However, the SVD is computationally expensive and Van Huffel and Zha [795, 1993] have developed a more efficient algorithm, which uses instead a rank revealing complete orthogonal decomposition.

4.6.5. Generalized TLS problems. In many parameter estimation problems, some of the columns are known exactly. In Van Huffel and Vandewalle [791, 1989] the TLS problem and the TLS algorithm are generalized for this case. It is no restriction to assume that the error-free columns are in leading positions in A. In the multivariate version of this **mixed LS-TLS problem** one has a linear relation

$$(A_1, A_2 + E_2)X = B + F, \quad A_1 \in \mathbf{R}^{m \times n_1},$$

where $A = (A_1, A_2) \in \mathbf{R}^{m \times n}$, $n = n_1 + n_2$. It is assumed that the rows of the errors (E_2, F) are independently and identically distributed with zero mean and the same variance. The mixed LS-TLS problem can then be expressed

(4.6.16) $$\min_{E_2, F} \|(E_2, F)\|_F, \quad (A_1, A_2 + E_2)X = B + F.$$

When A_2 is empty, this reduces to solving an ordinary least squares problem. When A_1 is empty this is the standard TLS problem. Hence this mixed problem includes both extreme cases.

Using a generalization of the Eckhart–Young Theorem 1.2.3, Golub, Hoffman, and Stewart [369, 1987] proved that the solution can be obtained by computing

4.6. TOTAL LEAST SQUARES

a QR factorization of A and then solving a TLS problem of reduced dimension. The following algorithm is based on that given by Van Huffel and Vandewalle [792, 1991, Sec. 3.6.3].

ALGORITHM 4.6.1. SOLVING MIXED LS-TLS PROBLEMS. Let $A = (A_1, A_2) \in \mathbf{R}^{m \times n}$, $n = n_1 + n_2$, $m \geq n$, and $B \in \mathbf{R}^{m \times d}$. Assume that the columns of A_1 are linearly independent. Then the following algorithm solves the mixed LS-TLS problem (4.6.16).

Step 1. Compute the QR decomposition

$$(A_1, A_2, B) = Q \begin{pmatrix} R \\ 0 \end{pmatrix}, \quad R = \begin{pmatrix} R_{11} & R_{12} \\ 0 & R_{22} \end{pmatrix},$$

where Q is orthogonal, and $R_{11} \in \mathbf{R}^{n_1 \times n_1}$, $R_{22} \in \mathbf{R}^{(n_2+d) \times (n_2+d)}$ are upper triangular. If $n_1 = n$, then the solution X is obtained by solving $R_{11} X = R_{12}$ (usual least squares); otherwise continue (solve a reduced TLS problem).

Step 2. Compute the SVD of R_{22}

$$R_{22} = U \Sigma V^T, \quad \Sigma = \mathrm{diag}\,(\sigma_1, \ldots, \sigma_{n_2+d}),$$

where the singular values are ordered in decreasing order of magnitude.

Step 3a. Determine $k \leq n_2$ such that

$$\sigma_k > \sigma_{k+1} = \cdots = \sigma_{n_2+d} = 0,$$

and set $V_{22} = (v_{k+1}, \ldots, v_{n_2+d})$. If $n_1 > 0$ then compute V_2 by back-substitution from

$$R_{11} V_{12} = -R_{12} V_{22}, \quad V_2 = \begin{pmatrix} V_{12} \\ V_{22} \end{pmatrix},$$

else set $V_2 = V_{22}$.

Step 3b. Perform Householder transformations such that

$$V_2 Q = \begin{pmatrix} Y & Z \\ 0 & \Gamma \end{pmatrix},$$

where $\Gamma \in \mathbf{R}^{d \times d}$ is upper triangular. If Γ is nonsingular then the solution X is obtained from

$$X \Gamma = -Z.$$

Otherwise the TLS problem is nongeneric and has no solution.

A more general version of this algorithm is given in [791, 1989], where it is assumed that the covariance matrix $C = \mathcal{E}(\Delta^T \Delta)$ of the errors $\Delta \in \mathbf{R}^{m \times (n_2+d)}$ is known. This algorithm uses the GSVD of the matrix pair (R_{22}, R_C), where $C = R_C^T R_C$, i.e., R_C is the Cholesky factor of C.

An even more general problem is the **restricted TLS problem** introduced by Van Huffel and Zha in [794, 1991]. Here it is assumed that

$$(A, B) = (A_0, B_0) + E^*,$$

where A_0, B_0 are the error-free data and E^* represents a restricted perturbation of the form
$$E^* = DEC, \quad D \in \mathbf{R}^{m \times p}, \quad C \in \mathbf{R}^{q \times (n+d)},$$
where C and D are known and $E \in \mathbf{R}^{p \times q}$ is unknown. The problem then considered is that of finding a matrix $(\Delta \hat{A}, \Delta \hat{B}) = D\hat{E}C$ such that
$$\mathcal{R}(B + \Delta \hat{B}) \subset \mathcal{R}(A + \Delta \hat{A})$$
and $\|\hat{E}\|_F$ is minimal.

De Moor [208, 1993] studies more general structured and weighted total least squares problems. He shows that applications in systems and control include total least squares with relative errors and/or fixed elements. These problems can be solved via a nonlinear GSVD.

We finally mention that Watson [820, 1984] has considered the approximation problem where $\|(E, r)\|_p$ is minimized for the class of matrix norms defined by
$$\|B\|_p = \Big(\sum_{i,j} \|b_{ij}\|^p\Big)^{1/p}, \quad 1 \leq p < \infty.$$

4.6.6. Linear orthogonal distance regression. A geometrical interpretation of the TLS problem (4.6.3) can be obtained as follows; cf. Golub and Van Loan [389, 1989, Sec. 12.3.3]. The quantity
$$d_i^2(x) = \|a_i^T x - b_i\|^2 / (\|x\|_2^2 + 1),$$
where a_i^T is the ith row of A, is the square of the orthogonal distance from the point $\begin{pmatrix} a_i \\ b_i \end{pmatrix} \in \mathbf{R}^{n+1}$ to the plane through the origin
$$P_x = \left\{ \begin{pmatrix} a \\ b \end{pmatrix} \mid a \in \mathbf{R}^n, b \in \mathbf{R}, \quad b = a^T x \right\}.$$
Hence the TLS solution minimizes the sum of squares orthogonal distances
$$\sum_{i=1}^m d_i^2(x),$$
and therefore is a special case of orthogonal distance regression; see Section 9.4.3.

We now consider a slightly more general orthogonal distance regression problem. Let $y_i \in \mathbf{R}^n$, $i = 1, 2, \ldots, m$, be $m > n$ given points. We want to determine a hyperplane in $\mathbf{R}^n$,
$$M = \{c^T z = h \mid c \in \mathbf{R}^n, \ \|c\|_2 = 1, \ h \in \mathbf{R}\},$$
such that the sum of squares of the orthogonal distances from the given points to M is minimized. For any hyperplane M the orthogonal projections of the points y_i onto M are given by
(4.6.17) $$z_i = y_i - (c^T y_i - h)c.$$

4.6. TOTAL LEAST SQUARES

It is readily verified that the point z_i lies on M and the residual $z_i - y_i$ is parallel to c and hence orthogonal to M. It follows that the problem is equivalent to minimizing

$$\psi(c, h) = \sum_{i=1}^{m}(c^T y_i - h)^2, \quad \text{subject to} \quad \|c\|_2 = 1.$$

This problem can be written in matrix form

(4.6.18) $$\min_{c,h} \left\| (Y^T, -e) \begin{pmatrix} c \\ h \end{pmatrix} \right\|, \quad \|c\|_2 = 1,$$

where

$$Y = (y_1, \ldots, y_m), \quad e = (1, \ldots, 1)^T.$$

For fixed c, this expression is minimized when the residual $Y^T c - he$ is orthogonal to e, i.e., $e^T(Y^T c - he) = 0$. Since $e^T e = m$

(4.6.19) $$h = c^T Y e / m = c^T \bar{y},$$

where

$$\bar{y} = \sum_{i=1}^{m} y_i / m$$

is the mean value of the given points. This shows that h is determined by the condition that the mean value $\bar{y}$ lies on the optimal plane M.

We now subtract the mean value $\bar{y}$ from the points and form the matrix

$$\tilde{Y} = (\tilde{y}_1, \ldots, \tilde{y}_m), \quad \tilde{y}_i = y_i - \bar{y}, \quad i = 1, \ldots, m.$$

Since by (4.6.19)

$$(Y^T, -e) \begin{pmatrix} c \\ h \end{pmatrix} = Y^T c - ey^T c = (Y^T - e\bar{y}^T)c = \tilde{Y}^T c,$$

we have reduced the problem (4.6.18) to

(4.6.20) $$\min \|\tilde{Y}^T c\|_2, \quad \text{subject to} \quad \|c\|_2 = 1.$$

By the min-max characterization, Theorem 1.2.6, the solution to (4.6.20) is $c = u_n$, where u_n is the left singular vector of $\tilde{Y}$ corresponding to the smallest singular value σ_n. The minimum is given by

$$\sum_{i=1}^{m}(c^T y_i - h)^2 = \sigma_n.$$

The fitted points z_i can be obtained by

(4.6.21) $$z_i = \tilde{y}_i - (c^T y_i)c + \bar{y},$$

i.e., by first orthogonalizing the shifted points $\tilde{y}_i$ against $c = u_n$, and then adding the mean value back.

As an example we consider the problem of fitting by orthogonal regression m pairs of points $(x_i, y_i) \in \mathbf{R}^2$, $i = 1, \ldots, m$, to a straight line

$$cx + sy = h, \qquad c^2 + s^2 = 1.$$

We compute the SVD of the matrix

$$\tilde{Y} = \begin{pmatrix} \tilde{x}_1 & \tilde{x}_2 & \cdots & \tilde{x}_m \\ \tilde{y}_1 & \tilde{y}_2 & \cdots & \tilde{y}_m \end{pmatrix} = (u_1 \; u_2) \begin{pmatrix} \sigma_1 & 0 \\ 0 & \sigma_2 \end{pmatrix} \begin{pmatrix} v_1^T \\ v_2^T \end{pmatrix},$$

where $\sigma_2 \geq \sigma_1 \geq 0$. Then the coefficients in the equation of the straight line are given by

$$(c \; s) = u_2^T, \qquad h = u_2^T \begin{pmatrix} \bar{x} \\ \bar{y} \end{pmatrix}.$$

If $\sigma_2 = 0$ but $\sigma_1 > 0$ the matrix $\tilde{Y}$ has rank one. In this case the given points lie on a straight line. If $\sigma_1 = \sigma_2 = 0$, then $\tilde{Y} = 0$, and $x_i = \bar{x}$, $y_i = \bar{y}$ for all $i = 1, \ldots, m$. Note that u_2 is uniquely determined if and only if $\sigma_1 \neq \sigma_2$. It is left to the reader to discuss the case $\sigma_1 = \sigma_2 \neq 0$!

Note that in contrast to the TLS problem the orthogonal regression problem always has a solution. The solution is unique when $\sigma_{n-1} \neq \sigma_n$, and the minimum sum of squares equals σ_n^2. We have $\sigma_n = 0$, if and only if the given points y_i, $i = 1, \ldots, m$, all lie on the hyperplane M. In the extreme case, all points coincide. Then $\tilde{Y} = 0$, and any plane going through $\bar{y}$ is a solution.

The above method solves the problem of fitting an $(n-1)$-dimensional manifold to a given set of points in R. It is readily generalized to the fitting of an $(n-p)$-dimensional manifold by orthogonalizing the shifted points y against the p left singular vectors of Y corresponding to the p smallest singular values.

Chapter 5
Constrained Least Squares Problems

5.1. Linear Equality Constraints

5.1.1. Introduction. In this section we consider least squares problems in which the unknowns are required to satisfy a subsystem of linear equations exactly. Such problems arise, e.g., in curve and surface fitting where the curve may be required to interpolate certain data points. Least squares problems with inequality constraints are treated in Section 5.2, where it is shown how these can be reduced to solving a sequence of problems with equality constraints.

PROBLEM LSE. Least Squares with Equality Constraints. Given matrices $A \in \mathbf{R}^{m \times n}$ and $B \in \mathbf{R}^{p \times n}$ find a vector $x \in \mathbf{R}^n$ which solves

$$(5.1.1) \qquad \min_x \|Ax - b\|_2 \quad \text{subject to} \quad Bx = d.$$

Problem LSE obviously has a solution if and only if the linear system $Bx = d$ is consistent. A sufficient condition for this is that B has linearly independent rows, i.e., $\text{rank}(B) = p$, since then $Bx = d$ is consistent for any right-hand side d.

A solution to (5.1.1) is unique if and only if

$$(5.1.2) \qquad \text{rank}(B) = p \quad \text{and} \quad \text{rank}\begin{pmatrix} A \\ B \end{pmatrix} = n.$$

The second condition in (5.1.2) is equivalent to

$$(5.1.3) \qquad \mathcal{N}(A) \cap \mathcal{N}(B) = \{0\},$$

i.e., the nullspaces of A and B intersect only trivially. If the condition (5.1.3) is not satisfied, then there is a vector $z \neq 0$ such that $Az = Bz = 0$. Then if x solves (5.1.1) $x + z$ is a different solution and it follows that the condition (5.1.3) is necessary for uniqueness. A constructive proof that (5.1.2) is sufficient to ensure that the problem (5.1.1) has a unique solution is given in Section 5.1.3. In the case of nonuniqueness there always is a unique solution of minimum norm.

A robust algorithm for problem LSE should check for possible inconsistency of the constraint equations $Bx = d$. If it is not known a priori that the constraints are consistent, then (5.1.1), as suggested by Leringe and Wedin [525, 1970], may be reformulated as a **sequential least squares problem**

$$(5.1.4) \qquad \min_{x \in S} \|Ax - b\|_2, \qquad S = \{x \mid \|Bx - d\|_2 = \min\}.$$

This problem always has a unique solution of minimum norm. Most of the methods described in the following for solving problem LSE can be adapted to solve (5.1.4) with little modification.

The most natural way to solve problem LSE is to derive an equivalent unconstrained least squares problem of lower dimension. There are basically two different ways to perform this reduction: **direct elimination** and the **nullspace method**. These methods are described below.

5.1.2. Method of direct elimination. In the method of direct elimination we start by reducing the matrix B to upper trapezoidal form. It is essential that column pivoting be used in this step. In order to be able to also solve the more general problem (5.1.4) we will compute a QR decomposition of B. By Theorem 1.3.4 there is an orthogonal matrix $Q_B \in \mathbf{R}^{p \times p}$ and a permutation matrix Π_B such that

$$(5.1.5) \qquad Q_B^T B \Pi_B = \begin{pmatrix} R_{11} & R_{12} \\ 0 & 0 \end{pmatrix} \begin{matrix} \}r \\ \}p-r \end{matrix},$$

where $r = \mathrm{rank}(B) \le p$ and R_{11} is upper triangular and nonsingular.

If we also apply Q_B^T to the vector d the constraints become

$$(5.1.6) \qquad (R_{11}, R_{12})\bar{x} = \bar{d}_1, \qquad \bar{d} = Q_B^T d = \begin{pmatrix} \bar{d}_1 \\ \bar{d}_2 \end{pmatrix},$$

where $\bar{x} = \Pi_B^T x$ and $\bar{d}_2 = 0$ if and only if the constraints are consistent.

We also apply the permutation Π_B to the columns of A and partition the resulting matrix conformably with (5.1.5):

$$(5.1.7) \qquad Ax - b = \bar{A}\bar{x} - b = (\bar{A}_1, \bar{A}_2)\begin{pmatrix} \bar{x}_1 \\ \bar{x}_2 \end{pmatrix} - b,$$

where $\bar{A} = A\Pi_B$. We now eliminate the variables $\bar{x}_1$ from (5.1.7) using (5.1.6). Substituting $\bar{x}_1 = R_{11}^{-1}(\bar{d}_1 - R_{12}\bar{x}_2)$ we get

$$Ax - b = \hat{A}_2 x_2 - \hat{b},$$

where
$$(5.1.8) \qquad \hat{A}_2 = \bar{A}_2 - \bar{A}_1 R_{11}^{-1} R_{12}, \qquad \hat{b} = b - \bar{A}_1 R_{11}^{-1} \bar{d}_1.$$

Hence the reduced unconstrained least squares problem

$$(5.1.9) \qquad \min_{\bar{x}_2} \|\hat{A}_2 \bar{x}_2 - \hat{b}\|_2, \qquad \hat{A}_2 \in \mathbf{R}^{m \times (n-r)},$$

is equivalent to the original problem LSE.

5.1. LINEAR EQUALITY CONSTRAINTS

The solution to the unconstrained problem (5.1.9) can be obtained from the QR decomposition of $\hat{A}_2$. We now show that if the condition (5.1.2) holds then rank($\hat{A}_2$) = $n - r$ and (5.1.9) has a unique solution. If rank($\hat{A}_2$) < $n - r$ then there is a vector $v \neq 0$ such that

$$\hat{A}_2 v = \bar{A}_2 v - \bar{A}_1 R_{11}^{-1} R_{22} v = 0.$$

If we let $u = -R_{11}^{-1} R_{12} v$, then

$$R_{11} u + R_{12} v = 0, \qquad \bar{A}_1 u + \bar{A}_2 v = 0.$$

Hence the vector

$$w = \Pi_B \begin{pmatrix} u \\ v \end{pmatrix} \neq 0$$

is a null vector to both B and A, and (5.1.2) does not hold.

If (5.1.2) holds then we can compute the QR decomposition

$$Q_A^T \hat{A}_2 = \begin{pmatrix} R_{22} \\ 0 \end{pmatrix}, \qquad Q_A^T b = \begin{pmatrix} c_1 \\ c_2 \end{pmatrix},$$

where $R_{22} \in \mathbf{R}^{(n-r) \times (n-r)}$ is upper triangular and nonsingular.

We then compute $\bar{x}$ from the triangular system

(5.1.10) $$\begin{pmatrix} R_{11} & R_{12} \\ 0 & R_{22} \end{pmatrix} \bar{x} = \begin{pmatrix} \bar{d}_1 \\ c_1 \end{pmatrix}.$$

Then $x = \Pi_B \bar{x}$ solves problem LSE.

The coding of the algorithm outlined above can be kept remarkably compact, as is illustrated by the Algol program given in Björck and Golub [110, 1967]. Note that the reduction in (5.1.8) can be interpreted as performing r steps of Gaussian elimination on the system

$$\begin{pmatrix} R_{11} & R_{12} \\ \bar{A}_1 & \bar{A}_2 \end{pmatrix} \begin{pmatrix} \bar{x}_1 \\ \bar{x}_2 \end{pmatrix} = \begin{pmatrix} \bar{d}_1 \\ b \end{pmatrix}.$$

REMARK 5.1.1. *The set of vectors* $x = \Pi_B \bar{x}$, *where* $\bar{x}$ *satisfies* (5.1.6) *is exactly the set of vectors which minimize* $\|Bx - d\|_2$. Thus, the algorithm outlined above actually solves the more general problem (5.1.4). If condition (5.1.2) is not satisfied, then the reduced problem (5.1.9) does not have a unique solution. Then column permutations are also needed in the QR decomposition of $\hat{A}_2$. In this case we can compute either a minimum norm solution or a basic solution to (5.1.4); see Section 1.3.3.

5.1.3. The nullspace method. We assume here that rank(B) = p. First compute an orthogonal matrix $Q_B \in \mathbf{R}^{n \times n}$ such that

(5.1.11) $$Q_B^T B^T = \begin{pmatrix} R_B \\ 0 \end{pmatrix},$$

where $R_B \in \mathbf{R}^{p \times p}$ is upper triangular and nonsingular. Let

$$Q_B = (Q_1, Q_2), \qquad Q_1 \in \mathbf{R}^{n \times p}, \qquad Q_2 \in \mathbf{R}^{n \times (n-p)}.$$

Then $\mathcal{N}(B) = \mathcal{R}(Q_2)$, i.e., Q_2 gives an orthogonal basis for the nullspace of B. Any vector $x \in \mathbf{R}^n$ which satisfies $Bx = d$ can then be represented as

(5.1.12) $$x = x_1 + Q_2 y_2, \qquad x_1 = B^\dagger d = Q_1 R_B^{-T} d.$$

Hence
$$Ax - b = Ax_1 + AQ_2 y_2 - b, \qquad y_2 \in \mathbf{R}^{n-p},$$

and it remains to solve the reduced system

(5.1.13) $$\min_{y_2} \|(AQ_2)y_2 - (b - Ax_1)\|_2.$$

Let y_2 be the minimum length solution to (5.1.13),

$$y_2 = (AQ_2)^\dagger (b - Ax_1),$$

and let x be defined by (5.1.12). Then, since $x_1 \perp Q_2 y_2$, it follows that

$$\|x\|_2^2 = \|x_1\|_2^2 + \|Q_2 y_2\|_2^2 = \|x_1\|_2^2 + \|y_2\|_2^2$$

and x is the minimum norm solution to problem LSE.

Now assume that the condition (5.1.2) is satisfied. Then the matrix

$$C = \begin{pmatrix} B \\ A \end{pmatrix} Q_B = \begin{pmatrix} R_B^T & 0 \\ AQ_1 & AQ_2 \end{pmatrix}$$

must have rank n. But then all columns in C must be linearly independent, and it follows that $\operatorname{rank}(AQ_2) = n - p$. Then we can compute the QR decomposition

$$Q_A^T (AQ_2) = \begin{pmatrix} R_A \\ 0 \end{pmatrix},$$

where R_A is upper triangular and nonsingular. The unique solution to (5.1.13) can then be computed from

(5.1.14) $$R_A y_2 = c_1, \qquad c = \begin{pmatrix} c_1 \\ c_2 \end{pmatrix} = Q_A^T (b - Ax_1),$$

and we finally obtain $x = x_1 + Q_2 y_2$, the unique solution to problem LSE.

The representation (5.1.12) of the solution x has been used as a basis for a perturbation theory by Leringe and Wedin [525, 1970], which generalizes the results given in Section 1.4.2. The corresponding bounds for problem LSE are more complicated, but show that the problem LSE is well-conditioned if $\kappa(B)$ and $\kappa(AQ_2)$ are small. It is important to note that these two condition numbers can be small even when $\kappa(A)$ is large. Any method which starts with minimizing $\|Ax - b\|_2$ will give inaccurate results in such a case. Eldén [268, 1980] has given a

5.1. Linear Equality Constraints

less complicated and more complete perturbation theory for problem LSE based on the concept of a weighted pseudoinverse. This theory is generalized to rank deficient problems in Wei [830, 833, 1992]. A perturbation theory based on the augmented system formulation is given by Wedin in [828, 1985].

The method of direct elimination and the nullspace method both have good numerical stability. In a numerical comparison by Leringe and Wedin [525, 1970] they gave almost identical results. The operation count for the method of direct elimination is slightly lower, because Gaussian elimination is used to derive the reduced unconstrained problem.

5.1.4. Problem LSE by generalized SVD. Following Van Loan [799, 1985] we now analyze problem LSE in terms of the generalized singular value decomposition (GSVD). For simplicity we assume that the conditions in (5.1.2) are satisfied. This ensures that the problem has a unique solution, and the GSVD can be written (see Section 4.2.2):

$$(5.1.15) \qquad U^T A = \begin{pmatrix} D_A \\ 0 \end{pmatrix} Z, \qquad V^T B = (D_B \; 0) Z,$$

where $D_A = \mathrm{diag}\,(\alpha_1, \ldots, \alpha_n)$, $D_B = \mathrm{diag}\,(\beta_1, \ldots, \beta_p)$, and

$$(5.1.16) \qquad 0 = \alpha_1 = \cdots = \alpha_q < \alpha_{q+1} \leq \cdots \leq \alpha_p < \alpha_{p+1} = \cdots = \alpha_n,$$
$$\beta_1 \geq \cdots \geq \beta_p > 0.$$

We can also assume, without loss of generality, that $\alpha_i^2 + \beta_i^2 = 1$, and

$$\sigma_i(Z) = \sigma_i \begin{pmatrix} A \\ B \end{pmatrix} > 0, \qquad i = 1, \ldots, n.$$

Using the GSVD the problem (5.1.1) is transformed into diagonal form:

$$(5.1.17) \qquad \min_y \left\| \begin{pmatrix} D_A \\ 0 \end{pmatrix} y - c \right\|_2^2 \quad \text{subject to} \quad (D_B \; 0) y = d,$$

where $Zx = y$, $c = U^T b$, $\tilde{d} = V^T d$. It is easily verified that (5.1.17) has the solution

$$(5.1.18) \qquad y_i = \begin{cases} \tilde{d}_i/\beta_i, & i = 1, \ldots, p, \\ c_i, & i = p+1, \ldots, n, \end{cases}$$

where we have used the fact that $\alpha_i = 1$, $i = p+1, \ldots, n$. Hence, letting $W = (w_1, \ldots, w_n) = Z^{-1}$, the solution x_{LSE} to (5.1.1) can be written

$$(5.1.19) \qquad x_{LSE} = \sum_{i=1}^{p} (\tilde{d}_i/\beta_i) w_i + \sum_{i=p+1}^{n} c_i w_i.$$

It is interesting to compute how much the residual $r_{LS} = b - Ax_{LS}$ increases as a result of the constraints $Bx = d$. A solution to the unconstrained problem to minimize $\|Ax - b\|_2$ is

$$x_{LS} = \sum_{i=q+1}^{p} (c_i/\alpha_i) w_i + \sum_{i=p+1}^{n} c_i w_i.$$

We have $r_{LSE} - r_{LS} = A(x_{LS} - x_{LSE})$, and using the relations $Ax_i = \alpha_i u_i$, $i = 1, \ldots, n$, we can show that

$$\text{(5.1.20)} \quad \|r_{LSE}\|_2^2 = \|r_{LS}\|_2^2 + \sum_{i=q+1}^{p} \rho_i^2, \quad \rho_i = c_i - (\alpha_i/\beta_i) v_i^T d.$$

5.1.5. The method of weighting. The method of weighting for solving problem LSE is based on the following simple observation. Assume that in a least squares problem we want some equations to be exactly satisfied. We can achieve that by giving these equations a large weight γ and solving the resulting unconstrained least squares problem. Hence, to solve (5.1.1), we would compute the solution $x(\gamma)$ to the problem

$$\text{(5.1.21)} \quad \min_x \left\| \begin{pmatrix} \gamma B \\ A \end{pmatrix} x - \begin{pmatrix} \gamma d \\ b \end{pmatrix} \right\|_2^2.$$

Note that if (5.1.2) holds, then (5.1.21) is a full rank least squares problem.

Updating methods for solving weighted problems were considered in Section 4.4.4. There it was shown (cf. (4.4.10)) that, provided that $\text{rank}(B) = p$ in (5.1.21), the residual $d - Bx(\gamma)$ will be proportional to γ^{-2} for large values of γ, and hence

$$\lim_{\gamma \to \infty} x(\gamma) = x_{LSE}.$$

A more general analysis of the problem (5.1.21) is given below in terms of the GSVD of A and B.

The method of weighting is attractive for its apparent simplicity. It allows the use of a subroutine or program for *unconstrained* least squares problems to be used to solve problem LSE. However, as was emphasized in Section 4.4, for large values of γ care must be exercised in the way (5.1.21) is solved, because then the matrix in (5.1.21) is poorly conditioned. In particular, except in special cases, the method of normal equations will fail for large values of γ.

Accurate solutions to (5.1.21) can be computed even for large values of γ using the QR decomposition methods described in Section 4.4.3. In particular, the self-scaling fast rotations developed by Anda and Park [15, 1996] can be used without any risk of overshooting the optimal weights γ. Indeed, consider the application of Givens rotations to a pair of weighted and unweighted rows in

$$\frac{1}{\sqrt{\gamma^2 a_{pi}^2 + a_{qi}^2}} \begin{pmatrix} \gamma a_{pi} & a_{qi} \\ -a_{qi} & \gamma a_{pi} \end{pmatrix} \begin{pmatrix} \gamma a_{pi} \\ a_{qi} \end{pmatrix} = \begin{pmatrix} \gamma a'_{pi} \\ 0 \end{pmatrix}.$$

As $\gamma \to \infty$, this becomes, after a row and column scaling,

$$\begin{pmatrix} 1 & 0 \\ -a_{qi}/a_{pi} & 1 \end{pmatrix} \begin{pmatrix} a_{pi} \\ a_{qi} \end{pmatrix} = \begin{pmatrix} a'_{pi} \\ 0 \end{pmatrix},$$

which is just a Gaussian elimination step in the method of direct elimination.

5.1. LINEAR EQUALITY CONSTRAINTS

It is important to note that it is essential that column permutations are used in the first p steps to ensure that the corresponding submatrix B_1 in

$$\begin{pmatrix} \gamma B \\ A \end{pmatrix} \Pi = \begin{pmatrix} \gamma B_1 & \gamma B_2 \\ A_1 & A_2 \end{pmatrix}$$

is well-conditioned. Powell and Reid [670, 1969] recommend that first the pivot column be selected by the standard column pivoting strategy (see Section 2.7.3) and then the largest element in the pivot column be permuted to the top before the Householder transformation is applied. However, it usually suffices in practice to order the constraint rows first, as is done in (5.1.21).

The following example from Van Loan [798, 1985] shows that column pivoting is necessary for accuracy.

EXAMPLE 5.1.1. Consider the problem $\min_x \|Ax - b\|_2$, subject to $Bx = d$, where

$$A = \begin{pmatrix} 1 & 1 & 1 \\ 1 & 3 & 1 \\ 1 & -1 & 1 \\ 1 & 1 & 1 \end{pmatrix}, \quad b = \begin{pmatrix} 1 \\ 2 \\ 3 \\ 4 \end{pmatrix},$$

$$B = \begin{pmatrix} 1 & 1 & 1 \\ 1 & 1 & -1 \end{pmatrix}, \quad d = \begin{pmatrix} 7 \\ 4 \end{pmatrix}.$$

This problem is well-conditioned and has the solution

$$x_{LSE} = \tfrac{1}{4}(23, -1, 6)^T.$$

The weighted problem (5.1.21) was solved in VAX double precision arithmetic, $u = 10^{-17}$, with $\gamma = 10^9 \approx u^{-\frac{1}{2}}$. With column pivoting, full double precision accuracy was obtained, whereas without column pivoting the error was of the order 10^{-9}. The trouble arises because the first two columns of the matrix B are linearly dependent. The submatrix B_1 consisting of the first and third columns of B is, however, well-conditioned. ∎

Van Loan [798, 1985] used the GSVD analysis in Section 5.1.4 to analyze the method of weighting. The solution $x(\gamma)$ to (5.1.21) satisfies the normal equations

$$(A^T A + \gamma^2 B^T B) x(\gamma) = A^T b + \gamma^2 B^T d.$$

Using (5.1.15)–(5.1.17) these can be transformed into

(5.1.22) $$\left(D_A^2 + \gamma^2 \begin{pmatrix} D_B^2 & 0 \\ 0 & 0 \end{pmatrix} \right) y(\gamma) = (D_A, 0) \tilde{b} + \gamma^2 \begin{pmatrix} D_B \\ 0 \end{pmatrix} \tilde{d},$$

where we have put $Zx(\gamma) = y(\gamma)$. From (5.1.22) we deduce that

$$y_i(\gamma) = \begin{cases} \dfrac{\alpha_i \tilde{b}_i + \gamma^2 \beta_i \tilde{d}_i}{\alpha_i^2 + \gamma^2 \beta_i^2}, & i = 1, \ldots, p, \\ \tilde{b}_i, & i = p+1, \ldots, n. \end{cases}$$

Hence from (5.1.18) we find that $y_i(\gamma) = y_i, i = 1, \ldots, q, p+1, \ldots, n$, and with ρ_i and μ_i defined by (5.1.20),

$$y_i(\gamma) - y_i = \frac{\mu_i^2}{\mu_i^2 + \gamma^2} \frac{\rho_i}{\alpha_i}, \quad i = q+1, \ldots, p.$$

This suggests that if μ_p is large then a large weight γ may be required for $x(\gamma)$ to well approximate the constrained solution.

A detailed analysis of the method of weighting has also been given by Lawson and Hanson [520, 1974, Sec. 22].

5.1.6. Solving LSE problems by updating. Problem LSE can also be solved by an updating technique. Assume that the conditions (5.1.2) are satisfied. Then the problem

(5.1.23) $$\min_x \left\| \begin{pmatrix} A \\ B \end{pmatrix} x - \begin{pmatrix} b \\ d \end{pmatrix} \right\|_2$$

has a unique solution x_u, which can be computed, e.g., by the QR decomposition of the matrix in (5.1.23). Let R be the R-factor of this matrix and compute the matrix $C = BR^{-1} \in \mathbf{R}^{p \times n}$ by back-substitution in $R^T C^T = B^T$. Then the solution to the constrained problems (5.1.1) is given by

$$x = x_u + R^{-1} w, \qquad w = C^T (CC^T)^{-1} (d - B x_u).$$

Since $BR^{-1} = C$, it holds that

$$Bx = Bx_u + BR^{-1}w = Bx_u + d - Bx_u = d,$$

so the constraints are satisfied exactly. However, for this approach to be stable it is necessary that R not be worse conditioned than the constrained problem.

Note that w is the solution to the minimum norm problem

$$\min \|w\|, \quad \text{subject to} \quad Cw = d - Bx_u,$$

and can also be computed from the QR decomposition of C; see Algorithm 2.4.6. If $p \ll n$ the cost of the updating step is small. This and similar updating techniques are especially useful in *sparse problems*; see also Section 6.7.3.

5.2. Linear Inequality Constraints

5.2.1. Classification of problems. This section is concerned with methods for linear least squares problems subject to different types of inequality constraints.

PROBLEM LSI.

(5.2.1) $$\min_x \|Ax - b\|_2 \quad \text{subject to} \quad l \leq Cx \leq u,$$

where $A \in \mathbf{R}^{m \times n}$ and $C \in \mathbf{R}^{p \times n}$, and the inequalities are to be interpreted componentwise.

5.2. Linear Inequality Constraints

If c_i^T denotes the ith row of the constraint matrix C then the constraints can also be written

$$l_i \leq c_i^T x \leq u_i, \quad i = 1, \ldots, p.$$

It is convenient to allow the elements $l_i = -\infty$ and $u_i = \infty$, which correspond to cases where the lower and upper bounds, respectively, are not present.

Note that if linear equality constraints are present, then these can be eliminated using one of the methods given in Section 5.1. Both the direct elimination method and the nullspace method will reduce the problem to a lower-dimensional problem without equality constraints. An equality constraint can also be specified by setting the corresponding bounds equal, i.e., $l_i = u_i$, but this is generally not efficient.

An important special case is when the inequalities are simple bounds.

PROBLEMS BLS.

$$(5.2.2) \qquad \min_x \|Ax - b\|_2, \quad \text{subject to} \quad l \leq x \leq u.$$

Bound-constrained least squares problems arise in many practical applications, e.g., reconstruction problems in geodesy and tomography, contact problems for mechanical systems, and the modeling of ocean circulation. Sometimes it can be argued that the linear model is only realistic when the variables are constrained within meaningful intervals. For reasons of computational efficiency it is essential that such constraints be considered separately from more general constraints in (5.2.1). If the matrix A has full column rank, Problem BLS is a strictly convex optimization problem, and then problem BLS has a unique solution for any vector b. Also, the problem is known to be solvable in polynomial time when A has full column rank.

In the special case when only one-sided bounds on x are specified it is no restriction to assume that these are nonnegativity constraints, and we have the following problem.

PROBLEM NNLS.

$$(5.2.3) \qquad \min_x \|Ax - b\|_2 \quad \text{subject to} \quad x \geq 0.$$

Nonnegativity constraints are natural for problems where the variables, by definition, can only take on nonnegative values.

Another special case of problem LSI is the least distance problem.

PROBLEM LSD.

$$(5.2.4) \qquad \min_x \|x\|_2, \text{ subject to } g \leq Gx \leq h,$$

or more generally,

$$(5.2.5) \qquad \min_x \|x_1\|_2, \text{ subject to } g \leq G_1 x_1 + G_2 x_2 \leq h,$$

where

$$x = \begin{pmatrix} x_1 \\ x_2 \end{pmatrix}, \quad x_1 \in \mathbf{R}^k, \quad x_2 \in \mathbf{R}^{n-k}.$$

5.2.2. Basic transformations of problem LSI. We first remark that (5.2.1) is equivalent to the quadratic programming problem

(5.2.6) $$\min_x (x^T B x + c^T x), \quad \text{subject to} \quad l \le Cx \le u,$$

where
$$B = A^T A, \quad c = -2 A^T b.$$

The problem (5.2.6) also arises as a subproblem in general nonlinear programming algorithms. Therefore, it has been studied extensively, and many algorithms have been proposed for it. In the application to problem LSI the matrix B is positive semidefinite, and hence (5.2.6) is a convex program. In general, to use algorithms for quadratic programming for solving (5.2.1) is not a numerically safe approach, since working with the explicit cross-product matrix B should be avoided. However, methods for quadratic programming can often be adapted to work directly with A.

In some important applications, problem LSI can be defined with a triangular matrix A; see Schittkowski [706, 1983]. When this is not the case it is often advantageous to perform an initial transformation of A in (5.2.1) to triangular form. Using standard column pivoting (see Section 2.7.3), we compute a rank revealing QR decomposition

(5.2.7) $$Q^T A P = R = \begin{pmatrix} R_{11} & R_{12} \\ 0 & 0 \end{pmatrix} \begin{matrix} \}r \\ \}m-r \end{matrix},$$

where Q is orthogonal, P a permutation matrix, and R_{11} upper triangular and nonsingular. We assume that the numerical rank r of A is determined using some specified tolerance, as discussed in Section 2.7. The objective function in (5.2.1) then becomes

$$\|(R_{11}, R_{12}) x - c_1\|_2, \quad c = \begin{pmatrix} c_1 \\ c_2 \end{pmatrix} = Q^T b,$$

since we can delete the last $(m-r)$ rows in R and c.

By a further transformation, discussed by Cline [173, 1975], problem LSI can be brought into a least distance problem. We perform orthogonal transformations from the right in (5.2.7) to obtain a complete orthogonal decomposition (see Section 2.7.5)

$$Q^T A P V = \begin{pmatrix} T & 0 \\ 0 & 0 \end{pmatrix} \begin{matrix} \}r \\ \}m-r \end{matrix},$$

where T is triangular and nonsingular. Then problem LSI (5.2.1) can be written

$$\min_y \|T y_1 - c_1\|_2, \quad \text{subject to} \quad l \le Ey \le u,$$

where
$$E = (E_1, E_2) = CPV, \quad y = \begin{pmatrix} y_1 \\ y_2 \end{pmatrix} = V^T P x,$$

5.2. LINEAR INEQUALITY CONSTRAINTS

E and y are conformally partitioned, and $y_1 \in \mathbf{R}^r$. We now make the further change of variables

$$z_1 = Ty_1 - c_1, \qquad z_2 = y_2.$$

Substituting $y_1 = T^{-1}(z_1 + c_1)$ in the constraints we arrive at an equivalent least distance problem:

(5.2.8) $$\min_z \|z_1\|_2, \text{ subject to } \tilde{l} \leq G_1 z_1 + G_2 z_2 \leq \tilde{u},$$

where

$$G_1 = E_1 T^{-1}, \qquad G_2 = E_2,$$
$$\tilde{l} = l - G_1 c_1, \qquad \tilde{u} = u - G_1 c_1.$$

We note that if A has full column rank, then $r = n$ and $z = z_1$, so we get a least distance problem of the form (5.2.4).

Methods for solving problem LSI based on the above transformation to a least distance problem have been given by Lawson and Hanson [520, 1974, Chap. 23], Haskell and Hanson [441, 1981], and Schittkowski [706, 1983]. The method proposed by Schittkowski for solving the least distance problem is a primal method, as opposed to the dual approach used in [520] and [441]. The dual approach is based on the following equivalence between a least distance problem and a nonnegativity-constrained problem.

THEOREM 5.2.1. *Consider the least distance problem with lower bounds*

(5.2.9) $$\min_x \|x\|_2, \quad \text{subject to} \quad g \leq Gx.$$

Let $u \in \mathbf{R}^{m+1}$ be the solution to the nonnegativity constrained problem

(5.2.10) $$\min_u \|Eu - f\|_2, \quad \text{subject to} \quad u \geq 0,$$

where

$$E = \begin{pmatrix} G^T \\ g^T \end{pmatrix}, \qquad f = \begin{pmatrix} 0 \\ 1 \end{pmatrix} \begin{matrix} \}n \\ \}1 \end{matrix}.$$

Let the residual corresponding to the solution be

$$r = (r_1, \ldots, r_{n+1})^T = Eu - f,$$

and put $\sigma = \|r\|_2$. If $\sigma = 0$, then the constraints $g \leq Gx$ are inconsistent and (5.2.9) has no solution. If $\sigma \neq 0$, then the vector x defined by

(5.2.11) $$x = (x_1, \ldots, x_n)^T, \qquad x_j = -r_j/r_{n+1}, \qquad j = 1, \ldots, n$$

is the unique solution to (5.2.9).

Proof. See Lawson and Hanson [520, 1974, pp. 165–167]. ∎

Cline [173, 1975] and Haskell and Hanson [441, 1981] describe how the modified LSD problem (5.2.5) with only upper bounds can be solved in two steps, each of which requires a solution of a problem of type NNLS, the first of these having additional linear equality constraints.

5.2.3. Active set algorithms for problem LSI.
We now consider methods for solving problem LSI, which do not use a transformation into a least distance problem.

In general, methods for problems with linear inequality constraints are iterative in nature. We consider here so-called **active set algorithms**, which are based on the following observation. At the solution to (5.2.1) a certain subset of constraints $l \leq Cx \leq u$ will be active, i.e., satisfied with equality. If this subset was known a priori, the solution to the LSI problem would also be the solution to a problem with equality constraints only, for which efficient solution methods are known; see Section 5.1.

In active set algorithms a sequence of equality-constrained problems are solved corresponding to a prediction of the correct active set, called the working set. The working set includes only constraints which are satisfied at the current approximation, but not necessarily all such constraints. In each iteration the value of the objective function is decreased and the optimum is reached in finitely many steps. A general description of active set algorithms for linear inequality-constrained optimization is given in Gill, Murray, and Wright [360, 1981, Chap. 5.2].

Any point x which satisfies all constraints in (5.2.1) is called a **feasible point**. In general, a feasible point from which to start the active set algorithm is not known. (A trivial exception is the case when all constraints are simple bounds, as in (5.2.2) and (5.2.3).) Therefore, the active set algorithm consists of two phases, where in the first phase a feasible point is determined as follows. For any point x denote by $I = I(x)$ the set of indices of constraints violated at x. Introduce an artificial objective function as the sum of all infeasibilities,

$$\phi(x) = \sum_{i \in I} \max\{(c_i^T x - u_i), (l_i - c_i^T x)\}.$$

In the first phase $\phi(x)$ is minimized subject to the constraints

$$l_i \leq c_i^T x \leq u_i, \qquad i \notin I.$$

If the minimum of $\phi(x)$ is positive, then the inequalities are inconsistent, and the problem has no solution. Otherwise, when a feasible point has been found, the objective function is changed to $\|Ax - b\|_2$. Except for that, the computations in phases one and two use the same basic algorithm.

We now briefly outline an active set algorithm for solving the LSI problem. In the case when A has full column rank the algorithm described below is essentially equivalent to the algorithm given by Stoer [759, 1971].

Let x_k, the iterate at the kth step, satisfy the working set of n_k linearly independent constraints with the associated matrix C_k. We take

$$x_{k+1} = x_k + \alpha_k p_k,$$

where p_k is a search direction and α_k a nonnegative step length. The search direction is constructed so that the working set of constraints remains satisfied

5.2. LINEAR INEQUALITY CONSTRAINTS

for all values of α_k. This will be the case if $C_k p_k = 0$. In order to satisfy this condition we compute a decomposition

$$(5.2.12) \qquad C_k Q_k = (0, \ T_k), \qquad T_k \in \mathbf{R}^{n_k \times n_k},$$

where T_k is triangular and nonsingular, and Q_k is a product of orthogonal transformations. (This is essentially the QR decomposition of C_k^T.) If we partition Q_k conformally,

$$(5.2.13) \qquad Q_k = (\underbrace{Z_k}_{n-n_k}, \underbrace{Y_k}_{n_k}),$$

then the $n - n_k$ columns of Z_k form a basis for the nullspace of C_k. Hence the condition $C_k p_k = 0$ is satisfied if we take

$$(5.2.14) \qquad p_k = Z_k q_k, \qquad q_k \in \mathbf{R}^{n-n_k}.$$

We now determine q_k so that $x_k + Z_k q_k$ minimizes the objective function, i.e., in phase two q_k solves the unconstrained least squares problem

$$(5.2.15) \qquad \min_{q_k} \|A Z_k q_k - r_k\|_2, \qquad r_k = b - A x_k.$$

To simplify the discussion we assume in the following that the matrix AZ_k is of full rank so that (5.2.15) has a unique solution. To compute this solution we need the QR decomposition of the matrix AZ_k. This is obtained from the QR decomposition of the matrix AQ_k, where

$$(5.2.16) \qquad P_k^T A Q_k = P_k^T (A Z_k, A Y_k) = \begin{pmatrix} R_k & S_k \\ 0 & U_k \\ 0 & 0 \end{pmatrix} \begin{matrix} \}n - n_k \\ \}n_k \end{matrix}.$$

The advantage of computing this larger decomposition is that then the orthogonal matrix P_k need not be saved and can be discarded after being applied also to the residual vector r_k. The solution q_k to (5.2.15) can now be computed from the triangular system

$$R_k q_k = c_k, \qquad P_k^T r_k = \begin{pmatrix} c_k \\ d_k \end{pmatrix} \begin{matrix} \}n - n_k \\ \}n_k \end{matrix}.$$

The next approximate solution is then taken to be $x_{k+1} = x_k + \alpha_k p_k$, where α_k is a step length to be chosen.

We now determine $\bar{\alpha}$, the maximum nonnegative step length along p_k for which x_{k+1} remains feasible with respect to the constraints not in the working set. If $\bar{\alpha} \leq 1$ we take $\alpha_k = \bar{\alpha}$, and then add the constraints which are hit to the working set for the next iteration.

If $\bar{\alpha} > 1$ we take $\alpha_k = 1$. In this case x_{k+1} will minimize the objective function when the constraints in the working set are treated as equalities, and the orthogonal projection of the gradient onto the subspace of feasible directions will be zero:

$$Z_k^T g_{k+1} = 0, \qquad g_{k+1} = -A^T r_{k+1}.$$

In this case we check the optimality of x_{k+1} by computing Lagrange multipliers for the constraints in the working set. At x_{k+1} these are defined by the equation

$$(5.2.17) \qquad C_k^T \lambda = g_{k+1} = -A^T r_{k+1}.$$

The residual vector to the new unconstrained problem r_{k+1} satisfies

$$P_k^T r_{k+1} = \begin{pmatrix} 0 \\ d_k \end{pmatrix}.$$

Hence, multiplying (5.2.17) by Q_k^T and using (5.2.12) we obtain

$$Q_k^T C_k^T \lambda = \begin{pmatrix} 0 \\ T_k^T \end{pmatrix} \lambda = -Q_k^T A^T P_k \begin{pmatrix} 0 \\ d_k \end{pmatrix},$$

so from (5.2.16)

$$T_k^T \lambda = -(U_k^T,\ 0)\, d_k.$$

The Lagrange multiplier λ_i for the constraint $l_i \leq c_i^T x \leq u_i$ in the working set is said to be optimal if $\lambda_i \leq 0$ at un upper bound and $\lambda_i \geq 0$ at a lower bound. If all multipliers are optimal then we have found an optimal point and are finished. If a multiplier is not optimal then the objective function can be decreased by deleting the corresponding constraint from the working set. If more than one multiplier is not optimal, then it is usual to delete that constraint whose multiplier deviates most from optimality.

At each iteration step the working set of constraints is changed, which leads to a change in the matrix C_k. If a constraint is dropped, a row in C_k is deleted; if a constraint is added, a new row in C_k is introduced. An important feature of an active set algorithm is the efficient solution of the sequence of unconstrained problems (5.2.15). Using techniques described in Section 3.2 methods can be developed to update the matrix decompositions (5.2.12) and (5.2.15). In (5.2.12) the matrix Q_k is modified by a sequence of orthogonal transformations from the right. These transformations are then applied to Q_k in (5.2.16) and this decomposition, together with the vector $P_k^T r_{k+1}$, is similarly updated. Since these updatings are quite intricate they will not be described in detail here.

For the case when A has full rank the problem LSI always has a unique solution. If A is rank deficient there will still be a unique solution if all active constraints at the solution have nonzero Lagrange multipliers. Otherwise there is an infinite manifold M of optimal solutions with a unique optimal value. In this case we can seek the unique solution of minimum norm, which satisfies

$$\min \|x\|_2, \qquad x \in M.$$

This is a least distance problem.

In the rank deficient case it can happen that the matrix AZ_k in (5.2.15) is rank deficient, and hence R_k is singular. Note that if some R_k is nonsingular it can become singular during later iterations only when a constraint is deleted from the working set, in which case only its last diagonal element can become zero. This simplifies the treatment of the rank deficient case. To make the initial R_k nonsingular one can add artificial constraints to ensure that the matrix AZ_k has full rank. For a further discussion of the treatment of the rank deficient case, see Gill et al. [355, 1986].

A possible further complication is that the working set of constraints can become linearly dependent. This can cause possible cycling in the algorithm, so that its convergence cannot be ensured. A simple remedy that is often used is to enlarge the feasible region of the offending constraint by a small quantity; see also Gill, Murray, and Wright [360, 1981, Chap. 5.8.2].

Stoer's algorithm [759, 1971] for problem LSI has been implemented by Pazelt [654, 1973], and an English version of this code is given by Eichhorn and Lawson [262, 1975]. Schittkowski and Stoer [708, 1979] give an implementation of the same method using Gram–Schmidt decompositions. An advantage of this implementation is that it is relatively easy to take data changes into account. The implementation described by Crane et al. [192, 1980] is based on this work. There is a restrictive assumption in these realizations that A is of full column rank.

A robust and general set of Fortran subroutines for problem LSI and convex quadratic programming is given by Gill et al. [355, 1986]. The method is a two-phase active set method. It also allows a linear term in the objective function and handles a mixture of bounds and general linear constraints.

5.2.4. Active set algorithms for BLS. As remarked earlier the problem LSI simplifies considerably when the only constraints are simple bounds. This problem is important in its own right and also serves as a good illustration of the general algorithm. Hence we now consider the algorithm for problem BLS in more detail.

We first note that feasibility of the bounds is resolved by simply checking whether $l_i \leq u_i, i = 1, \ldots, p$. Further, the specification of the working set is equivalent to a partitioning of x into free and fixed variables. During an iteration the fixed variables will not change and can be effectively removed from the problem.

We divide the index set of x according to

$$\{1, 2, \ldots, n\} = \mathcal{F} \cup \mathcal{B},$$

where $i \in \mathcal{F}$ if x_i is a free variable and $i \in \mathcal{B}$ if x_i is fixed at its lower or upper bound. The matrix C_k will now consist of the rows $e_i, i \in \mathcal{B}$, of the unit matrix I_n. We let $C_k = E_{\mathcal{B}}^T$, and if $E_{\mathcal{F}}$ is similarly defined we can write

$$Q_k = (E_{\mathcal{F}}, E_{\mathcal{B}}), \qquad T_k = I_{n_k}.$$

This shows that Q_k is simply a permutation matrix, and the product

$$AQ_k = (AE_{\mathcal{F}}, AE_{\mathcal{B}}) = (A_{\mathcal{F}}, A_{\mathcal{B}})$$

corresponds to a permutation of the columns of A. Assume now that the bound corresponding to x_q is to be dropped. This can be achieved by

$$AQ_{k+1} = AQ_k P_R(k,q),$$

where $P_R(k,q)$, $q > k+1$, is a permutation matrix which performs a right circular shift in which the columns are permuted:

$$k+1, \ldots, q-1, q \quad \Rightarrow \quad q, k+1, \ldots, q-1.$$

Similarly, if the bound corresponding to x_q becomes active it can be added to the working set by

$$AQ_{k+1} = AQ_k P_L(q,k),$$

where $P_L(q,k)$, $q < k-1$, is a permutation matrix, which performs a left circular shift in which the columns are permuted:

$$q, q+1, \ldots, k \quad \Rightarrow \quad q+1, \ldots, k, q.$$

Subroutines for updating the QR decomposition (QRD) after right or left circular shifts are included in LINPACK and are described in Dongarra et al. [228, 1979, Chap. 10].

For problem BLS the equation (5.2.17) for the Lagrange multipliers simplifies to

$$\lambda = E_\mathcal{B}^T g_{k+1} = -E_\mathcal{B}^T A^T r_{k+1} = -(U_k^T, 0) d_k.$$

Hence the Lagrange multipliers are simply equal to the corresponding components of the gradient vector $-A^T r_{k+1}$. This leads to Algorithm 5.2.1 given below. (It is assumed that rank $(A) = n$.)

Several implementations of varying generality of active set methods for problem BLS have been developed. Lawson and Hanson [520, 1974] give a Fortran implementation of an algorithm for problem NNLS, which is similar to Algorithm 5.2.1. They also give a Fortran subroutine based on this algorithm for problem LSD with lower bounds only. Zimmermann [856, 1977] gives a special implementation of Stoer's method for problem BLS based on Gram–Schmidt decompositions.

Haskell and Hanson [441, 1981] give an algorithm for problems with nonnegativity constraints on selected variables and equality constraints (NNLSE), where the equality constraints are handled by the method of weighting; see Section 5.1.5. In this algorithm no assumption on the rank of A is made. They describe several methods of transforming problems of type LSI with added linear equality constraints, into the form NNLSE. In Hanson [436, 1986] further developments of this algorithm are described.

The active set algorithms usually restrict the change in dimension of the working set by dropping or adding only one constraint at each iteration. For large scale problems this can force many iterations to be taken when, e.g., many variables have to leave their bound. Hence an active set algorithm can be slow to converge when the set of active constraints cannot be guessed well by the user. Moré and Toraldo [590, 1989] have presented an algorithm for bound-constrained quadratic programming problems which combines the standard active set strategy with the gradient projection method. They report that on nondegenerate problems the gradient projection algorithm requires considerably fewer iterations and less time than an active set algorithm. Similar advantages are claimed for the block principal pivoting methods developed by Portugal, Júdice, and Vicente [665, 1994] for problem NNLS.

ALGORITHM 5.2.1. ACTIVE SET ALGORITHM FOR PROBLEM BLS.

Initialization:
$\mathcal{F} := \{1, 2, \ldots, n\}; \quad \mathcal{B} := \emptyset; \quad Q := Z := I_n;$
$x := (l + u)/2;$
Compute QRD $P^T A Q = \begin{pmatrix} R \\ 0 \end{pmatrix}; \quad P^T b = \begin{pmatrix} c \\ d \end{pmatrix};$

Main loop:
  begin repeat
    Compute unconstrained optimum in free variables:
    $q := R^{-1}(c - SE_{\mathcal{B}}x); \quad z := E_{\mathcal{F}}q;$
    if $l_i \leq z_i \leq u_i$ for all $i \in \mathcal{F}$ **then**
    begin comment: Check for optimality.
      Compute Lagrange multipliers $\lambda := U^T(d - UE_{\mathcal{B}}x)$
      if $\mathcal{B} = \emptyset$ or $\text{sign}(\gamma_i)\lambda_i \leq 0$ for all $i \in \mathcal{B}$
      go to finished;
      Find index t such that $\text{sign}(\gamma_t)\lambda_t = \max_{i \in \mathcal{B}} sign(\gamma_i)\lambda_i;$
      Move index t from $\mathcal{B}$ to $\mathcal{F}$, i.e. free variable x_t.
      Update $P^T A(E_{\mathcal{F}}, E_{\mathcal{B}}) = \begin{pmatrix} R & S \\ 0 & U \end{pmatrix}$ and $P^T b = \begin{pmatrix} c \\ d \end{pmatrix}.$
    end
    else
    begin For all $i \in \mathcal{F}$: $\alpha_i := \begin{cases} (x_i - l_i)/(x_i - z_i), & z_i < l_i \\ (u_i - x_i)/(z_i - x_i); & z_i > u_i \end{cases}$
      $\alpha := \min_{i \in \mathcal{F}} \alpha_i; \quad x := x + \alpha(z - x); \quad (0 \leq \alpha < 1)$
      Move from $\mathcal{F}$ to $\mathcal{B}$ all indices q for which $x_q = l_q$ or $x_q = u_q$.
      Put $\gamma_q := 1$ **if** $x_q = l_q$ **and** $\gamma_q := -1$ **if** $x_q = u_q;$
      Update $P^T A(E_{\mathcal{F}}, E_{\mathcal{B}}) = \begin{pmatrix} R & S \\ 0 & U \end{pmatrix}$ and $P^T b = \begin{pmatrix} c \\ d \end{pmatrix}.$
    end
  end main loop
finished.

5.3. Quadratic Constraints

5.3.1. Ill-posed problems. Least squares problems with quadratic constraints arise in a variety of applications, such as smoothing of noisy data, and in trust region methods for nonlinear least squares problems; see Section 9.2.3. However, the most important source is the solution of discretized ill-posed problems. Such problems arise naturally from inverse problems where one tries to determine, e.g., the structure of a physical system from its behavior. As an example, consider the integral equation of the first kind,

$$\int K(s,t)f(t)dt = g(s), \tag{5.3.1}$$

where the operator K is compact. It is well known that this is an ill-posed problem in the sense that the solution f does not depend continuously on the data g. This is because there are rapidly oscillating functions $f(t)$ which come arbitrarily close to being annihilated by the integral operator.

Let the integral equation (5.3.1) be discretized into a corresponding least squares problem

$$\min_f \|Kf - g\|_2. \tag{5.3.2}$$

The singular values of $K \in \mathbf{R}^{m \times n}$ will decay exponentially to zero. Hence K will not have a well-defined numerical δ-rank r, since by (2.8.3) this requires that $\sigma_r > \delta \geq \sigma_{r+1}$ holds with a distinct gap between the singular values σ_r and σ_{r+1}. Therefore, most of the methods for rank deficient least squares problems given in Section 2.7 are not very useful in this context. In general, any attempt to solve (5.3.2) without restricting f will give a meaningless result.

One of the most successful methods for solving ill-conditioned problems of this type is **Tikhonov regularization**; see Section 2.7.2. In this method the solution space is restricted by imposing an a priori bound on $\|Lf\|_2$ for a suitably chosen matrix $L \in \mathbf{R}^{p \times n}$. Typically L is taken to be a discrete approximation to some derivative operator, e.g.,

$$L = \begin{pmatrix} 1 & -1 & & & \\ & 1 & -1 & & \\ & & \ddots & \ddots & \\ & & & 1 & -1 \end{pmatrix} \in \mathbf{R}^{(n-1) \times n}, \tag{5.3.3}$$

which approximates the first derivative operator except for a scaling factor.

The above approach leads us to take f as the solution to the problem

$$\min_f \|Kf - g\|_2 \quad \text{subject to} \quad \|Lf\|_2 \leq \gamma. \tag{5.3.4}$$

Here the parameter γ governs the balance between a small residual and a smooth solution. The determination of a suitable γ is often a major difficulty in the solution process. Alternatively, we can consider the related problem

$$\min_f \|Lf\|_2 \quad \text{subject to} \quad \|Kf - g\|_2 \leq \rho. \tag{5.3.5}$$

The problems (5.3.4) and (5.3.5) are called **regularization methods** for the ill-conditioned problem (5.3.2) in the terminology of Tikhonov [775, 1963]; in the statistical literature the solution of problem (5.3.4) is called a **ridge estimate**.

5.3.2. Quadratic inequality constraints.
Problems (5.3.4) and (5.3.5) are special cases of the general problem LSQI.

PROBLEM LSQI. Least Squares with Quadratic Inequality Constraint.

$$(5.3.6) \qquad \min_x \|Ax - b\|_2 \quad \text{subject to} \quad \|Cx - d\|_2 \leq \gamma,$$

where $A \in \mathbf{R}^{m \times n}$, $C \in \mathbf{R}^{p \times n}$, $\gamma > 0$.

A particularly simple but important case is when

$$(5.3.7) \qquad C = I_n \quad \text{and} \quad d = 0.$$

We call this the **standard form** of LSQI.

Conditions for existence and uniqueness and properties of solutions to problem LSQI have been given by Gander [317, 1981]. Clearly, problem LSQI has a solution if and only if

$$(5.3.8) \qquad \min_x \|Cx - d\|_2 \leq \gamma,$$

and in the following we assume that this condition is satisfied. We define a C-generalized solution $x_{A,C}$ to the problem $\min_x \|Ax - b\|_2$ to be a solution to the problem (cf. Section 2.7.4)

$$(5.3.9) \qquad \min_{x \in S} \|Cx - d\|_2, \qquad S = \{x \in \mathbf{R}^n \mid \|Ax - b\|_2 = \min\}.$$

Notice that for $C = I$ and $d = 0$ we have $x_{A,I} = A^I b$. Then the constraint in problem LSI is binding only if

$$(5.3.10) \qquad \|Cx_{A,C} - d\|_2 > \gamma.$$

This observation gives rise to the following theorem.

THEOREM 5.3.1. *Assume that problem LSQI has a solution. Then either $x_{A,C}$ is a solution or (5.3.10) holds and the solution occurs on the boundary of the constraint region. In the latter case the solution $x = x_\lambda$ satisfies the generalized normal equations*

$$(5.3.11) \qquad (A^T A + \lambda C^T C) x_\lambda = A^T b + \lambda C^T d,$$

where λ is determined by the secular equation

$$(5.3.12) \qquad f_\lambda = \|Cx_\lambda - d\|_2 - \gamma = 0.$$

Proof. Using the method of Lagrange multipliers we minimize $\psi(x)$ where

$$\psi(x)A = \|Ax - b\|_2^2 + \lambda(\|Cx - d\|_2^2 - \gamma^2).$$

A necessary condition for a minimum is that the gradient of $\psi(x)$ equals zero, which gives (5.3.11). ∎

In the following we assume that (5.3.10) holds so that the constraint is binding. Then there is a unique solution to problem LSQI if and only if

$$\text{(5.3.13)} \qquad \operatorname{rank}\begin{pmatrix} A \\ C \end{pmatrix} = n.$$

As we shall see, only positive values of λ are of interest. Then (5.3.11) are the normal equations for the least squares problem

$$\text{(5.3.14)} \qquad \min_x \left\| \begin{pmatrix} A \\ \mu C \end{pmatrix} x - \begin{pmatrix} b \\ \mu d \end{pmatrix} \right\|_2, \quad \mu = \lambda^{1/2}.$$

Hence, to solve (5.3.11) for a given value of λ, it is not necessary to form the cross-product matrices $A^T A$ and $C^T C$.

A numerical method for solving problem LSQI can be based on applying, e.g., Newton's method for solving the secular equation (5.3.12). However, it is preferable to use the alternative formulation due to Hebden:

$$\text{(5.3.15)} \qquad h_\lambda = \frac{1}{\|Cx_\lambda - d\|_2} - \frac{1}{\gamma} = 0,$$

where x_λ is computed from (5.3.14). Reinsch [683, 1971] has shown that h_λ is convex, and hence that Newton's method is monotonically convergent to the solution λ^* if started within $[0, \lambda^*]$. If derivatives cannot be computed, the secant method can be used, and it can be shown that if the initial iterates are both nonnegative this method also is monotonically convergent.

For every function value h_λ we have to compute a new QR decomposition of (5.3.14) for computing x_λ. Methods which avoid this have been given by Eldén [267, 1977] and will be described later in this section.

5.3.3. Problem LSQI by GSVD. We now consider the use of the GSVD in Theorem 4.2.2 for analyzing problem LSQI. For ease of notation we assume that $m \geq n$ and put $q = \min(p, n)$. Then we have

$$\text{(5.3.16)} \qquad U^T A = \begin{pmatrix} D_A \\ 0 \end{pmatrix} Z, \quad V^T C = \begin{pmatrix} D_C & 0 \\ 0 & 0 \end{pmatrix} Z,$$

where $U \in \mathbf{R}^{m \times m}$ and $V \in \mathbf{R}^{p \times p}$ are orthogonal, Z nonsingular, and $D_A = \operatorname{diag}(\alpha_1, \ldots, \alpha_n)$, $D_C = \operatorname{diag}(\beta_1, \ldots, \beta_q)$ are the generalized singular values.

The rank condition (5.3.13) implies that

$$\text{(5.3.17)} \qquad \alpha_i^2 + \beta_i^2 > 0, \quad i = 1, \ldots, r, \qquad \alpha_i > 0, \quad i = r+1, \ldots, n,$$

where $r = \operatorname{rank}(C)$. Using the orthogonality of U and V problem LSQI becomes

$$\text{(5.3.18)} \qquad \min_y \|D_A y - \tilde{b}_1\|_2 \quad \text{subject to} \quad \left\| \begin{pmatrix} D_C y_1 - \tilde{d}_1 \\ \tilde{d}_2 \end{pmatrix} \right\|_2 \leq \gamma,$$

5.3. QUADRATIC CONSTRAINTS

where

$$Zx = y = \begin{pmatrix} y_1 \\ y_2 \end{pmatrix} \begin{matrix} \}q \\ \}n-q \end{matrix}, \quad U^T b = \begin{pmatrix} \tilde{b}_1 \\ \tilde{b}_2 \end{pmatrix} \begin{matrix} \}n \\ \}m-n \end{matrix}, \quad V^T d = \begin{pmatrix} \tilde{d}_1 \\ \tilde{d}_2 \end{pmatrix} \begin{matrix} \}q \\ \}p-q \end{matrix}.$$

Clearly a solution exists if and only if $\|\tilde{d}_2\|_2 \leq \gamma$, which is condition (5.3.8). Further, the vector y defined by

$$y_i = \begin{cases} \tilde{b}_i/\alpha_i, & \alpha_i \neq 0, \quad i = 1, \ldots, n, \\ \tilde{d}_i/\beta_i, & \alpha_i = 0, \quad i = 1, \ldots, r, \end{cases}$$

is a solution to (5.3.9). Hence the condition (5.3.10) becomes

$$(5.3.19) \qquad \sum_{\substack{i=1 \\ \alpha_i \neq 0}}^{q} (\beta_i \tilde{b}_i/\alpha_i - \tilde{d}_i)^2 + \|\tilde{d}_2\|_2^2 > \gamma^2,$$

and we assume that this condition is satisfied. The generalized normal equations (5.3.11) can now be written

$$(5.3.20) \qquad \begin{cases} (\alpha_i^2 + \lambda \beta_i^2) y_{i,\lambda} = \alpha_i \tilde{b}_i + \lambda \beta_i \tilde{d}_i, & i = 1, \ldots q, \\ \alpha_i y_{i,\lambda} = \tilde{b}_i, & i = q+1, \ldots, n. \end{cases}$$

A simple calculation shows that the secular equation becomes

$$(5.3.21) \qquad f_\lambda^2 = \sum_{\substack{i=1 \\ \alpha_i \neq 0}}^{q} \left(\alpha_i \frac{\beta_i \tilde{b}_i - \alpha_i \tilde{d}_i}{\alpha_i^2 + \lambda \beta_i^2} \right)^2 + \|\tilde{d}_2\|_2^2 = \gamma^2.$$

From (5.3.19) it follows that $f(0) > \gamma$. Since f_λ is monotone decreasing for $\lambda > 0$ there exists a unique positive root to (5.3.21). It can be shown that this is the desired root; see Gander [317, 1981]. From (5.3.21) we can cheaply compute function values and derivatives of f_λ for given numerical values of λ.

For LSQI problems of standard form the above algorithm simplifies. If we let the SVD of A be

$$U^T A = \begin{pmatrix} D_A \\ 0 \end{pmatrix} V^T,$$

where U and V are orthogonal, we have $\beta_i = 1$, $i = 1, \ldots, n$. Assume that the singular values $\alpha_i = \sigma_i(A)$ are ordered so that

$$\alpha_1 \geq \alpha_2 \geq \cdots \geq \alpha_n \geq 0.$$

The rank condition (5.3.13) is now trivially satisfied, and the condition (5.3.10) simplifies to

$$\|A^I b\|_2^2 = \sum_{i=1}^{n} (\tilde{b}_i/\alpha_i)^2 > \gamma^2.$$

We assume that this condition is satisfied and determine λ by solving the secular equation

$$(5.3.22) \qquad f_\lambda^2 = \sum_{i=1}^n y_{i,\lambda}^2 = \gamma^2, \qquad y_{i,\lambda} = \alpha_i \tilde{b}_i / (\alpha_i^2 + \lambda).$$

We finally obtain the solution from

$$(5.3.23) \qquad x = Vy = \sum_{i=1}^n y_{i,\lambda^*} v_i, \qquad V = (v_1, \ldots, v_n),$$

where λ^* is the unique positive solution to (5.3.22). This algorithm requires $mn^2 + 17/3 n^3$ flops.

5.3.4. Problem LSQI by QR decomposition. The GSVD and SVD, respectively, are the proper decompositions to analyze problem LSQI in general and standard form. These decompositions also lead to the most stable computational algorithms for the numerical solution of these problems. However, as for the unconstrained least squares problem more efficient algorithms which are almost as satisfactory can be devised which make use of simpler matrix decompositions. For the problem LSQI such methods have been given by Eldén [267, 1977].

We first consider the problem in standard form with $B = I_n$ and $d = 0$. In order to solve the secular equation $\|x_\lambda\|_2 - \gamma = 0$, we have to compute the solution $x = x_\mu$ to the least squares problem

$$(5.3.24) \qquad \min_x \left\| \begin{pmatrix} A \\ \mu I_n \end{pmatrix} x - \begin{pmatrix} b \\ 0 \end{pmatrix} \right\|_2$$

for a sequence of values of $\mu = \lambda^{1/2}$. To do this efficiently we first transform A to upper bidiagonal form (see Section 2.6.2)

$$(5.3.25) \qquad A = Q \begin{pmatrix} B \\ 0 \end{pmatrix} P^T,$$

where P and Q are orthogonal matrices and B is upper diagonal

$$B = \begin{pmatrix} \gamma_1 & \delta_1 & & & \\ & \gamma_2 & \delta_2 & & \\ & & \ddots & \ddots & \\ & & & \gamma_{n-1} & \delta_{n-1} \\ & & & & \gamma_n \end{pmatrix}.$$

This decomposition can be computed in only $mn^2 + n^3$ flops using Householder transformations. If we put

$$(5.3.26) \qquad x = Py, \qquad \tilde{b} = Q^T b = \begin{pmatrix} \tilde{b}_1 \\ \tilde{b}_2 \end{pmatrix} \begin{matrix} \}n \\ \}m-n \end{matrix},$$

5.3. Quadratic Constraints

the problem (5.3.24) is transformed into

$$(5.3.27) \qquad \min_x \left\| \begin{pmatrix} B \\ \mu I_n \end{pmatrix} y - \begin{pmatrix} \tilde{b}_1 \\ 0 \end{pmatrix} \right\|_2.$$

Since P is orthogonal the secular equation becomes $\|y_\lambda\|_2 = \gamma$. For a given value of μ we can now determine two sequences of Givens transformations

$$G_k = R_{k,n+k}, \qquad k = 1, \ldots, n,$$
$$J_k = R_{n+k,n+k+1}, \qquad k = 1, \ldots, n-1$$

so that B_μ is again upper bidiagonal,

$$(5.3.28) \qquad G_n J_{n-1} \cdots G_2 J_1 G_1 \begin{pmatrix} B & \tilde{b}_1 \\ \mu I_n & 0 \end{pmatrix} = \begin{pmatrix} B_\mu & g_1 \\ 0 & g_2 \end{pmatrix}.$$

Then y_μ is computed from the bidiagonal system $B_\mu y_\mu = g_1$ by back-substitution.

The construction of the Givens rotations G_k and J_k is sufficiently well demonstrated by the case $n = 3$ below. The first transformation G_1 is chosen to zero the element in position (4,1). This creates a new nonzero element in position (4,2), which is then annihilated by the transformation J_1. This step reduces the dimension of the problem by one and the transformation proceeds recursively. The first step is shown below ($n = 3$), where transformed elements are denoted by a prime:

$$J_1 G_1 \begin{pmatrix} \gamma_1 & \delta_1 & \\ & \gamma_2 & \delta_2 \\ & & \gamma_3 \\ \mu & & \\ & \mu & \\ & & \mu \end{pmatrix} = \begin{pmatrix} \gamma_1' & \delta_1' & \\ & \gamma_2 & \delta_2 \\ & & \gamma_3 \\ 0 & 0 & \\ & \mu' & \\ & & \mu \end{pmatrix}.$$

The transformation in (5.3.28), and the computation of y_λ takes only about $11n$ flops. Eldén [267, 1977] gives a more detailed operation count and also shows how to compute derivatives of the function

$$f_\mu = \|y_\mu\|_2 - \mu.$$

We now consider the more general form of problem LSQI, where $d = 0$ and $B = L \in \mathbf{R}^{(n-t) \times n}$ is a banded matrix. We now have to solve

$$(5.3.29) \qquad \min_x \left\| \begin{pmatrix} A \\ \mu L \end{pmatrix} x - \begin{pmatrix} b \\ 0 \end{pmatrix} \right\|_2$$

for a sequence of values of μ. We can no longer start by transforming A to bidiagonal form since the necessary transformations from the right would destroy the sparsity of L. Instead we use the QR decomposition of A

$$Q_1^T A = \begin{pmatrix} R_1 \\ 0 \end{pmatrix}, \qquad Q_1^T b = \begin{pmatrix} d_1 \\ d_2 \end{pmatrix},$$

to transform (5.3.29) into the equivalent problem

(5.3.30) $$\min_x \left\| \begin{pmatrix} R_1 \\ \mu L \end{pmatrix} x - \begin{pmatrix} d_1 \\ 0 \end{pmatrix} \right\|_2.$$

Some problems give rise to a matrix A which has a banded structure. Then it can be shown (see Theorem 6.2.1) that the matrix R_1 will be an upper triangular band matrix with the same bandwidth w_1 as A, and the complete matrix in (5.3.30) will be of banded form. In many cases the banded matrix L is also upper triangular. If not, it is convenient to reduce it to this form by computing the QR decomposition

$$Q_2^T L = R_2,$$

where R_2 has bandwidth w_2. Since Q_2 is orthogonal we have reduced (5.3.30) to the form

(5.3.31) $$\min_x \left\| \begin{pmatrix} R_1 \\ \mu R_2 \end{pmatrix} x - \begin{pmatrix} d_1 \\ 0 \end{pmatrix} \right\|_2.$$

This problem can be efficiently solved by one of the orthogonalization methods of Sections 6.2.3–6.2.4. Note that this involves a reordering of the rows of the matrix in (5.3.31) so that the column indices of the first nonzero element in each row form a nondecreasing sequence. The resulting algorithm has been described in detail by Eldén [271, 1984]. The number of operations for each value of μ is $O(n(w_1^2 + w_2^2))$.

We now describe an algorithm for the case when R_1 does not have band structure. The idea is to transform (5.3.29) to a regularization problem of standard form. Note that if L was nonsingular we could achieve this by the change of variables $y = Lx$. However, normally $L \in \mathbf{R}^{(n-t) \times n}$ and is of rank $n - t < n$. The transformation to standard form can then be achieved using the pseudoinverse of L by a technique due to Eldén [267, 1977]. We compute the QR decomposition of L^T:

$$L^T = (V_1, V_2) \begin{pmatrix} R_2 \\ 0 \end{pmatrix},$$

where V_2 spans the nullspace of L. If we set $y = Lx$, then

(5.3.32) $$x = L^I y + V_2 w, \qquad L^I = V_1 R_2^{-T},$$

where L^I is the pseudoinverse of L, and

$$Ax - b = AL^I y - b + AV_2 w.$$

We form AV_2 and compute its QR decomposition

$$AV_2 = (Q_1, Q_2) \begin{pmatrix} U \\ 0 \end{pmatrix}, \qquad U \in \mathbf{R}^{t \times t}.$$

Then

$$Q^T(Ax - b) = \begin{pmatrix} Q_1^T(AL^I y - b) + Uw \\ Q_2^T(AL^I y - b) \end{pmatrix} = \begin{pmatrix} r_1 \\ r_2 \end{pmatrix}.$$

5.3. QUADRATIC CONSTRAINTS

Now, if A and L have no nullspace in common, then AV_2 has rank t and U is nonsingular. Thus, we can always determine w so that $r_1 = 0$ and (5.3.29) is equivalent to

$$(5.3.33) \quad \min_y \left\| \begin{pmatrix} \tilde{A} \\ \mu I \end{pmatrix} y - \begin{pmatrix} \tilde{b} \\ 0 \end{pmatrix} \right\|_2, \quad \tilde{A} = Q_2^T A L^I, \quad \tilde{b} = Q_2^T b,$$

which is of standard form. We then retrieve x from (5.3.32).

We remark that if m is substantially larger than n it is better to apply the above reduction to (5.3.30) instead of (5.3.29). Since the reduction involves the pseudoinverse of L it is numerically less stable than GSVD or the direct solution of (5.3.29) if

$$\kappa(L) \gg \kappa \begin{pmatrix} A \\ \mu L \end{pmatrix}.$$

However, in practice it seems to give very similar results; see Varah [803, 1979].

An important special case is when in LSQI we have $A = K$, $C = L$, and both K and L are upper triangular Toeplitz matrices, i.e.,

$$K = \begin{pmatrix} k_1 & k_2 & \cdots & k_{n-1} & k_n \\ & k_1 & k_2 & & k_{n-1} \\ & & \ddots & \ddots & \vdots \\ & & & k_1 & k_2 \\ & & & & k_1 \end{pmatrix}$$

and L is as in (5.3.3). Such systems arise when convolution-type Volterra integral equations of the first kind,

$$\int_0^t K(t-s) f(t) dt = g(s), \quad 0 \leq t \leq T,$$

are discretized. Eldén [271, 1984] has developed a method for solving problems of this kind which only uses $\frac{9}{2}n^2$ flops for each value of μ. It can be modified to handle the case when K and L also have a few nonzero diagonals below the main diagonal. Although K can be represented by n numbers this method uses $n^2/2$ storage locations. A modification of this algorithm which uses only $O(n)$ storage locations is given in Bojanczyk and Brent [120, 1986].

5.3.5. Cross-validation. So far we have assumed that the parameter $\lambda = \mu^2$ is determined by solving the secular equation (5.3.12), where γ is known a priori or somehow determined from additional information about the solution. We now describe a method for determining the smoothing parameter μ directly from the data. The underlying statistical model is that the components of b are subject to random errors of zero mean and covariance matrix $\sigma^2 I_m$, where σ^2 may or may not be known. We take $d = 0$ in (5.3.11) and write the solution as a function of μ:

$$(5.3.34) \quad x_\mu = M_\mu^{-1} A^T b, \quad M_\mu = A^T A + \mu^2 C^T C.$$

The predicted values of b can then be written

(5.3.35) $$Ax_\mu = P_\mu b, \qquad P_\mu = AM_\mu^{-1}A^T,$$

where the symmetric matrix P_μ is called the influence matrix.

Craven and Wahba [193, 1979] have suggested that when σ^2 is known then μ should be chosen to minimize an unbiased estimate of the expected true mean square error given by

$$T(\mu) = \frac{1}{m}\|Ax_\mu - b\|_2^2 + \frac{2}{m}\sigma^2 \text{trace}(I_m - P_\mu) + \sigma^2.$$

Here trace(A) denotes the trace of the matrix A. When σ^2 is not known then μ may be chosen to minimize the generalized cross-validation (GCV) function given by

(5.3.36) $$V(\mu) = \frac{\frac{1}{m}\|Ax_\mu - b\|_2^2}{\frac{1}{m^2}\left[\text{trace}(I_m - P_\mu)\right]^2},$$

since the minimizer of $V(\mu)$ is asymptotically the same as the minimizer of $T(\mu)$, when m is large; see Golub, Heath, and Wahba [368, 1979].

EXAMPLE 5.3.1. (See Golub and Van Loan [389, 1989, Problem 12.1-5].) Let $A = (1, 1, \ldots, 1)^T \in \mathbf{R}^{m \times 1}$, $b \in \mathbf{R}^m$, and put

$$\bar{b} = \sum_{i=1}^m \frac{b_i}{m}, \qquad s^2 = \sum_{i=1}^m \frac{(b_i - \bar{b})^2}{m-1}.$$

Then the cross-validation function becomes

$$V(\lambda) = \frac{m(m-1)s^2 + \nu^2 m^2 \bar{b}^2}{(m-1+\nu)^2}, \qquad \nu = \frac{\lambda}{m+\lambda}.$$

It is readily verified that $V(\lambda)$ is minimized for $\nu = s^2/(m\bar{b}^2)$, which leads to an optimal λ given by

$$\lambda = \left[\left(\frac{\bar{b}}{s}\right)^2 - \frac{1}{m}\right]^{-1}. \quad \blacksquare$$

Minimization of either $T(\mu)$ or $V(\mu)$ requires that $\|Ax_\mu - b\|_2$ and trace $(I_m - P_\mu)$ can be accurately and efficiently computed. For a problem in standard form, i.e., when $B = I_n$, these quantities can be computed from the SVD of A:

$$A = U \begin{pmatrix} \Sigma \\ 0 \end{pmatrix} V^T, \qquad \Sigma = \text{diag}(\sigma_1, \ldots, \sigma_n).$$

We get

$$P_\mu = A(A^T A + \mu^2 I_n)A^T = U \begin{pmatrix} \Omega & 0 \\ 0 & 0 \end{pmatrix} U^T,$$

where

$$\Omega = \text{diag}(\omega_1, \ldots, \omega_n), \qquad \omega_i = \frac{\sigma_i^2}{\sigma_i^2 + \mu^2}.$$

5.3. QUADRATIC CONSTRAINTS

From an easy calculation it follows that with $c = U^T b$

(5.3.37) $$\|(I_m - P_\mu)b\|_2^2 = \mu^2 \sum_{i=1}^{n} \frac{c_i^2}{\sigma_i^2 + \mu^2} + \sum_{i=n+1}^{m} c_i^2.$$

Since ω_i are the eigenvalues of P we further have

(5.3.38) $$\text{trace}(I_m - P_\mu) = m - \sum_{i=1}^{n} \omega_i = m - n + \sum_{i=1}^{n} \frac{1}{\sigma_i^2 + \mu^2}.$$

Using the GSVD (5.3.16), formulas similar to (5.3.37) and (5.3.38) are easily derived for the general case $B \neq I_n$.

Eldén [270, 1984] has given a method for computing the cross-validation function C_μ for the standard problem $(C = I)$ based on the bidiagonalization of A, which is more efficient than that based on the SVD. Using the decomposition (5.3.26)–(5.3.27) the norm of the residual vector can be expressed

$$\|Ax_\mu - b\|_2^2 = \|By_\mu - \tilde{b}_1\|_2^2 + \|\tilde{b}_2\|_2^2.$$

Given y_μ this computation can be performed in $O(n)$ operations. However, for small values of μ cancellation occurs, and there may be large errors. Cancellation can be avoided by basing the computations on the identity

$$B(B^T B + \mu^2 I)^{-1} - I = -\mu^2 (BB^T + \mu^2 I)^{-1}.$$

A decomposition of $BB^T + \mu^2 I$ can be obtained using a variant of the algorithm of Eldén described in Section 5.3.4.

To compute the trace term in (5.3.36) we write using (5.3.35)

$$\begin{aligned} \text{trace}\,(I - P_\mu) &= m - \text{trace}\,(AM_\mu^{-1}A^T) \\ &= m - \text{trace}\,(I - \mu^2 M_\mu^{-1}) = m - n + \mu^2 \text{trace}\,M_\mu^{-1}. \end{aligned}$$

We further have, using properties of the trace function

$$\text{trace}\,(M_\mu)^{-1} = \text{trace}\,(B^T B + \mu^2)^{-1} = \text{trace}\,(B_\mu^T B_\mu)^{-1}.$$

A recursive procedure for computing the diagonal elements of the inverse of a band matrix can now be used to compute the last trace term in $O(n)$ operations; see Eldén [271, 1984].

In many important applications the matrices $A^T A$ and $C^T C$ have band structure. For example, when fitting a polynomial smoothing spline of degree $2k - 1$ to m data points the half-bandwidth will be k and $k + 1$, respectively; see Reinsch [683, 1971]. Then computing the cross-validation function using the singular value decomposition will require $O(m^3)$ operations and is not efficient. Hutchinson and de Hoog [479, 1985] give a method which requires only $\mathcal{O}(k^2 m)$ operations, and which generalizes Eldén's algorithm. It is based on the observation that to compute $\text{trace}(P)_\mu$ only the central $2k + 1$ bands of the inverse M_μ^{-1} are needed. These can be efficiently computed from the Cholesky factor R_μ of M_μ by the algorithm in Section 6.7.4.

Chapter 6
Direct Methods for Sparse Problems

6.1. Introduction

In this chapter we review methods for solving the linear least squares problem

(6.1.1) $$\min_x \|Ax - b\|_2, \quad A \in \mathbf{R}^{m \times n}, \quad m \geq n,$$

which are effective when the matrix A is sparse, i.e., when the matrix A has "relatively few" nonzero elements. A more precise definition is difficult to give. J. H. Wilkinson defined a sparse matrix to be "any matrix with enough zeros that it pays to take advantage of them." Often the gain in operations and storage can be huge, making otherwise intractable problems possible to solve. Note that very large problems are by necessity sparse (or structured), since otherwise they would be intractable.

In Rice [685, 1983] sources of least squares problems of large, and sometimes of enormous, size are identified and discussed. The following sources are mentioned:

1. the geodetic survey problem,
2. the photogrammetry problem,
3. the molecular structure problem,
4. the gravity field of the earth,
5. tomography,
5. the force method in structural analysis,
6. the very long base line problem,
7. surface fitting,
8. cluster analysis and pattern matching.

A sparse least squares problem of spectacular size is described in Kolata [507, 1978]. This is the problem of least squares adjustment of coordinates of the geodetic stations comprising the North American Datum. It consists of about 6.5 million equations in 540,000 unknowns (= twice the number of stations). The equations are mildly nonlinear, so it suffices to solve two or three linearized problems of this size. The adjustment of the geodetic measurements for the entire earth is planned for the future!

A natural distinction is between sparse matrices with regular zero pattern (e.g., banded structure) and matrices with an irregular pattern of nonzero elements. An example of a symmetric irregular pattern arising from a structural problem in aerospace is illustrated in Figure 6.1.1. Other application areas can give patterns with quite different characteristics.

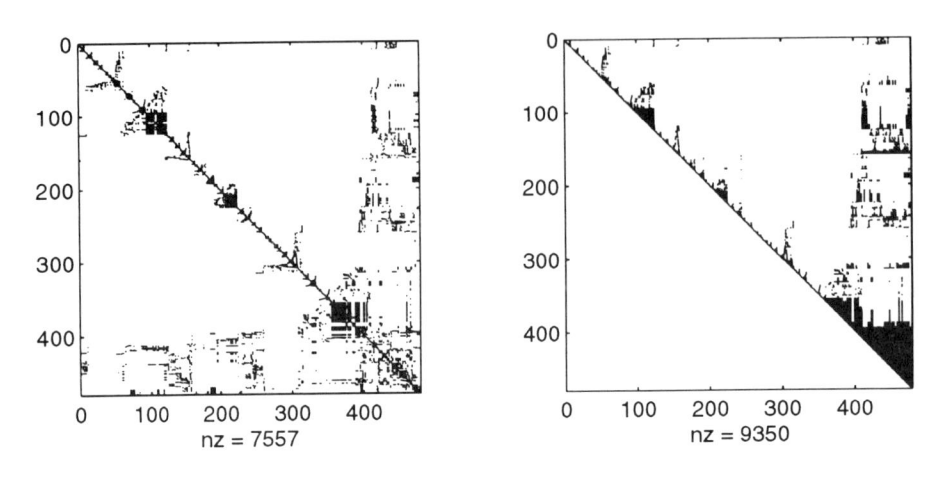

FIG. 6.1.1. *Nonzero pattern of a matrix and its Cholesky factor.*

In order to solve large sparse matrix problems efficiently it is important that only nonzero elements of the matrices be stored and operated on. One must also try to keep the **fill** small as the computation proceeds, which is the term used to denote the creation of new nonzeros. Only in the last 20 years or so have direct methods been developed for general sparse problems. For an early survey of sparse matrix research, see Duff [236, 1977]. An excellent modern introduction to the practice of sparse matrix computation is given by Duff, Erisman, and Reid [240, 1986]. For special surveys on methods for sparse least squares problems, see Björck [87, 1976], Heath [443, 1984], and Ikramov [482, 1985].

We will initially assume that A has full column rank, i.e., rank$(A) = n$. However, problems where rank$(A) = m < n$ or rank$(A) < \min(m, n)$ occur in practice. Other important variations include weighted problems, problems with linear equality constraints, and problems with upper and lower bounds on the variables. It may be possible to take advantage of a sparse right-hand side b. Also, if only a part of the solution vector x is needed savings can be achieved.

Sparse least squares problems may be solved either by direct or iterative methods. Preconditioned iterative methods can often be considered as hybrids between these two classes of solution. Solving sparse least squares problems using the method of normal equations or by QR decomposition is closely related to solving sparse positive definite systems by Cholesky factorization. An excellent introduction to theory and methods for the latter class of problems is given in the monograph by George and Liu [336, 1981]. A different approach is based on the augmented system (2.3) and uses methods for computing a sparse decomposition of a sparse, symmetric, and *indefinite* matrix.

The choice of solution method for large sparse least squares problems depends partly on the computing environment. Important considerations are whether we are using a virtual memory machine or need to consider the use of auxiliary storage. The efficiency of an algorithm may vary considerably depending on whether vector or parallel computers (either of shared-memory or local-memory type) are used.

6.2. Banded Least Squares Problems

The simplest class of sparse rectangular matrices is the class of matrices which have a banded structure. A banded matrix $A \in \mathbf{R}^{m \times n}$ has the property that in each row the nonzero elements lie in a narrow band. We define

(6.2.1) $\qquad f_i(A) = \min\{j \mid a_{ij} \neq 0\}, \qquad l_i(A) = \max\{j \mid a_{ij} \neq 0\}$

to be column subscripts of the first and last nonzeros in the ith row of A.

DEFINITION 6.2.1. *The rectangular matrix $A \in \mathbf{R}^{m \times n}$ is said to have row bandwidth w if*
(6.2.2) $\qquad w(A) = \max_{1 \leq i \leq m} (l_i(A) - f_i(A) + 1).$

For this structure to have practical significance we need to have $w \ll n$. Note that, although the row bandwidth is independent of the row ordering, it will depend on the column ordering. Methods to permute the columns in A in order to achieve a small bandwidth are discussed in Section 6.5.1. We define the **bandwidth** of a symmetric matrix to be the bandwidth of its strictly upper (lower) triangular part

DEFINITION 6.2.2. *The bandwidth (or half-bandwidth) of a symmetric matrix $C \in \mathbf{R}^{n \times n}$ is given by*
$$p = \max_{1 \leq i \leq n} (l_i(C) - i).$$

We now prove a relationship between the row bandwidth w of A and the bandwidth p of the corresponding matrix of normal equations $C = A^T A$.

THEOREM 6.2.1. *Assume that the matrix $A \in \mathbf{R}^{m \times n}$ has row bandwidth w. Then the symmetric matrix $C = A^T A$ has at most bandwidth $p = w - 1$.*

Proof. From Definition 6.2.1 it follows that $a_{ij} a_{ik} \neq 0 \Rightarrow |j - k| < w$. Hence,

$$|j - k| \geq w \Rightarrow (A^T A)_{jk} = \sum_{i=1}^{m} a_{ij} a_{ik} = 0.$$

But then $C = A^T A$ must have bandwidth $p = w - 1$. ∎

It is often advantageous to sort the rows of A so that the column indices $f_i(A), i = 1, 2, \ldots, m$, of the first nonzero element in each row form a nondecreasing sequence, i.e.,
$$i \leq k \Rightarrow f_i(A) \leq f_k(A).$$

A matrix whose rows are sorted in this way is said to be in **standard form**. A row ordering within the blocks A_k may be specified by sorting the rows so that the column indices $l(A_k)$ form a nondecreasing sequence.

If we take into account that A may have a variable row bandwidth the relationship between the structure of A and that of $A^T A$ becomes more complicated. The structure of $A^T A$ can now be generated as follows.

THEOREM 6.2.2. *Assume that $A \in \mathbf{R}^{m \times n}$ is a banded matrix in standard form, and define*

$$(6.2.3) \qquad w_k = \max_{\{i \mid f_i(A) = k\}} (l_i(A) - f_i(A) + 1), \qquad k = 1, \ldots, q,$$

i.e., w_k is the bandwidth of the block A_k in (6.2.5). Then the banded structure of $C = A^T A$ is given by $l_1(C) = w_1$,

$$l_k(C) = \max\{w_k + k - 1, l_{k-1}(C)\}, \qquad k = 2, \ldots, q.$$

Proof. The proof is by induction. ∎

We now consider the resulting structure of the Cholesky factor R of $A^T A$. It can be shown that $R + R^T$ inherits the envelope of $A^T A$ where we have the following definition.

DEFINITION 6.2.3. *The **envelope** of a symmetric matrix C is the set of indices*

$$\text{Env}(C) = \{(i, j) \mid i \leq j \leq l_i(C)\}.$$

THEOREM 6.2.3. *Let R be the Cholesky factor of the positive definite matrix $C = A^T A$. Then it holds that*

$$\text{Env}(R + R^T) = \text{Env}(C).$$

Proof. The proof is by induction in the dimension n of C. The theorem obviously holds for $n = 1$. Assume it holds for $n - 1$ and partition $C \in \mathbf{R}^{n \times n}$ and its Cholesky factor as

$$C = \begin{pmatrix} C_1 & u \\ u^T & s \end{pmatrix}, \qquad R = \begin{pmatrix} R_1 & v \\ 0 & t \end{pmatrix},$$

where $C_1 = R_1^T R_1$ is positive definite and of dimension $n - 1$. By the induction assumption we have $\text{Env}(R_1 + R_1^T) = \text{Env}(C_1)$. From $C = R^T R$ it follows that v satisfies the lower triangular system $R_1^T v = u$. Hence if $u_i = 0$, $i = 1, \ldots, l$, it follows that also $v_i = 0$, $i = 1, \ldots, l$. ∎

6.2.1. Storage schemes for banded matrices. When the row bandwidth is not constant we can use a **compressed row storage** scheme, in which elements are stored contiguously row by row in one vector. All elements within the band of A are stored. For example, the matrix

$$(6.2.4) \qquad A = \begin{pmatrix} 0 & a_{12} & a_{13} & 0 \\ 0 & a_{22} & 0 & a_{24} \\ a_{31} & a_{32} & 0 & 0 \\ a_{41} & a_{42} & 0 & 0 \\ 0 & 0 & a_{53} & a_{54} \\ 0 & 0 & 0 & a_{64} \end{pmatrix}$$

6.2. Banded Least Squares Problems

would be stored as

$$AC = (a_{12}, a_{13} \mid a_{22}\ 0,\ a_{24} \mid a_{31}, a_{32} \mid a_{41}, a_{42} \mid a_{53}, a_{54} \mid a_{64}),$$
$$IA = (1,\ 3,\ 6,\ 8,\ 10,\ 12,\ 13),$$
$$FA = (2,\ 2,\ 1,\ 1,\ 3,\ 4).$$

Here FA contains the column indices $f_i(A)$ for each row, and IA the position in the array AC of the first element in the ith row of A. Note that zeros within the band are stored. The last element in IA equals the number of elements in the envelope plus one. In the simplest case when A has *constant bandwidth*, i.e., when

$$l_i(A) - f_i(A) = w - 1, \qquad i = 1, 2, \ldots m,$$

the vector IA is not needed.

According to Theorem 6.2.1 the matrix of normal equations $A^T A$ corresponding to a banded matrix has a *symmetric* banded structure. For example, for the matrix in (6.2.4) we have

$$A^T A = \begin{pmatrix} c_{11} & c_{12} & & \\ c_{21} & c_{22} & c_{23} & c_{24} \\ & c_{32} & c_{33} & c_{34} \\ & c_{42} & c_{43} & c_{44} \end{pmatrix}.$$

Symmetric band matrices can be stored row-wise using a band (envelope) storage scheme, where all elements in a row from the diagonal to the last nonzero are stored.

6.2.2. Normal equations for banded problems.

In the method of normal equations we first form

$$C = A^T A = \sum_{i=1}^{m} a_i a_i^T, \qquad d = A^T b = \sum_{i=1}^{m} b_i a_i,$$

where a_i^T, $i = 1, \ldots, m$, are the rows of A. We then perform the Cholesky factorization $C = R^T R$, and finally solve the two triangular systems resulting from $R^T(Rx) = d$.

In the case when A has constant row bandwidth w, forming C and d requires about $\frac{1}{2}mw(w+3)$ multiplications (b is assumed to have no zero elements). From Theorems 6.2.1 and 6.2.3 it follows that C and $R + R^T$ both have bandwidth $p = w - 1$, and from Algorithm 2.2.2 we can derive a row-wise band Cholesky algorithm. When the Cholesky factor is available, the solution of $R^T R x = A^T b = d$ can be computed by solving the two banded triangular systems $R^T y = d$ and $Rx = y$ by forward- and back-substitution.

ALGORITHM 6.2.1. BAND CHOLESKY ALGORITHM.

for $i = 1 : n$
$$r_{ii} = \left(c_{ii} - \sum_{k=\max(1,i-p)}^{i-1} r_{ki}^2 \right)^{1/2};$$
for $j = i+1 : \min(i+p, n)$
$$r_{ij} = \left(c_{ij} - \sum_{k=\max(1,j-p)}^{i-1} r_{ki} r_{kj} \right)/r_{ii};$$
end
end

ALGORITHM 6.2.2. BANDED FORWARD- AND BACK-SUBSTITUTION.

for $i = 1 : n$
$$y_i := \left(d_i - \sum_{k=max\{1,i-p\}}^{i-1} r_{ki} y_k \right)/r_{ii}$$
end

for $i = n : (-1) : 1$
$$x_i := \left(y_i - \sum_{k=i+1}^{\min\{i+p,n\}} r_{ik} x_k \right)/r_{ii};$$
end

In the constant bandwidth case the upper triangular part of C can be stored in an $n \times (p+1)$ array. The elements of the upper triangular part of C (and R) may be stored by diagonals in a two-dimensional $n \times (p+1)$ array as illustrated below ($n = 5, p = 2$):

$$\begin{pmatrix} c_{11} & c_{12} & c_{13} \\ c_{22} & c_{23} & c_{24} \\ c_{33} & c_{34} & c_{35} \\ c_{44} & c_{45} & \times \\ c_{55} & \times & \times \end{pmatrix}.$$

This means that the (i, j) entry of C (and R) is found in the position $(i, j - i + 1)$ entry of the storage array. The elements marked $\times$ in the lower right corner may be arbitrary since they are not used. In an implementation the subscript computation should be made as efficient as possible. For a detailed discussion see "Contribution I/4" in Wilkinson and Reinsch [843, 1971], and Dongarra et al. [228, 1979 Chap. 4].

If $p \ll n$ the Cholesky factorization requires about $\frac{1}{2}n(p+1)(p+2)$ flops and n square roots. Together the forward- and back-substitution require $(2n - p - 1)(p + 2)$ multiplications. The algorithm is easily modified so that y and x overwrite d in storage. It follows that if full advantage is taken of the band

6.2. BANDED LEAST SQUARES PROBLEMS

structure of the matrices involved, the solution of a least squares problem where A has bandwidth w requires a total of

$$\frac{1}{2}(mw(w+3) + n(w-1)(w+2)) + n(2w-1) + O(m+n) \text{ flops.}$$

6.2.3. Givens QR decomposition for banded problems. In this section we consider methods based on Givens QR decomposition for the banded least squares problem $\min_x \|Ax - b\|_2$. From Theorems 6.2.1 and 6.2.3 it follows that R and the upper triangular part of $A^T A$ are banded matrices with nonzero elements only in the first $p = w - 1$ superdiagonals. Hence R again has w nonzeros in each row.

We now describe a sequential row orthogonalization scheme using Givens rotations. In this method we initialize an upper triangular matrix R to zero and then update R adding a row of A at a time. We will assume that the rows of A have been ordered so that A is in standard form. The orthogonalization proceeds row-wise, and in the ith step row a_i^T is merged with the triangular matrix R_{i-1} to produce the triangular matrix R_i. It uses the algorithm *givrot* defined in Section 2.3.2.

ALGORITHM 6.2.3. ROW ORTHOGONALIZATION FOR BAND MATRICES. Let $A \in \mathbf{R}^{m \times n}$ be a matrix of row bandwidth w. Initialize $R = R_0$ to be an upper triangular matrix of bandwidth w with all elements zero.

$$\begin{aligned}
&\text{for } i = 1, 2, \ldots, m, \\
&\quad \text{for } j = f_i(A), \ldots, \min\{f_i(A) + w - 1, n\} \\
&\qquad \text{if } a_{ij} \neq 0 \text{ then} \\
&\qquad\quad [c, s] = \text{givrot}(r_{jj}, a_{ij}); \\
&\qquad\quad \begin{pmatrix} r_j^T \\ a_i^T \end{pmatrix} := \begin{pmatrix} c & s \\ -s & c \end{pmatrix} \begin{pmatrix} r_j^T \\ a_i^T \end{pmatrix}; \\
&\qquad \text{end} \\
&\quad \text{end} \\
&\text{end}
\end{aligned}$$

By initializing R to zero the description above is also valid for the processing of the initial n rows of A. If at some stage $r_{jj} = 0$, then the whole jth row in R_{i-1} must be zero and the remaining part of the current row a_i^T can just be inserted in row j of R_{i-1}. (Note that a row permutation is just a special case of a Givens rotation.) The number of rotations needed to process row a_i^T is at most equal to $\min(i - 1, w)$.

In Figure 6.2.1 we show the situation before the elimination of the ith row. The updating of R by Givens method when a new row is added is seen to be basically identical to updating a *full triangular matrix* formed by rows and columns $f_i(A)$ to $l_i(A)$ of R by the full row formed by elements $f_i(A)$ to $l_i(A)$. For a detailed discussion see Cox [189, 1981]. Note that only the indicated $w \times w$ upper triangular part of R_{i-1} is involved in this step.

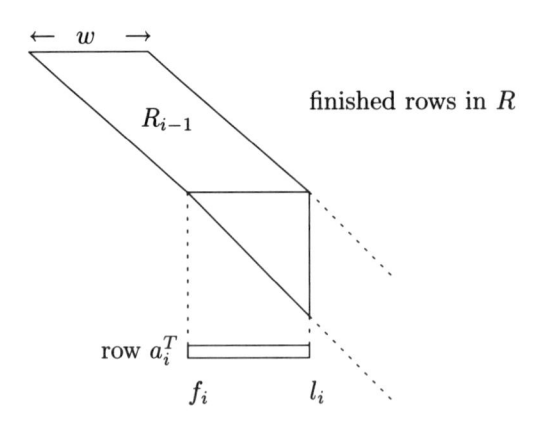

FIG. 6.2.1. *The ith step of reduction of a banded matrix.*

If A has constant bandwidth and is in standard form then the last $(n - l_i(A))$ columns of R have not been touched and are still zero as initialized. Further, the first $(f_i(A) - 1)$ rows of R are already finished at this stage and can be read out to secondary storage. Thus, very large problems can be handled since primary storage is needed only for the active triangular part in Figure 6.2.1

It is clear from the above that the processing of row a_i^T requires at most $2w^2$ flops if 4-multiply Givens rotations are used. Thus the complete orthogonalization requires about $2mw^2$ flops, and can be performed in $\frac{1}{2}w(w + 3)$ locations of primary storage. We remark that if the rows of A are processed in random order, then we can only bound the operation count by $2mnw$ flops, which is a factor of n/w worse (see Cox [189, 1981]). Thus, it almost invariably pays to sort the rows. The algorithm can be modified to also handle problems with variable row bandwidth w_i. In this case an envelope data structure for $A^T A$ is set up using Theorem 6.2.2, in which the R-factor will fit.

In Algorithm 6.2.3 the Givens rotations could also be applied to one or several right-hand sides b to produce

$$c = Q^T b = \begin{pmatrix} c_1 \\ c_2 \end{pmatrix}, \quad c_1 \in \mathbf{R}^n.$$

The least squares solution is then obtained from $Rx = c_1$ by back-substitution. The vector c_2 is not stored, but used to accumulate the residual sum of squares $\|r\|_2^2 = \|c_2\|_2^2$. If we have to treat several right-hand sides, which are not available in advance, then the Givens rotations should be saved. As described in Section 2.3.1 each Givens rotation can be represented by a single floating point number; see (2.3.12). Since at most w rotations are needed to process each row it follows that Q can be stored in no more space than that allocated for A.

6.2.4. Householder QR decomposition for banded problems. For the case when $m \gg n$ more efficient schemes based on Householder transformations may be developed. Such a scheme for banded systems was first given by Reid

6.2. BANDED LEAST SQUARES PROBLEMS

[679, 1967]. Lawson and Hanson [520, 1974, Chap. 11] give a similar algorithm and also provide Fortran subroutines implementing their algorithm.

Assume that the rows of A have been sorted so that A is in standard form. Then we can write

$$(6.2.5) \qquad A = \begin{pmatrix} A_1 \\ A_2 \\ \vdots \\ A_q \end{pmatrix}, \qquad q \leq n,$$

where the block A_k consists of all rows of A for which the first nonzero element is in column k, $k = 1, \ldots, q$. Initialize $R = R_0$ to be an upper triangular matrix of bandwidth w. The Householder algorithm proceeds in q steps, $k = 1, \ldots, q$. After the first $k-1$ steps the first $k-1$ blocks have been reduced by a sequence of Householder transformations to an upper trapezoidal banded matrix R_{k-1}. In step k we merge the kth block A_k, and compute

$$Q_k^T \begin{pmatrix} R_{k-1} \\ A_k \end{pmatrix} = R_k,$$

where Q_k is a product of Householder transformations and R_k is again upper trapezoidal. Note that because the block A_k has its first nonzero element in column k, this and later steps will not involve the first $k-1$ rows and columns of R_{k-1}. Hence the first $k-1$ rows of R_{k-1} are rows in the final matrix R.

The reduction using this algorithm takes about $w(w+1)(m+3n/2)$ flops, which is approximately half as many as for the Givens method. As in the Givens algorithm the Householder transformations can also be applied to one or several right-hand sides b to produce $c = Q^T b$, from which the least squares solution is then obtained from $Rx = c_1$ by back-substitution.

It is essential that the Householder transformations be subdivided as outlined above, otherwise intermediate fill will occur and the operation count will increase greatly. The reader is encouraged to work through the example below.

EXAMPLE 6.2.1. The least squares approximation of a discrete set of data by a linear combination of cubic B-splines gives rise to a banded linear least squares problem. Let

$$s(t) = \sum_{j=1}^{n} x_j B_j(t),$$

where $B_j(t), j = 1, 2, \ldots, n$, are the normalized cubic B-splines, and let (y_i, t_i), $i = 1, \ldots, m$, be given data points. If we determine x to minimize

$$\sum_{i=1}^{m} (s(t_i) - y_i)^2 = \|Ax - y\|_2^2,$$

then since the only B-splines with nonzero values for $t \in [\lambda_{k-1}, \lambda_k]$ are B_j, $j = k, k+1, k+2, k+3$, the matrix A will be a banded matrix with $w = 4$. In particular, assume that $m = 17$, $n = 10$, and that A consists of blocks A_k^T, $k = 1, \ldots, 7$.

$$\begin{pmatrix}
\times & \times & \times & \times & & & & & & & & & \\
1 & \times & \times & \times & + & & & & & & & & \\
1 & 2 & \times & \times & + & + & & & & & & & \\
 & 3 & 4 & \times & \times & + & & & & & & & \\
 & 3 & 4 & 5 & \times & + & & & & & & & \\
 & & & 6 & 7 & 8 & \times & & & & & & \\
 & & & 6 & 7 & 8 & 9 & & & & & & \\
 & & & 6 & 7 & 8 & 9 & & & & & & \\
 & & & & & & \times & \times & \times & \times & & & \\
 & & & & & & \times & \times & \times & \times & & & \\
 & & & & & & \times & \times & \times & \times & & & \\
 & & & & & & \times & \times & \times & \times & & & \\
 & & & & & & \times & \times & \times & \times & & & \\
 & & & & & & & \times & \times & \times & \times & & \\
 & & & & & & & \times & \times & \times & \times & & \\
 & & & & & & & & \times & \times & \times & \times & \\
 & & & & & & & & \times & \times & \times & \times &
\end{pmatrix}$$

FIG. 6.2.2. *The matrix A after reduction of the first $k = 3$ blocks using nine Householder transformations.*

In Figure 6.2.2 we show that the matrix after the first three blocks have been reduced by Householder transformations $P_1, \ldots, P_9$. Elements which have been zeroed by P_j are denoted by j and fill elements by $+$. In step $k = 4$ only the indicated part of the matrix is involved. ∎

Some problems (for example, periodic spline approximation) lead to matrices which have an augmented band structure,

$$A = (A_1, \; A_2),$$

where A_1 is a band matrix and A_2 a generally full matrix with a small number of columns. The band matrix algorithm is easily extended to matrices of this structure. Note also that this form is a special case of block angular form, which is treated in the next section.

6.3. Block Angular Least Squares Problems

6.3.1. Block angular form. As noted in Rice [685, 1983] there is a substantial similarity in the structure of several large sparse least squares problems. The matrices possess a block structure, perhaps at several levels, which reflects a "local connection" structure in the underlying physical problem. In particular, the problem can often be put in the following bordered block diagonal or **block angular form**:

$$(6.3.1) \quad A = \begin{pmatrix} A_1 & & & & B_1 \\ & A_2 & & & B_2 \\ & & \ddots & & \vdots \\ & & & A_M & B_M \end{pmatrix}, \quad x = \begin{pmatrix} x_1 \\ x_2 \\ \vdots \\ x_{M+1} \end{pmatrix}, \quad b = \begin{pmatrix} b_1 \\ b_2 \\ \vdots \\ b_M \end{pmatrix},$$

where

$$A_i \in \mathbf{R}^{m_i \times n_i}, \quad B_i \in \mathbf{R}^{m_i \times n_{M+1}}, \quad i = 1, 2, \ldots, M,$$

6.3. BLOCK ANGULAR LEAST SQUARES PROBLEMS

and
$$m = m_1 + m_2 + \cdots + m_M, \qquad n = n_1 + n_2 + \cdots + n_{M+1}.$$

From (6.3.1) we see that the variables $x_1, \ldots, x_M$ are coupled only to the variables x_{M+1}. Some applications where the form (6.3.1) arises naturally are in photogrammetry (see Golub, Luk, and Pagano [375, 1979]), Doppler radar positioning (see Manneback, Murigande, and Toint [561, 1985]), and geodetic survey problems (see Golub and Plemmons [380, 1980]). Weil and Kettler [834, 1971] have given a heuristic algorithm for permuting a general sparse matrix into this form.

The normal matrix of A in (6.3.1) is of doubly bordered block diagonal form,

$$A^T A = \begin{pmatrix} A_1^T A_1 & & & & A_1^T B_1 \\ & A_2^T A_2 & & & A_2^T B_2 \\ & & \ddots & & \vdots \\ & & & A_M^T A_M & A_M^T B_M \\ B_1^T A_1 & B_2^T A_2 & \cdots & B_M^T A_M & C \end{pmatrix},$$

where
$$C = \sum_{k=1}^{M} B_k^T B_k = R_{M+1}^T R_{M+1},$$

and R_{M+1} is the Cholesky factor of C. We assume in the following that $\mathrm{rank}(A) = n$, which implies that the matrices $A_i^T A_i$, $i = 1, 2, \ldots, M$, and C are positive definite. It is easily seen that then the Cholesky factor of $A^T A$ will have a block structure similar to that of A,

(6.3.2)
$$R = \begin{pmatrix} R_1 & & & & S_1 \\ & R_2 & & & S_2 \\ & & \ddots & & \vdots \\ & & & R_M & S_M \\ & & & & R_{M+1} \end{pmatrix},$$

where $R_i \in \mathbf{R}^{n_i \times n_i}$, and the Cholesky factor of $A_i^T A_i$, $i = 1, \ldots, M+1$ is nonsingular.

A number of problems have two levels of sparsity structure, i.e., the blocks A_i and/or B_i are themselves large and sparse matrices, often of the same general sparsity pattern as A. There may also be more than two levels of structure. There is a wide variation in the number and sizes of blocks. Some problems have large blocks with M of moderate size (10–100), while others have many more but smaller blocks.

6.3.2. QR methods for block angular problems. An algorithm for least squares problems of block angular form based on QR decomposition of A has been given by Golub, Luk, and Pagano [375, 1979]. This proceeds in three steps.

1. For $i = 1, 2, \ldots, M$ reduce the diagonal block A_i to upper triangular form by a sequence of orthogonal transformations to (A_i, B_i) and the right-hand side b_i, yielding

$$Q_i^T(A_i, B_i) = \begin{pmatrix} R_i & S_i \\ 0 & T_i \end{pmatrix}, \qquad Q_i^T b_i = \begin{pmatrix} c_i \\ d_i \end{pmatrix}.$$

   Any sparse structure in the blocks A_i should be exploited.

2. To obtain R_{M+1} and c_{M+1} compute the QR decomposition

$$\tilde{Q}_{M+1}^T (T \quad d) = \begin{pmatrix} R_{M+1} & c_{M+1} \\ 0 & d_{M+1} \end{pmatrix},$$

   where

$$T = \begin{pmatrix} T_1 \\ \vdots \\ T_M \end{pmatrix}, \qquad d = \begin{pmatrix} d_1 \\ \vdots \\ d_M \end{pmatrix}.$$

   The residual norm is given by $\rho = \|d_{M+1}\|_2$.

3. Find the solution to $\min_{x_{M+1}} \|T x_{M+1} - d\|_2$ from the triangular system

$$R_{M+1} x_{M+1} = c_{M+1},$$

   and compute $x_M, \ldots, x_1$ by back-substitution in the sequence of triangular systems

$$R_i x_i = c_i - S_i x_{M+1}, \qquad i = M, \ldots, 1.$$

There are several ways to organize this algorithm. In steps 1 and 3 the computations can be performed in parallel on the M subsystems. It is then advantageous to continue the reduction in step 1 so that the matrices T_i, $i = 1, \ldots, M$, are brought into upper trapezoidal form.

Cox [190, 1990] considers two modifications of this algorithm, for which the storage requirement is reduced. First he notices that by merging steps 1 and 2 it is not necessary to hold all blocks T_i simultaneously in memory. He shows that even more storage can be saved by discarding R_i and S_i after they have been computed in step 1, and recomputing R_i and S_i for step 3. Indeed, only R_i needs to be recomputed, since when x_{M+1} has been computed in step 2, we have that x_i, $i = 1, \ldots, M$, is the solution to the least squares problem

$$\min_{x_i} \|A_i x_i - g_i\|_2, \qquad g_i = b_i - B_i x_{M+1}.$$

Hence to determine x_i we only need to (re-)compute the QR decompositions of (A_i, g_i). In some practical problems this modification can reduce the storage requirement by an order of magnitude, while the recomputation of R_i only increases the operation count by a few percent. This is true, e.g., for large problems with dense blocks where

$$n_i \ll n_{M+1} \ll M n_i, \qquad m_i \gg n_i.$$

6.4. Tools for General Sparse Problems

Using the structure of the R-factor in (6.3.2), the diagonal blocks of the variance-covariance matrix $C = (R^T R)^{-1} = R^{-1} R^{-T}$ can be written (see Golub, Plemmons, and Sameh [381, 1988])

$$
\begin{aligned}
& C_{M+1,M+1} = R_{M+1}^{-1} R_{M+1}^{-T}, \\
(6.3.3) \quad & C_{i,i} = R_i^{-1}(I + W_i^T W_i) R_i^{-T}, \quad W_i^T = S_i R_{M+1}^{-1}, \quad i = 1, \ldots, M.
\end{aligned}
$$

Hence, if we compute the QR decompositions

$$Q_i \begin{pmatrix} W_i \\ I \end{pmatrix} = \begin{pmatrix} U_i \\ 0 \end{pmatrix}, \qquad i = 1, \ldots, M,$$

we have $I + W_i^T W_i = U_i^T U_i$ and

$$C_{i,i} = (U_i R_i^{-T})^T (U_i R_i^{-T}), \qquad i = 1, \ldots, M.$$

This method assumes that all the matrices R_i and S_i have been retained. For a discussion of computing variances and covariances using the modified algorithm see Cox [190, 1990].

In some applications the matrices R_i will be sparse but a lot of fill occurs in the blocks B_i in step 1. Then the triangular matrix R_{M+1} will be full, and expensive to compute. For such problems a block preconditioned iterative method may be more efficient. Here an iterative method, e.g., the conjugate gradient method, is applied to the problem

$$\min_y \|(AM^{-1})y - b\|_2, \qquad y = Mx,$$

where $M = \operatorname{diag}(R_1, \ldots, R_M, I)$. It may be possible to also compute a sparse QR decomposition of the last block column in (6.3.1),

$$Q_B^T B = \begin{pmatrix} R_B \\ 0 \end{pmatrix}, \qquad B = (B_1^T, \ldots, B_M^T)^T.$$

Then it is advantageous to use the preconditioner $M = \operatorname{diag}(R_1, \ldots, R_M, R_B)$; see Golub, Manneback, and Toint [376, 1986]. Such preconditioners are further discussed in Section 7.3.2.

6.4. Tools for General Sparse Problems

One of the main objectives of a sparse matrix data structure is to economize on storage while at the same time facilitating subsequent operations on the matrix. We first describe some suitable storage schemes for sparse vectors and matrices.

6.4.1. Storage schemes for general sparse matrices. We first consider a scheme for storing a sparse vector x. We store the nonzero elements[1] of x in

---
[1] All zeros are not necessarily exploited, and some stored "nonzero" elements may have the numerical value zero.

compressed form in a vector xc with dimension nnz, where nnz is the number of nonzero elements in x. Further, we store in an integer vector ix the indices of the corresponding nonzero elements in xc. Hence, the sparse row vector x can be represented by the triple (ix, xc, nnz), where

$$xc_k = x_{ix(k)}, \qquad k = 1, \ldots, nnz.$$

So, for example, the vector $x = (0, 4, 0, 0, 1, 0, 0, 0, 6, 0)$ is stored in compressed form as

$$xc = (1, 4, 6), \qquad ix = (5, 2, 9), \qquad nnz = 3.$$

Note that it is not necessary to store the nonzero elements in any particular order in xc. This makes it very easy to add further nonzero elements in x, since these can just be appended in the last positions of xc.

Operations on sparse vectors are simplified if one of the vectors is first **uncompressed**, i.e., stored as a dense vector. This can be done in time proportional to the number of nonzeros, and allows direct random access to specified elements in the vector. Vector operations, e.g., adding a multiple a of a sparse vector x to an uncompressed sparse vector y, or computing the inner product $x^T y$, can then be performed in constant time per nonzero element. Assume, for example, that the vector x is held in compressed form and y in a full length array. Then the operation $y := a * x + y$ may be expressed as

$$\text{for } k = 1, \ldots, nnz, \quad y(ix(k)) := a * x(k) + y(ix(k)).$$

REMARK 6.4.1. In the design of sparse matrix algorithms in MATLAB the above scheme is formalized by introducing a sparse accumulator SPA. The SPA consists of a dense vector of real (or complex) values, a dense vector of true/false "occupied flags," and an unordered list of the indices whose occupied flags are true. For a more complete description see Gilbert, Moler, and Schreiber [351, 1992].

There are many different ways to generalize this storage scheme to store a sparse rectangular matrix in compressed form. A simple scheme is to store the nonzero elements in "coordinate form" as an unordered one-dimensional array AC together with two integer vectors ix and jx containing the corresponding row and column indices,

$$ac(k) = a_{i,j}, \qquad i = ix(k), \qquad j = jx(k), \qquad k = 1, \ldots, nnz.$$

Hence, A is stored as an unordered set of triples consisting of a numerical value and two indices. We denote the number of nonzero entries in A by nnz=nnz(A). For the initial representation of a general sparse matrix this **coordinate scheme** is very convenient, and further nonzero elements are easily added to the structure. It has the drawback that storage overhead is large, since two extra integer vectors of length nnz are needed. More important is that it is difficult to access the matrix A by rows or by columns, which is needed, e.g., when implementing the Cholesky factorization.

6.4. Tools for General Sparse Problems

Another possibility is to store the matrix as a collection of sparse row (or column) vectors, where for each vector its nonzero elements are stored in AC in compressed form. The corresponding column subscripts are stored in the integer vector JA, i.e., the column subscript of the element ac_k is given in $ja(k)$. Finally, we need a third vector $IA(i)$, which gives the position in the array AC of the first element in the ith row of A. Alternatively a similar scheme storing A as a collection of column vectors may be used.

For example, in this **compressed row storage** scheme the matrix

$$A = \begin{pmatrix} a_{11} & 0 & a_{13} & 0 & 0 \\ a_{21} & a_{22} & 0 & 0 & 0 \\ 0 & 0 & a_{33} & 0 & 0 \\ 0 & a_{42} & 0 & a_{44} & 0 \\ 0 & 0 & 0 & a_{54} & a_{55} \\ 0 & 0 & 0 & 0 & a_{65} \end{pmatrix}$$

would be stored by rows as

$$\begin{aligned} AC &= (a_{11}, a_{13} \mid a_{22}, a_{21} \mid a_{33} \mid a_{42}, a_{44} \mid a_{54}, a_{55} \mid a_{65}), \\ IA &= (1, 3, 5, 6, 8, 10, 11), \\ JA &= (1, 3, 2, 1, 3, 2, 4, 4, 5, 5). \end{aligned}$$

The components in each row need not be ordered; indeed, there is often little advantage in ordering them. To access a nonzero a_{ij} there is no direct method of calculating the corresponding index in the vector AC. Some testing on the subscripts in ja has to be done. However, more frequently a complete row of A has to be retrieved, and this can be done quite efficiently.

In the general sparse storage scheme only nonzero elements are stored. This saving is, however, bought at the cost of storage for the vector ja of column subscripts. This overhead storage can be decreased by using a clever compressed scheme due to Sherman; see George and Liu [336, 1981, pp. 139–142].

If the matrix is stored as a collection of sparse row vectors, the entries in a particular column cannot be retrieved without a search of nearly all elements. These entries are needed, for instance, to find the rows which are involved in a stage of Gaussian elimination. One possibility would then be to also store the matrix as a set of column vectors.

An important distinction is between **static** storage structures that remain fixed and **dynamic** structures that can accommodate fill. A static structure like the general sparse storage scheme can be used when the location of the nonzeros in the matrix can be predicted in advance. Otherwise the data structure for the factors must dynamically allocate space for the fill during the elimination. Such storage schemes often use linked lists.

Storage schemes similar to the ones given above can be used for storing a sparse symmetric positive definite matrix $B \in \mathbf{R}^{n \times n}$. Obviously it is sufficient to store the upper (or lower) triangular part of B, including the main diagonal, and the compressed row storage scheme above can be used unchanged. However, since for a positive definite matrix all diagonal elements are positive it is convenient to store these in a separate vector; see George and Liu [336, 1981, pp. 79–80].

6.4.2. Graph representation of sparse matrices. In the method of normal equations for solving sparse linear least squares problems an important step is to determine a column permutation P such that the matrix $P^T A^T A P$ has a sparse Cholesky factor R, and to then generate a storage structure for R. This should be done symbolically using only the nonzero structure of A (or $A^T A$) as input. To perform such tasks the representation of the structure of a sparse matrix as a directed or undirected graph is a powerful tool.

A useful way to represent the structure of a symmetric matrix is by an **undirected graph** $G = (X, E)$, consisting of a set of **nodes** X and a set of **edges** E (unordered pairs of nodes). A graph is **ordered** (labeled) if its nodes are labeled. The ordered graph $G(A) = (X, E)$, representing the structure of a symmetric matrix $A \in \mathbf{R}^{n \times n}$, consists of nodes labeled $1, \ldots, n$ and edges $(x_i, x_j) \in E$ if and only if $a_{ij} = a_{ji} \neq 0$. Thus there is a direct correspondence between nonzero elements and edges in its graph; see Figure 6.4.1.

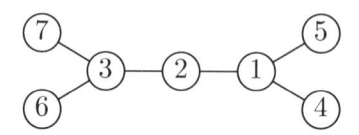

FIG. 6.4.1. *The matrix A and its labeled graph.*

Two nodes, x and y, are said to be **adjacent** if there is an edge $(x, y) \in E$. The adjacency set of x in G is defined by

$$\mathrm{Adj}_G(x) = \{y \in X \mid x \text{ and } y \text{ are adjacent}\}.$$

The number of nodes adjacent to x is denoted by $|\mathrm{Adj}_G(x)|$, and is called the **degree** of x. A **path** of length $l \geq 1$ between two nodes, u_1 and u_{l+1}, is an ordered set of distinct nodes $u_1, \ldots, u_{l+1}$, such that

$$(u_i, u_{i+1}) \in E, \quad i = 1, \ldots, l.$$

If there is such a chain of edges between two nodes, then they are said to be **connected**. If there is a path between every pair of distinct nodes, then the graph is connected. A disconnected graph consists of at least two separate connected subgraphs. ($\bar{G} = (\bar{X}, \bar{E})$ is a subgraph of $G = (X, E)$ if $\bar{X} \subset X$ and $\bar{E} \subset E$.) If

6.4. TOOLS FOR GENERAL SPARSE PROBLEMS

$G = (X, E)$ is a connected graph, then $Y \subset X$ is called a separator if G becomes disconnected after the removal and the nodes Y.

A symmetric matrix A is said to be **reducible** if there is a permutation matrix P such that $P^T A P$ is block diagonal. Such a symmetric permutation $P^T A P$ of A corresponds to a reordering of the nodes in $G(A)$ without changing the graph. It follows that the graph $G(P^T A P)$ is connected if and only if $G(A)$ is connected. It is then easy to prove that A is reducible if and only if its graph $G(A)$ is disconnected.

The structure of an unsymmetric matrix can similarly be represented by a **directed graph** $G = (X, E)$, where the edges now are ordered pairs of nodes. A directed graph is **strongly connected** if there is a path between every pair of distinct nodes along directed edges.

6.4.3. Predicting the structure of $A^T A$. In the method of normal equations we form the matrix $C = A^T A$ and apply the Cholesky algorithm to compute R such that $C = R^T R$. To apply the graph algorithms of the previous section, we first need to determine the structure of the matrix $A^T A$.

Partitioning A by rows we have (cf. (2.2.5))

$$(6.4.1) \qquad A^T A = \sum_{i=1}^{m} a_i a_i^T,$$

where a_i^T now denotes the ith row of A. This expresses $A^T A$ as the sum of m matrices of rank one. In the following we will often appeal to a **no-cancellation** assumption, i.e., whenever two nonzero numerical quantities are added or subtracted, the result is assumed to be nonzero. Invoking such a no-cancellation assumption, it follows that the nonzero structure of $A^T A$ is the direct sum of the nonzero structures of $a_i a_i^T$, $i = 1, 2, \ldots, m$.

The graph $G(A^T A)$ can be constructed directly from the structure of the matrix A. Appealing to (6.4.1) and the no-cancellation assumption the graph of $A^T A$ is the direct sum of all the graphs $G(a_i a_i^T)$, $i = 1, 2, \ldots, m$, i.e., we form the union of all nodes and edges not counting multiple edges. Note that the nonzeros in any row a_i^T will generate a subgraph where all pairs of nodes are connected. Such a subgraph is called a **clique** and corresponds to a full submatrix in $A^T A$. Clearly, the structure of $A^T A$ is not changed by dropping any row of A whose nonzero structure is a subset of another row, which can be used to speed up the algorithm.

Alternatively, we can use the following characterization to construct the graph $G(A^T A)$: under the no-cancellation assumption it holds that

$$(6.4.2) \qquad (A^T A)_{jk} \neq 0 \iff a_{ij} \neq 0 \text{ and } a_{ik} \neq 0,$$

for at least one row $i = 1, 2, \ldots, m$, i.e., when columns j and k intersect. Because of this characterization the graph $G(A^T A)$ is also known as the **column intersection** graph of A.

6.4.4. Predicting the structure of R.

From the graph $G(A^TA)$ the structure of the Cholesky factor R can be predicted by using a graph model of Gaussian elimination. The use of graphs to symbolically model Gaussian elimination for a symmetric positive definite matrix $A \in \mathbf{R}^{n \times n}$ is due to Parter [653, 1961], although the details were later given by Rose [688, 1972].

The fill under the factorization process can be analyzed by considering a sequence of undirected graphs $G_i = G(A^{(i)})$, $i = 0, \ldots, n - 1$, where $A^{(0)} = A$. These **elimination graphs** can be recursively formed in the following way.

Take $G_0 = G(A)$, and form G_i from $G_{(i-1)}$ by removing the node i and its incident edges and adding fill edges. The fill edges in eliminating node v in the graph G are

$$\{(j,k) \mid (j,k) \in \mathrm{Adj}_G(v), j \neq k\}.$$

Thus the fill edges correspond to the set of edges required to make the adjacent nodes of v pairwise adjacent. The elimination graphs for the matrix in Figure 6.4.1 are pictured in Figure 6.4.2. It can be verified that the number of fill-in elements (edges) is ten. The filled graph $G_F(A)$ of A is a graph with n

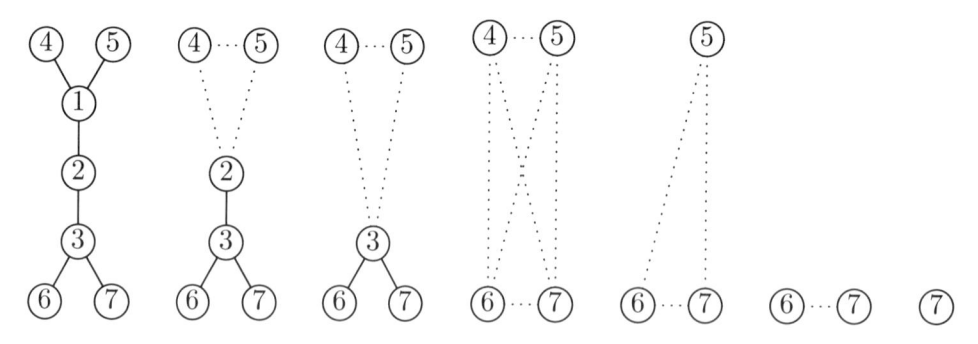

FIG. 6.4.2. *Sequence of elimination graphs of the matrix in Figure* 6.4.1.

vertices and edges corresponding to all the elimination graphs G_i, $i = 0, \ldots, n-1$. The filled graph bounds the structure of the Cholesky factor R,

(6.4.3) $$G(R^T + R) \subset G_F(A).$$

Under a no-cancellation assumption, the relation (6.4.3) holds with equality.

The following characterization of the filled graph describes how it can be computed directly from $G(A)$.

THEOREM 6.4.1. *Let $G(A) = (X, E)$ be the undirected graph of A. Then (x_i, x_j) is an edge of the filled graph $G_F(A)$ if and only if $(x_i, x_j) \in E$, or there is a path in $G(A)$ from node i to node j passing only through nodes with numbers less than $\min(i, j)$.*

We mention that it is possible to predict the structure of R working directly with $G(A)$ without forming $G(A^TA)$. Gilbert, Moler, and Schreiber [351, 1992] describe such an algorithm implemented in MATLAB, which makes the step of determining the structure of A^TA redundant.

6.4. Tools for General Sparse Problems

Under the no-cancellation assumption it follows that if A contains at least one full row then $A^T A$ will be full even if the rest of the matrix is sparse. An example is the matrix

$$A = \begin{pmatrix} \times & \times & \times & \times & \times \\ \times & \times & & & \\ & & \times & \times & \\ & & & \times & \times \\ & & & & \times & \times \\ & & & & & \times \end{pmatrix}.$$

(Sparse problems with only a few dense rows can be treated by updating the solution to the corresponding problem where the dense rows have been deleted; see Section 6.7.4.) If the no-cancellation assumption is not satisfied this may considerably overestimate the number of nonzeros in $A^T A$. For example, if A is orthogonal then $A^T A = I$ and is sparse even when A is dense.

The matrix R in the QR decomposition mathematically equals the Cholesky factor of $A^T A$; see Theorem 1.3.2. R is uniquely determined apart from possible sign differences in its rows. Hence, in particular, its nonzero structure is unique. Thus it seems that the same symbolic algorithm as for the Cholesky factor can be used to determine the structure of R in the QR decomposition. However, this method may in fact be too generous in allocating space for nonzeros in R. To see this, consider the matrix in Figure 6.4.3. For this matrix $R = A$ since A is already upper triangular. However, since $A^T A$ is full the algorithm above will predict R to be full. Note that this can occur because we begin not with the structure of $A^T A$, but with the structure of A. Hence the elements in $A^T A$ are not independent, and cancellation can occur in the Cholesky factorization irrespective of the numerical values of the nonzero elements in A. We call this **structural cancellation**, in contrast to numerical cancellation, which occurs only for certain values of the nonzero elements in A.

Another approach to predicting the structure of R is to perform the Givens or Householder algorithm symbolically working from the structure of A. George and Heath [333, 1980] proved the following result.

THEOREM 6.4.2. *The structure of R as predicted by symbolic factorization of $A^T A$ includes the structure of R as predicted by the symbolic Givens method, which includes the structure of R.*

Manneback [560, 1985] has proved that the structure predicted by a symbolic Householder algorithm is also strictly included in the structure predicted from $A^T A$. However, both the Givens and Householder rules can also overestimate the structure of R. Gentleman [331, 1976] gives an example where structural cancellation occurs for the Givens rule.

Coleman, Edenbrandt, and Gilbert [177, 1986] exhibited a class of matrices for which symbolic factorization of $A^T A$ correctly predicts the structure of R, since it can be proved that structural cancellation will not occur. From the above it follows that for this class the Givens and Householder rules will also give the correct result.

DEFINITION 6.4.1. *A matrix $A \in \mathbf{R}^{m \times n}$, $m \geq n$, is said to have the strong **Hall property** if for every subset of k columns, $0 < k < m$, the corresponding submatrix has nonzeros in at least $k + 1$ rows. (Thus, when $m > n$, every subset of $k \leq n$ has the required property, and when $m = n$, every subset of $k < n$ columns has the property.)*

THEOREM 6.4.3. *Let $A \in \mathbf{R}^{m \times n}$, $m \geq n$, have the strong Hall property. Then the structure of $A^T A$ will correctly predict that of R, excluding numerical cancellations.*

Obviously the matrix A in Figure 6.4.3 strong Hall property since the first column has only one nonzero element. However, the matrix $\tilde{A}$ obtained by deleting the first column has this property. Although both $A^T A$ and $\tilde{A}^T \tilde{A}$ are full, only the Cholesky factor of $\tilde{A}^T \tilde{A}$ is structurally full.

$$
\begin{array}{cc}
\begin{matrix}
\times & \times & \times & \times & \times \\
& \times & & & \\
& & \times & & \\
& & & \times & \\
& & & & \times \\
\end{matrix}
&
\begin{matrix}
\times & \times & \times & \times \\
\times & & & \\
& \times & & \\
& & \times & \\
& & & \times \\
\end{matrix}
\\
A & \tilde{A}
\end{array}
$$

FIG. 6.4.3. *Strong Hall property of matrix $\tilde{A}$.*

6.4.5. Block triangular form of a sparse matrix. An arbitrary rectangular matrix $A \in \mathbf{R}^{m \times n}$, $m \geq n$, can by row and column permutations be brought into the block triangular form

$$(6.4.4) \qquad PAQ = \begin{pmatrix} A_h & U_{hs} & U_{hv} \\ & A_s & U_{sv} \\ & & A_v \end{pmatrix}.$$

Here the diagonal block A_h is underdetermined (i.e., has more columns than rows), A_s is square and A_v is overdetermined (has more rows than columns), and all three blocks have a nonzero diagonal; see the example in Figure 6.4.4. The submatrices A_v and A_h^T both have the strong Hall property. The off-diagonal blocks denoted by U are possibly nonzero matrices of appropriate dimensions. This block triangular form (6.4.4) of a sparse matrix is based on a canonical decomposition of bipartite graphs discovered by Dulmage, Mendelsohn, and Johnson in a series of papers [255, 1958], [256, 1959], [257, 1963], and [491, 1963].

Following the notations in Pothen and Fan [668, 1990], we call the decomposition of A into the submatrices A_h, A_s, and A_v the **coarse decomposition**. One or two of the diagonal blocks may be absent in the coarse decomposition. It may be possible to further decompose the diagonal blocks in (6.4.4) to obtain

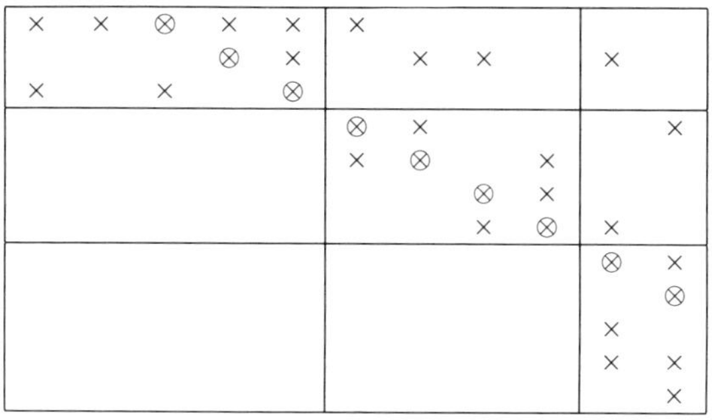

6.4. Tools for General Sparse Problems

FIG. 6.4.4. *The coarse block triangular decomposition of A.*

the **fine decompositions** of these submatrices. Each of the blocks A_h and A_v may be further decomposable into block diagonal form,

$$A_h = \begin{pmatrix} A_{h1} & & \\ & \ddots & \\ & & A_{hp} \end{pmatrix}, \quad A_v = \begin{pmatrix} A_{v1} & & \\ & \ddots & \\ & & A_{vq} \end{pmatrix},$$

where each $A_{h1}, \ldots, A_{hp}$ is underdetermined and each $A_{v1}, \ldots, A_{vq}$ is overdetermined. The submatrix A_s may be decomposable in block upper triangular form

$$(6.4.5) \qquad A_s = \begin{pmatrix} A_{s1} & U_{12} & \ldots & U_{1,t} \\ & A_{s2} & \ldots & U_{2,t} \\ & & \ddots & \vdots \\ & & & A_{st} \end{pmatrix}$$

with square diagonal blocks $A_{s1}, \ldots, A_{st}$ which have nonzero diagonal elements. The resulting decomposition can be shown to be essentially unique. Any one block triangular form can be obtained from any other by applying row permutations that involve the rows of a single block row, column permutations that involve the columns of a single block column, and symmetric permutations that reorder the blocks.

A square matrix which can be permuted to the form (6.4.5), with $t > 1$, is said to be **reducible**; otherwise it is called **irreducible**. (Some authors reserve the terms reducible for the case $Q = P^T$, and use the terms bireducible and bi-irreducible.) All the diagonal blocks in the fine decomposition are irreducible; this implies that $A_{s1}, \ldots, A_{st}$ all have the strong Hall property; see Coleman, Edenbrandt, and Gilbert [177, 1986].

For the case when A is a square and structurally nonsingular matrix there is a two-stage algorithm for permuting A to block upper triangular form; see Tarjan [773, 1972], Gustavson [415, 1976], and Duff [235, 1977], [237, 1981]. The program MC13D by Duff and Reid [248, 1978], included in the Harwell subroutine library, implements the fine decomposition of A_s. An algorithm for the more general block triangular form described above has been given by Pothen and Fan [668, 1990]; see also [666, 1984]. This algorithm depends on the concept of matchings in bipartite graphs.

The **bipartite graph** associated with A is denoted $G(A) = \{R, C, E\}$. Here $R = (r_1, \ldots, r_m)$ is a set of vertices corresponding to the rows of A, $C = (c_1, \ldots, c_m)$ is a set of vertices corresponding to the columns of A, E is the set of edges, and $\{r_i, c_j\} \in E$ if and only if a_{ij} is nonzero. A **matching** in $G(A)$ is a subset of its edges with no common end points. In the matrix A this corresponds to a subset of nonzeros, no two of which belong to the same row or column. A maximum matching is a matching with a maximum number $r(A)$ of edges. The **structural rank** of A equals $r(A)$. (In Figure 6.4.4 the structural rank of $A \in \mathbf{R}^{12 \times 11}$ equals 9.) (Note that the mathematical rank is always less than or equal to its structural rank.)

The algorithm of Pothen and Fan [668, 1990] consists of the following steps:

1. Find a maximum matching in the bipartite graph $G(A)$ with row set R and column set C.

2. According to the matching, partition R into the sets VR, SR, HR and C into the sets VC, SC, HC corresponding to the horizontal, square, and vertical blocks.

3. Find the diagonal blocks of the submatrix A_v and A_h from the connected components in the subgraphs $G(A_v)$ and $G(A_h)$. Find the block upper triangular form of the submatrix A_s from the strongly connected components in the associated directed subgraph $G(A_s)$, with edges directed from columns to rows.

The reordering to block triangular form in a preprocessing phase can save work and intermediate storage in solving least squares problems. If A has structural rank equal to n, then the first block row in (6.4.4) must be empty, and the original least squares problem can after reordering be solved by a form of block back-substitution. First compute the solution of

$$(6.4.6) \qquad \min_{\tilde{x}_v} \|A_v \tilde{x}_v - \tilde{b}_v\|_2,$$

where $\tilde{x} = Q^T x$ and $\tilde{b} = Pb$ have been partitioned conformally with PAQ in (6.4.4). The remaining part of the solution $\tilde{x}_k, \ldots, \tilde{x}_1$ is then determined by

$$(6.4.7) \qquad A_{si} \tilde{x}_i = \tilde{b}_i - \sum_{j=i+1}^{k} U_{ij} \tilde{x}_j, \quad i = k, \ldots, 2, 1.$$

Finally, we have $x = Q\tilde{x}$. We can solve the subproblems in (6.4.6) and (6.4.7) by computing the QR decompositions of A_v and $A_{s,i}$, $i = 1, \ldots, k$. Since $A_{s1}, \ldots, A_{sk}$ and A_v have the strong Hall property the structures of the matrices R_i are correctly predicted by the structures of the corresponding normal matrices.

If A has structural rank n but is *numerically* rank deficient it will not be possible to factorize all the diagonal blocks in (6.4.5). In this case the block triangular structure given by the Dulmage–Mendelsohn form cannot be preserved, or some blocks may become severely ill-conditioned.

If the matrix A has structural rank less than n, then we have an underdetermined block A_h. In this case we can still obtain the form (6.4.5) with a square block A_{11} by permuting the extra columns in the first block to the end. The least squares solution is then not unique, but a unique solution of minimum length can be found as outlined in Section 2.7.

The block triangular form of the matrices in the Harwell–Boeing test collection (Duff, Grimes, and Lewis [242, 1989]) and the time required to compute them are given in Pothen and Fan [668, 1990]. For results on savings achieved by using this form in QR decomposition codes for sparse matrices see Puglisi [672, 1993, Chap. 9]. Note that for some applications, e.g., for matrices arising from discretizations of partial differential equations, it may be known a priori that the matrix is irreducible. In other applications the block triangular decomposition may be known in advance from the underlying physical structure. In both these cases the algorithm discussed above is not useful.

6.5. Fill Minimizing Column Orderings

A reordering of the columns of AP of A corresponds to a symmetric reordering of the rows and columns of $A^T A$. Although this will not affect the number of nonzeros in $A^T A$, only their positions, it may greatly affect the number of nonzeros in the Cholesky factor R. Before carrying out the Cholesky factorization numerically, it is therefore important to find a permutation matrix P such that $P^T A^T AP$ has a sparse Cholesky factor R. (Note that the ordering of the rows in A has no effect on the matrix $A^T A$.)

Ideally it would be desirable to find an ordering which minimized the number of nonzero elements in R. However, it is known that to find an ordering optimal in this sense is an NP-complete problem, i.e., it cannot be solved in polynomial time. Hence, most ordering algorithms are heuristic, and will in general only give a suboptimal solution.

The simplest ordering methods use a priori information, such as ordering the columns in increasing column count. Such orderings are usually inferior to local ordering methods which use information from successively reduced submatrices. By far the most important local ordering method is the **minimum degree ordering** and various **nested dissection** orderings.

6.5.1. Bandwidth reducing ordering methods.

Some reordering methods have the objective of minimizing the bandwidth, or rather the area of the envelope of $A^T A$. Note that by Theorem 6.2.3 zeros outside the envelope will not suffer fill-in during the Cholesky factorization. Such ordering methods often perform well for matrices that come from one-dimensional problems or problems that are in some sense long and thin.

The most widely used ordering algorithm for reducing bandwidth or envelope is based on the **Cuthill–McKee** method. This method tries to minimize the envelope of the permuted symmetric matrix $P^T AP$ by gathering the nonzero elements close to the main diagonal. Cuthill [197, 1972] noticed that it is then important to label adjacent nodes x and y as close as possible to each other. Their method, given as Algorithm 6.5.1, is based on a local minimization criterion.

ALGORITHM 6.5.1. CUTHILL–MCKEE ORDERING.

> Determine a starting node r and put $x_1 := r$;
> for $i = 1, \ldots, n$
>   Find all unnumbered nodes in Adj (x_i)
>   and number them in increasing order;
> end

It was observed by George [332, 1971] that the ordering obtained by *reversing* the Cuthill–McKee ordering often gives less fill-in, although the bandwidth remains the same.

The performance of the Cuthill–McKee ordering method strongly depends on the choice of the starting node. George and Liu [336, 1981] recommend a strategy where a node of maximal or nearly maximal eccentricity $l(x) = \max_{y \in X}(d(x,y))$, is chosen as a starting node. Here $d(x,y)$ denotes the length of the shortest path between the two nodes x and y in the graph $G(A)$.

6.5.2. Minimum degree ordering. The **minimum degree ordering** is one of the most effective ordering algorithms. It is a symmetric analogue of an ordering algorithm proposed by Markowitz [567, 1957] for linear programming applications. The same strategy was employed by Tinney and Walker [776, 1967] for symmetric matrices.

In terms of the Cholesky factorization the minimum degree algorithm is equivalent to choosing the ith pivot column as one with the minimum number of nonzero elements in the unreduced part of $A^T A$. This will minimize the number of entries that will be modified in the next elimination step, and hence tend to minimize the arithmetic and amount of fill that occurs in this step. Although this, of course, will not in general provide a minimization of global arithmetic or fill, it has proved to be very effective in reducing both of these objectives. The name "minimum degree" comes from the graph-theoretic formulation of the algorithm, which was first given by Rose [688, 1972].

ALGORITHM 6.5.2. MINIMUM DEGREE ORDERING.

> Let $G^{(0)} = G(A)$.
> for $i = 1, \ldots, n-1$
>   Select a node y in $G^{(i-1)}$ of minimal degree.
>   Choose y as the next pivot.
>   Update the elimination graph to get $G^{(i)}$.
> end

6.5. FILL MINIMIZING COLUMN ORDERINGS

The minimum degree ordering for the matrix in Figure 6.4.1 will choose the pivots, e.g., in order $4, 5, 6, 7, 1, 2, 3$. No fill-in occurs, compared to ten fill-in elements with the initial ordering! Note that this ordering is not unique since several nodes in the initial graph have degree 1. The way tie-breaking is done may have an important influence on the goodness of the ordering. One can, e.g., choose the minimum degree node at random or as the first node in a candidate set of nodes. Examples are known where minimum degree will give very bad orderings if the tie-breaking is systematically done badly.

An example due to Duff, Erisman, and Reid [240, 1986] which shows that the minimum degree algorithm is not optimal is given in Figure 6.5.1. Here the initial ordering will give no fill. However, node 5 has minimum degree equal to 2, and if this node is eliminated first fill will occur in position $(4, 6)$.

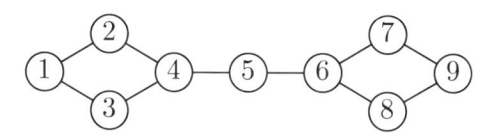

$$A = \begin{pmatrix} \times & \times & \times & \times & & & & & \\ \times & \times & \times & \times & & & & & \\ \times & \times & \times & \times & & & & & \\ \times & \times & \times & \times & \times & & & & \\ & & & \times & \times & \times & & & \\ & & & & \times & \times & \times & \times & \times \\ & & & & & \times & \times & \times & \times \\ & & & & & \times & \times & \times & \times \\ & & & & & \times & \times & \times & \times \end{pmatrix}$$

FIG. 6.5.1. *A matrix A for which minimum degree is not optimal.*

Remarkably fast symbolic implementations of the minimum degree algorithm exist, which use refinements of the elimination graph model of the Cholesky factorization described above. George and Liu [338, 1989] survey the extensive development of efficient versions of the minimum degree algorithm. One way is to decrease the number of degree updates as follows. The nodes $Y = \{y_1, \ldots, y_p\}$ are called **indistinguishable** if they have the same adjacency sets (including the node itself), i.e.,

$$\text{Adj}(v_i) \cup v_i = \text{Adj}(v_j) \cup v_j, \quad 1 \leq i, j \leq p.$$

If one of these nodes is eliminated the degree of the remaining nodes in the set will decrease by one, and they all become of minimum degree. This allows us to eliminate all nodes in Y simultaneously and perform the graph transformation and node update only once. Indeed, indistinguishable nodes can in the minimum degree algorithm be merged and treated as one **supernode**. In the matrix in Figure 6.5.1 there are two sets of indistinguishable nodes $\{1, 2, 3\}$ and $\{7, 8, 9\}$.

An implementation problem with the minimum degree algorithm is that in the graph updating, the space required to represent $G^{(i)}$ may be larger than for the previous graph $G^{(i-1)}$, since edges are added. An efficient technique for handling the storage of elimination graphs is to represent the graph as a number of cliques, $\{K_1, \ldots, K_q\}$. Indeed the original graph $G(A) = (X, E)$ can be regarded as consisting of $|E|$ cliques, each having two nodes (i.e., an edge). The degree of a node s then equals the number of different nodes $v \neq s$ in all clique elements $\{K_{s1}, \ldots, K_{st}\}$ to which s belongs. Updating the elimination graph when s is eliminated then requires two steps: first the cliques $\{K_{s1}, \ldots, K_{st}\}$ are removed from $\{K_1, \ldots, K_q\}$; second the new clique $K = (K_{s1} \cup \cdots \cup K_{st}) - \{s\}$ is added into the clique set. The key point now is that using this generalized element approach the amount of storage will never exceed the amount of storage needed to represent the original graph, since it is easily shown that $|K| < \sum_{i=1}^{t} |K_{si}|$.

We showed above that the structure of each row in $A \in \mathbf{R}^{m \times n}$ corresponds to a clique in the graph $G(C)$, $C = A^T A$. Therefore we can use the generalized element approach to represent C as a sequence of cliques. This allows an implementation of the minimum degree algorithm for $A^T A$ which bypasses the step of forming the structure of the matrix $C = A^T A$, with resulting savings in work and storage.

6.5.3. Nested dissection orderings. We now discuss a general procedure, called substructuring or dissection, for obtaining a block angular form as defined in Section 6.3. As an example, consider a geodetic position network consisting of geodetic stations connected through observations. To each station corresponds a set of unknown coordinates to be determined. A technique for breaking down such geodetic problems into geographically defined subproblems connected in a well-defined way has been applied for more than a century and dates back to Helmert [445, 1880]. The idea is to choose a set of stations $\mathcal{B}$, which separates

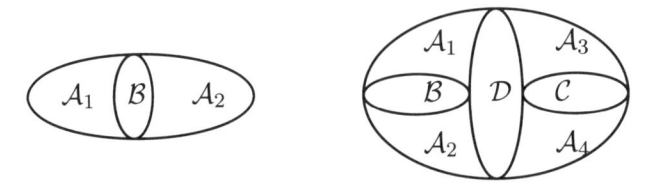

FIG. 6.5.2. *One and two levels of dissection of a region.*

the other stations into two regional blocks $\mathcal{A}_1$ and $\mathcal{A}_2$ so that station variables in $\mathcal{A}_1$ are not connected by observations to station variables in $\mathcal{A}_2$. The variables are then ordered so that those in $\mathcal{A}_1$ appear first, those in $\mathcal{A}_2$ second, and those in $\mathcal{B}$ last. Finally we order the equations so that those including $\mathcal{A}_1$ come first, those including $\mathcal{A}_2$ next, and those only involving variables in $\mathcal{B}$ come last. The dissection can be continued by dissecting the regions $\mathcal{A}_1$ and $\mathcal{A}_2$ each into two subregions, and so on in a recursive fashion. The blocking of the region for one and two levels of dissection is pictured in Figure 6.5.2.

Figure 6.5.3 shows the block structure in A induced by one and two levels of

6.5. Fill Minimizing Column Orderings

dissection, and the structure of the corresponding elimination trees. The block of rows corresponding to $\mathcal{A}_i$, $i = 1, 2, \ldots$, can be processed independently. The variables in $\mathcal{B}_i$ are then eliminated, etc.; compare the block angular structure in (6.3.1). There is a finer structure in A not shown. For example, in one level of dissection most of the equations involve variables in $\mathcal{A}_1$ or $\mathcal{A}_2$ only, but not in $\mathcal{B}$.

$$A = \begin{pmatrix} A_1 & & B_1 \\ & A_2 & B_2 \end{pmatrix}, \quad A = \begin{pmatrix} A_1 & & & & B_1 & & D_1 \\ & A_2 & & & B_2 & & D_2 \\ & & A_3 & & & C_3 & D_3 \\ & & & A_4 & & C_4 & D_4 \end{pmatrix}$$

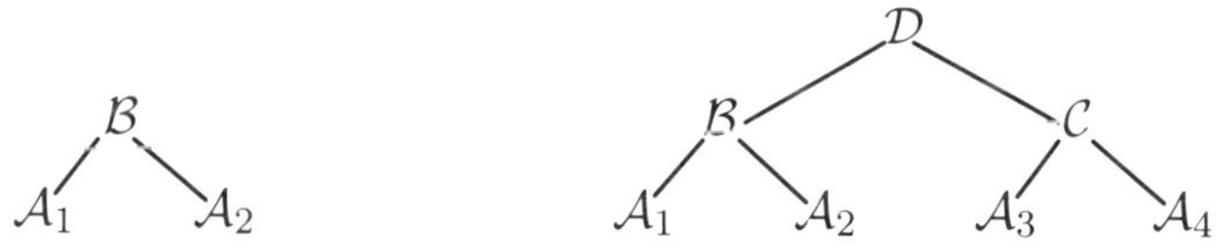

FIG. 6.5.3. *Block structure induced in A by one and two levels of dissection and the corresponding elimination trees.*

It is advantageous to perform the dissection in such a way that in each stage of the dissection the number of variables in the two partitions are roughly the same. Also, the number of variables in the separator nodes should be as small as possible. In particular, if each dissection is done so that the variables contained in the two partitions are at least halved, then after at most $\log^2 n$ levels each partition contains only one variable. Of course, it is usually preferable to stop before this point.

For a detailed discussion of dissection and orthogonal decompositions in geodetic survey problems, see Golub and Plemmons [380, 1980] Avila and Tomlin [30, 1979] discuss the solution of very large least squares problems by nested dissection on a parallel processor using the method of normal equations.

The dissection procedure described above is a variation of nested dissection orderings developed for general sparse positive definite systems; see George and Liu [336, 1981, Chap. 7–8]. Hence this approach applies to general sparse least squares problems. The column ordering and partitioning of A are then determined from the graph $G(A^T A)$; see George, Poole, and Voigt [348, 1978]. The use of such orderings for sparse least squares problems is treated in George, Heath, and Plemmons [335, 1981] and George and Ng [343, 1983]. It is known that **planar graphs**, i.e., graphs which can be drawn in the plane without two edges crossing, have small balanced separators. In Lipton, Rose, and Tarjan [539, 1979] it is shown that for any planar graph G with n nodes there exists a separator with $O(\sqrt{n})$ nodes such that each subgraph has at most $n/2$ nodes.

6.6. The Numerical Cholesky and QR Decompositions

As is well known, mathematically the Cholesky factor of $A^T A$ is equal to the factor R in the QR decomposition of A. Hence, the ordering methods discussed above, which work on the structure of these matrices, apply equally well to both the Cholesky and QR decompositions. In this section we discuss the numerical phase of sparse factorization methods.

6.6.1. The Cholesky factorization. An algorithm using the normal equations for solving sparse linear least squares problems is often split up in a symbolical and a numerical phase as follows. (We assume that rank$(A) = n$; for modifications needed to treat the case when rank$(A) < n$, see Section 6.7.1.)

ALGORITHM 6.6.1. SPARSE NORMAL EQUATIONS.

1. Determine symbolically the structure of $A^T A$.

2. Determine a column permutation P_c such that $P_c^T A^T A P_c$ has a sparse Cholesky factor R.

3. Perform the Cholesky factorization of $P_c^T A^T A P_c$ symbolically to generate a storage structure for R.

4. Compute $B = P_c^T A^T A P_c$ and $c = P_c^T A^T b$ numerically, storing B in the data structure of R.

5. Compute the Cholesky factor R numerically and solve $R^T z = c$, $Ry = z$, giving the solution $x = P_c y$.

Here steps 1, 2, and 3 involve only symbolic computation. It should be emphasized that the reason why the ordering algorithm in step 2 can be done symbolically only working on the structure of $A^T A$ is that pivoting is not required for numerical stability of the Cholesky algorithm.

For details of the implementation of the numerical factorization in step 5 we refer to George and Liu [336, 1981, Chap. 5]. Some available software packages are surveyed in Section 6.9.1. For well-conditioned problems the method of normal equations is quite satisfactory, and often provides a solution of sufficient accuracy. However, for ill-conditioned or stiff problems this method may lead to substantially less accurate solutions than methods based on the QR decomposition. For moderately ill-conditioned problems using the normal equations with iterative refinement may be a good choice; see Section 6.6.5. Meissl [576] gives an analysis of roundoff errors using normal equations for a super-large geodetic problem.

6.6.2. Row sequential QR decomposition. The potential numerical instability of the method of normal equations is due to loss of information in explicitly forming $A^T A$ and $A^T b$, and to the fact that the condition number of $A^T A$ is the square of that of A (see Section 2.2). Orthogonalization methods avoid both of these sources of inaccuracy by working directly with A. The main steps of a sparse QR algorithm are outlined below.

6.6. The Numerical Cholesky and QR Decompositions

ALGORITHM 6.6.2. SPARSE QR ALGORITHM.

1. Same as steps 1–3 in Algorithm 6.6.1.

2. Find a suitable row permutation P_r and reorder the rows to obtain $P_r A P_c$ (see Section 6.6.3).

3. Compute R and c numerically by applying orthogonal transformations to $(P_r A P_c, P_r b)$ (e.g., as described in Algorithm 6.6.3).

4. Solve $Ry = c$ and take $x = P_c y$.

For dense problems the most effective serial method for computing the QR decomposition is to use a sequence of Householder reflections; see Algorithm 2.3.2. In this algorithm we put $A^{(1)} = A$, and compute $A^{(k+1)} = P_k A^{(k)}$, $k = 1, \ldots, n$, where P_k is chosen to annihilate the subdiagonal elements in the kth column of $A^{(k)}$. In the sparse case this method will cause each column in the remaining unreduced part of the matrix, which has a nonzero inner product with the column being reduced, to take on the sparsity pattern of their union. In this way, even though the final R may be sparse, a lot of intermediate fill-in will take place with consequent cost in operations and storage. However, as was shown in Section 6.2.4, the Householder method can be modified to work efficiently for sparse banded problems, by applying the Householder reductions to a sequence of small dense subproblems. The generalization of this leads to multifrontal sparse QR methods; see Section 6.6.4. Here we first consider a row sequential algorithm by George and Heath [333, 1980], in which the problem with intermediate fill-in in the orthogonalization method is avoided by using a row-oriented method employing Givens rotations.

ALGORITHM 6.6.3. ROW SEQUENTIAL QR ALGORITHM. Assume that R_0 is initialized to have the structure of the final R and has all elements equal to zero. The rows a_k^T of A are processed sequentially, $k = 1, 2, \ldots, m$, and we denote by $R_{k-1} \in \mathbf{R}^{n \times n}$ the upper triangular matrix obtained after processing rows $a_1^T, \ldots, a_{k-1}^T$. The kth row $a_k^T = (a_{k1}, a_{k2}, \ldots, a_{kn})$ is processed as follows: we uncompress this row into a full vector and scan the nonzeros from left to right. For each $a_{kj} \neq 0$ a Givens rotation involving row j in R_{k-1} is used to annihilate a_{kj}. This may create new nonzeros both in R_{k-1} and in the row a_k^T. We continue until the whole row a_k^T has been annihilated. Note that if $r_{jj} = 0$, this means that this row in R_{k-1} has not yet been touched by any rotation and hence the entire jth row must be zero. When this occurs the remaining part of row k is inserted as the jth row in R.

To illustrate the algorithm we use the example in Figure 6.6.1, taken from George and Ng [343, 1983]. We assume that the first k rows of A have been processed to generate $R^{(k)}$. In Figure 6.6.1 nonzero elements of $R^{(k-1)}$ are denoted by ×, nonzeros introduced into $R^{(k)}$ and a_k^T during the elimination of a_k^T are denoted by +, and all the elements involved in the elimination of a_k^T are circled. Nonzero elements created in a_k^T during the elimination are of course ultimately annihilated. The sequence of row indices involved in the elimination are $\{2, 4, 5, 7, 8\}$, where 2 is the column index of the first nonzero in a_k^T.

Note that unlike the Householder method intermediate fill now only takes place in the row that is being processed. It follows from Theorem 6.4.2 that if the structure of R has been predicted from that of $A^T A$, then any intermediate matrix R_{i-1} will fit into the predicted structure.

$$\begin{bmatrix} R_{k-1} \\ a_k^T \end{bmatrix} = \begin{bmatrix} \times & 0 & \times & 0 & 0 & \times & 0 & 0 & 0 & 0 \\ & \otimes & 0 & \oplus & \otimes & 0 & 0 & 0 & 0 & 0 \\ & & \times & 0 & \times & 0 & 0 & 0 & \times & 0 \\ & & & \otimes & \oplus & 0 & \otimes & 0 & 0 & 0 \\ & & & & \otimes & \oplus & 0 & 0 & 0 & 0 \\ & & & & & \times & 0 & 0 & \times & 0 \\ & & & & & & \otimes & \otimes & 0 & 0 \\ & & & & & & & \otimes & 0 & 0 \\ & & & & & & & & \times & \times \\ & & & & & & & & 0 & \times \\ 0 & \otimes & 0 & \otimes & \oplus & 0 & \oplus & \oplus & 0 & 0 \end{bmatrix}$$

FIG. 6.6.1. *Circled elements $\otimes$ in R_{k-1} are involved in the elimination of a_k^T; fill elements are denoted by $\oplus$.*

For simplicity we have not included the right-hand side in Figure 6.6.1, but the Givens rotations should be applied simultaneously to b to form $Q^T b$. In the implementation by George and Heath [333, 1980] the Givens rotations are not stored but discarded after use. Hence, only enough storage to hold the final R and a few extra vectors for the current row and right-hand side(s) is needed in main memory. Discarding Q creates a problem if we later wish to solve additional problems having the same matrix A but a different right-hand side b since we cannot form $Q^T b$. In most cases a satisfactory method to deal with this problem is to use the corrected seminormal equations; see Section 6.6.6.

If Q is required, then the Givens rotations should be saved separately using the scheme outlined in Section 2.3.2. This in general requires far less storage and fewer operations than computing and storing Q itself; see Gilbert, Ng and Peyton [352, 1993].

6.6.3. Row orderings for sparse QR decomposition. Assuming that the columns have been permuted by some ordering method, the final R is independent of the ordering of the rows in A. However, the number of operations needed to compute the QR decomposition may depend on the row ordering. This fact was stressed already in the discussion of algorithms for the QR decomposition of banded matrices; see Section 6.2. Another illustration is given by the contrived example (adapted from George and Heath [333, 1980]) in Figure 6.6.2. Here the cost for reducing A is $O(mn^2)$, but that for PA is only $O(n^2)$.

6.6. The Numerical Cholesky and QR Decompositions

Assuming that the rows of A do not have widely differing norms, the row ordering does not affect numerical stability and can be chosen based on sparsity consideration only. We consider the following heuristic algorithm for determining a row ordering, which is an extension of the row ordering recommended for banded sparse matrices.

$$A = \begin{pmatrix} \times & \times & \times & \times & \times \\ \times & & & & \\ \vdots & & & & \\ \times & & & & \\ \hline \times & & & & \\ & \times & & & \\ & & \times & & \\ & & & \times & \end{pmatrix} \begin{array}{l} \left.\vphantom{\begin{matrix}1\\1\\1\\1\end{matrix}}\right\} m \\ \left.\vphantom{\begin{matrix}1\\1\\1\\1\end{matrix}}\right\} n \end{array}, \quad PA = \begin{pmatrix} \times & & & & \\ \times & & & & \\ \vdots & & & & \\ \times & & & & \\ \hline \times & \times & \times & \times & \times \\ & \times & & & \\ & & \times & & \\ & & & \times & \end{pmatrix} \begin{array}{l} \left.\vphantom{\begin{matrix}1\\1\\1\\1\end{matrix}}\right\} m \\ \left.\vphantom{\begin{matrix}1\\1\\1\\1\end{matrix}}\right\} n \end{array}.$$

FIG. 6.6.2. *A bad and a good row ordering.*

ALGORITHM 6.6.4. ROW ORDERING ALGORITHM. Denote the column index for the first and last nonzero elements in the ith row of A by $f_i(A)$ and $l_i(A)$, respectively. First sort the rows after increasing $f_i(A)$, so that $f_i(A) \le f_k(A)$ if $i < k$. Then for each group of rows with $f_i(A) = k$, $k = 1, \ldots, \max_i f_i(A)$, sort all the rows after increasing $l_i(A)$.

We note that using this row ordering algorithm on the matrix A in Figure 6.6.2 will produce the good row ordering PA. This rule does not in general determine a unique ordering. One way to resolve ties is to use a strategy by Duff [234, 1974], and consider the cost of symbolically rotating a row a_i^T into all other rows with a nonzero element in column $l_i(A)$. Here, by cost we mean the total number of new nonzero elements created. The rows are then ordered according to ascending cost. For this ordering it follows that the rows $1, \ldots, f_i(A) - 1$ in R_{i-1} will not be affected when the remaining rows are processed. These rows therefore are the final first $f_i(A) - 1$ rows in R and may, e.g., be transferred to auxiliary storage.

An alternative row ordering is obtained by ordering the rows after increasing values of $l_i(A)$. This ordering has been found to work well in some contexts; see George and Heath [333, 1980]. With this ordering it holds that when row a_i^T is being processed only the columns $f_i(A)$ to $l_i(A)$ of R_{i-1} will be involved, since all the previous rows only have nonzeros in columns up to at most $l_i(A)$. Hence R_{i-1} will have zeros in column $l_{i+1}(A), \ldots, n$, and no fill will be generated in row a_i^T in these columns.

6.6.4. Multifrontal QR decomposition.
A significant advance in direct methods for sparse matrix factorization is the **multifrontal method** by Duff and Reid [251, 1983]. This method reorganizes the factorization of a sparse matrix into a sequence of partial factorizations of small dense matrices, and is well suited for parallelism. A multifrontal algorithm for the QR decomposition

was first developed by Liu [540, 1986]. Liu generalized the row-oriented scheme of George and Heath by using submatrix rotations and remarked that this scheme is essentially equivalent to a multifrontal method. He showed that his algorithm can give a significant reduction in QR decomposition time at a modest increase in working storage. George and Liu [337, 1987] presented a modified version of Liu's algorithm which uses Householder transformations instead of Givens rotations.

There are several advantages with the multifrontal approach. The solution of the dense subproblems can more efficiently be handled by vector machines. Also, it leads to independent subproblems which can be solved in parallel. The good data locality of the multifrontal method gives fewer page faults on paging systems, and out-of-core versions can be developed. Multifrontal methods for sparse QR decompositions have been extensively studied and several codes developed by Lewis, Pierce, and Wah [527, 1989], Puglisi [672, 1993], Matstoms [573, 1994], and C. Sun [765, 1995].

$$A = \begin{pmatrix} \times & & & & & \times & & \times & \times & & \\ \times & & & & & \times & & \times & \times & & \\ \times & & & & & \times & & \times & \times & & \\ & \times & & & & \times & & & \times & \times \\ & \times & & & & \times & & & \times & \times \\ & \times & & & & \times & & & \times & \times \\ & & \times & & & & \times & \times & \times & & \\ & & \times & & & & \times & \times & \times & & \\ & & \times & & & & \times & \times & \times & & \\ & & & \times & & \times & & & \times & \times \\ & & & \times & & \times & & & \times & \times \\ & & & \times & & \times & & & \times & \times \end{pmatrix}.$$

FIG. 6.6.3. *A matrix A corresponding to a 3×3 mesh.*

We first describe the multiple front idea on the small 12×9 example in Figure 6.6.3, adopted from Liu [540, 1986]. This matrix arises from a 3×3 mesh problem using a nested dissection ordering, and its graph $G(A)$ is given in Figure 6.6.4.

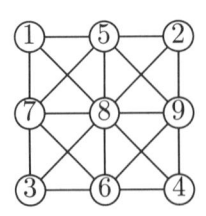

FIG. 6.6.4. *The graph $G(A^T A)$ and a nested dissection ordering.*

We first perform a QR decomposition of rows 1–3. Since these rows have nonzeros only in columns $\{1, 5, 7, 8\}$ this operation can be carried out as a QR decomposition of a small dense matrix of size 3×4 by leaving out the zero

6.6. THE NUMERICAL CHOLESKY AND QR DECOMPOSITIONS

columns. The first row equals the first of the final R of the complete matrix and can be stored away. The remaining two rows form an **update matrix** F_1 and will be processed later. The other three block rows 4–6, 7–9, and 10–12 can be reduced in a similar way. Moreover, these tasks are independent and can be done in parallel. After this first stage the matrix has the form shown in Figure 6.6.5. The first row in each of the four blocks are final rows in R and can be removed, which leaves four upper trapezoidal update matrices, F_1–F_4.

$$\begin{pmatrix} \times & & & & & & \times & \times & \times \\ & & & & & & \times & \times & \times \\ & & & & & & \times & \times & \times \\ & \times & & & \times & & & & \times & \times \\ & & & & \times & & & & \times & \times \\ & & & & & & & & \times & \times \\ & & \times & & & \times & \times & \times & & \\ & & & & & \times & \times & \times & & \\ & & & & & \times & \times & & & \\ & & & \times & & \times & & & \times & \times \\ & & & & & \times & & & \times & \times \\ & & & & & & & & \times & \times \end{pmatrix}.$$

FIG. 6.6.5. *The reduced matrix after the first elimination stage.*

In the second stage we can simultaneously merge F_1, F_2 and F_3, F_4 into two upper trapezoidal matrices by eliminating columns 5 and 6. In merging F_1 and F_2 we need to consider only the set of columns $\{5, 7, 8, 9\}$. We first reorder the rows after the index of the first nonzero element, and then perform a QR decomposition:

$$Q^T \begin{pmatrix} \times & \times & \times & \\ \times & & \times & \times \\ & \times & \times & \\ & \times & & \times \end{pmatrix} = \begin{pmatrix} \times & \times & \times & \times \\ & \times & \times & \times \\ & & \times & \times \\ & & & \times \end{pmatrix}.$$

The merging of F_3 and F_4 is performed similarly. Again, the first row in each reduced matrix is a final row in R, and is removed. In the final stage we merge the remaining two upper trapezoidal (in this example triangular) matrices and produce the final factor R. This corresponds to eliminating columns 7, 8, and 9.

The scheme described here can also be viewed as a special type of variable row pivoting method as studied by Gentleman [329, 1973], Duff [234, 1974], and Zlatev [857, 1982]. However, as observed by Liu [540, 1986], variable row pivoting schemes have never become very popular because of the difficulty of generating good orderings for the rotations and because these schemes are complicated to implement. Also, the dynamic storage structure needed tends to reduce the efficiency.

A basic concept for multifrontal methods is the **elimination tree**, which captures the row dependencies in the Cholesky factor R.

DEFINITION 6.6.1. *Let $R \in \mathbf{R}^{n \times n}$ be the Cholesky factor of $C = A^T A$. The elimination tree of C, denoted by $T(C)$, is a rooted tree with n nodes labeled from 1 to n, where node p is the **parent** of node i, if and only if*

$$p = \min_{j}\{j > i \mid r_{ij} \neq 0\}.$$

The elimination tree $T(C)$ can be obtained from the filled graph $G_F(C)$ in the following way. If directed edges from lower- to higher-numbered nodes are introduced in the filled graph, then a directed edge from node j to node $i > j$ indicates that row i depends on row j. To exhibit this row dependency relation this directed graph is reduced by a **transitive reduction**: if there is a directed path from j to i of greater length than one, then the edge from j to i is redundant and is removed. The removal of all such redundant edges generates precisely the elimination tree. For the matrix A in Figure 6.6.3 the filled graph $G_F(A^T A)$ equals $G(A^T A)$ with an additional edge between nodes 7 and 9. The result of the transitive reduction and the elimination tree is shown in Figure 6.6.6. Liu [541, 1990] gives an algorithm for determining the elimination tree in time proportional to nnz(R) and in space proportional to nnz(A).

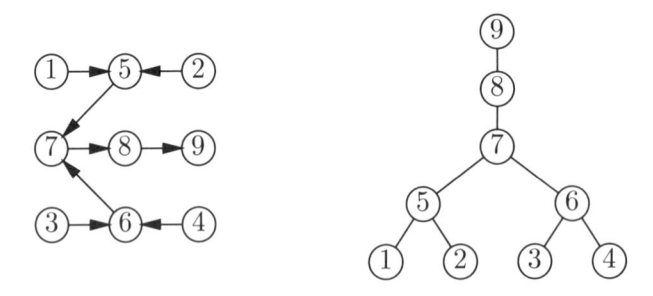

FIG. 6.6.6. *The transitive reduction and elimination tree $T(A^T A)$.*

The elimination tree provides in compact form all information about the row dependencies. The tree can be uniquely represented by the parent vector $PARENT[i]$, $i = 1, \ldots, n$, of all n nodes of the tree. The following theorem, which is the basis for the multifrontal method, is proved in Duff [238, 1986].

THEOREM 6.6.1. *Let $T[j]$ denote the subtree rooted in node j. The columns k and j can be eliminated independently of each other if $k \notin T[j]$.*

It follows that if $T[i]$ and $T[j]$ are two disjoint subtrees of $T(C)$ and $s \in T[i]$, $t \in T[j]$, then columns s and t can be eliminated in any order. The elimination tree prescribes an order relation for the elimination of columns in the QR factorization, namely, a column associated with a child node must be eliminated before the parent column. Columns associated with different subtrees of $T(C)$ are, on the other hand, independent and can be eliminated in parallel. An excellent treatment of elimination trees and their role in sparse factorization is given by Liu [541, 1990].

The organization of the multifrontal method is based on the elimination tree, and nodes in the tree are visited in turn given by the ordering. Each node x_j in the tree is associated with a frontal matrix F_j which consists of the set of rows A_j in A with the first nonzero in location j, together with one update matrix contributed by each child node of x_j. After eliminating the variable j in the frontal matrix, the first row in the reduced matrix is the jth row of the upper triangular factor R. The remaining rows form a new update matrix U_j, and is stored in a stack until needed. Hence a formal description of the method is as follows.

ALGORITHM 6.6.5. MULTIFRONTAL SPARSE QR ALGORITHM.

For $j := 1$ to n do

1. Form the frontal matrix F_j by combining the set of rows A_j and the update matrix U_s for each child x_s of the node x_j in the elimination tree $T(A^T A)$;

2. By an orthogonal transformation, eliminate variable x_j in F_j to get $\bar{U}_j$. Remove the first row in $\bar{U}_j$, which is the jth row in the final matrix R. The rest of the matrix is the update matrix U_j;

end.

The node ordering of an elimination tree is such that children nodes are numbered before their parent node. Such orderings are called **topological orderings**. All topological orderings of the elimination tree are equivalent in the sense that they give the same triangular factor R. A **postordering** is a topological ordering in which a parent node j always has node $j-1$ as one of its children. Postorderings are particularly suitable for the multifrontal method, and can be determined by a depth-first search; see Liu [541, 1990]. For example, the ordering of the nodes in the tree in Figure 6.6.6 can be made into a postordering by exchanging labels 3 and 5. The important advantage of using a postordering in the multifrontal method is that data management is simplified since the update matrices can be managed in a stack on a last-in–first-out basis. This also reduces the storage requirement.

The frontal matrices in the multifrontal method are often too small to make it possible to efficiently utilize vector processors and matrix-vector operations in the solution of the subproblems. A useful modification of the multifrontal method, therefore, is to amalgamate several nodes into one supernode. Instead of eliminating one column in each node, the decomposition of the frontal matrices now involves the elimination of several columns, and it may be possible to use Level 2 or even Level 3 BLAS; see Dongarra et al. [229, 1990].

In general, nodes can be grouped together to form a supernode if they correspond to a block of contiguous columns in the Cholesky factor, where the diagonal block is full triangular and these rows all have identical off-block diagonal column structures. Because of the computational advantages of having large supernodes, it is advantageous to relax this condition and also amalgamate nodes which satisfy this condition if some local zeros are treated as nonzeros. A practical restriction is that if too many nodes are amalgamated then the frontal matrices become sparse. (In the extreme case when all nodes are amalgamated into one supernode, the frontal matrix becomes equal to the original sparse matrix!) Note also that non-numerical operations often make up a large part of the total decomposition time, which limits the possible gain. For a discussion of supernodes and other modifications of the multifrontal method, see Liu [541, 1990] and Matstoms [572, 1994]; the latter gives a detailed description of the implementation of a supernodal multifrontal sparse QR algorithm.

In many implementations of the multifrontal algorithms the orthogonal transformations are not stored, and the seminormal equations (Section 6.6.5) are used for treating additional right-hand sides. If Q is needed then it should not be stored explicitly, but instead be represented by the Householder vectors of the frontal orthogonal transformations. For a K by K grid problem with $n = K^2$, $m = s(K-1)^2$ it is known (see George and Ng [347, 1988]) that $\mathrm{nnz}(R) = O(n \log n)$ but Q has $O(n\sqrt{n})$ nonzeros. Lu and Barlow [548, 1993] show that storing the frontal Householder matrices only requires $O(n \log n)$ storage.

6.6.5. Iterative refinement and seminormal equations. Due to storage considerations the matrix Q in a sparse QR decomposition is often discarded. This creates a problem if later an additional right-hand side b is to be treated, since we cannot form $Q^T b$. If the original matrix A is saved one can use the **seminormal equations** (SNE)

$$(6.6.1) \qquad R^T R x = A^T b.$$

However, the numerical stability of this method is no better than the method of normal equations. This is true even though we are using a factor R computed by QR decomposition and thus of better "quality" than that obtained from a Cholesky factorization of $A^T A$. This is related to the fact that already the rounding errors in computing $A^T b$ will give rise to an error δx, for which

$$\|\delta x\|_2 \leq mu\|(A^T A)^{-1}\|_2 \|A\|_2 \|b\|_2 \leq mu\kappa^2(A)\left(\frac{\|b\|_2}{\|A\|_2}\right),$$

where u is the unit roundoff. This error usually dominates.

By adding a correction step to (6.6.1) we obtain a solution of much better accuracy. The method of **corrected seminormal equations** (CSNE) is obtained as follows. Let $\bar{R}$ denote the computed R-factor and $\bar{x}$ the computed solution using (6.6.1). A corrected solution x_c is then determined as follows:

$$(6.6.2) \qquad \bar{r} = fl(b - A\bar{x}), \quad \bar{R}^T \bar{R} w = A^T \bar{r}, \quad x_c = \bar{x} + w.$$

The correction step is similar to doing one step of iterative refinement in fixed precision (see Sections 2.9.1 and 2.9.2).

A detailed error analysis of the CSNE method is given in Björck [91, 1987]. The factor $\bar{R}$ computed by a QR decomposition (see Section 2.4) is the exact R-factor of a perturbed matrix $A + E$, where $\|E\|_F \leq cu\|A\|_F$ and c a constant

6.6. The Numerical Cholesky and QR Decompositions

depending on m and n. Neglecting terms of higher order in $u\kappa$, it can be shown that the error in x_c from CSNE is

$$\|x - \bar{x}_c\|_2 \leq \sigma u\kappa \left(c_2\|x\|_2 + n^{1/2}m\frac{\|b\|_2}{\|A\|_2}\right) + mn^{1/2}u\kappa\left(\|x\|_2 + \kappa\frac{\|r\|_2}{\|A\|_2}\right),$$

where (see Remark 2.9.1)

$$\sigma = c_3 u\kappa\kappa', \quad c_3 \leq 2n^{1/2}(c_1 + 2n + m/2), \quad \kappa' = \min_{D>0} \kappa(AD).$$

A comparison with the bounds for a backward stable method shows that if $\sigma < 1$, then the error bound for the seminormal equations with one refinement step is no worse than the error bound for a backward stable method. The condition $\sigma < 1$ is roughly equivalent to requiring that the solution $\bar{x}$ from the seminormal equation has at least one correct digit, which is usually the case in practical applications. However, we caution that for problems with widely differing row scalings (stiff problems), CSNE is less satisfactory.

For more ill-conditioned problems several refinement steps may be used. Denote by x_s the computed solution after s refinement steps. With R from QR the error $\|x - x_s\|$ initially behaves as

$$\|x - x_s\| \sim cu\kappa\kappa'(cu\kappa')^s.$$

Assuming that $c \approx 1$, $\kappa' = \kappa$, acceptable-error stable level is achieved in p steps if $\kappa(A) < u^{-p/(p+1)}$. With $u = 10^{-16}$, the maximum condition number for which acceptable-error stable results are obtained after p refinements is

$$\kappa_{\max}(p) = 10^8, 10^{10.7}, 10^{16}, \quad p = 1, 2, \infty.$$

Fixed precision iterative refinement is also an efficient method for improving a solution obtained from the normal equations. In this method a sequence of improved approximations is computed as follows (see Algorithm 2.9.2).

Set $x_0 = 0$, and for $s = 0, 1, 2, \ldots$ until convergence do

(6.6.3) $\qquad r_s = b - Ax_s, \quad \bar{R}^T\bar{R}\delta x_s = A^T r_s, \quad x_{s+1} = x_s + \delta x_s.$

Here $\bar{R}$ is the computed Cholesky factor of $A^T A$. Each step of this algorithm requires two matrix-vector multiplications with A and A^T, and the solution of two triangular systems. The first step, $i = 0$, is identical to the normal equations. The rate of convergence of the error norm in this iteration can be shown to be approximately equal to

$$\bar{\rho} = cu\kappa'(A)^2.$$

If several steps of refinement are carried out, this will give good accuracy for a large class of problems (see Foster [310, 1991]), and we have

$$\|x - x_s\| \sim cu\kappa\kappa'(cu(\kappa')^2)^s.$$

Assuming that $c \approx 1$, $\kappa' = \kappa$, acceptable-error stable level is achieved in p steps if $\kappa(A) \leq u^{-p/(2p+1)}$. For example, with $u = 10^{-16}$, the maximum condition number for which acceptable-error stable results are obtained after p refinements is

$$\kappa_{\max}(p) = 10^{5.3}, 10^{6.4}, 10^{8}, \qquad p = 1, 2, \infty.$$

This can be compared with the result from fixed point iterative refinement when R from a QR decomposition is used. We conclude that for moderately ill-conditioned problems the normal equations combined with iterative refinement can give very good accuracy, but for more ill-conditioned problems QR and the seminormal equations are much superior.

6.7. Special Topics

6.7.1. Rank revealing sparse QR decomposition.

So far we have assumed that A is not rank deficient. In the dense case possibly rank deficient problems are handled by introducing column pivoting in the QR decomposition of A; see Sections 2.7.3–2.7.5. In sparse QR decomposition the column ordering is chosen to produce a sparse R-factor and fixed in advance of any numerical computation. Column pivoting cannot be used, since then R will generally not fit into the previously generated fixed storage structure.

Heath [442, 1982] suggested the following modification of the row sequential QR algorithm described in Section 6.6.2. Suppose this algorithm is applied to a matrix A of rank $r < n$ using *exact arithmetic*. Then it follows that the resulting R-factor must have $n - r$ zero diagonal elements. In this algorithm a row is only inserted into R when it makes the diagonal entry nonzero. Further processing of this row can only increase the diagonal element. It follows that if a row has a zero diagonal element then *all* its elements are zero, and hence the final R will have the form depicted in Figure 6.7.1. By permuting the zero rows of R to the bottom, and the columns of R corresponding to the zero diagonal elements to the right, we obtain R in rank revealing form.

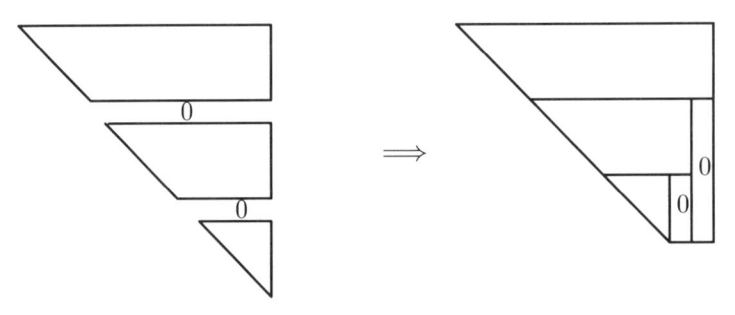

FIG. 6.7.1. *Structure of upper triangular matrix R.*

In finite precision we will usually end up with an R with no zero diagonal elements even when rank $(A) < n$. Although this is not always the case, the rank is often revealed by the presence of small diagonal elements. However, since a small diagonal element no longer implies that the rest of the row is negligible Heath suggests the following postprocessing of R. Starting from the top, the diagonal of R is examined for small elements. In each row whose diagonal element falls below a certain tolerance the diagonal element is put equal to zero. The rest of the row is then reprocessed, zeroing out all its other nonzero elements. This might increase some previously small diagonal elements in rows below, which is why one has to start from the top. After this we again end up with a matrix of the form shown in Figure 6.7.1.

In the test for small diagonal elements a relative tolerance can be used based on the largest diagonal element in R. Heath [442, 1982] reports that on a typical test batch the rank determined by this algorithm agreed with that determined by QR decomposition with column pivoting. However, this way of determining rank is in general not satisfactory. It may be that R is numerically rank deficient and yet has no small diagonal element; see Example 6.7.1.

EXAMPLE 6.7.1. Consider a matrix of the form (Jordan block):

$$R_n = \begin{pmatrix} \delta & 1 & & & \\ & \delta & 1 & & \\ & & \ddots & \ddots & \\ & & & \delta & 1 \\ & & & & \delta \end{pmatrix} \in \mathbf{R}^{n \times n}.$$

From inspection of the Gershgorin circles of $R_n^T R_n$, it follows that $(n-1)$ of the singular values are close to unity. Since their product equals $\det^{1/2}(R_n^T R_n) = \delta^n$, the remaining singular value approximately equals $\sigma_{\min} = \delta^n$. For $\delta = 0.1$ and $n = 20$ we thus have $\sigma_{\min} \approx 10^{-20}$, and yet no diagonal element is small! This ill-conditioning is much more severe than for the matrix in Example 2.7.1. ∎

More reliable rank revealing algorithms for sparse QR decompositions have been developed by Bischof and Hansen [80, 1992], Hwang, Lin, and Pierce [480, 1993], and Pierce and Lewis [659, 1995]. These are based on the techniques introduced in Section 2.7.5 using inverse iteration for determining rank and ill-conditioning in triangular matrices. Denote the first j columns of the final R by R_j. Assume that R_j is not too ill-conditioned, but when the next column is added $R_{j+1} = (R_j, r_{j+1})$ is found to be almost rank deficient. We then permute the column r_{j+1} to the end of the columns, and continue. This may happen several times during the numerical factorization, and at the end we have a QR factorization

$$(A_1, A_2) = Q \begin{pmatrix} R_1 & R_2 \\ 0 & S \end{pmatrix},$$

where $R_1 \in \mathbf{R}^{(n-r) \times (n-r)}$ is well-conditioned. An important fact stated in the theorem below is that R_1 will always fit into the storage structure predicted for R. In general R_2 and S will be dense, but provided $r \ll n$ this is often acceptable.

The following theorem is implicit in a paper by Foster [308, 1986].

THEOREM 6.7.1. Let $A = [a_1, a_2, \ldots, a_n]$ and let

$$A_{\mathcal{F}_k} = [a_{j_1}, a_{j_2}, \ldots, a_{j_r}], \qquad 1 \leq j_1 < j_2 < \cdots < j_r \leq n$$

be a submatrix of A. Denote the Cholesky factors of $A^T A$ and $A_{\mathcal{F}_k}^T A_{\mathcal{F}_k}$ by R and $R_{\mathcal{F}_k}$, respectively. Then the nonzero structure of $R_{\mathcal{F}_k}$ is included in the nonzero structure predicted for R under the no-cancellation assumption.

Proof. Consider the ordered graph $G = G(X, E)$ of $A^T A$. The ordered graph $G_{\mathcal{F}_k} = G_{\mathcal{F}_k}(X_{\mathcal{F}_k}, E_{\mathcal{F}_k})$ of $A_{\mathcal{F}_k}^T A_{\mathcal{F}_k}$ is obtained by deleting all nodes in G not in $\mathcal{F}_k = [j_1, j_2, \ldots, j_r]$ and all edges leading to the deleted nodes. It holds that $(R_\mathcal{F})_{ij} \neq 0$ only if there exists a path in $G_{\mathcal{F}_k}$ from node i to node j ($i < j$) through nodes with numbers less than i. If such a path exists in $G_{\mathcal{F}_k}$ it must exist also in G and hence we will have predicted $R_{ij} \neq 0$. ∎

6.7.2. Updating sparse least squares solutions.

We remarked earlier that a single dense row in A will lead to a full matrix $A^T A$ and therefore, invoking the no-cancellation assumption, to a full Cholesky factor R. Problems where the matrix A is sparse except for a few dense rows can be treated by first solving the problem with the dense rows deleted. The effect of the dense rows is then incorporated into the solution by updating. We stress that only the solution is updated, *not* the Cholesky factor.

Consider the problem

$$(6.7.1) \qquad \min_x \left\| \begin{pmatrix} A_s \\ A_d \end{pmatrix} x - \begin{pmatrix} b_s \\ b_d \end{pmatrix} \right\|_2,$$

where $A_s \in \mathbf{R}^{m_1 \times n}$ is sparse and $A_d \in \mathbf{R}^{m_2 \times n}$, $m_2 \ll n$, contains the dense rows. We assume for simplicity that $\operatorname{rank}(A_s) = n$. Denote by x_s the solution to the sparse problem

$$\min_x \|A_s x - b_s\|_2,$$

and let the corresponding Cholesky factor be R_s. The residual vectors in (6.7.1) corresponding to x_s are

$$(6.7.2) \qquad r_s(x_s) = b_s - A_s x_s, \quad r_d(x_s) = b_d - A_d x_s.$$

We now wish to compute the solution $x = x_s + z$ to the full problem (6.7.1), and hence to choose z to minimize $\|r_s(x)\|_2^2 + \|r_d(x)\|_2^2$, where

$$r_s(x) = r_s(x_s) - A_s z, \quad r_d(x) = r_d(x_s) - A_d z.$$

Since $A_s^T r_s(x_s) = 0$, this is equivalent to the problem

$$\min_z \{\|A_s z\|_2^2 + \|A_d z - r_d(x_s)\|_2^2\}.$$

Letting $u = R_s z$ and $B_d = A_d R_s^{-1}$, we have $\|A_s z\|_2 = \|u\|_2$, and the problem reduces to

$$(6.7.3) \qquad \min_u \left\| \begin{pmatrix} B_d \\ I_n \end{pmatrix} u - \begin{pmatrix} r_d(x_s) \\ 0 \end{pmatrix} \right\|_2^2.$$

This is equivalent to the minimum norm problem (cf. Theorem 2.5.1)

$$(6.7.4) \qquad \min \left\| \begin{pmatrix} u \\ v \end{pmatrix} \right\|_2 \quad \text{subject to} \quad (B_d \ \ I_{m_2}) \begin{pmatrix} u \\ v \end{pmatrix} = r_d(x_s),$$

where $v = r_d(x_s) - B_d u$. Since $C = (\, B_d \quad I_{m_2} \,) \in \mathbf{R}^{m_2 \times n}$ has full row rank we can compute the QR decomposition

$$C^T = \begin{pmatrix} B_d^T \\ I_{m_2} \end{pmatrix} = Q_C \begin{pmatrix} R_C \\ 0 \end{pmatrix},$$

and $R_C \in \mathbf{R}^{m_2 \times m_2}$ is nonsingular. Hence u can be obtained from

$$\begin{pmatrix} u \\ v \end{pmatrix} = Q_C \begin{pmatrix} R_C^{-T} r_d \\ 0 \end{pmatrix}.$$

Finally, solving $R_s z = u$ for z by back-substitution, we obtain the solution to (6.7.1) as $x = x_s + z$.

The updating scheme can be generalized to the case where the sparse subproblem has rank less than n; see Heath [442, 1982]. A general scheme for updating equality-constrained linear least squares solutions, where the constraints are also split into a sparse and a dense set, has been developed by Björck [90, 1984]. It is important to point out that these updating algorithms cannot be expected to be stable in all cases. Stability will be a problem whenever the sparse subproblem is more ill-conditioned than the full problem.

There are problems where even though A is fairly sparse in all rows and columns, the matrix $A^T A$ will be practically full. Large problems which have these characteristics occur, e.g., in image reconstruction and certain other inverse problems. These problems usually have to be solved by iterative methods.

6.7.3. Partitioning for out-of-core solution. Many large sparse linear least squares problems are so large that it is impossible to store even the Cholesky factor R. We now briefly describe a simple automatic partitioning scheme by George, Heath, and Plemmons [335, 1981] for solving such problems.

Assume that an appropriate ordering and a partitioning of the columns of A have been found, e.g., by the method of nested dissection in Section 6.5.3. Denote by Y_i the set of column indices in the ith partition, $i = 1, \ldots, p$. We now order first the rows having nonzero elements with column indices in Y_1, giving us a set of row indices Z_1. Among the unordered rows we now order the rows having nonzero elements with column indices in Y_2, and so on. This induces a partitioning of the row indices $\{Z_1, Z_2, \ldots, Z_p\}$ of A which can be defined as follows. Let $Z_0 = \emptyset$ and

$$Z_i = \{k \mid \exists j \in Y_i, \text{ with } a_{kj} \neq 0\} - \bigcup_{l=1}^{i-1} Z_l, \quad i = 1, 2, \ldots, p.$$

If the rows of A are permuted to appear in the order $Z_1, Z_2, \ldots, Z_p$ then A will have a block upper trapezoidal form with (in general) rectangular diagonal blocks A_{ii}, $i = 1, \ldots, p$, as depicted below:

$$(6.7.5) \qquad \begin{pmatrix} A_{11} & A_{12} & \cdots & A_{1p} \\ & A_{22} & \cdots & A_{2p} \\ & & \ddots & \vdots \\ & & & A_{pp} \end{pmatrix}.$$

For a matrix of block upper triangular form the sequential orthogonalization method can be applied to a *block row* at a time. In the first step only the blocks $A_{11}, \ldots, A_{1p}$ are processed, transforming A_{11} to upper triangular form. The first $|Y_1|$ rows of the resulting matrix are the first $|Y_1|$ rows of the final R and can be stored away. The remaining rows are adjoined to the next block row $A_{22} \ldots A_{2p}$ to give $\bar{A}_{22}, \ldots, \bar{A}_{2p}$, and now $\bar{A}_{22}$ is transformed into upper triangular form, etc. Hence we do a chain of sparse QR decompositions. If we assume that the rows of A are stored on auxiliary storage at all times, then the only main storage required is that for holding the $|Y_i|$ rows of R generated at step i, $i = 1, \ldots, p$. A slightly more efficient way to carry out this process is described in George, Heath, and Plemmons [335, 1981] where the data management is outlined in more detail.

The multifrontal method can also be adopted for out-of-core solution of large linear systems. Reid [682, 1984] describes a multifrontal method for the Cholesky factorization, which can be adopted also for the QR decompositions. Note that the scheme outlined above is a special case of a multifrontal method where the elimination tree is just a chain.

6.7.4. Computing selected elements of the covariance matrix. In Section 2.8.3 we discussed methods for computing the covariance matrix $\sigma^2 C$, where $C = (R^T R)^{-1}$, for the least squares solution x. When the matrix R is sparse, Golub and Plemmons [380, 1980] have shown that the algorithm (2.8.12)–(2.8.14) can be used to very efficiently compute *all elements in C, which are associated with nonzero elements in R*. Since R has a nonzero diagonal this includes the diagonal elements of C giving the variance of x. If R has bandwidth w, then the corresponding elements in C can be computed in only nw^2 flops by the algorithm below.

We define the index set K by

$$r_{ij} \neq 0, \quad (i,j) \in K; \quad r_{ij} = 0, \text{ otherwise.}$$

We will compute all elements c_{ij}, $(i,j) \in K$, in the upper triangular part of C.

Let f_k be the row index for the first nonzero element in the kth column of R, i.e., $f_k = \min_{1 \leq i \leq k-1}\{i \mid r_{ik} \neq 0\}$. We start with the last column of C and compute

$$c_{nn} = r_{nn}^{-2},$$

(6.7.6) $$c_{in} = -r_{ii} \sum_{\substack{j=i+1 \\ (i,j) \in K}}^{n} r_{ij} c_{jn}, \quad i = n-1, \ldots, f_n, \quad (i,n) \in K.$$

Assume now that we have computed all elements c_{ij}, $j = n, \ldots, k+1$, $i \leq j$, $(i,j) \in K$. Then from (2.8.13),

(6.7.7) $$c_{kk} = r_{kk}^{-1}\left[r_{kk}^{-1} - \sum_{\substack{j=k+1 \\ (k,j) \in K}}^{n} r_{kj} c_{kj}\right],$$

and similarly from (2.8.14), for $i = k-1, \ldots, f_k$,

$$(6.7.8) \qquad c_{ik} = -r_{ii}^{-1} \left[\sum_{\substack{j=i+1 \\ (i,j) \in K}}^{k} r_{ij} c_{kj} + \sum_{\substack{j=k+1 \\ (i,j) \in K}}^{n} r_{ij} c_{kj} \right], \qquad (i,k) \in K.$$

It can be shown that since R is the Cholesky factor of $A^T A$ its structure is such that $(i,j) \in K$ and $(i,k) \in K$ implies that $(j,k) \in K$ if $j < k$ and $(k,j) \in K$ if $j > k$. Hence all elements needed in (6.7.6)–(6.7.8) have been computed.

6.8. Sparse Constrained Problems

6.8.1. An active set method for problem BLS.
In Section 5.2 we considered active set methods for solving problem LSI

$$(6.8.1) \qquad \min_x \|Ax - b\|_2 \quad \text{subject to} \quad l \leq Cx \leq u.$$

It is often the case that the matrices A and C in this problem are sparse. Unfortunately, it is difficult to take advantage of this sparsity, since it is usually destroyed by the sequence of transformations applied to A and C during the iterations in an active set method. However, for the least squares problem with simple bounds (BLS),

$$(6.8.2) \qquad \min_x \|Ax - b\|_2, \quad \text{subject to} \quad l \leq x \leq u,$$

it is possible to preserve sparsity in the active set Algorithm 5.2.1

An application of problem BLS occurs in polishing large optics. Calculating the amount of material to be removed requires the solution of a linear least squares problem. The nonnegativity constraints come in because material cannot be added to the surface by polishing. A typical problem might have about 8,000 to 20,000 equations and the same number of unknowns, with a few percent of the matrix elements being nonzero. In general the problem is rank deficient, and the nonnegativity constraints are active at a significant fraction of the elements of the solution vector.

We now describe the algorithm given in Björck [93, 1988] for problem BLS. We assume that initially a sparse QR decomposition of A is computed. If the orthogonal transformations are applied also to b, we have

$$(6.8.3) \qquad Q^T A P_c = \begin{pmatrix} R \\ 0 \end{pmatrix}, \qquad Q^T b = \begin{pmatrix} c \\ d \end{pmatrix},$$

where P_c is the column permutation performed for sparsity. This decomposition can be obtained by the sequential QR algorithm or the multifrontal algorithm described in Section 6.6.4. This reduces the problem to the upper triangular form

$$(6.8.4) \qquad \min_x \|Rx - c\|_2, \quad \text{subject to} \quad l \leq x \leq u,$$

where R is sparse.

We divide the index set of x according to $[1, 2, \ldots, n] = \mathcal{F} \cup \mathcal{B}$, where $i \in \mathcal{F}$ if x_i is a free variable, and $i \in \mathcal{B}$ if x_i is fixed at its lower or upper bound. We will assume that $\mathcal{F}$ and $\mathcal{B}$ are ordered sets, with indices ordered in increasing order. To this partitioning corresponds a permutation matrix $P = (E_\mathcal{F}, E_\mathcal{B})$, where $E_\mathcal{F}$ and $E_\mathcal{B}$ consist of the columns e_i of the unit matrix for which $i \in \mathcal{F}$ and $i \in \mathcal{B}$, respectively.

Following Algorithm 5.2.1 we choose an initial solution $x^{(0)}$ satisfying $l < x^{(0)} < u$, and take
$$\mathcal{F}_0 = [1, 2, \ldots, n], \qquad R_{\mathcal{B}_0} = \emptyset,$$
so that $R_{\mathcal{F}_0} = R$. (The reason for this is that, as will become apparent, it is a cheaper and more stable operation to fix a free variable than the opposite operation.)

Let $x^{(k)}$ be the iterate at the kth step ($k = 0, 1, \ldots$) and write

(6.8.5) $$x^{(k)}_{\mathcal{F}_k} = E^T_{\mathcal{F}_k} x^{(k)}, \qquad x^{(k)}_{\mathcal{B}_k} = E^T_{\mathcal{B}_k} x^{(k)},$$

for the free and fixed parts of the solution. The unconstrained problem (6.8.4) with the variables $x^{(k)}_{\mathcal{B}_k}$ fixed becomes

(6.8.6) $$\min_{x_{\mathcal{F}_k}} \|R_{\mathcal{F}_k} x_{\mathcal{F}_k} - c_k\|_2, \qquad c_k = c - R_{\mathcal{B}_k} x^{(k)}_{\mathcal{B}_k},$$

where $RP_k = (RE_{\mathcal{F}_k}, RE_{\mathcal{B}_k}) = (R_{\mathcal{F}_k}, R_{\mathcal{B}_k})$. To simplify the discussion we assume in the following that the matrix $R_{\mathcal{F}_k}$ has full column rank, so that (6.8.6) has a unique solution. This is always the case if $\text{rank}(A) = n$.

To solve (6.8.6) we need the QR decomposition of $R_{\mathcal{F}_k}$. We obtain this by considering the first block of columns of the QR decomposition

(6.8.7) $$Q_k^T (R_{\mathcal{F}_k}, R_{\mathcal{B}_k}) = \begin{pmatrix} U_k & S_k \\ 0 & V_k \end{pmatrix}, \qquad Q_k^T c = \begin{pmatrix} d_k \\ e_k \end{pmatrix}.$$

The solution to (6.8.6) is now given by
$$U_k x^{(k)}_{\mathcal{F}_k} = d_k - S_k x^{(k)}_{\mathcal{B}_k},$$
and we take
$$x^{(k+1)} = x^{(k)} + \alpha(z^{(k)} - x^{(k)}),$$
where $z^{(k)} = E_{\mathcal{F}_k} x^{(k)}_{\mathcal{F}_k} + E_{\mathcal{B}_k} x^{(k)}_{\mathcal{B}_k}$ and α is a nonnegative step length. (Note that $z^{(0)}$ is just the solution to the unconstrained problem (6.8.2).)

Let $\bar{\alpha}$ be the maximum value of α, for which $x^{(k+1)}$ remains feasible. There are now two possibilities.

If $\bar{\alpha} < 1$, then $z^{(k)}$ is not feasible. We then take $\alpha = \bar{\alpha}$, and move all indices $q \in \mathcal{F}_k$ for which $x_q^{(k+1)} = l_q$ or u_q from $\mathcal{F}_k$ to $\mathcal{B}_k$. Thus the free variables which hit their lower or upper bounds will be fixed for the next iteration step.

If $\bar{\alpha} \geq 1$, then we take $\alpha = \bar{\alpha}$. Then $x^{(k+1)} = z^{(k)}$ equals the unconstrained minimum when the variables $x_{\mathcal{B}_k}$ are kept fixed. The Lagrange multipliers are

6.8. Sparse Constrained Problems

then checked to see if the objective function can be decreased further by freeing one of the fixed variables. If not, we have found the global minimum.

At each iteration step the sets $\mathcal{F}_k$ and $\mathcal{B}_k$ are changed. If a constraint is dropped a column from $R_{\mathcal{B}_k}$ is moved to $R_{\mathcal{F}_k}$; if a constraint is added a column is moved from $R_{\mathcal{F}_k}$ to $R_{\mathcal{B}_k}$. The solution of the sequence of unconstrained problems (6.8.6) and the computation of the corresponding Lagrange multipliers can be efficiently achieved, provided that the QR decomposition (6.8.7) can be updated. We now consider the feasibility in the sparse case of computing this sequence of decompositions as the active set algorithm proceeds. We use a simple example to illustrate the problem encountered.

EXAMPLE 6.8.1. Assume that the matrix R in (6.8.3) is a banded upper triangular matrix of bandwidth $w = 3$. In (6.8.8) we show the structure of the matrix $\bar{R} = (R_{\mathcal{F}_k}, R_{\mathcal{B}_k})$ for $n = 11$ and the partitioning $\mathcal{F}_k = [2,3,4,6,9,10,11]$, $\mathcal{B}_k = [1,5,7,8]$:

$$(6.8.8) \qquad \bar{R} = \begin{pmatrix} * & * & & & & & * & & & & \\ * & * & * & & & & & & & & \\ & * & * & & & & & * & & & \\ & & * & * & & & & * & & & \\ & & & * & & & & * & * & & \\ & & & & * & & & & * & * & \\ & & & & & * & & & * & * & \\ & & & & & * & * & & & & * \\ & & & & & & * & * & * & & \\ & & & & & & & * & * & & \\ & & & & & & & & * & & \end{pmatrix}.$$

If this matrix $\bar{R}$ is transformed into upper triangular form by symbolically performing a sequence of plane rotations the structure becomes

$$(6.8.9) \qquad \begin{pmatrix} U & S \\ 0 & V \end{pmatrix} = \begin{pmatrix} * & * & * & & & & & * & & & \\ & * & * & & & & & * & * & & \\ & & * & * & & & & * & * & & \\ & & & * & & & & * & * & * & * \\ & & & & * & * & * & & * & * & \\ & & & & & * & * & & * & * & \\ & & & & & & * & & * & * & \\ & & & & & & & * & * & * & * \\ & & & & & & & & * & * & * \\ & & & & & & & & & * & * \\ & & & & & & & & & & * \end{pmatrix}.$$

We observe that although the block U is still sparse, the lower right block V has filled in completely. Since we do not know in advance the set B of active constraints at the solution the above approach of updating the full decomposition (6.8.7) is feasible only if for all possible partitionings $[1,2,\ldots,n] = \mathcal{F}_k \cup \mathcal{B}_k$ the R-factor of $(R_{\mathcal{F}_k}, R_{\mathcal{B}_k})$ remains sparse. Note that we are free to order the columns only within the sets $\mathcal{F}$ and $\mathcal{B}_k$, and all columns in $\mathcal{B}_k$ must be ordered after those in $\mathcal{F}_k$. ∎

We now pursue and generalize the above example. Let the matrix R be banded upper triangular of bandwidth $w = 3$. Then $R^T R$ is a penta-diagonal symmetric matrix. We take (n even)

$$\mathcal{F}_k = [1, 3, \ldots, n-1], \qquad \mathcal{B}_k = [2, 4, \ldots, n],$$

and hence eliminate all odd nodes first. The fill-in in the elimination can be determined from the elimination graph model; see Section 6.4.4. It is easy to verify that the block V in the factorization becomes a full upper triangular matrix. Similarly, the block S will fill in its lower triangular part. The total number of new nonzero elements created in the decomposition will be $(n/2 - 2)(n/2 - 1) = n^2/4 - (3/2)n + 2$. Clearly this is not acceptable.

We conclude that even when R has a simple band structure it is not feasible to recur the full QR decomposition (6.8.7). We now show that the alternative of keeping only the factor U_k associated with $R_{\mathcal{F}_k}$ can be implemented in a stable and efficient way. The key to this is the fact that by Theorem 6.7.1 the structure of the factor U_k corresponding to $R_{\mathcal{F}_k}$ will always be contained in the predicted structure of the initial matrix R.

We now describe in more detail an algorithm for in-place updating of $U_{\mathcal{F}_k}$ in the factorization

$$Q_{\mathcal{F}_k}^T R_{\mathcal{F}_k} = \begin{pmatrix} U_{\mathcal{F}_k} \\ 0 \end{pmatrix}.$$

(i) DELETING COLUMN q FROM $R_{\mathcal{F}_k}$: We perform a sequence of plane rotations to annihilate the nonzero elements in the qth row from left to right. A nonzero element in position (q, j), $j > q$, is annihilated by a rotation of rows q and j. This creates a nonzero element in position (q, j), which need not be computed since we are going to delete the qth column. It may also create intermediate fill-in in the qth row outside the given data structure. However, there can be no fill-in outside the data structure in the jth row. At the end, by deleting the qth column, we get the updated $U_{\mathcal{F}_k}$ stored in the original data structure.

We illustrate this algorithm by a simple example. Let $U_{\mathcal{F}_k}$ be upper triangular with bandwidth $w = 3$, and take $n = 8, q = 4$. In Figure 6.8.1 we denote by + fill-in elements (not computed), by (*) canceled elements, and by (#) intermediate fill-in. Note that there is no space in the data structure for the elements (#) Therefore it is convenient to store the qth row in a dense working vector.

(ii) ADDING COLUMN q TO $R_{\mathcal{F}_k}$: Note that the column has to be inserted in its proper place for Theorem 6.7.1 to apply. We will do the updating in two steps. First we update adding the new column after the last column in $U_{\mathcal{F}_k}$. This column will then in general be a full column. Second, we permute it in place and apply transformations to reduce the resulting matrix to upper triangular form. The first step will be discussed in the next section. The second step is achieved by performing the algorithm for deleting a column in reverse order, and we now describe this step in more detail.

The elements in the qth column above the main diagonal are elements in the final updated $U_{\mathcal{F}_k}$ and thus can be placed directly in the fixed data structure.

6.8. SPARSE CONSTRAINED PROBLEMS

$$\begin{pmatrix} * & * & * & & & & & \\ & * & * & * & & & & \\ & & * & * & * & & & \\ & & & * & (*) & (*) & (\#) & (\#) \\ & & & + & * & * & * & \\ & & & + & & * & * & * \\ & & & + & & & * & * \\ & & & + & & & & * \end{pmatrix}$$

FIG. 6.8.1. *Deleting column q from $R_{\mathcal{F}_k}$.*

The elements below the main diagonal are in general nonzero, and we annihilate these from bottom up. The element $(j,q), j > q$ is annihilated by a rotation in the rows q and j, which will create fill-in in the qth row. (Note that the qth row is initially zero except for its diagonal element.) *No fill-in outside the data structure can occur in the jth row.* After the elements below the diagonal in the qth column have been eliminated we have obtained an upper triangle, which by Theorem 6.7.1 will fit into the fixed data structure. Hence, any intermediate fill-in which may occur in row q must eventually cancel out.

We again use Figure 6.8.1 to illustrate the process. Column q and row q are stored in two dense working vectors and at the end inserted into the fixed sparse data structure. The intermediate fill-in (#) created in the first two steps must eventually cancel.

When the column r_{j_q} is adjoined after the last column in $R_{\mathcal{F}_k}$:

$$(R_{\mathcal{F}_k}, r_q) = (r_j, \ldots, r_{j_r}, r_{j_q}),$$

the updated upper triangular factor has the form

$$\begin{pmatrix} U_{\mathcal{F}_k} & u \\ 0 & \gamma \end{pmatrix},$$

where u and γ are to be determined. If $Q_{\mathcal{F}_k}$ is available this updated factor can be stably computed. However, since the matrix $Q_{\mathcal{F}_k}$ in general is not sparse, it is not feasible to keep and update $Q_{\mathcal{F}_k}$. The alternative, to store the sequence of rotations defining $Q_{\mathcal{F}_k}$, is also not attractive, since then the amount of storage needed cannot be predicted.

Since $Q_{\mathcal{F}_k}$ is not available we use the method of corrected seminormal equations (CSNE) (see Section 6.6.6) to compute u and γ as follows. Let z be the solution to the least squares problem

(6.8.10) $$\min_z \|R_{\mathcal{F}_k} z - r_{j_q}\|_2^2.$$

(It is assumed that a copy of the initial factor R in (6.8.3) is saved so that r_{j_q} can be retrieved.) We solve (6.8.10) using the CSNE. Then

$$u = U_{\mathcal{F}_k} z, \qquad \gamma = \|r_{j_q} - R_{\mathcal{F}_k} z\|_2.$$

Numerical results illustrating the stability of this algorithm are given in Björck [93, 1988].

In the algorithm above the set of free and fixed variables only change in one element in each iteration. If many variables are fixed at the solution, then this strategy may be inefficient. Portugal, Júdice, and Vicente [665, 1994] consider more general, so-called block pivoting strategies, which allow the set of free and fixed elements to change by many elements in each step. For these algorithms it may no longer be efficient to update the QR decomposition. Instead, the unconstrained subproblem in each iteration step is solved by computing a new sparse QR decomposition from scratch. Computational experience with a block pivoting algorithm is reported in [665].

6.8.2. Interior point methods for problem BLS. We consider here only problem NNLS (see Section 5.2.1)

$$\min_x \|Ax - b\|_2 \quad \text{subject to} \quad x \geq 0,$$

which is a special case of problem BLS. The algorithm can be generalized to solve problems with both lower and upper bounds in a fairly straightforward way.

The Karush–Kuhn–Tucker optimal conditions for NNLS give rise to an equivalent monotone linear complementarity problem (LCP)

(6.8.11) $\qquad y = A^T(Ax - b), \quad y \geq 0, \quad x \geq 0, \quad x^T y = 0.$

These conditions can be written as a system of nonlinear equations

$$F_1(x, y) = XYe = 0,$$

where $y = A^T(Ax - b)$, $x, y \geq 0$, and

$$X = \text{diag}(x), \quad Y = \text{diag}(y), \quad e = (1, 1, \ldots, 1)^T.$$

This system is the basis of the first of two primal-dual interior point methods for problem NNLS, developed by Portugal, Júdice, and Vicente [665, 1994]. The second uses the Newton directions for the nonlinear system

$$F_2(x, y) = \begin{pmatrix} XYe \\ A^T(Ax - b) - y \end{pmatrix} = 0,$$

where the iterands are not forced to satisfy the linear constraints $y = A^T(Ax - b)$.

In the algorithms a sequence of points $\{x_k\}$ are computed by

$$(x_{k+1}, y_{k+1}) = (x_k, y_k) + \theta_k(u_k, v_k),$$

6.8. SPARSE CONSTRAINED PROBLEMS

where θ_k is a positive step size. The Newton direction (u_k, v_k) for the second algorithm satisfies the linear system

$$(6.8.12) \quad \begin{pmatrix} Y_k & X_k \\ A^T A & -I \end{pmatrix} \begin{pmatrix} u_k \\ v_k \end{pmatrix} = \begin{pmatrix} -X_k y_k + \mu_k e \\ A^T r_k + y_k \end{pmatrix},$$

where $r_k = b - Ax_k$, $X_k = \text{diag}(x_k)$, $Y_k = \text{diag}(y_k)$, and μ_k is a centralization parameter; see Lustig, Marsden, and Shanno [554, 1991]. The step size θ_k is chosen to satisfy

$$\theta_k \leq \begin{cases} \theta_k^{\max} & \text{if } u_k^T v_k \leq 0; \\ \min(\theta_k^{\max}, \hat{\theta}_k) & \text{if } u_k^T v_k > 0, \end{cases}$$

where $\theta_k^{\max}$ is the largest value such that $x_{k+1} \geq 0$, $y_{k+1} \geq 0$, and

$$\hat{\theta}_k = (x_k^T y_k - n\mu_k)/(u_k^T v_k).$$

Choosing θ_k in this way can be shown to guarantee a monotonic decrease of $g(x, y) = x^T y$ in each iteration. Computational experience has shown that the condition $\theta_k^{\max} < \hat{\theta}_k/2$ is usually satisfied, and in practice one takes $\theta_k = 0.99995 \cdot \theta_k^{\max}$.

From (6.8.12) it can be seen that u_k is the solution to the least squares problem

$$(6.8.13) \quad \min_{u_k} \left\| \begin{pmatrix} A \\ (X_k Y_k)^{-1/2} \end{pmatrix} u_k - \begin{pmatrix} r_k \\ (X_k Y_k)^{-1/2} \mu_k e \end{pmatrix} \right\|_2.$$

After u_k has been calculated v_k is determined from the first block equation in (6.8.12),

$$v_k = -y_k + X_k^{-1}(\mu_k e - Y_k u_k).$$

The above approach can be improved by using a predictor corrector scheme. In this scheme one first determines a first direction (u_k, v_k) by taking $\mu_k = 0$ in (6.8.12). This direction is corrected by

$$(z_k, w_k) = (u_k, v_k) + (\bar{u}_k, \bar{v}_k),$$

where $(\bar{u}_k, \bar{v}_k)$ satisfies

$$(6.8.14) \quad \begin{pmatrix} Y_k & X_k \\ A^T A & -I \end{pmatrix} \begin{pmatrix} \bar{u}_k \\ \bar{v}_k \end{pmatrix} = \begin{pmatrix} -U_k V_k e + \mu_k e \\ 0 \end{pmatrix},$$

and $U_k = \text{diag}(u_k)$, $V_k = \text{diag}(v_k)$. When u_k and v_k have been computed, z_k can be computed as the solution of the least squares problem

$$\min_{z_k} \left\| \begin{pmatrix} A \\ (X_k Y_k)^{-1/2} \end{pmatrix} u_k - \begin{pmatrix} r_k \\ (X_k Y_k)^{-1/2}(\mu_k - U_k V_k)e \end{pmatrix} \right\|_2.$$

Finally w_k is found from

$$w_k = -y_k + X_k^{-1}(\mu_k e - Y_k z_k - U_k v_k).$$

Following [554] the parameter μ_k is taken as

$$\mu_k = (x_k + \theta_k u_k)^T (y_k + \theta_k v_k)/n^2,$$

with θ_k as above. This choice does not guarantee a decrease in $g(x,y)$, but works well in practice. Implementation issues are discussed and computational experience presented in [665].

In this method the subproblems (6.8.13) and (6.8.14) have to be solved from scratch at each iteration, since no reliable updating methods are available. Portugal, Júdice, and Vicente [665, 1994] report computational experience with the predictor-corrector algorithm for problem NNLS both for random matrices and matrices from the Harwell–Boeing collection in Duff, Grimes, and Lewis [242, 1989]. They found this method to give high accuracy even when, for better efficiency, the subproblems were solved by forming the normal equation and computing a sparse Cholesky factorization.

6.9. Software and Test Results

6.9.1. Software for sparse direct methods.
Below we list and briefly comment on some often-used software packages for solving sparse least squares problems with direct methods. Problem-related issues such as dimension, sparsity and structure, and conditioning should be considered when determining the choice of direct method to be used. Possible rank deficiency, occurrence of weighted rows, and the number of right-hand sides are other issues to be considered. An overview of available codes is given in Table 6.9.1.

Several of the subroutines are in the public domain and available, e.g., from netlib. Access to netlib is via the Internet address *netlib@ornl.gov*, which refers to a gateway machine at Oak Ridge National Laboratory in Oak Ridge, Tennessee. This address should be understood on all the major networks. For access from Europe, there is a duplicate collection in Oslo at *netlib@nac.no*. For the Pacific, try *netlib@draci.cs.uow.edu.au*, located at the University of Wollongong, NSW, Australia.

MATLAB, distributed by The MathWorks, has been extended to include sparse matrix storage and operations. The operations included are described in Gilbert, Moler, and Schreiber [351, 1992]. In particular, a minimum degree preordering algorithm and a sparse Cholesky decomposition are included. For example, to solve a least squares problem by the method of normal equations with one step of fixed precision iterative refinement (see Section 6.6.5), one writes

```
q=colmmd(A);          % Find minimum degree ordering of A
A=A(:,q);             % Permute column of A
R=chol(A'*A);         % Sparse Cholesky decomposition
x=R\(A'*b);           % Least squares solution
x=x+R\(A'*(b-A*x))    % Perform one correction step
x(q)=x;               % Permute the solution
```

There is also a built-in sparse least squares solver in MATLAB. This currently uses the augmented system formulation with the scaling parameter chosen to be

6.9 SOFTWARE AND TEST RESULTS

TABLE 6.9.1
Survey of some commonly used software packages.

Package	Purpose	Author	Distribution
MATLAB	LU and Cholesky	Gilbert et al. [351]	The MathWorks Inc.
SQR	Matlab QR	Matstoms [572]	netlib
MA27	LDL^T	Duff and Reid [251]	HSL
MA47	LDL^T	Duff et al. [241]	HSL
QR27	Householder QR	Matstoms [573, 574]	*qr27@math.liu.se*
SPARSPAK-A	Cholesky	Chu et al. [167]	Univ. of Waterloo
SPARSPAK-B	Givens QR	George and Ng [344]	Univ. of Waterloo
YSMP	Cholesky	Eisenstat et al.[265]	Yale
LLSS01	Incomplete QR	Zlatev and Nielsen [859]	Tech. Univ. Denm.

$\alpha = 10^{-3} \max |a_{ij}|$. The solution is computed using the minimum degree ordering and the built-in sparse LU decomposition.

Matstoms [572, 1994] has developed a multifrontal sparse QR algorithm to be used with MATLAB. This is implemented as four M-files, which are available from netlib. The main routine is called sqr, and the statements q=colmmd(A); [R,p,c]=sqr(A(:,q),b) will compute the factor R in a sparse QR decomposition of $A(:,q)$, and $c = Q^T b$. For further details we refer to [572].

More recently C. Sun[2] has developed another software package for computing a sparse QR decomposition This package is implemented in C and designed to be used within a MATLAB environment. C. Sun has also developed a parallel multifrontal algorithm for sparse QR decomposition on distributed-memory multiprocessors; see [763, 1996].

Pierce and Lewis [659, 1995] at Boeing have implemented a multifrontal sparse rank revealing QR decomposition/least squares solution module. This code has some optimization for vector computers in general, but it also works very well on a wide variety of scientific workstations. It is included in the collection of sparse matrix, and very large dense matrix codes, which are available in the commercial software package, BCSLIB-EXT, from Boeing Information and Support Services, Seattle. This library of FORTRAN callable routines is also given to researchers in laboratories and academia for testing and comparing and as a professional courtesy.

The Harwell Subroutine Library (HSL) has a subroutine MA45 for solving the normal equations. If the least squares problem is written in the augmented system form (2.5.10), then the multifrontal subroutine MA27 for solving symmetric indefinite linear systems can be used. However, the MA27 code does not exploit the special structure of the augmented system, and there is a new routine MA47 which caters explicitly to this kind of system.

---
[2]Advanced Computing Research Institute, Cornell University, *csun@cs.cornell.edu*.

Closely related to the Harwell MA27 code is a subroutine QR27 that has been developed by Matstoms [571, 1992]. This code is available for academic research and can be ordered from *pomat@math.liu.se*. A parallel version of QR27 has been developed for shared-memory MIMD computers; see Matstoms [574, 1995].

SPARSPAK is a collection of routines for solving sparse systems of linear systems developed at the University of Waterloo. It is divided into two portions: SPARSPAK-A deals with sparse symmetric positive definite systems and SPARSPAK-B handles sparse linear least squares problems, including linear equality constraints. For solving least squares problems, both A and B parts are needed. The original reference is George and Heath [333, 1980]. A more recent paper by Heath [443, 1984] details some of the extensions to this algorithm as well as some alternatives.

SPARSPAK-B has the feature that dense rows of A, which would cause R to fill, can be withheld from the decomposition and the final solution updated to incorporate them at the end. Only the upper triangular factor is maintained, and the Givens rotations are not saved. SPARSPAK is licensed and distributed by the University of Waterloo, Canada.[3]

Zlatev and Nielsen [859, 1979] have developed a Fortran subroutine called LLSS01, which uses fast Givens rotations to perform the QR decomposition. The orthogonal matrix Q is not stored, and elements in R smaller than a user-specified tolerance are dropped. The solution is computed using fixed precision iterative refinement or, alternatively, preconditioned conjugate gradient, with the computed matrix R as preconditioner; see [860, 1988].

We finally mention the Sparse Matrix Manipulation System (SMMS), developed by Alvarado, which is described in [10, 1990]. SMMS is a collection of directly executable sparse matrix commands, and includes routines for sparse orthogonal decomposition. This package is available from *eceserv0.ece.wisc.edu* in directories *pub/smms93* and *pub/smmspc*.

6.9.2. Test results. We report here on a comparison of accuarcy and execution time for four different numerical methods for sparse least squares problems made by Matstoms [572, 1994]. The experiments are carried out on nine of the matrices from the Harwell–Boeing test collection (Duff, Grimes, and Lewis [242, 1989]), together with five matrices (ARTFnnnn) formed by the merging of two Harwell–Boeing matrices,

$$A = \begin{pmatrix} WELLnnnn \\ 0 \quad ASHnnn \end{pmatrix}.$$

Table 6.9.2 summarizes dimensions and other chracteristics of these test problems. The numerical values of the ABB and ASH matrices are random numbers uniformly distributed in $[-1, 1]$, while WELL and ILLC have their

---

[3]Detailed information can be obtained from Mr. Peter Sprung, Department of Computing Services, University of Waterloo, Waterloo, Ontario N2L 3G1, Canada.

TABLE 6.9.2
Characteristics of test problems.

Matrix	m	n	nnz	$\kappa(A)$ Set 1	$\kappa(A)$ Set 2	Description
ABB313	313	176	1,557	$1.5 \cdot 10^1$	$1.5 \cdot 10^7$	Sudan survey
ASH219	219	85	438	$7.8 \cdot 10^0$	$7.1 \cdot 10^6$	Geodesy problem
ASH331	331	104	662	$5.2 \cdot 10^0$	$5.4 \cdot 10^6$	Geodesy problem
ASH608	608	188	1,216	$6.3 \cdot 10^1$	$6.4 \cdot 10^6$	Geodesy problem
ASH958	958	292	1,916	$6.9 \cdot 10^1$	$6.9 \cdot 10^6$	Geodesy problem
WELL1033	1,033	320	4,732	$1.7 \cdot 10^2$	$1.1 \cdot 10^8$	Gravity-meter
WELL1850	1,850	712	8,758	$1.1 \cdot 10^2$	$2.0 \cdot 10^7$	Gravity-meter
ILLC1033	1,033	320	4,732	$1.9 \cdot 10^4$	$3.1 \cdot 10^9$	Gravity-meter
ILLC1850	1,850	712	8,758	$1.4 \cdot 10^3$	$1.3 \cdot 10^9$	Gravity-meter
ARTF1252	1,252	320	5,170	$3.6 \cdot 10^1$	$3.4 \cdot 10^7$	WELL1033,ASH219
ARTF1641	1,641	320	5,948	$4.1 \cdot 10^0$	$3.0 \cdot 10^6$	WELL1033,ASH608

original values. The WELL and ILLC matrices have the same nonzero structure but different numerical values. A second set of test matrices are constructed by down-weighting the rows $(n-1,\ldots,m)$ in the Harwell-Boeing matrices by the factor 16^{-5}. A set of consistent least squares problems is defined by taking the exact solution to be $x = (1,\ldots,1)^T$, and $b = Ax$.

The following four methods were compared using MATLAB. The first, aug, is the built-in least squares solver in MATLAB which uses a sparse LU decomposition of the augmented system. The second method, qls, uses the sparse QR decomposition sqr by Matstoms, which applies the transformation Q^T to the right-hand side. The third method, csne, uses the same QR decomposition but then applies the corrected seminormal equations. The fourth method, cne, the corrected normal equations, uses the built-in sparse Cholesky decomposition with one step of refinement.

Results are given in Table 6.9.3 for the unweighted test matrices. These problems are all well-conditioned or (ILLC1033, ILLC1850) moderately ill-conditioned. The error shown is the relative error $\|x - \bar{x}\|_2 / \|x\|_2$ in the computed solution $\bar{x}$. Here all four methods give high accuracy. The execution times vary a lot. The method cne is faster than the two methods using the QR decomposition by a factor of about 10. This significant difference is partly explained by the fact that the Cholesky decomposition routine is implemented in the core of MATLAB, while the QR decomposition routine is implemented using M-files. Normally we would expect the QR decomposition to be slower by a factor of 2–3. A comparison of execution times between methods aug and qls shows no consistent behavior. For the problems ABB313 and ASH219–ASH958 aug is about twice as fast. On the other hand, for problem WELL1850, aug is 5 times slower, and for ARTF1641, more than 30 times slower. This illustrates the fact that the fill-in in the factors computed in the augmented systems method is not linked to the fill-in in the

TABLE 6.9.3
Errors and execution times for test set 1.

Problem	aug	qls	csne	cne	aug	qls	csne	cne
ABB313	$5.4 \cdot 10^{-16}$	$9.9 \cdot 10^{-16}$	$1.6 \cdot 10^{-16}$	$1.9 \cdot 10^{-16}$	2.5	4.9	4.5	0.8
ASH219	$3.7 \cdot 10^{-16}$	$3.8 \cdot 10^{-16}$	$8.4 \cdot 10^{-17}$	$6.0 \cdot 10^{-17}$	0.9	2.2	2.0	0.2
ASH331	$2.7 \cdot 10^{-16}$	$4.0 \cdot 10^{-16}$	$7.0 \cdot 10^{-17}$	$6.4 \cdot 10^{-17}$	1.7	2.7	2.4	0.3
ASH608	$3.4 \cdot 10^{-16}$	$4.3 \cdot 10^{-16}$	$5.9 \cdot 10^{-17}$	$6.2 \cdot 10^{-17}$	3.2	5.0	4.5	0.6
ASH958	$3.5 \cdot 10^{-16}$	$5.2 \cdot 10^{-16}$	$6.6 \cdot 10^{-17}$	$6.4 \cdot 10^{-17}$	5.6	8.8	7.9	1.0
WELL1033	$3.4 \cdot 10^{-15}$	$2.5 \cdot 10^{-15}$	$1.0 \cdot 10^{-15}$	$6.6 \cdot 10^{-16}$	17	16	15	1.3
WELL1850	$2.6 \cdot 10^{-15}$	$2.1 \cdot 10^{-15}$	$3.5 \cdot 10^{-16}$	$4.4 \cdot 10^{-16}$	152	30	29	3.2
ILLC1033	$1.7 \cdot 10^{-13}$	$8.8 \cdot 10^{-14}$	$4.3 \cdot 10^{-14}$	$7.2 \cdot 10^{-14}$	17	16	15	1.3
ILLC1850	$5.8 \cdot 10^{-14}$	$2.1 \cdot 10^{-14}$	$6.7 \cdot 10^{-15}$	$5.6 \cdot 10^{-15}$	152	31	28	3.2
ARTF1252	$1.2 \cdot 10^{-15}$	$1.5 \cdot 10^{-15}$	$1.3 \cdot 10^{-16}$	$1.7 \cdot 10^{-16}$	51	22	19	1.8
ARTF1641	$1.1 \cdot 10^{-15}$	$8.3 \cdot 10^{-16}$	$1.0 \cdot 10^{-16}$	$8.8 \cdot 10^{-17}$	774	26	24	2.5

R-factor in any simple way.

Since the sparsity structures of the second set of test problems are the same, the execution times are identical to those for the first set. For these ill-conditioned problems the accuracy of the four methods varies widely; see Table 6.9.4. In all cases method csne was the most accurate, followed by qls. The method cne did well on some of the problems, but gave poor accuracy or failed on the most ill-conditioned ones. These results are consistent with the expected behavior, see Section 6.6.5. That the accuracy of aug was also worse on some of the problems is related to the less-than-optimal scaling parameter used.

TABLE 6.9.4
Errors for test set 2.

Test problem	aug	qls	csne	cne
ABB313	$1.5 \cdot 10^{-14}$	$4.0 \cdot 10^{-11}$	$1.3 \cdot 10^{-16}$	$3.9 \cdot 10^{-9}$
ASH219	$4.0 \cdot 10^{-16}$	$1.6 \cdot 10^{-11}$	$6.6 \cdot 10^{-17}$	$1.4 \cdot 10^{-16}$
ASH331	$1.2 \cdot 10^{-14}$	$5.1 \cdot 10^{-12}$	$1.1 \cdot 10^{-16}$	$1.0 \cdot 10^{-16}$
ASH608	$8.8 \cdot 10^{-16}$	$5.2 \cdot 10^{-12}$	$8.4 \cdot 10^{-17}$	$1.9 \cdot 10^{-16}$
ASH958	$4.2 \cdot 10^{-15}$	$1.3 \cdot 10^{-11}$	$6.7 \cdot 10^{-17}$	$7.2 \cdot 10^{-17}$
WELL1033	$7.2 \cdot 10^{-7}$	$5.2 \cdot 10^{-11}$	$3.0 \cdot 10^{-13}$	$7.4 \cdot 10^{-3}$
WELL1850	$1.5 \cdot 10^{-8}$	$5.4 \cdot 10^{-11}$	$1.4 \cdot 10^{-13}$	$9.8 \cdot 10^{-7}$
ILLC1033	$9.9 \cdot 10^{-10}$	$9.5 \cdot 10^{-10}$	$5.6 \cdot 10^{-10}$	Failed
ILLC1850	$3.1 \cdot 10^{-5}$	$5.6 \cdot 10^{-10}$	$9.5 \cdot 10^{-12}$	$6.5 \cdot 10^{-8}$
ARTF1252	$6.4 \cdot 10^{-8}$	$1.4 \cdot 10^{-9}$	$1.7 \cdot 10^{-13}$	$5.3 \cdot 10^{-6}$
ARTF1641	$1.0 \cdot 10^{-10}$	$1.0 \cdot 10^{-10}$	$2.7 \cdot 10^{-14}$	$1.3 \cdot 10^{-9}$

Chapter 7
Iterative Methods For Least Squares Problems

7.1. Introduction

In this chapter we consider the iterative solution of large sparse least squares problems

$$\min_x \|Ax - b\|_2.$$

We assume in the following, unless otherwise stated, that A has full column rank, so that the problem has a unique solution.

In principle any iterative method for symmetric positive definite linear systems can be applied to the system of normal equations $A^T A x = A^T b$. The explicit formation of the matrix $A^T A$ can be avoided by using the **factored form** of the normal equations

(7.1.1) $$A^T(Ax - b) = 0.$$

Working only with A and A^T separately has two important advantages. First, as has been much emphasized for direct methods, a small perturbation in $A^T A$, e.g., by roundoff, may change the solution much more than perturbations of similar size in A itself. Second, we avoid the fill which can occur in the formation of $A^T A$.

We also consider iterative methods for computing a minimum norm solution of a consistent underdetermined system,

$$\min \|y\|_2, \qquad A^T y = c.$$

If A^T has full row rank the unique solution satisfies the normal equations of the second kind (see Section 1.1.4),

(7.1.2) $$y = Az, \qquad A^T(Az) = c.$$

Also in this case, the explicit formation of the cross-product matrix $A^T A$ can be avoided.

Another approach is to use iterative methods applied to the augmented system

(7.1.3) $$\begin{pmatrix} I & A \\ A^T & 0 \end{pmatrix} \begin{pmatrix} y \\ x \end{pmatrix} = \begin{pmatrix} b \\ c \end{pmatrix};$$

see Theorem 1.1.5. This avoids forming the normal equations, but has the drawback that since the augmented system is symmetric indefinite many standard iterative methods cannot be applied.

7.1.1. Iterative versus direct methods. In iterative methods an initial approximate solution is successively improved until an acceptable solution is obtained. The matrix A need not be stored, but can instead be defined by the action of A and A^T on vectors. This makes iterative methods especially attractive for problems where the elements A are easily generated on demand.

The main weakness of iterative methods is their poor robustness and often narrow range of applicability. Often a particular iterative solver may be found to be very efficient for a specific class of problems, but if used for other cases it may be excessively slow or break down. Preconditioned iterative methods, which are based on an approximate factorization, can be considered as a compromise between direct and iterative solvers.

For some classes of sparse problems fill-in will make sparse direct methods prohibitively costly in terms of storage and operations. An example is the case when $A \in \mathbf{R}^{m \times n}$ is large and sparse but $A^T A$ is almost dense. Note that if $A^T A$ is dense and A has the strong Hall property (see Section 6.4.4), then the Cholesky factor R will also be dense, and this also rules out methods based on the QR decomposition. This is in contrast to the dense case where the elements in the upper triangular part of $A^T A$ are always fewer than the mn $(m \geq n)$ nonzero elements of A.

As an example of a problem where direct methods are not suitable we consider the case when A has a random sparsity structure, such that an element a_{ij} is nonzero with probability $p < 1$. Ignoring numerical cancellation it follows that $(A^T A)_{jk} \neq 0$ with probability

$$q = 1 - (1 - p^2)^m \approx 1 - e^{-mp^2}.$$

Therefore $A^T A$ will be almost dense when $mp \approx m^{1/2}$, i.e., when the average number of nonzero elements in a column equals about $m^{1/2}$. This type of structure often occurs in reconstruction problems. An example is the inversion problem for the velocity structure for the Central California Microearthquake Network, for which (in 1980) $m = 500,000$, $n = 20,000$, and A has about 10^7 nonzero elements with a very irregular structure. The matrix $A^T A$ will be almost dense. A similar situation occurs in image reconstruction by X-ray. The structure of the matrix A^T in a very small model problem for image reconstruction is illustrated in Figure 7.1.1. It is easily verified that, except for the diagonal blocks, the matrix $A^T A$ will be completely dense.

7.1.2. Computing sparse matrix-vector products. The efficient implementation of iterative methods depends to a large extent on the performance of sparse matrix-vector products Av and $A^T u$. In some applications it may be possible to express the elements in these matrix-vector products by relatively

7.1. INTRODUCTION

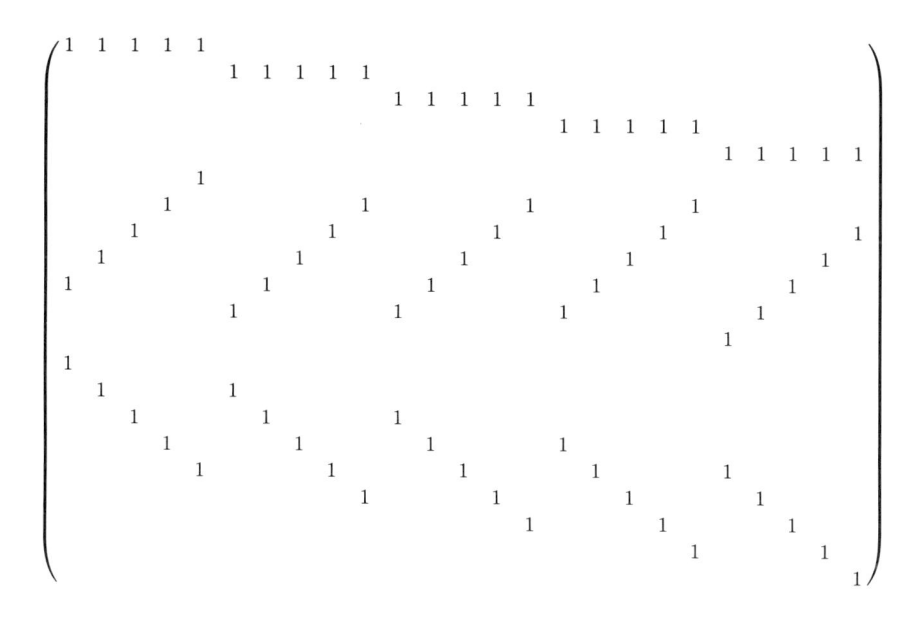

$$\begin{pmatrix}
1 & 1 & 1 & 1 & 1 & & & & & & & & & & & & & & & \\
 & & & & & 1 & 1 & 1 & 1 & 1 & & & & & & & & & & \\
 & & & & & & & & & & 1 & 1 & 1 & 1 & 1 & & & & & \\
 & & & & & & & & & & & & & & & 1 & 1 & 1 & 1 & 1 \\
\end{pmatrix}$$

FIG. 7.1.1. *The structure of A^T in a model image reconstruction problem.*

simple formulas, thus eliminating the need for explicitly storing A. When this is not the case, the choice of the data structure used to store the sparse matrix A is crucial for efficiency. There are several requirements that may be conflicting. As mentioned in Section 6.4.1, ideally only nonzero elements in A should be stored and operated on. Additionally the data structure should be chosen so that hardware features like vector registers can be exploited. Such requirements have led to a great many different data structures to be used, and a consequent lack of portability. Some of the more common structures are described below. We mention that a proposal for standard computational kernels (BLAS) aimed at iterative solvers have been given by Duff et al. [244, 1995].

In Section 6.4.1 we introduced the **compressed row storage** scheme for storing sparse matrices. In this scheme the nonzero elements of A are stored row-wise in a one-dimensional array AC of length nz, the number of nonzero elements of A. The column indices of these elements are stored in an integer vector ja, and the vector ip of length $m+1$ contains pointers to the beginning of each row in AC. With this storage scheme the matrix

$$A = \begin{pmatrix} a_{11} & 0 & a_{13} & 0 \\ a_{21} & a_{22} & 0 & a_{24} \\ 0 & a_{32} & a_{33} & 0 \\ 0 & a_{42} & 0 & a_{44} \\ 0 & 0 & a_{53} & a_{54} \end{pmatrix}$$

is stored in one real and two integer vectors as

$$AC = (a_{11}, a_{13} \mid a_{21}, a_{22}, a_{24} \mid a_{32}, a_{33} \mid a_{42}, a_{44} \mid a_{53}, a_{54}),$$
$$ja = (1, 3, 1, 2, 4, 2, 3, 2, 4, 3, 4),$$
$$ip = (1, 3, 6, 8, 10, 12).$$

The matrix-vector product $u = Av$ is then implemented as follows:

>**for** $i = 1 : m$
>>$u(i) = 0$;
>>**for** $j = ip(i) : ip(i+1) - 1$
>>>$u(i) = u(i) + v(ja(j)) * AC(j)$;
>>
>>**end**
>
>**end**

This is very efficient on *scalar* processors, but does not vectorize well. This is because the vector length in this scheme is equal to the number of nonzero elements in a row, which is usually too small for efficiency. The same comment applies to the **compressed column storage** scheme.

For the transpose product $v = A^T u$ we should not use the equation

$$v_i = \sum_j a_{j,i} u_j, \quad i = 1, \ldots, n,$$

since this accesses elements column by column, which is very inefficient with this storage scheme. We instead perform it as follows:

>**for** $i = 1 : n$ $v(i) = 0$; **end**
>**for** $j = 1 : n$
>>**for** $i = ip(j) : ip(j+1) - 1$
>>>$v(ja(i)) = v(ja(i)) + u(j) * AC(i)$;
>>
>>**end**
>
>**end**

An alternative scheme more suitable for vector processors is the **compressed matrix storage mode**. Here the matrix A is stored in two rectangular arrays AC and ka with m rows and k columns, where k is the maximum number of nonzero elements per row of A. Each row of AC contains the nonzero elements of the corresponding row of A. If a row of A has fewer than k elements, the corresponding row is padded with zeros to length k. ka is an integer array, which contains the column number of the corresponding elements of AC. If the corresponding element is zero any index in the range $1 : n$ can be used. Hence

7.1. Introduction

the matrix A in the example above would be stored as

$$AC = \begin{pmatrix} a_{11} & a_{13} & 0 \\ a_{22} & a_{21} & a_{24} \\ a_{33} & a_{32} & 0 \\ a_{44} & a_{42} & 0 \\ a_{53} & a_{54} & 0 \end{pmatrix}, \quad ka = \begin{pmatrix} 1 & 3 & * \\ 2 & 1 & 4 \\ 3 & 2 & * \\ 4 & 2 & * \\ 3 & 4 & * \end{pmatrix}.$$

The matrix-vector product $u = Av$ is here implemented as follows:

> **for** $i = 1 : m$
> $\quad u(i) = 0;$
> $\quad$ **for** $j = 1 : k$
> $\quad\quad u(i) = u(i) + v(ka(i,j)) * AC(i,j);$
> $\quad$ **end**
> **end**

This code will vectorize on the outer loop with vectors of length m. The code for the transpose product $v = A^T u$ is very similar.

The **compressed diagonals** storage mode is suitable for problems in which the nonzero matrix elements all lie along few diagonals. Here the matrix A is stored in two rectangular arrays AD and a vector la of pointers. The array AD has n rows and nd columns, where nd is the number of diagonals. AD contains the diagonals of A that have at least one nonzero entry, and la contains the corresponding diagonal numbers. The superdiagonals are padded to length n with k trailing zeros, where k is the diagonal number. The subdiagonals are padded to length n with $|k|$ leading zeros. The matrix A in the example above would be stored as

$$AD = \begin{pmatrix} a_{11} & a_{13} & a_{21} & 0 \\ a_{22} & a_{24} & a_{32} & a_{42} \\ a_{33} & 0 & 0 & a_{53} \\ a_{44} & 0 & a_{54} & 0 \end{pmatrix},$$

$$la = (0 \ 2 \ -1 \ -2).$$

The matrix-vector product $u = Av$ is here implemented as follows:

> **for** $i = 1 : m \ \ u(i) = 0;$ **end**
> **for** $j = 1 : nd$
> $\quad k = la(j);$
> $\quad$ **for** $i = \max(1, 1 - k) : \min(n, n - k)$
> $\quad\quad u(i) = u(i) + v(k + i) * AD(i,j);$
> $\quad$ **end**
> **end**

This code will vectorize with vectors of length n. The transpose matrix-vector product $v = A^T u$ is a minor variation of this code.

7.2. Basic Iterative Methods

For general treatments of iterative methods for linear systems see Varga [805, 1962] and Young [847, 1971]. Surveys of iterative methods for least squares problems are given by Björck [87, 1976] and Heath [443, 1984].

7.2.1. General stationary iterative methods. The simplest class of iterative methods for solving the normal equations (7.1.1) is the class of **stationary iterative methods**. These have the form

$$(7.2.1) \qquad Mx^{(k+1)} = Nx^{(k)} + b, \qquad k = 0, 1, \ldots,$$

where $x^{(0)}$ is an initial approximation. Here $A^T A = M - N$, M nonsingular, is a **splitting** of the matrix of normal equations. For the iteration to be practical it must be easy to solve linear systems with matrix M.

To analyze the convergence of the method (7.2.1) we define

$$(7.2.2) \qquad G = M^{-1}N = I - M^{-1}A^T A, \qquad c = M^{-1}b,$$

where G is the iteration matrix. The iterations (7.2.1) can then be written

$$(7.2.3) \qquad x^{(k+1)} = Gx^{(k)} + c, \qquad k = 0, 1, \ldots.$$

We call the iterative method (7.2.3) convergent if the generated sequence $\{x^{(k)}\}$ converges for all initial vectors $x^{(0)}$. It follows that if the method converges and $\lim_{k \to \infty} = x$, then x satisfies $x = Gx + c$, and hence x is a least squares solution. Of fundamental importance in the study of convergence of stationary iterative methods are conditions for a sequence of powers of a matrix to converge to the null matrix. The **spectral radius** of a matrix $G \in \mathbf{R}^{n \times n}$ is the nonnegative number

$$\rho(G) = \max_{1 \le i \le n} |\lambda_i(G)|.$$

For any consistent matrix norm it holds that $\rho(G) \le \|G\|$, which can be used to derive *sufficient* conditions for convergence. The following theorems give *necessary and sufficient* conditions for convergence.

THEOREM 7.2.1. *Let $G \in \mathbf{R}^{n \times n}$ be a given matrix. Then the following conditions are equivalent:*

1. $\lim_{k \to \infty} G^k = 0$,
2. $\lim_{k \to \infty} G^k x = 0, \quad \forall x \in \mathbf{C}^n$,
3. $\rho(G) < 1$,
4. $\|G\| < 1 \qquad$ *for at least one matrix norm.*

In the following we assume that $A^T A$ is positive definite. For the case when rank$(A^T A) < n$, convergence of the iteration (7.2.1) has been investigated by Keller [503, 1965] and Young [847, 1971].

7.2. Basic Iterative Methods

THEOREM 7.2.2. *The stationary iterative method $x^{(k+1)} = Gx^{(k)} + c$ is convergent for all initial vectors $x^{(0)}$ if and only if $\rho(G) < 1$.*

Proof. Subtracting $x = Gx + c$ from (7.2.3) it follows that

$$x^{(k)} - x = G(x^{(k-1)} - x) = \cdots = G^k(x^{(0)} - x).$$

Hence $\lim_{k \to \infty} x^{(k)} = x$ for *all* initial vectors $x^{(0)}$ if and only if $\lim_{k \to \infty} G^k = 0$. The theorem now follows from Theorem 7.2.1. ∎

Usually, we are not only interested in convergence, but also in the *rate* of convergence. For the error $x^{(k)} - x$ at step k it holds for any consistent pair of norms

$$\|x^{(k)} - x\| \leq \|G^k\| \|x^{(0)} - x\| \leq \|G\|^k \|x^{(0)} - x\|.$$

Thus $\|G^k\|$ measures the factor by which the norm of the error is reduced after k iterations. We make the following definition.

DEFINITION 7.2.1. *Assume that the iterative method defined by (7.2.3) is convergent. For any given matrix norm $\|\cdot\|$ we define the* **average rate** *$R_k(G)$ and the* **asymptotic rate** *$R_\infty(G)$ of convergence by*

$$R_k(G) = -\frac{1}{k} \ln \|G^k\|, \quad \text{and} \quad R_\infty(G) = -\ln \rho(G),$$

respectively.

The norm of the error is reduced by a fixed factor δ provided that $\|G^k\| \leq \delta$. Hence we need at most k iterations, where k satisfies

$$k \geq -\ln \delta / R_k(G).$$

In many cases it is desirable that the iteration matrix G has real eigenvalues. This will be the case if the iterative method is **symmetrizable**, where we have the following definition.

DEFINITION 7.2.2. *The stationary iterative method (7.2.1) is said to be* **symmetrizable** *if there is a nonsingular matrix W such that the matrix*

$$W(I - G)W^{-1} = WM^{-1}A^TAW^{-1}$$

is symmetric and positive definite.

A stationary iterative method for the normal equations is symmetrizable if the splitting matrix M is symmetric and positive definite. Then with $W = R$, where R is the Cholesky factor of A^TA, we have

$$RM^{-1}A^TAR^{-1} = RM^{-1}R^TRR^{-1} = RM^{-1}R^T,$$

which is symmetric, positive definite.

7.2.2. Splittings of rectangular matrices. The concept of splitting has been extended to rectangular matrices by Plemmons [661, 1972]. Berman and Plemmons [65, 1974] define $A = M - N$ to be a **proper splitting** if the ranges and nullspaces of A and M are equal. They show that for a proper splitting the iteration

(7.2.4) $$x^{(k+1)} = M^\dagger(Nx^{(k)} + b)$$

converges to the pseudoinverse solution $x = A^\dagger b$ for every $x^{(0)}$ if and only if the spectral radius $\rho(M^\dagger N) < 1$. The iterative method (7.2.4) avoids the explicit recourse to the normal system.

Tanabe [770, 1971] considers stationary iterative methods of the form (7.2.4) for computing more general solutions $x = A^- b$, where A^- is any generalized inverse of A, $(AA^-A = I)$. He shows that the iteration can always be written in the form

$$x^{(k+1)} = x^{(k)} + B(b - Ax^{(k)})$$

for some matrix B, and characterizes the solution in terms of the $\mathcal{R}(AB)$ and $\mathcal{N}(BA)$.

Splittings of rectangular matrices have also been investigated by Chen [163, 1975]. Chen shows that if $\operatorname{rank}(A) = n$, then for any consistent iterative method (7.2.1) there exists a splitting such that the method can be written in the form (7.2.4), and for every such method the iteration matrix equals $G = I - C^{-T}A^TA$ for some nonsingular matrix C. Hence the most general iterative method of the form (7.2.1) for the least squares problem is equivalent to Richardson's first-order method (see below) applied to the linear system $C^{-T}A^T(Ax - b) = 0$.

7.2.3. Classical iterative methods. Consider the splitting

$$M = \frac{1}{\alpha}I, \qquad N = \frac{1}{\alpha}I - A^TA,$$

where $\alpha > 0$ is a parameter. This gives the iteration matrix $G = I - \alpha A^T A$, which is **Richardson's first-order method**. Since $I - G = \alpha A^T A$ is symmetric and positive definite this iteration method is symmetrizable. It can be written in the form

(7.2.5) $$x^{(k+1)} = x^{(k)} + \alpha A^T(b - Ax^{(k)}),$$

which does not require the explicit formation of $A^T A$. The eigenvalues of G equal

$$\lambda_k(G) = 1 - \alpha \sigma_k^2, \qquad k = 1, \ldots, n,$$

where σ_k are the singular values of A. From this it can be shown that Richardson's method converges to the least squares solution $x = A^\dagger b$ if

$$x^{(0)} \in \mathcal{R}(A^T), \qquad 0 < \alpha < 2/\sigma_1^2(A).$$

Assume that all columns in A are nonzero, and let

(7.2.6) $$A = (a_1, \ldots, a_n) \in \mathbf{R}^{m \times n}, \qquad d_j = a_j^T a_j = \|a_j\|_2^2 > 0.$$

7.2. BASIC ITERATIVE METHODS

In **Jacobi's** method a sequence of approximations

$$x^{(k)} = (x_1^{(k)}, \ldots, x_n^{(k)})^T, \qquad k = 1, 2, \ldots,$$

is computed from

(7.2.7) $\qquad x_j^{(k+1)} = x_j^{(k)} + a_j^T(b - Ax^{(k)})/d_j, \quad j = 1, 2, \ldots, n.$

We define the **standard splitting** of $A^T A$ to be

$$A^T A = L_A + D_A + L_A^T, \qquad D_A > 0,$$

where $D_A = \text{diag}(d_1, \ldots, d_n)$ is a diagonal matrix and L_A is strictly lower triangular. Then Jacobi's method corresponds to the splitting $M = D_A$, and we can write (7.2.7) in matrix form as

(7.2.8) $\qquad x^{(k+1)} = x^{(k)} + D_A^{-1} A^T (b - Ax^{(k)}).$

Jacobi's method is symmetrizable since

$$D_A^{1/2}(I - G_J) D_A^{-1/2} = D_A^{-1/2} A^T A D_A^{-1/2}.$$

Jacobi's method can also be used to solve the normal equations of second type (7.1.2). This method can be written in the form

(7.2.9) $\qquad y^{(k+1)} = y^{(k)} + A D_A^{-1}(c - A^T y^{(k)}).$

The **Gauss–Seidel method** is obtained from the standard splitting above by taking $M = L_A + D_A$, and can be written

$$x^{(k+1)} = x^{(k)} + D_A^{-1}(A^T b - L_A x^{(k+1)} + (D_A + L_A^T) x^{(k)}).$$

In Björck and Elfving [105, 1979] it is shown that the Gauss–Seidel method is a special case of a class of **residual reducing** methods, and can be implemented without forming the matrix $A^T A$ explicitly.

Let $p_j \notin \mathcal{N}(A)$, $j = 1, 2, \ldots$, be a sequence of nonzero n-vectors and compute a sequence of approximations of the form

(7.2.10) $\qquad x^{(j+1)} = x^{(j)} + \alpha_j p_j, \qquad \alpha_j = p_j^T A^T (b - Ax^{(j)}) / \|Ap_j\|_2^2.$

It is easily verified that $r^{(j+1)} \perp Ap_j = 0$, where $r_j = b - Ax^{(j)}$, and hence

$$\|r^{(j+1)}\|_2^2 = \|r^{(j)}\|_2^2 - |\alpha_j|^2 \|Ap_j\|_2^2 \leq \|r^{(j)}\|_2^2,$$

i.e., the class of methods (7.2.10) is residual reducing. For a square matrix A, method (7.2.11) was developed by de la Garza [207, 1951]. This class of residual reducing projection methods was studied by Householder and Bauer [477, 1960].

If A has linearly independent columns we obtain the Gauss–Seidel method for the normal equations by taking p_j in (7.2.10) equal to the unit vectors e_j in cyclic order. Then if $A = (a_1, a_2, \ldots, a_n)$, we have $Ap_j = Ae_j = a_j$. An iteration step in the Gauss–Seidel method consists of n minor steps where we put $z^{(1)} = x^{(k)}$, and $x^{(k+1)} = z^{(n+1)}$ is computed by

$$(7.2.11) \qquad z^{(j+1)} = z^{(j)} + e_j a_j^T r^{(j)}/d_j, \qquad r^{(j)} = b - Az^{(j)},$$

$j = 1, 2, \ldots, n$. Note that in the jth minor step only the jth component of $z^{(j)}$ is changed, and hence the residual $r^{(j)}$ can be cheaply updated. With $r^{(1)} = b - Ax^{(k)}$ we obtain the recursions

$$(7.2.12) \qquad z^{(j+1)} = z^{(j)} + \delta_j e_j, \quad r^{(j+1)} = r^{(j)} - \delta_j a_j,$$
$$\delta_j = a_j^T r^{(j)}/d_j, \quad j = 1, \ldots, n.$$

Note that in the jth minor step only the jth column of A is accessed. In contrast to the Jacobi method the Gauss–Seidel method is not symmetrizable and the ordering of the columns of A will influence the convergence.

The Gauss–Seidel method for solving the normal equations of second kind $y = Az$, $A^T Az = c$ (see (7.1.2)) can also be implemented without forming $A^T A$. Following Björck and Elfving [105, 1979] we define a class of **error reducing** methods. Let $p_i \notin \mathcal{N}(A)$, $i = 1, 2, \ldots$, be a sequence of nonzero n-vectors and compute approximations of the form

$$(7.2.13) \qquad y^{(j+1)} = y^{(j)} + \alpha_j A p_j, \qquad \alpha_j = p_j^T (c - A^T y^{(j)})/\|Ap_j\|_2^2.$$

If the system $A^T y = c$ is consistent there is a unique solution y of minimum norm. If we denote the error by $d^{(j)} = y - y^{(j)}$, then by construction $d^{(j+1)} \perp Ap_i$, and thus

$$\|d^{(j+1)}\|_2^2 = \|d^{(j)}\|_2^2 - |\alpha_j|^2 \|Ap_j\|_2^2 \leq \|d^{(j)}\|_2^2.$$

Hence this class of methods is error reducing. Taking p_j to be the unit vectors e_j in cyclic order, we have $Ap_j = a_j$, the jth column in A. Then the iterative method (7.2.13) takes the form

$$(7.2.14) \qquad y^{(j+1)} = y^{(j)} + a_j(c_j - a_j^T y^{(j)})/d_j, \qquad j = 1, \ldots, n.$$

In the Gauss–Seidel method for $A^T A z = c$ the approximation $z^{(j)}$ is updated by

$$\Delta z^{(j)} = e_j(c_j - a_j^T A z^{(i)})/d_j,$$

and with $y^{(j)} = Az^{(j)}$ and $y^{(j+1)} = y^{(j)} + A\Delta z^{(j)}$ we recover (7.2.14). This shows that if we take $y^{(0)} = Az^{(0)}$, then for an arbitrary $z^{(0)}$ (7.2.14) is equivalent to the Gauss–Seidel method for (7.1.2). For the case of a square matrix A this method was originally devised by Kaczmarz [493, 1937]. The convergence properties of Kaczmarz's method for general $m \times n$ matrices has been studied by Tanabe [770, 1971]. We remark that Kaczmarz's method has been rediscovered and used successfully in image reconstruction; see [448, 1973]. In this context the method is known as the unconstrained ART algorithm (algebraic reconstruction technique).

The Jacobi method has the advantage over Gauss–Seidel that it is more easily adapted to parallel computation, since (7.2.9) just requires a matrix-vector multiplication. Further, it does not require A to be stored (or generated) columnwise, since products of the form Ax and $A^T r$ can conveniently be computed also if A can only be accessed by rows. In this case, if $a_1^T, \ldots, a_m^T$ are the rows of A, then we have

$$(Ax)_i = a_i^T x, \quad i = 1, \ldots, n, \quad A^T r = \sum_{i=1}^{m} a_i r_i.$$

That is, for Ax we use an inner product formulation, and for $A^T r$, an outer product formulation.

The Jacobi and Gauss–Seidel methods can be generalized to block matrices. We refer to Section 7.3 for a discussion of block iterative methods.

7.2.4. Successive overrelaxation methods. The **successive overrelaxation (SOR) method** for the normal equations $A^T A x = A^T b$ is obtained by introducing an **relaxation parameter** ω in the Gauss–Seidel method (7.2.13),

$$(7.2.15) \quad z^{(j+1)} = z^{(j)} + \delta_j e_j, \quad r^{(j+1)} = r^{(j)} - \delta_j a_j,$$
$$\delta_j = \omega a_j^T r^{(j)} / d_j, \quad j = 1, \ldots, n.$$

The SOR method always converges when $A^T A$ is positive definite and ω satisfies $0 < \omega < 2$. The SOR shares with the Gauss–Seidel the advantage of simplicity and small storage requirements.

Similarly, one step of the SOR method applied to the normal equations of the second kind ($y = Az$, $A^T A z = c$) can be written as

$$(7.2.16) \quad y^{(j+1)} = y^{(j)} + \omega a_j (c_j - a_j^T y^{(j)}) / d_j, \quad j = 1, \ldots, n,$$

i.e., by introducing an acceleration parameter ω in the generalized Kaczmarz's method.

One wants to choose ω such that the asymptotic rate of convergence is maximized. If the matrix $A^T A$ has special properties, this optimal ω can be expressed in closed form. We make the following definitions.

DEFINITION 7.2.3. *The matrix $A^T A$ is said to have "property A" if there exists a permutation matrix P such that PAP^T has the form*

$$(7.2.17) \quad \begin{pmatrix} D_1 & U_1 \\ L_1 & D_2 \end{pmatrix},$$

where D_1, D_2 are diagonal matrices.

DEFINITION 7.2.4. *A matrix A^TA with the decomposition $A^TA = D_A(I - L - U)$, where D_A is nonsingular, and L (and U) strictly lower (upper) triangular, is said to be* **consistently ordered** *if it has the property that the eigenvalues of*

$$J(\alpha) = \alpha L + \alpha^{-1} U, \qquad \alpha \neq 0,$$

are independent of α.

It is easy to show that a matrix of the form (7.2.17) is consistently ordered. However, a consistently ordered matrix need not have this form. An important example of a consistently ordered matrix is any block tridiagonal matrix, whose diagonal blocks are nonsingular *diagonal* matrices. The following result is due to Young.

THEOREM 7.2.3. *Let A^TA be a consistently ordered matrix, and assume that the eigenvalues μ of $B_J = L + U$ are real and $\rho_J = \rho(B_J) < 1$. Then the optimal relaxation parameter ω in SOR is given by*

$$\text{(7.2.18)} \qquad \omega_{\text{opt}} = \frac{2}{1 + \sqrt{1 - \rho_J^2}},$$

and for this optimal value we have $\rho(B_{\omega_{\text{opt}}}) = \omega_{\text{opt}} - 1$.

Proof. See Young [847, 1971].

For consistently ordered matrices, using ω_{opt} in SOR gives a great increase in the rate of convergence. However, when the assumptions in the theorem are not satisfied, then the SOR method may not be effective for any choice of ω. Then the symmetric SOR (SSOR) method is often advantageous to use. In the SSOR method the forward sweep $j = 1, \ldots, n$ is followed by a backward sweep $j = n, \ldots, 1$ in (7.2.15) or (7.2.16). The SSOR method is less sensitive to the choice of ω, and often $\omega = 1$ is close to the optimum. Further, it can be shown that the SSOR method is symmetrizable and hence can be accelerated further by, e.g., the Chebyshev semi-iterative method described in the following subsection.

7.2.5. Semi-iterative methods. Consider the stationary iterative method

$$\text{(7.2.19)} \qquad \tilde{x}^{(k+1)} = B\tilde{x}^{(k)} + c, \qquad k = 0, 1, \ldots,$$

for solving the normal equations $A^TAx = A^Tb$, which corresponds to a symmetrizable matrix splitting $A^TA = M - N$, i.e.,

$$B = M^{-1}N = I - M^{-1}A^TA, \qquad c = M^{-1}A^Tb.$$

Then the eigenvalues $\{\lambda_i\}_{i=1}^n$ of $M^{-1}A^TA$ are real. Assume that lower and upper bounds are known such that

$$\text{(7.2.20)} \qquad 0 < a \leq \lambda_i < b.$$

Then the eigenvalues $\{\rho_i\}_{i=1}^n$ of $B = I - M^{-1}A^TA$ satisfy

$$c < \rho_i \leq d < 1, \qquad c = 1 - b, \quad d = 1 - a.$$

7.2. Basic Iterative Methods

(Note that we allow $c \leq -1$, even though then $\rho(B) \geq 1$, and the basic method is not convergent!) In the simplest case $M = I$ and $B = I - A^T A$.

For the iterative method (7.2.19) we have the error equation

$$\tilde{x}^{(k)} - x = B^k(\tilde{x}^{(0)} - x).$$

To accelerate the convergence of the sequence $\{\tilde{x}^{(k)}\}$ we take linear combinations of the first k approximations

$$x^{(0)} = \tilde{x}^{(0)}, \qquad x^{(k)} = \sum_{i=0}^{k} c_{ki} \tilde{x}^{(i)}, \quad k = 1, 2, \ldots,$$

where $\sum_{i=0}^{k} c_{ki} = 1$. The resulting method is known as a **semi-iterative method**; see Varga [805, 1962].

Introducing the generating polynomial

$$P_k(t) = \sum_{i=0}^{k} c_{ki} t^i, \quad P_k(1) = 1,$$

it follows from the error equation that

$$x^{(k)} - x = P_k(B)(x^{(0)} - x),$$

where $P_k(B)$ is a polynomial in the matrix B. Therefore, this procedure is also known as **polynomial acceleration**.

A measure of the rate of convergence after k steps for the accelerated sequence is $\rho(P_k(B)) \leq \max_{t \in [c,d]} |P_k(t)|$. To minimize this quantity we solve

$$\min_{P_k \in \Pi_k^1} \max_{t \in [c,d]} |P_k(t)|,$$

where Π_k^1 denotes the set of polynomials P_k of degree $\leq k$ such that $P_k(1) = 0$. The solution to this minimization problem is given by

$$P_k(t) = \frac{T_k(z(t))}{T_k(z(1))}, \qquad z(t) = \frac{2t - (d+c)}{d-c},$$

where $T_k(z)$ denotes the Chebyshev polynomial of degree k. Here $z(t)$ is a linear transformation which maps the interval $t \in [c,d]$ onto $z \in [-1,1]$. (For a proof of this and a review of properties of Chebyshev polynomials we refer to Young [847, 1971, pp. 302–303].) For this choice we have

(7.2.21) $$\rho(P_k(B)) \leq \frac{1}{T_k(\mu^{-1})}, \qquad \mu^{-1} = z(1) = \frac{b+a}{b-a},$$

and it holds that

$$T_k(\mu^{-1}) = \cosh(k\gamma), \qquad \cosh(\gamma) = \mu^{-1}.$$

Solving the quadratic equation in e^γ,
$$e^\gamma + e^{-\gamma} = 2\cosh(\gamma) = 2(\kappa^2 + 1)/(\kappa^2 - 1)$$
for γ gives
$$\gamma = \log\left(\frac{\kappa+1}{\kappa-1}\right) > 2/\kappa.$$

It follows that k iterations will reduce the error norm by at least a factor $1/\cosh(k\gamma)$, where
$$\cosh(k\gamma) = \frac{1}{2}(e^{k\gamma} + e^{-k\gamma}) > \frac{1}{2}e^{k\gamma} > \frac{1}{2}e^{2k/\kappa}.$$

Thus to reduce the error norm by at least a factor of $1/\delta$ it is sufficient to perform k iterations, where
$$(7.2.22) \qquad k > \frac{1}{2}\kappa \ln\frac{2}{\delta}.$$

This is an order of magnitude improvement of the asymptotic convergence rate compared to the basic stationary method. Note that if the splitting matrix M is symmetric and positive definite then $\kappa = (b/a)^{1/2} > 1$ is an upper bound for the condition number of $AM^{-1/2}$. (In particular, if $M = I$ then $\kappa = \kappa(A)$.) For achieving a close-to-optimal acceleration of convergence it is necessary that the upper and lower bounds in (7.2.20) for the eigenvalues λ_i are sufficiently accurate.

The vectors $x^{(k)}$ can be computed by means of the three-term recurrence relation for the Chebyshev polynomials, without computing the $\tilde{x}^{(k)}$ first. The iteration can then be written (see Golub and Varga [390, 1961]):

$$(7.2.23) \quad x^{(k+1)} = \begin{cases} x^{(0)} + \alpha s^{(0)}, & k = 0; \\ x^{(k-1)} + \omega_{k+1}(\alpha s^{(k)} + x^{(k)} - x^{(k-1)}), & k = 1, 2, \ldots, \end{cases}$$

where
$$(7.2.24) \qquad r^{(k)} = b - Ax^{(k)}, \quad Ms^{(k)} = A^T r^{(k)}, \quad \alpha = 2/(a+b),$$

and
$$(7.2.25) \qquad \omega_1 = 2, \quad \mu = \frac{b-a}{b+a}, \quad \omega_{k+1} = \left(1 - \frac{\mu^2}{4}\omega_k\right)^{-1}.$$

This is the **Chebyshev semi-iterative method**. Note that in each iteration we have to perform two matrix-vector multiplications $Ax^{(k)}$ and $A^T r^{(k)}$, and solve one linear system $Ms^{(k)} = A^T r^{(k)}$.

The **second-order Richardson method** can also be described by equation (7.2.23). Here one takes α and μ as above and
$$(7.2.26) \qquad \omega_k = \hat{\omega} = \frac{2}{1 + \sqrt{1-\mu^2}}.$$

Interestingly enough, it can be shown that in the Chebyshev method we have $\omega_k \to \hat{\omega}$.

7.2. Basic Iterative Methods

The eigenvalues of the iteration matrix of the SOR method $B_{\omega_{\text{opt}}}$ are all complex and have modulus $|\omega_{\text{opt}}|$. In this case convergence acceleration is of no use. (A precise formulation is given in Young [847, 1971, p. 375].) However, Chebyshev acceleration can be applied to the SSOR method, often with a substantial gain in convergence rate.

7.2.6. Preconditioning. Because of potentially slow convergence one main emphasis in the development of iterative methods is on convergence acceleration. A general technique to improve convergence is by **preconditioning**, which for the linear least squares problem is equivalent to a transformation of variables. Let $S \in \mathbf{R}^{n \times n}$ be a nonsingular matrix and consider the problem

$$(7.2.27) \qquad \min_y \|(AS^{-1})y - b\|_2, \quad Sx = y.$$

If S is chosen so that AS^{-1} has a more favorable spectrum than A, this will improve convergence of an iterative method applied to (7.2.27).

It is important to note that the product AS^{-1} should not be explicitly formed, but treated as a product of two operators. In an iterative method preconditioned by the matrix S, matrix-vector products of the forms $AS^{-1}y$ and $S^{-T}A^T r$ will occur. Thus the extra cost of preconditioning will be in solving linear systems of the form $Sx = y$ and $S^T q = s$. Hence S has to be chosen so that such systems can be easily solved. To summarize, a good preconditioner S should have the following properties, which are partly contradictory:

- AS^{-1} should be better conditioned than A and/or have only a few distinct singular values.

- S should have about the same number of nonzeros as A.

- It should be cheap to solve equations with matrices S and S^T.

A simple and cheap preconditioner corresponds to a diagonal scaling of the columns of A,

$$(7.2.28) \qquad S = D^{1/2} = \text{diag}\,(d_1^{1/2}, \ldots, d_n^{1/2}), \quad d_j = \|a_j\|_2^2.$$

Since AS^{-1} has columns of unit length it follows from Theorem 1.4.7 that this approximately minimizes $\kappa(AS^{-1})$ over all diagonal scalings.

In Section 7.2.5 we developed the Chebyshev semi-iterative method which can be used to accelerate the convergence of the SSOR method. A different interpretation is to view the SSOR method as a preconditioner for the Chebyshev method. The SSOR preconditioner corresponds to taking

$$(7.2.29) \qquad S = D_A^{-1/2}(D_A + \omega L_A^T), \quad 0 \leq \omega < 2,$$

with the standard splitting

$$A^T A = L_A + D_A + L_A^T,$$

where L_A is strictly lower triangular. Note that taking $\omega = 0$ in (7.2.29) gives (7.2.28).

In Björck and Elfving [105, 1979] it is shown how to implement the preconditioner (7.2.29) without actually forming $A^T A$ or L. We use the notation

$$A = (a_1, \ldots, a_n), \qquad d_j = a_j^T a_j = \|a_j\|_2^2.$$

We can compute the vectors $t = (t_1, \ldots, t_n)^T = S^{-1} p$ and $q = q_0 = A S^{-1} p$ simultaneously as follows. Set $q_n = 0$, and for $j = n, n-1, \ldots, 1$ compute

$$(7.2.30) \qquad t_j = (d_j^{1/2} p_j - \omega a_j^T q_j)/d_j, \qquad q_{j-1} = q_j + t_j a_j.$$

The vector $s = (s_1, \ldots, s_n)^T = S^{-T} A^T r$ is computed as follows. Set $h_1 = r$, and for $j = 1, 2, \ldots, n$ compute

$$(7.2.31) \qquad s_j = a_j^T h_j / d_j^{1/2}, \qquad h_{j+1} = h_j - \omega (d_j^{-1/2} s_j) a_j.$$

Hence, in order to use the SSOR preconditioner, one only needs to access one column of A at a time. The number of operations per step approximately doubles when $\omega \neq 0$, compared to diagonal scaling ($\omega = 0$).

Theory and numerical experiments indicate that $\omega = 1$ is often close to the optimum choice of ω. Convergence can be affected also by reordering the columns. Such a reordering can also be used to introduce parallelism into the scheme. These topics are addressed in Section 7.5.1. Some tests using the SSOR preconditioner for solving nonsymmetric systems arising from partial differential equations are given in Saad [698, 1988].

7.3. Block Iterative Methods

In many large sparse least squares problems arising from multidimensional models the matrix A has a natural column block structure

$$(7.3.1) \qquad A = (A_1, A_2, \ldots, A_N),$$

where

$$A_j \in \mathbf{R}^{m \times n_j}, \qquad n_1 + \cdots + n_N = n.$$

A special example of such a block structure is the block angular form described in Section 6.3.1.

7.3.1. Block column preconditioners.
For problems of the structure (7.3.1) block versions of the preconditioners (7.2.28) and (7.2.29) are particularly suitable. Let the QR decompositions of the blocks be

$$(7.3.2) \qquad A_j = Q_j R_j, \qquad Q_j \in \mathbf{R}^{m \times n_j}, \qquad j = 1, \ldots, N.$$

Then to (7.2.28) corresponds the block diagonal preconditioner

$$(7.3.3) \qquad S = R_B = \mathrm{diag}\,(R_1, R_2, \ldots, R_N).$$

7.3. Block Iterative Methods

For this choice we have $AS^{-1} = (Q_1, Q_2, \ldots, Q_N)$, i.e., the columns of each block are mutually orthogonal.

If we split $A^T A$ according to

$$A^T A = L_B + D_B + L_B^T,$$

where L_B is strictly lower block triangular, then the block SSOR preconditioner becomes

(7.3.4) $$S = R_B^{-1}(D_B + \omega L_B^T).$$

This preconditioner was introduced for least squares problems by Björck [89, 1979]. As with the corresponding point preconditioner it can be implemented without actually forming $A^T A$.

We now consider the use of the block diagonal preconditioner (7.3.3). We will partition x and $y = Sx$ conformably with (7.3.1):

$$x = (x_1, x_2, \ldots, x_N)^T, \quad y = (y_1, y_2, \ldots, y_N)^T.$$

Jacobi's method (7.2.8) applied to the preconditioned problem (7.2.27) can then be written

$$y_j^{(k+1)} = y_j^{(k)} + Q_j^T(b - AS^{-1}y^{(k)}), \quad j = 1, \ldots, N,$$

or in terms of the original variables

(7.3.5) $$x_j^{(k+1)} = x_j^{(k)} + R_j^{-1}Q_j^T(b - Ax^{(k)}), \quad j = 1, \ldots, N.$$

This is the block Jacobi method for the normal equations. Note that the correction $z_j = x_j^{(k+1)} - x^{(k)}$ equals the solution to the problem

(7.3.6) $$\min_{z_j} \|A_j z_j - r^{(k)}\|_2, \quad r^{(k)} = b - Ax^{(k)},$$

and that these corrections can be computed in parallel. Often Q_j is not available and we have to use $Q_j = A_j R_j^{-1}$. This is equivalent to using the method of seminormal equations (3.4.6) for solving (7.3.6). This can lead to some loss of accuracy, and a correction step is recommended unless all the blocks A_j are well-conditioned.

Similarly, we can derive a block SOR method for the normal equations. Let $r_1^{(k)} = b - Ax^{(k)}$ and for $j = 1, \ldots, N$ compute

(7.3.7) $$x_j^{(k+1)} = x_j^{(k)} + \omega z_j, \quad r_{j+1}^{(k)} = r_j^{(k)} - \omega A_j z_j,$$

where z_j is the solution to

(7.3.8) $$\min_{z_j} \|A_j z_j - r_j^{(k)}\|_2.$$

Taking $\omega = 1$ in (7.3.7) gives the block Gauss–Seidel method.

In order to use the block SSOR preconditioner (7.3.4) for the conjugate gradient method we have to be able to compute vectors $q = AS^{-1}p$ and $s = S^{-T}A^T r$ efficiently, given $p = (p_1, \ldots, p_N)^T$ and r. The following algorithms for this, derived in Björck [89, 1979], generalize the point SSOR algorithms (7.2.30) and (7.2.31).

Put $q^{(N)} = 0$ and compute $q = q^{(0)}$ and $S^{-1}p = (z_1, \ldots, z_N)^T$:

$$(7.3.9) \quad z_j = R_j^{-1}(p_j - R_j^{-T}A_j^T q^{(j)}), \quad q^{(j-1)} = q^{(j)} - A_j z_j,$$
$$j = N, \ldots, 2, 1.$$

Put $r^{(1)} = r$, and compute $s = (s_1, \ldots, s_N)^T$:

$$(7.3.10) \quad s_j = R_j^{-T}A_j^T r^{(j)}, \quad r^{(j+1)} = r^{(j)} - A_j R_j^{-1} s_j,$$
$$j = 1, 2, \ldots, N.$$

The choice of partitioning A into blocks is important for the storage and computational efficiency of the methods. An important criterion is that it should be possible to compute the factorizations $A_j = Q_j R_j$ (or at least the factors R_j) without too much fill. Note that if A_j is the direct sum of blocks, then in the SOR method the computation of z_j splits up into independent subproblems. This makes it possible to achieve efficiency through parallelism.

For grid problems a high degree of parallelism can be achieved. For example, Kamath and Sameh [496, 1989] give a scheme for a three-dimensional $n \times n \times n$ mesh and a seven-point difference star, for which $N = 9$ and each block consists of $n^2/9$ separate subblocks of columns. Hence each subproblem can be solved with a parallelism of $n^2/9$. For some ordering algorithms based on graph-theoretic algorithms see Dennis and Steihaug [224, 1986].

7.3.2. The two-block case. The case $N = 2$, $A = (A_1, A_2)$ is of special interest. For the block diagonal preconditioner (7.3.3) we have $AS^{-1} = (Q_1, Q_2)$ and the matrix of normal equations for the preconditioned system becomes

$$(7.3.11) \quad (AS^{-1})^T AS^{-1} = \begin{pmatrix} I & K \\ K^T & I \end{pmatrix}, \quad K = Q_1^T Q_2.$$

This matrix has "property A"; see Definition 7.2.3. This means that it is possible to reduce the work required per iteration by approximately one-half for many iterative methods. This preconditioner is also called the **cyclic Jacobi preconditioner**.

For matrices with "property A" the SOR theory holds. As shown by Elfving [279, 1980] it follows that the optimal ω in the block SOR method (7.3.7) for $N = 2$ is given by

$$\omega_{\text{opt}} = 2/(1 + \sin\theta_{\min}), \quad \cos\theta_{\min} = \sigma_{\max}(Q_1^T Q_2).$$

Here $\theta_{\min}$ is the smallest principal angle between $\mathcal{R}(A_1)$ and $\mathcal{R}(A_2)$; see Definition 1.2.2. Using ω_{opt} in the block SOR method reduces the number of iterations by a factor of $2/\sin\theta_{\min}$ compared to using $\omega = 1$.

7.3. BLOCK ITERATIVE METHODS

For $N = 2$, the preconditioner (7.3.4) with $\omega = 1$ also has special properties; see Golub, Manneback, and Toint [376, 1986]. We have

$$S = \begin{pmatrix} R_1 & \omega Q_1^T A_2 \\ 0 & R_2 \end{pmatrix},$$

and it follows that for $\omega = 1$

(7.3.12) $\qquad (A_1, A_2) S^{-1} = (Q_1, (I - P_1) Q_2),$

where $P_1 = Q_1 Q_1^T$ is the orthogonal projector onto Range(A_1). It follows that the two blocks in (7.3.12) are mutually orthogonal, and thus the preconditioned problem (7.2.27) can be split into two problems, which are

(7.3.13) $\qquad \min_{y_2} \|(I - P_1) Q_2 y_2 - b\|_2, \quad y_1 = Q_1^T b.$

This effectively reduces the original system to a system of size n_2. Hence this preconditioner is also called the **reduced system preconditioner**. The matrix of normal equations becomes

$$(AS^{-1})^T AS^{-1} = \begin{pmatrix} I & 0 \\ 0 & Q_2^T (I - P_1) Q_2 \end{pmatrix} = I - \begin{pmatrix} 0 & 0 \\ 0 & K^T K \end{pmatrix},$$

where $K = Q_1^T Q_2$. This reduction of the variables corresponding to the first block columns can also be performed in the case when $N > 2$.

For the case $N = 2$ the SSOR preconditioned conjugate gradient method has been carefully studied by Manneback [560, 1985]. It is shown that the choice $\omega = 1$, i.e., using the reduced system preconditioning, is optimal with respect to the number of iterations. Further, Hageman, Luk, and Young [418, 1980] have shown the equivalence of reduced system preconditioning and cyclic Jacobi preconditioning ($\omega = 0$) for Chebyshev semi-iteration and the conjugate gradient method. The reduced system preconditioning will essentially generate the same approximations in half the number of iterations. Since the work per iteration is about doubled for $\omega \neq 0$, this means that cyclic Jacobi preconditioning is optimal for the conjugate gradient method in the class of SSOR preconditioners.

Golub, Manneback, and Toint [376, 1986] have applied the block SSOR preconditioned conjugate gradient method to the Doppler positioning problem. Here the matrix is of block angular form (see Section 6.3.1) but is partitioned into two blocks (A, B) where A consists of all the diagonal blocks, and B, the last block column.

Some experimental results of block SSOR preconditionings for the case $N > 2$ are given by Björck [89, 1979]. In this case, $\omega = 1$ is not necessarily optimal. However, in the tests the optimal value of ω was close to 1, and the number of iterations required was not sensitive to the choice of ω around $\omega = 1$.

7.4. Conjugate Gradient Methods

The conjugate gradient (CG) method was developed in the early fifties by Hestenes and Stiefel [451, 1952]. Because in exact arithmetic it converges in at most n steps it was first considered as a direct method. However, the finite termination does not hold with roundoff, and the method came into wide use first in the mid-seventies, when it was realized that it should be regarded as an iterative method. It has now become a standard tool for solving large sparse linear systems and linear least squares problems. In the original paper by Hestenes and Stiefel [451, 1952], and in the subsequent paper by Stiefel [755, 1952] a version of CG for the solution of normal equations was given. Läuchli [516, 1959] was the first to discuss a preconditioned CG method and to apply it to solving least squares geodetic network problems. Other early discussions of the conjugate gradient method for linear least squares are found in Lawson [519, 1973] and Chen [163, 1975].

7.4.1. CGLS and variants. The CG method is a special case of Krylov space methods. We make the following definition.

DEFINITION 7.4.1. *Given a matrix $B \in \mathbf{R}^{n \times n}$ and a vector $c \in \mathbf{R}^n$ the* **Krylov subspace** $\mathcal{K}_k(B, c)$ *is*

$$\mathcal{K}_k(B, c) = \mathrm{span}\{c, Bc, \ldots, B^{k-1}c\}.$$

The kth iterate in the CG method is uniquely determined by the following variational property. Let $\hat{x} = A^\dagger b$ be the pseudoinverse solution and $\hat{r} = b - A\hat{x}$ the corresponding residual. Then $x^{(k)}$ minimizes the error functional

(7.4.1) $$E_\mu(x^{(k)}) = (\hat{x} - x^{(k)})^T (A^T A)^\mu (\hat{x} - x^{(k)})$$

over all vectors $x^{(k)}$ in the affine subspace

(7.4.2) $$x^{(k)} \in x^{(0)} + \mathcal{K}_k(A^T A, s^{(0)}), \qquad s^{(0)} = A^T(b - Ax^{(0)}).$$

Only the values $\mu = 0, 1, 2$ are of practical interest, and they correspond to the cases

(7.4.3) $$\begin{aligned} \mu = 0 \quad &\text{minimizes} \quad \|\hat{x} - x^{(k)}\|_2, \\ \mu = 1 \quad &\text{minimizes} \quad \|\hat{r} - r^{(k)}\|_2^2 = \|r^{(k)}\|_2^2 - \|\hat{r}\|_2^2, \\ \mu = 2 \quad &\text{minimizes} \quad \|A^T(\hat{r} - r^{(k)})\|_2^2 = \|s^{(k)}\|_2^2. \end{aligned}$$

Taking $\mu = 0$ is only feasible when the system $Ax = b$ is consistent, and the resulting method is equivalent to a method due to Craig [191, 1955]. In this case, denoting by $s^{(k)} = A^T r^{(k)}$ the residual of the normal equations, we have

$$(s^{(k)})^T (A^T A)^{-1} s^{(k)} = (b - Ax^{(k)}) A(A^T A)^{-1} A^T (b - Ax^{(k)}) = (r^{(k)})^T r^{(k)},$$

since $A(A^T A)^{-1} A^T$ is the orthogonal projection onto $\mathcal{R}(A)$. For $\mu = 1$ the method is denoted CGLS in [639] and CGNR in [315]. The second expression for $E_1(x^{(k)})$ in (7.4.3) follows from the fact that $\hat{r} \perp \hat{r} - r^{(k)}$.

7.4. CONJUGATE GRADIENT METHODS

The variational property of the CG method implies that the error functional $E_\mu(x^{(k)})$ decreases monotonically. For $\mu = 1$, also, $\|r^{(k)}\|$ will decrease monotonically. The following estimate of the rate of convergence is known to hold:

$$(7.4.4) \qquad E_\mu(x^{(k)}) < 2\left(\frac{\kappa-1}{\kappa+1}\right)^k E_\mu(x^{(0)}),$$

where $\kappa = \kappa(A) = \sqrt{\kappa(A^T A)}$. Further, we have the following result.

LEMMA 7.4.1. *Let $\{x^{(k)}\}$ be the sequence of conjugate gradient approximations, which minimize $E_\mu(x)$ subject to (7.4.2). Then for $\mu = 1,2$ the sequences $E_\nu(x^{(k)})$, $\nu \leq \mu$, all decrease monotonically.*

Proof. For $\mu = 1$ see Hestenes and Stiefel [451, 1952, p. 416]. For $\mu = 2$ see [451, Sec. 7]. ∎

For $\mu = 1$ we are assured that both $\|\hat{r} - r^{(k)}\|$ and $\|\hat{x} - x^{(k)}\|$ decrease monotonically, but $\|A^T r^{(k)}\|$ will often exhibit large oscillations when $\kappa(A)$ is large. We stress that this behavior is *not* a result of rounding errors. For $\mu = 2$, $\|A^T r^{(k)}\|$ will also decrease monotonically, but this choice $\mu = 2$ gives slower convergence and lower final accuracy in both $\|\hat{r} - r^{(k)}\|_2$ and $\|\hat{x} - x^{(k)}\|_2$. It also requires more operations and storage; see Table 7.4.1.

TABLE 7.4.1

Comparison of storage and operations.

	μ	Storage	Flops/step
CGLS	0	$2n+3m$	$2n+3m$
	1	$2n+2m$	$3n+2m$
	2	$4n+3m$	$5n+3m$

In the following we only consider the case $\mu = 1$, which is of most practical interest. The preferred version is that originally given by Hestenes and Stiefel [451, 1952].

ALGORITHM 7.4.1. CGLS. Let $x^{(0)}$ be an initial approximation, set

$$r^{(0)} = b - Ax^{(0)}, \quad p^{(0)} = s^{(0)} = A^T r^{(0)}, \quad \gamma_0 = \|s^{(0)}\|_2^2,$$

and for $k = 0, 1, 2, \ldots$ while $\gamma_k > tol$ compute

$$q^{(k)} = Ap^{(k)},$$
$$\alpha_k = \gamma_k / \|q^{(k)}\|_2^2,$$
$$x^{(k+1)} = x^{(k)} + \alpha_k p^{(k)},$$
$$r^{(k+1)} = r^{(k)} - \alpha_k q^{(k)},$$
$$s^{(k+1)} = A^T r^{(k+1)},$$
$$\gamma_{k+1} = \|s^{(k+1)}\|_2^2,$$
$$\beta_k = \gamma_{k+1} / \gamma_k,$$
$$p^{(k+1)} = s^{(k+1)} + \beta_k p^{(k)}.$$

CGLS requires the storage of two n-vectors, x and p, and two m-vectors, r and q. (Note that s can share storage with q.) Each iteration requires about $2nz(A) + 3n + 2m$ flops, where $nz(A)$ are the number of nonzero elements in A. A comparison with other choices of μ is given in Table 7.4.1.

7.4.2. Convergence properties of CGLS. For convenience we assume in the following that $x^{(0)} = 0$, and hence $r^{(0)} = b$. We also assume in this subsection that *exact* arithmetic is used. The effect of roundoff will be discussed in the next subsection.

An upper bound for the rate of convergence of CGLS can be derived from the fact that the approximation $x^{(k)}$ minimizes the quadratic form

$$(7.4.5) \qquad \min_{x \in \mathcal{K}_k(A^T A, A^T b)} s^T (A^T A)^{-1} s, \qquad s = A^T(b - Ax).$$

For the residual vector $s^{(k)} = A^T r^{(k)}$ produced by CGLS we have

$$s^{(k)} \in T_k, \qquad T_k = \{A^T(b - Ax) \mid x \in \mathcal{K}_k(A^T A, A^T b)\}.$$

Hence any vector $s \in T_k$ can be written

$$s_k = (I - A^T A P_{k-1}(A^T A))A^T b = R_k(A^T A)A^T b,$$

where P_{k-1} is a polynomial of degree $k-1$. Hence the **residual polynomial** R_k is of degree $\leq k$ and satisfies $R_k(0) = 1$. If we let $\tilde{\Pi}_k^1$ denote the set of all such polynomials, then by the minimizing property (7.4.5)

$$\|s^{(k)}\|_{(A^T A)^{-1}} = \min_{R_k \in \tilde{\Pi}_k^1} \|R_k(A^T A) A^T b\|_{(A^T A)^{-1}}.$$

Let σ_i, u_i, v_i be the singular values and left and right singular vectors of A. Then we can expand the right-hand sides b and $A^T b$ as

$$b = \sum_{i=1}^m \gamma_i u_i, \qquad A^T b = \sum_{i=1}^n \gamma_i \sigma_i v_i,$$

and for CGLS we have

$$\|s^{(k)}\|^2_{(A^T A)^{-1}} = \min_{R_k \in \tilde{\Pi}_k^1} \sum_{i=1}^n \gamma_i^2 R_k^2(\sigma_i^2).$$

An upper bound can be obtained by substituting *any* polynomial $R_k \in \tilde{\Pi}_k^1$. In particular, taking

$$R_n(\sigma^2) = \left(1 - \frac{\sigma^2}{\sigma_2}\right)\left(1 - \frac{\sigma^2}{\sigma_1^2}\right) \cdots \left(1 - \frac{\sigma^2}{\sigma_n^2}\right),$$

we get $\|s_n\|_{(A^T A)^{-1}} = 0$, which proves that CGLS terminates after at most n steps. Similarly, if A only has t distinct singular values, then CGLS converges in

7.4. CONJUGATE GRADIENT METHODS

at most t steps. Hence, CGLS is particularly effective when A has low rank! If $\operatorname{rank}(A) = p$, then p steps suffice to get the exact solution. Fewer than n steps are needed also if in the expansion of b, $\gamma_i = 0$ for some $i \leq n$.

Let the set S contain all the nonzero singular values of A and assume that for some $\tilde{R}_k \in \tilde{\Pi}_k^1$ we have

$$\max_{\sigma \in S} |\tilde{R}_k(\sigma^2)| \leq M_k.$$

Then it follows that

$$\|s^{(k)}\|_{(A^T A)^{-1}}^2 \leq M_k^2 \sum_{i=1}^n \gamma_i^2 \sigma_i^{-2} = M_k^2 \|s^{(0)}\|_{(A^T A)^{-1}}^2$$

and since $\|s^{(k)}\|_{(A^T A)^{-1}}^2 = \|r - r^{(k)}\|_2^2$ we get

$$\|r - r^{(k)}\| \leq M_k \|r - r^{(0)}\|.$$

We can now select a set S on the basis of some assumption regarding the eigenvalue distribution of A and seek a polynomial $\tilde{R}_k \in \tilde{\Pi}_k^1$ such that $M_k = \max_{\sigma \in S} |\tilde{R}_k(\sigma^2)|$ is small. A simple choice is to take $S = [\sigma_n^2, \sigma_1^2]$ and seek the polynomial $\tilde{R}_k \in \tilde{\Pi}_k^1$ which minimizes

$$\max_{\sigma_n^2 \leq \sigma^2 \leq \sigma_1^2} |R_k(\sigma^2)|.$$

The solution to this problem is known to be the shifted Chebyshev polynomials introduced in the semi-iterative methods. Hence, for CGLS, the residual error is reduced according to

$$(7.4.6) \qquad \|r - r^{(k)}\|_2 < 2\Big(\frac{\kappa - 1}{\kappa + 1}\Big)^k \|r - r^{(0)}\|_2.$$

From this it follows that an upper bound for the number of iterations k to reduce the relative error by a factor of ϵ is given by

$$(7.4.7) \qquad \frac{\|r - r^{(k)}\|_2}{\|r - r^{(0)}\|_2} < \epsilon \iff k < \frac{1}{2}\kappa(A) \log \frac{2}{\epsilon}.$$

This is the same rate of convergence as for the Chebyshev semi-iterative method and the second-order Richardson method. However, these methods require that accurate lower and upper bounds for the singular values of A be known. For the CG method no such bounds are needed.

Provided that $x^{(0)} \in \mathcal{R}(A^T)$, which holds, e.g., if $x^{(0)} = 0$, $x^{(k)}$ will converge to the pseudoinverse solution $A^\dagger b$; see also Hestenes [450, 1975]. In the absence of rounding errors CGLS will compute the exact pseudoinverse solution in at most t iterations, where $t \leq n$ is the number of distinct nonzero singular values of A.

When A is ill-conditioned and the least squares solution x has substantial components along singular vectors corresponding to small singular values, then many more than n iterations may be needed. However, more important is that in some practical applications one is satisfied with approximations that are obtained in far fewer than n iterations. This is true when A is well-conditioned, or when the right-hand side is such that the effective condition number of the system $Ax = b$ is small; see below.

In some least squares problems the underlying model is such that the exact solution will have only small projections along the singular vector directions corresponding to small singular values. Let $A = U\Sigma V^T$ be the singular value decomposition of A. Then the least squares solution can be written

$$x = Vz = \sum_{i=1}^{n}(c_i/\sigma_i)v_i, \qquad c_i = u_i^T b.$$

Let δ be an error tolerance for x and assume that $\|c_i/\sigma_i\| < \delta$, $i = p+1,\ldots,n$. Then, assuming $x^{(0)} = 0$, the errors along v_i, $i > p$, are sufficiently small at the start of the iterations. The **effective condition number** determining the rate of convergence of the error along the other singular vectors will then be

$$\kappa_e(A, b) = \sigma_1/\sigma_p.$$

In discrete ill-posed problems it is often the case that $\kappa_e(A, b) \ll \kappa(A)$; see Section 6.7.

7.4.3. The conjugate gradient method in finite precision. There are many ways, all mathematically equivalent, in which to implement the CG method for least squares problems. In *exact* arithmetic they will all generate the same sequence of approximations, but in finite precision the achieved accuracy may differ substantially. It is important to notice that an implementation of the CG method for symmetric positive definite systems should *not* be applied directly to the normal equations; cf. the discussion in Paige and Saunders [639, 1982, Sec. 7.1]. Often the following variant of Algorithm 7.4.1, where the residual of the normal equations $s = A^T(b - Ax)$ is recurred instead of $r = b - Ax$, occurs in the literature.

ALGORITHM 7.4.2. Initialize

$$p^{(0)} = s^{(0)} = A^T(b - Ax^{(0)}), \qquad \gamma_0 = \|s^{(0)}\|_2^2,$$

and for $k = 0, 1, 2, \ldots$ while $\gamma_k > tol$ compute

$$\begin{aligned}
q^{(k)} &= Ap^{(k)}, \\
\alpha_k &= \gamma_k/\|q^{(k)}\|_2^2, \\
x^{(k+1)} &= x^{(k)} + \alpha_k p^{(k)}, \\
s^{(k+1)} &= s^{(k)} - \alpha_k(A^T q^{(k)}), \\
\gamma_{k+1} &= \|s^{(k+1)}\|_2^2, \\
\beta_k &= \gamma_{k+1}/\gamma_k, \\
p^{(k+1)} &= s^{(k+1)} + \beta_k p^{(k)}.
\end{aligned}$$

7.4. Conjugate Gradient Methods

The lack of stability of this implementation has been analyzed in Björck, Elfving, and Strakos [109, 1995]. Assuming that $x_0 = 0$, the only information about the right-hand side b is in the initialization of $p_0 = s_0 = A^T b$, and $\theta_1 v_1 = A^T b$, respectively. For the computed $A^T b$ we have

$$(7.4.8) \qquad |fl(A^T b) - A^T b| \leq \gamma_m |A^T||b|, \qquad \gamma_m = mu/(1 - mu),$$

and this is almost sharp. The perturbed solution $x + \delta x$ corresponding to $c = fl(A^T b)$ satisfies

$$A^T A(x + \delta x) = A^T b + \delta c, \qquad |\delta c| \leq \gamma_m |A^T||b|.$$

Subtracting $A^T A x = A^T b$ and solving for δx we obtain $\delta x = (A^T A)^{-1} \delta c$, and from this we get the componentwise bound

$$|\delta x| \leq \gamma_m |(A^T A)^{-1}||A^T||b|.$$

Taking norms we obtain

$$(7.4.9) \qquad \frac{\|\delta x\|}{\|x\|} \leq \gamma_m \|(A^T A)^{-1}\| \frac{\|A\|\|b\|}{\|x\|} = \gamma_m \kappa^2(A) \left(\frac{\|b\|}{\|A\|\|x\|} \right).$$

Since no reference to b is made in the iterative phase of the algorithm it follows that roundoff errors that occur in computing the vector $A^T b$ cannot be compensated for. Hence, the best possible error bound has the form given above. In cases where $\|r\| \ll \|b\|$ this version can be expected to produce much less than optimal accuracy.

Paige and Saunders [639, 1982] have developed an algorithm LSQR based on the Lanczos bidiagonalization algorithm, which is described in Section 7.6.1. LSQR is mathematically equivalent to CGLS but converges somewhat more quickly when A is ill-conditioned and many iterations have to be carried out. However, the achievable accuracy with CGLS and LSQR seem to be the same.

7.4.4. Preconditioned CGLS. The normal equations in factored form for the preconditioned problem are

$$(7.4.10) \qquad S^{-T} A^T (A S^{-1} y - b) = S^{-T} A^T (Ax - b) = 0.$$

Note that the preconditioned CGLS still minimizes the error functional $\|\hat{r} - r^{(k)}\|_2$, where $r = b - Ax$, but over a different Krylov subspace. Hence for the rate of convergence, we have

$$(7.4.11) \qquad \|r - r^{(k)}\|_2 < 2 \left(\frac{\kappa - 1}{\kappa + 1} \right)^k \|r - r^{(0)}\|_2, \qquad \kappa = \kappa(A S^{-1}).$$

It is often convenient to formulate the preconditioned method in terms of the original variables x. For the preconditioned conjugate gradient method (7.4.1) (PCCGLS) we obtain the following algorithm.

ALGORITHM 7.4.3. PCCGLS. Let $x^{(0)}$ be an initial approximation, set

$$r^{(0)} = b - Ax^{(0)}, \quad p^{(0)} = s^{(0)} = S^{-T}(A^T r^{(0)}), \quad \gamma_0 = \|s^{(0)}\|_2^2,$$

and for $k = 0, 1, 2, \ldots$ while $\gamma_k > tol$ compute

$$t^{(k)} = S^{-1} p^{(k)},$$
$$q^{(k)} = A t^{(k)},$$
$$\alpha_k = \gamma_k / \|q^{(k)}\|_2^2,$$
$$x^{(k+1)} = x^{(k)} + \alpha_k t^{(k)},$$
$$r^{(k+1)} = r^{(k)} - \alpha_k q^{(k)},$$
$$s^{(k+1)} = S^{-T}(A^T r^{(k+1)}),$$
$$\gamma_{k+1} = \|s^{(k+1)}\|_2^2,$$
$$\beta_k = \gamma_{k+1}/\gamma_k,$$
$$p^{(k+1)} = s^{(k+1)} + \beta_k p^{(k)}.$$

7.5. Incomplete Factorization Preconditioners

7.5.1. Incomplete Cholesky preconditioners. Several different preconditioners have been suggested for the least squares problem. We first note that if we take $S = R$, where R is the Cholesky factor of $A^T A$, then we have (neglecting roundoff) $\kappa(AS^{-1}) = \kappa(Q_1) = 1$. Hence with this choice of preconditioner all the above iterative methods will converge in one iteration! Thus, a preconditioner S, which in some sense approximates R, can be expected to be efficient (or rather, S^{-1} should approximate R^{-1}).

One approach is then to take S to be an **incomplete Cholesky factor** of $A^T A$, i.e., $S = R$, where

$$C = A^T A = R^T R - E,$$

and E is a **defect matrix**. We want $\|E\|$ to be small and R sparse. One way to compute such a matrix R is to use a direct method for sparse Cholesky factorization, but to only keep those elements in R which lie within a predetermined sparsity structure. If we allow nonzero elements in R for the set of index positions $(i, j) \in P$, then the incomplete Cholesky factorization can be described as follows.

Note that there is no need to explicitly compute the matrix $C = A^T A$. All that is required is to be able to access one row at a time, and thus the nonzero elements in the ith row of C can be computed when needed and then discarded.

In the simplest case nonzero elements are only allowed in the positions (i, j) where $c_{ij} \neq 0$, $P_0 = P_C$. This is called a level-zero incomplete factorization and for this the structure of R is the same as for the upper triangular part of $A^T A$. If we let E_0 be the corresponding defect matrix we take for the level-one incomplete factorization $P_1 = P_0 \cup E_0$. Higher level incomplete factorizations are defined recursively.

7.5. INCOMPLETE FACTORIZATION PRECONDITIONERS

ALGORITHM 7.5.1. INCOMPLETE CHOLESKY FACTORIZATION.

for $i = 1, 2, \ldots, n$
$$r_{ii} := \left(c_{ii} - \sum_{k=1}^{i-1} r_{ki}^2 \right)^{1/2}$$
for $j = i+1, \ldots, n$
if $(i, j) \notin P$ then $r_{ij} = 0$ else
$$r_{ij} := \left(c_{ij} - \sum_{k=1}^{i-1} r_{ki} r_{kj} \right) / r_{ii}$$
end
end
end

Positive-definiteness of $C = A^T A$ alone is not sufficient to guarantee the existence of an incomplete factor R, since definiteness may be lost because of dropped elements. We have the following definition.

DEFINITION 7.5.1. *A matrix $C = (c_{ij}) \in \mathbf{R}^{n \times n}$ with $c_{ij} \leq 0$ for all $i \neq j$ is an M-matrix if C is nonsingular, and $C^{-1} \geq 0$.*

The following relationship between positive-definiteness and M-matrices was given by Stieltjes [756, 1887].

THEOREM 7.5.1. *Let C be a real square matrix, with $c_{ij} \leq 0$, $i \neq j$. Then C is an M-matrix if C is positive definite.*

Meijerink and van der Vorst [575, 1977] proved the existence and stability of the incomplete Cholesky factorization when C is an M-matrix.

THEOREM 7.5.2. *If C is a symmetric M-matrix, there exists for every symmetric set P, such that $(i,j) \in P$ for $i = j$, a uniquely defined upper triangular matrix R with $r_{ij} = 0$ if $(i,j) \notin P$, such that*
$$C = R^T R - E, \quad (R^T R)^{-1} \geq 0, \quad E \geq 0.$$

Manteuffel [563, 1980] extended the existence of incomplete factorizations to the class of H-matrices. A matrix C is an H-matrix if the comparison matrix $\hat{C}$ is an M-matrix, where
$$\hat{c}_{ij} = \begin{cases} -|c_{ij}|, & \text{if } i \neq j; \\ c_{ii}, & \text{if } i = j. \end{cases}$$
Note that a diagonally dominant matrix is an H-matrix.

In the general case when C is not an H-matrix the incomplete factorization may break down because of zero pivots, or the preconditioner may fail to be positive definite due to the presence of negative pivots. To avoid this, corrections can be added to the diagonal elements. An off-diagonal element c_{ij} can be deleted by adding an error matrix E_{ij} with nonzero elements
$$\begin{pmatrix} c_{ii} & -c_{ij} \\ -c_{ji} & c_{jj} \end{pmatrix}.$$

If we choose the corrections to the diagonal elements such that $c_{ii}c_{jj} - c_{ij}^2 \geq 0$, then the matrix E_{ij} is positive semidefinite, and the eigenvalues of the modified matrix $C + E_{ij}$ cannot be smaller than those of C. Hence if C is positive definite and E is the sum of such modifications it follows that also $C + E$ is positive definite, and the incomplete factorization cannot break down.

Instead of prescribing the sparsity structure of the incomplete factor R one can discard elements in the Cholesky factorization whose magnitude falls below a certain preset threshold level. This approach has been considered by Ajiz and Jennings (see [5, 1984], which also contains a Fortran program). In their algorithm the elements of R are computed row by row. After rows $1, \ldots, i-1$ have been computed the transformed off-diagonal elements in the ith row,

$$c_{ij}^* = c_{ij} - \sum_{k=1}^{i-1} r_{ki} r_{kj}, \qquad i < j \leq n,$$

are first computed. Each of these elements c_{ij}^* is tested for rejection against a tolerance. If the element is rejected, then additions are made to the corresponding ith and jth diagonal elements. The diagonal modifications are chosen so that equal *relative* changes are made to c_{ii} and c_{jj}. After all the off-diagonal elements in row i have been computed, all additions are made to c_{ii}, and we compute

$$r_{ii} = \left(c_{ii} - \sum_{k=1}^{i-1} r_{ki}^2 \right)^{1/2}, \qquad r_{ij} = c_{ij}^*/r_{ii}, \quad j > i.$$

ALGORITHM 7.5.2. INCOMPLETE THRESHOLD CHOLESKY FACTORIZATION.

$$\begin{aligned}
&\text{for } i = 1, 2, \ldots, n \\
&\quad \text{for } j = i+1, \ldots, n \\
&\qquad c_{ij} = c_{ij} - \sum_{k=1}^{i-1} r_{ki} r_{kj}; \\
&\qquad \text{if } c_{i,j}^2 \leq \tau^2 * c_{ii} * c_{jj} \text{ then} \\
&\qquad\quad c_{ij} = 0; \quad \rho = \sqrt{c_{ii}/c_{jj}}; \\
&\qquad\quad c_{ii} = c_{ii} + \rho * |c_{ij}|; \\
&\qquad\quad c_{jj} = c_{jj} + (1/\rho) * |c_{ij}|; \\
&\qquad \text{end} \\
&\quad \text{end} \\
&\quad r_{ii} = \left(c_{ii} - \sum_{k=1}^{i-1} r_{ki}^2 \right)^{1/2}; \\
&\quad \text{for } j = i+1, \ldots, n \\
&\qquad r_{ij} = c_{ij}/r_{ii}; \\
&\quad \text{end} \\
&\text{end}
\end{aligned}$$

We emphasize again that there is no need to explicitly compute the matrix $C = A^T A$, except its diagonal elements. All that is required is to be able to access C one row at a time, and thus the ith row can be computed and then discarded.

A choice of $\tau = 0$ will retain all elements giving a complete Cholesky factorization of C. It can also be shown that a choice of $\tau = 1$ will cause *all* off-diagonal elements to be rejected and give a diagonal preconditioner. In practice, an intermediate value of τ in the interval $[0.01, 0.02]$ is recommended by Ajiz and Jennings [5, 1984].

A problem with the threshold Cholesky factorization is that the amount of storage needed to hold the factorization for a given τ cannot be determined in advance. One solution is to stop and restart with a larger value of τ if the allocated memory should not suffice. However, a too-large threshold may lead to slow convergence in the preconditioned CG method.

In the study of direct solution methods for sparse problems the orderings of the unknowns was of great importance. Several different orderings, e.g., minimum degree, nested dissection, and reverse Cuthill–McKee, were used to decrease the amount of fill-in in the full Cholesky factorization. Duff and Meurant [245, 1989] have studied the effect of different ordering strategies on the convergence of the CG method when this is preconditioned by incomplete Cholesky factorizations. They conclude that the rate of convergence of the CG method is not related to the number of fill-ins that are dropped, but is almost directly related to $\|E\|$, the norm of the residual matrix. Several orderings which give a small number of fill-ins did not perform well when used with a level-zero or level-one incomplete factorization.

When a drop tolerance is used to compute the incomplete factorization, good orderings for direct methods like the minimum degree algorithm seem to perform well. With these orderings fewer elements need to be dropped; see Munksgaard [592, 1980].

7.5.2. Incomplete orthogonal decompositions. Alternatively, an incomplete Cholesky factor R can be generated by modifying algorithms for orthogonal decomposition. The aim here is not to reduce the effect of rounding errors, but possibly to achieve faster convergence in the CG iterations.

An incomplete modified Gram–Schmidt (IMGS) factorization has been described by Jennings and Ajiz [487, 1984]. Here a drop tolerance for elements in R is used, where the magnitude of each off-diagonal element r_{ij} is compared against a drop tolerance τ scaled by the norm of the corresponding column norm $\|a_i\|_2$, i.e., elements which satisfy $|r_{ij}| < \tau d_i$ are dropped, where

(7.5.1) $$A = (a_1, a_2, \ldots, a_n), \qquad d_i = \|a_i\|_2.$$

At each stage in the IMGS decomposition it holds that $A = \bar{Q}\bar{R}$, where $\bar{R}$ is upper triangular with positive diagonal. Hence we have

$$\text{span}\,\{\bar{q}_1, \ldots, \bar{q}_n\} = \text{span}\,\{a_1, \ldots, a_n\},$$

and if A has full column rank this algorithm cannot break down.

ALGORITHM 7.5.3. IMGS.

$$
\begin{aligned}
&\text{for } i = 1, 2, \ldots, n \\
&\quad r_{ii} := \|a_i\|_2; \quad q_i = a_i/r_{ii}; \\
&\quad \text{for } j = i+1, \ldots, n \\
&\quad\quad r_{ij} = q_i^T a_j; \\
&\quad\quad \text{if } r_{ij} < \tau * d_i \text{ then } r_{ij} = 0; \text{ end} \\
&\quad\quad a_j = a_j - r_{ij} q_i; \\
&\quad \text{end} \\
&\text{end}
\end{aligned}
$$

Saad [698, 1988] discusses the implementation of an alternative dropping criterion, where the p_R largest elements in a row of R and the p_R largest elements in a column of Q are kept. It is possible also to prescribe the sparsity structure of R in advance, as was done in the incomplete Cholesky algorithm. This version can be obtained from Algorithm 7.5.3 by modifying it so that $r_{ij} = 0$ when $(i, j) \notin P$.

Wang [818, 1993] (see also Wang, Gallivan, and Bramley [819, 1994]) has given an alternative, more compact algorithm (CIMGS) for computing the IMGS preconditioner; see Algorithm 7.5.4. By compressing the information in the column vectors of A in inner product form, one need not store Q, and an algorithm similar to a Cholesky factorization is obtained. In rank one update form, this leads to the algorithm CIMGS given below. It can be shown that in exact arithmetic this produces the same incomplete factor R as IMGS, and therefore inherits the robustness of IMGS.

ALGORITHM 7.5.4. CIMGS.

$$
\begin{aligned}
&\text{for } i = 1, 2, \ldots, n \\
&r_{ii} = \sqrt{c_{ii}}; \\
&\quad \text{for } j = i+1, \ldots, n \\
&\quad\quad c_{ij} = c_{ij}/r_{ii}; \\
&\quad\quad \text{if } (i, j) \notin P \text{ then } r_{ij} = 0 \text{ else } r_{ij} = c_{ij}; \text{end} \\
&\quad \text{end} \\
&\quad \text{for } j = i+1, \ldots, n \\
&\quad\quad \text{for } k = i+1, \ldots, n \\
&\quad\quad\quad \text{if } (i, j) \in P \text{ or } (i, k) \in P \text{ then} \\
&\quad\quad\quad\quad c_{kj} = c_{kj} - c_{ik} * c_{ij} \text{ end} \\
&\quad\quad \text{end} \\
&\quad \text{end} \\
&\text{end}
\end{aligned}
$$

An alternative method for computing an incomplete QR decomposition is to apply the sequential sparse Givens QR factorization described in Section 6.6.1. To get an incomplete factorization we avoid performing the rotation to eliminate any element in A if its magnitude is small compared to the norm of the column in which it lies. Jennings and Ajiz [487, 1984] use the test

$$|a_{ij}| < \tau \|a_j\|_2,$$

where a_j is the jth column of A and τ is a parameter. The rows a_i^T of A are processed sequentially, $i = 1, 2, \ldots, m$. The nonzero elements in the ith row $a_i^T = (a_{i1}, a_{i2}, \ldots, a_{in})$ are scanned, and each nonzero element with $|a_{ij}| \geq \tau \|a_j\|_2$ is annihilated by a Givens rotation involving row j in R. If elements with $|a_{ij}| < \tau \|a_j\|_2$ were just neglected, then the final incomplete factor R could become singular even if $A^T A$ is positive definite. Instead, these elements are rotated into the diagonal element taking

$$r_{jj} = \left(r_{jj}^2 + a_{ij}^2\right)^{1/2}.$$

This guarantees that R is nonsingular, and the residual matrix E of $A^T A = R^T R - E$ will have zero diagonal elements.

A similar approach is taken by Zlatev and Nielsen [860, 1988]. They compute a sparse incomplete orthogonal factorization of A, neglecting all elements which in the course of computations become smaller than a certain **drop-tolerance** $\tau \geq 0$. This can sometimes substantially reduce the number of nonzero elements during the factorization process. The drop-tolerance $\tau = 2^{-3}$ is recommended, and τ is successively reduced if the iterations fail to converge. This approach can be very efficient for some classes of problems, in particular when storage is a limiting factor.

7.5.3. Preconditioners based on LU factorization. Other choices of preconditioners result from using sparse direct factorization methods on a square subblock of A. Assume that A has full column rank and that the rows of A have been permuted so that

(7.5.2) $$A = \begin{pmatrix} A_1 \\ A_2 \end{pmatrix}, \quad A_1 \in \mathbf{R}^{n \times n},$$

where A_1 is nonsingular. We can then use A_1 as a right preconditioner, and consider the least squares problem

(7.5.3) $$\min_y \|AA_1^{-1}y - b\|_2, \quad y = A_1 x.$$

In some sparse least squares problems it may be possible to choose A_1 when the data are collected, e.g., in geodetic computations (see Plemmons [663, 1979]). A choice for A_1 might also result directly from the nature of the problem. In the general case, A_1 and its LU factorization have to be found by a sparse direct factorization method.

The use of this preconditioner was first suggested by Läuchli in [517, 1961]. Läuchli explicitly computes the matrices A_1^{-1} and $C = A_2 A_1^{-1}$ by performing n Jordan elimination steps with complete pivoting, and applies the conjugate gradient method to the normal equations for problem (7.5.3). Alternatively, he suggests use of the Chebyshev semi-iterative method (7.2.23), with parameters estimated from the elements of C

$$a = 1 \leq \lambda(I + CC^T) \leq 1 + \text{trace}(C^T C) = b.$$

Läuchli did not consider pivoting for sparsity. A more efficient way to implement the algorithm above is to compute an LU factorization of A using a sparse LU code. Such codes have been discussed by Gilbert, Moler, and Schreiber [351, 1992], Duff and Reid [249, 1979], Sherman [719, 1978], and George and Ng [345, 1985]. Such codes can be used to compute an LU factorization of a permuted A, such that we obtain $A_1 = L_1 U_1$.

Let b and the residual vector r be partitioned conformally with A. Then the preconditioned system becomes

$$\min_y \left\| \begin{pmatrix} I_n \\ C \end{pmatrix} y - \begin{pmatrix} b_1 \\ b_2 \end{pmatrix} \right\|_2, \qquad C = A_2 A_1^{-1}.$$

The normal equations can be written in augmented form as

$$\begin{pmatrix} I_n & 0 & I_n \\ 0 & I_{m-n} & C \\ I_n & C^T & 0 \end{pmatrix} \begin{pmatrix} r_1 \\ r_2 \\ y \end{pmatrix} = \begin{pmatrix} b_1 \\ b_2 \\ 0 \end{pmatrix},$$

from which, by eliminating y, we obtain

(7.5.4) $$\begin{pmatrix} I_n & C^T \\ -C & I_{m-n} \end{pmatrix} \begin{pmatrix} r_1 \\ r_2 \end{pmatrix} = \begin{pmatrix} 0 \\ b_2 - Cb_1 \end{pmatrix}.$$

Evans and Li [282, 1989] suggest solving this system by the CG method. Also eliminating r_2 in (7.5.4), we obtain the symmetric positive definite system

(7.5.5) $$(I_n + C^T C) r_1 = C^T (b_2 - Cb_1),$$

which is the system of normal equations for r_1. The algorithm PCGLS can be adapted to (7.5.5).

Whenever C or C^T multiplies a vector in the algorithm below we interpret these operations as a product of operators,

$$s = r + C^T r = r + L_1^{-T}(U_1^{-T}(A_2^T r)), \qquad q = Cp = A_2(U_1^{-1}(L_1^{-1} p)).$$

Hence we only need a sparse factorization of the matrix A_1. In each step matrix-vector products with A_2 and A_2^T, and two solutions of $n \times n$ linear systems of the form $A_1^{-1} v$ and $A_1^{-T} v$ are required. The system (7.5.5) has been generalized to the case of constrained least squares problems in [43, 1988].

7.5. Incomplete Factorization Preconditioners

ALGORITHM 7.5.5. LU PRECONDITIONED CGLS. Let $x^{(0)}$ be an initial approximation, set

$$r^{(0)} = b - Ax^{(0)}, \quad p^{(0)} = s^{(0)} = r_1^{(0)} + C^T r_2^{(0)}, \quad \gamma_0 = \|s^{(0)}\|_2^2,$$

and for $k = 0, 1, 2, \ldots$ while $\gamma_k > tol$ compute

$$q^{(k)} = Cp^{(k)},$$
$$\alpha_k = \gamma_k / (\|p^{(k)}\|_2^2 + \|q^{(k)}\|_2^2),$$
$$r_1^{(k+1)} = r_1^{(k)} - \alpha_k p^{(k)},$$
$$r_2^{(k+1)} = r_2^{(k)} - \alpha_k q^{(k)},$$
$$s^{(k+1)} = r_1^{(k+1)} + C^T r_2^{(k+1)},$$
$$\gamma_{k+1} = \|s^{(k+1)}\|_2^2,$$
$$\beta_k = \gamma_{k+1}/\gamma_k,$$
$$p^{(k+1)} = s^{(k+1)} + \beta_k p^{(k)}.$$

Solve x from

$$A_1 x = b_1 - r_1.$$

The convergence of Algorithm 7.5.5 has been studied by Freund [314, 1987]. The eigenvalues of $(I_n + C^T C)$ are

$$\lambda_i = 1 + \sigma_i^2(C), \quad i = 1, \ldots, n,$$

where $\sigma_i(C)$ are the singular values of C. It follows that

$$\kappa(AA_1^{-1}) = \sqrt{1 + \sigma_1^2(C)} \Big/ \sqrt{1 + \sigma_n^2(C)} \leq \sqrt{1 + \alpha^2},$$

where
(7.5.6) $$\alpha = \sigma_1(C) = \|C\|_2 = \|A_2 A_1^{-1}\|_2.$$

Hence, for the rate of convergence we have

(7.5.7) $$\|r - (b - Ax^{(k)})\|_2 < 2 \left(\frac{\alpha}{1 + \sqrt{1 + \alpha^2}} \right)^{2k} \|r - (b - Ax^{(0)})\|_2.$$

To get fast convergence we want to have α small. This is achieved if the blocking (7.5.2) can be chosen so that $\|A_2\|_2$ is small and A_1 well-conditioned, because $\alpha \leq \sigma_{\max}(A_2)/\sigma_{\min}(A_1)$. Since C has at most $p = \min\{m-n, n\}$ distinct singular values, method PCGLS will in exact arithmetic converge in at most $\min\{p+1, n\}$ steps. Hence, if $p \ll n$ we can expect rapid convergence.

If $m > n$, then by eliminating $r_1 = -C^T r_2$ in (7.5.4) we obtain another symmetric positive definite system of dimension $(m - n) \times (m - n)$,

(7.5.8) $$(CC^T + I_{m-n})r_2 = b_2 - Cb_1.$$

If r_2 is known the solution can be obtained from

$$A_1 x = d, \qquad d = b_1 + C^T r_2.$$

The system (7.5.8) can be interpreted as the normal equations for

$$\min_{r_2} \left\| \begin{pmatrix} -C^T \\ I_{m-n} \end{pmatrix} r_2 - \begin{pmatrix} b_1 \\ b_2 \end{pmatrix} \right\|.$$

CGLS adapted to solve system (7.5.8) requires about the same storage and work as the previous algorithm. However, Yuan [849, 1993] shows that the CG method applied to (7.5.8) can more easily be adapted to weighted and generalized least squares problems with covariance matrix $W \neq I$.

With the partitioning (7.5.2) the normal equations for the original system can be written in augmented form as

$$(7.5.9) \qquad \begin{pmatrix} A_1 & 0 & I_n \\ A_2 & I_{m-n} & 0 \\ 0 & A_2^T & A_1^T \end{pmatrix} z = d, \qquad z = \begin{pmatrix} x \\ r_2 \\ r_1 \end{pmatrix}, \qquad d = \begin{pmatrix} b_1 \\ b_2 \\ 0 \end{pmatrix}.$$

Several block SOR methods have been developed for solving this system (7.5.9). A three-block SOR method, using a splitting $B = L_3 + D_3 + U_3$ with

$$D_3 = \begin{pmatrix} A_1 & 0 & 0 \\ 0 & I_{m-n} & 0 \\ 0 & 0 & A_1^T \end{pmatrix},$$

was studied in Niethammer, De Pillis, and Varga [601, 1984]. Markham, Neumann, and Plemmons [565, 1985] proposed a two-block SOR method using a splitting $B = L_2 + D_2 + U_2$ with

$$D_2 = \begin{pmatrix} A_1 & 0 & 0 \\ A_2 & I_{m-n} & 0 \\ 0 & 0 & A_1^T \end{pmatrix},$$

and showed that this was superior. The two-block SOR iteration can be expressed in the form

$$(7.5.10) \qquad z^{(k+1)} = \mathcal{L}_\omega z^{(k)} + (D_2 - \omega L_2)^{-1} d,$$

where ω is the SOR parameter and

$$\mathcal{L}_\omega = (D_2 - \omega L_2)^{-1} \Big[(1 - \omega) D_2 + U_2 \Big].$$

Hence the method requires the solution of a block triangular system with coefficient matrix

$$D_2 - \omega L_2 = \begin{pmatrix} A_1 & 0 & 0 \\ A_2 & I_{m-n} & 0 \\ 0 & \omega A_2^T & A_1^T \end{pmatrix}$$

at each step. The method converges for all ω in the interval $0 < \omega < 2/(1+\alpha)$. The optimal relaxation parameter and the spectral radius of the corresponding iteration matrix $\mathcal{L}_{\omega_b}$ are given by

$$\omega_b = \frac{2}{1 + \sqrt{1 + \alpha^2}}, \qquad \rho(\mathcal{L}_{\omega_b}) = \left(\frac{\alpha}{1 + \sqrt{1 + \alpha^2}}\right)^2,$$

where α is defined by (7.5.6).

It has been observed by Freund [314, 1987] that both these SOR methods generate approximations in the same affine Krylov subspace as the conjugate gradient method preconditioned with A_1. The bound (7.5.7) for the preconditioned CGLS method indicates the same asymptotic behavior as the two-block SOR with optimal ω. However since the CG method is optimal, the block SOR methods will converge more slowly.

Saunders [702, 1979] suggests using Gaussian elimination with row interchanges to compute a stable, sparse factorization $P_r A = LU$, where P_r is a permutation matrix, $L \in \mathbf{R}^{m \times n}$ is unit lower trapezoidal, and U is upper triangular. As preconditioner, one takes $S = U$. The rationale for this choice is that any ill-conditioning in A is usually reflected in U, and L tends to be well-conditioned. A preliminary pass through the rows of A are made to select a triangular subset with maximal diagonal elements. The matrix L is not saved, and subsequent use of the operator AU^{-1} involves back-substitution with U and multiplication with A. This approach has the advantage that often there is very little fill-in in U, and hence U is likely to contain fewer nonzeros than A.

7.6. Methods Based on Lanczos Bidiagonalization

7.6.1. Lanczos bidiagonalization. We consider here the Lanczos bidiagonalization (LBD) of a matrix $A \in \mathbf{R}^{m \times n}$, $m \geq n$. This has important applications for computing approximations to the large singular values and corresponding singular vectors, as well as for solving least squares problems.

In Section 2.5.1 we gave an algorithm for computing the factorization

$$(7.6.1) \qquad A = U \begin{pmatrix} B \\ 0 \end{pmatrix} V^T, \quad U^T U = I_m, \quad V^T V = I_n,$$

where $U = (u_1, \ldots, u_m)$ and $V = (v_1, \ldots, v_n)$ are chosen as products of Householder transformations and

$$B = B_n = \begin{pmatrix} \alpha_1 & \beta_1 & & & \\ & \alpha_2 & \beta_2 & & \\ & & \ddots & \ddots & \\ & & & \alpha_{n-1} & \beta_{n-1} \\ & & & & \alpha_n \end{pmatrix} \in \mathbf{R}^{n \times n}$$

is upper bidiagonal. An alternative approach to computing this factorization was given by Golub and Kahan [370, 1965]. Here the columns of U and V are generated sequentially, as in the Lanczos process.

If we set $U_1 = (u_1, \ldots, u_n)$ then from (7.6.1) we have

(7.6.2) $$AV = U_1 B, \qquad A^T U_1 = V B^T.$$

Equating the jth columns in these two equations yields

(7.6.3) $$\begin{aligned} Av_j &= \alpha_j u_j + \beta_{j-1} u_{j-1}, \\ A^T u_j &= \alpha_j v_j + \beta_j v_{j+1}, \quad j = 1, \ldots, n, \end{aligned}$$

where we have put $\beta_0 u_0 = \beta_n v_{n+1} \equiv 0$. Using the fact that $\|u_j\|_2 = \|v_j\|_2 = 1$, $j = 1, \ldots, n$, we can solve these equations for u_j and v_{j+1}, respectively, and obtain

(7.6.4) $$r_j = Av_j - \beta_{j-1} u_{j-1}, \quad \alpha_j = \|r_j\|_2, \quad u_j = r_j/\alpha_j,$$
(7.6.5) $$p_j = A^T u_j - \alpha_j v_j, \qquad \beta_j = \|p_j\|_2, \quad v_{j+1} = p_j/\beta_j,$$

$j = 1, \ldots, n$. (Note that we can always take $\alpha_j \geq 0$, $\beta_j \geq 0$.) Starting with an arbitrary vector $v_1 \in \mathbf{R}^n$, $\|v_1\|_2 = 1$, these equations can be used to recursively generate the vectors $u_1, v_2, u_2, \ldots, v_n, u_n$ and the corresponding elements in B_n, unless some α_j or β_j vanishes. It is easy to show by an inductive proof that these vectors are the first n columns of the desired matrices U and V.

Assume that v_j and u_j, $j \leq n$, can be computed without breakdown. It then follows directly from the recurrence relations (7.6.4)–(7.6.5) and Definition 7.4.1 that

$$v_j \in \mathcal{K}_j(A^T A, v_1), \qquad u_j \in \mathcal{K}_j(AA^T, Av_1).$$

Hence $v_1, \ldots, v_j$ and $u_1, \ldots, u_j$ form an orthogonal basis for these two Krylov spaces.

The process (7.6.4)–(7.6.5) terminates when $\alpha_j = 0$ or $\beta_j = 0$. (Since there are at most n orthogonal vectors in $\mathbf{R}^n$ it must terminate with $\beta_n = 0$ or earlier.) If it terminates with $\alpha_j = 0$, $j < n$, then from (7.6.4) it follows that $Av_j = \beta_{j-1} u_{j-1}$ and $\mathrm{span}\{Av_1, \ldots, Av_j\} \subset \mathrm{span}\{u_1, \ldots, u_{j-1}\}$. Hence this can only happen if $\mathrm{rank}(A) < n$. If the process terminates with $\beta_j = 0$, then it can be verified that

$$A(v_1, \ldots, v_j) = (u_1, \ldots, u_j) B_j, \qquad A^T(u_1, \ldots, u_j) = (v_1, \ldots, v_j) B_j^T,$$

and thus $\sigma(B_j) \subset \sigma(A)$.

In the Golub–Kahan bidiagonalization a starting vector $v_1 \in \mathbf{R}^n$ is chosen. A variant of this bidiagonalization algorithm where instead a starting vector $u_1 \in \mathbf{R}^m$ is used was described by Paige and Saunders [639, 1982]. This is the appropriate version for solving least squares problems. After n steps we obtain a transformation of A into **lower** bidiagonal form

(7.6.6) $$B_n = \begin{pmatrix} \alpha_1 & & & \\ \beta_2 & \alpha_2 & & \\ & \beta_3 & \ddots & \\ & & \ddots & \alpha_n \\ & & & \beta_{n+1} \end{pmatrix} \in \mathbf{R}^{(n+1) \times n}.$$

(Note that B_n is not square.) Again equating columns in (7.6.2) we obtain, setting $\beta_1 v_0 \equiv 0$, $\alpha_{n+1} v_{n+1} \equiv 0$, the recurrence relations

$$A^T u_j = \beta_j v_{j-1} + \alpha_j v_j,$$
(7.6.7)
$$A v_j = \alpha_j u_j + \beta_{j+1} u_{j+1}, \quad j = 1, \ldots, n.$$

Starting with a given vector $u_1 \in \mathbf{R}^m$, $\|u_1\|_2 = 1$, we can now recursively generate the vectors $v_1, u_2, v_2, \ldots, u_{m+1}$ and corresponding elements in B_n using, for $j = 1, 2, \ldots$, the formulas

(7.6.8) $\quad r_j = A^T u_j - \beta_j v_{j-1}, \quad \alpha_j = \|r_j\|_2, \quad v_j = r_j/\alpha_j,$
(7.6.9) $\quad p_j = A v_j - \alpha_j u_j, \quad \beta_{j+1} = \|p_j\|_2, \quad u_{j+1} = p_j/\beta_{j+1}.$

For this bidiagonalization scheme we have

$$u_j \in \mathcal{K}_j(AA^T, u_1), \qquad v_j \in \mathcal{K}_j(A^T A, A^T u_1).$$

We remark that both bidiagonalization algorithms are closely related to the Lanczos tridiagonalization scheme for the symmetric matrices $A^T A$, AA^T, and

$$C = \begin{pmatrix} 0 & A \\ A^T & 0 \end{pmatrix}.$$

There is a close relationship between the above bidiagonalization process and the Lanczos process applied to the two matrices AA^T and $A^T A$. Note that these matrices have the same nonzero eigenvalues σ_i^2, $i = 1, \ldots, n$, and that the corresponding eigenvectors equal the left and right singular vectors of A, respectively.

The LBD process (7.6.8)–(7.6.9) generates in exact arithmetic the same sequences of vectors $u_1, u_2, \ldots$ and $v_1, v_2, \ldots$ as are obtained by simultaneously applying the Lanczos process to AA^T with starting vector $u_1 = b/\|b\|_2$, and to $A^T A$ with starting vector $v_1 = A^T b/\|A^T b\|_2$.

The bidiagonalization of A is also related to the Lanczos process applied to the symmetric matrix

(7.6.10) $$C = \begin{pmatrix} 0 & A \\ A^T & 0 \end{pmatrix} \in \mathbf{R}^{(m+n) \times (m+n)},$$

whose eigenvalues equal $\pm \sigma_i$, $i = 1, \ldots, n$, and 0 with multiplicity $m - n$; see Theorem 1.2.2. Provided that the Lanczos process is started with the special vector $(b^T, 0)^T$ it generates in exact arithmetic the special tridiagonal matrix of dimension $(2n+1) \times (2n+1)$,

$$\begin{pmatrix} 0 & \alpha_1 & & & & & \\ \alpha_1 & 0 & \beta_2 & & & & \\ & \beta_2 & 0 & \ddots & & & \\ & & \ddots & \ddots & \alpha_k & & \\ & & & \alpha_n & 0 & \beta_{n+1} \\ & & & & \beta_{n+1} & 0 \end{pmatrix},$$

with zero entries in the diagonal. After an odd-even permutation of rows and columns this matrix equals
$$\begin{pmatrix} 0 & B_n \\ B_n^T & 0 \end{pmatrix},$$
where B_n is the matrix in (7.6.6).

The relations between LBD and these Lanczos methods are discussed in detail in Cullum, Willoughby, and Lake [196]. For block versions of LBD, see Golub, Luk, and Overton [374, 1981].

In floating point arithmetic the computed Lanczos vectors will lose orthogonality, and many of the relations in Section 7.6.1 will not hold to full precision. In spite of this the extreme (largest and smallest) singular values of the truncated bidiagonal matrix $B_k \in \mathbf{R}^{(k+1) \times k}$ tend to be quite good approximations to the corresponding singular values of A, even for $k \ll n$. Let the singular value decomposition of B_k be $B_k = P_{k+1} \Omega_k Q_k^T$. Then approximations to the singular vectors of A are
$$\hat{U}_k = U_k P_{k+1}, \qquad \hat{V}_k = V_k Q_k.$$

This is a simple way of realizing the Ritz–Galerkin projection process on the subspaces $\mathcal{K}_j(A^T A, v_1)$ and $\mathcal{K}_j(AA^T, Av_1)$. The corresponding approximations are called Ritz values and Ritz vectors. Results on the rate of convergence of the Ritz approximations are given in Section 7.6.4.

Lanczos algorithms for computing singular values and vectors are given in Cullum, Willoughby, and Lake [196, 1983] and Berry [67, 1992]. Berry has developed several codes for the partial singular value decompositions of sparse matrices; see [66, 1992], [68, 1993], and [69, 1994]. These include single vector or block Lanczos codes written in Fortran-77, which have been used in information retrieval problems and in seismic tomography. Typically, the 100–200 largest singular values and vectors for matrices having up to 30,000 rows and 20,000 columns are required.

7.6.2. Best approximation in the Krylov subspace. We now consider computing a sequence of approximate solutions to the linear least squares problem
$$\min_x \|Ax - b\|_2, \quad A \in \mathbf{R}^{m \times n}, \quad m \geq n.$$

Here we start the recursion (7.6.8)–(7.6.9) with the vector b, take

(7.6.11) $$\beta_1 u_1 = b, \quad \alpha_1 v_1 = A^T u_1,$$

and for $j = 1, 2, \ldots,$
$$\begin{aligned} \beta_{j+1} u_{j+1} &= A v_j - \alpha_j u_j, \\ \alpha_{j+1} v_{j+1} &= A^T u_{j+1} - \beta_{j+1} v_j, \end{aligned}$$

where $\alpha_{j+1} \geq 0$ and $\beta_{j+1} \geq 0$ are determined so that $\|u_{j+1}\|_2 = \|v_{j+1}\|_2 = 1$.

7.6. Methods Based on Lanczos Bidiagonalization

After k steps we have computed
$$V_k = (v_1, \ldots, v_k), \qquad U_{k+1} = (u_1, \ldots, u_{k+1})$$

and

(7.6.12) $$B_k = \begin{pmatrix} \alpha_1 & & & \\ \beta_2 & \alpha_2 & & \\ & \beta_3 & \ddots & \\ & & \ddots & \alpha_k \\ & & & \beta_{k+1} \end{pmatrix} \in \mathbf{R}^{(k+1) \times k}.$$

The recurrence relations (7.6.11)–(7.6.12) can now be written in matrix form as

(7.6.13) $$\beta_1 U_{k+1} e_1 = b,$$

(7.6.14) $$AV_k = U_{k+1} B_k, \qquad A^T U_{k+1} = V_k B_k^T + \alpha_{k+1} v_{k+1} e_{k+1}^T.$$

We now seek an approximate solution $x_k \in \mathcal{K}_k$, $\mathcal{K}_k = \mathcal{K}_k(A^T A, A^T b)$. Since $\mathcal{K}_k = \mathrm{span}(V_k)$, we can write
(7.6.15) $$x_k = V_k y_k.$$

Multiplying the first equation in (7.6.14) by y_k we obtain $Ax_k = AV_k y_k = U_{k+1} B_k y_k$, and then from (7.6.13)

(7.6.16) $$b - Ax_k = U_{k+1} t_{k+1}, \qquad t_{k+1} = \beta_1 e_1 - B_k y_k.$$

Using the orthogonality of U_{k+1} and V_k it follows that $\|b - Ax_k\|_2$ is minimized over all $x_k \in \mathrm{span}(V_k)$ by taking y_k to be the solution to the least squares problem

(7.6.17) $$\min_{y_k} \|B_k y_k - \beta_1 e_1\|_2.$$

Note the special form of the right-hand side, which holds because the starting vector was taken as b. Hence $x_k = V_k y_k$ solves $\min_{x_k \in \mathcal{K}_k} \|Ax - b\|_2$, and thus *mathematically* this method generates the same sequence of approximations as CGLS. Thus convergence properties for LSQR are the same as for CGLS, and under appropriate conditions the sequence x_k, $k = 1, 2, \ldots$, will converge quickly.

7.6.3. The LSQR algorithm. Since B_k is bidiagonal the subproblem (7.6.17) can be reliably solved by the QR decomposition of B_k,

(7.6.18) $$Q_k B_k = \begin{pmatrix} R_k \\ 0 \end{pmatrix}, \qquad Q_k(\beta_1 e_1) = \begin{pmatrix} f_k \\ \bar{\phi}_{k+1} \end{pmatrix},$$

where R_k is upper bidiagonal,

(7.6.19) $$R_k = \begin{pmatrix} \rho_1 & \theta_1 & & & \\ & \rho_2 & \theta_2 & & \\ & & \ddots & \ddots & \\ & & & \rho_{k-1} & \theta_k \\ & & & & \rho_k \end{pmatrix}, \qquad f_k = \begin{pmatrix} \phi_1 \\ \phi_2 \\ \vdots \\ \phi_{k-1} \\ \phi_k \end{pmatrix}.$$

The matrix Q_k is a product of plane rotations $Q_k = G_{k,k+1}G_{k-1,k}\cdots G_{12}$ chosen to eliminate the subdiagonal elements $\beta_2, \ldots, \beta_{k+1}$ of B_k. The solution vector y_k and the residual t_{k+1} can then be obtained from

$$(7.6.20) \qquad R_k y_k = f_k, \qquad t_{k+1} = Q_k^T \begin{pmatrix} 0 \\ \bar{\phi}_{k+1} \end{pmatrix}.$$

The factorization (7.6.18) need not be computed from scratch in each step. Instead a recurrence relation is developed as follows. Assume we have computed the factorization for B_{k-1}. In the next step the kth column is added, and a plane rotation $Q_k = G_{k,k+1}Q_{k-1}$ determined so that

$$(7.6.21) \quad G_{k,k+1}G_{k-1,k}\begin{pmatrix} 0 \\ \alpha_k \\ \beta_{k+1} \end{pmatrix} = \begin{pmatrix} \theta_k \\ \rho_k \\ 0 \end{pmatrix}, \quad G_{k,k+1}\begin{pmatrix} \bar{\phi}_k \\ 0 \end{pmatrix} = \begin{pmatrix} \phi_k \\ \bar{\phi}_{k+1} \end{pmatrix}.$$

(Note that the rotations $G_{k-2,k-1}, \ldots, G_{12}$ do not affect the kth column.)

If x_k would be formed as $x_k = V_k y_k$ as above, then it would be necessary to save (or recompute) the vectors $v_1, \ldots, v_k$. This can, however, be avoided, as shown by Paige and Saunders [639, 1982]. They derive a simple recursion for computing x_k from x_{k-1}, only storing one extra n-vector. The iterates x_k are formed as

$$x_k = (V_k R_k^{-1})f_k \equiv Z_k f_k.$$

Here Z_k satisfies the lower triangular system $R_k^T Z_k^T = V_k^T$, and hence the columns of $Z_k = (z_1, z_2, \ldots, z_k)$ can be found successively by forward substitution. With $z_0 = x_0 = 0$, we find using R_k in (7.6.19) and identifying the last columns in $Z_k R_k = V_k$,

$$z_k = \frac{1}{\rho_k}(v_k - \theta_k z_{k-1}), \qquad x_k = x_{k-1} + \phi_k z_k.$$

Some work can be saved by using the vectors $w_k \equiv \rho_k z_k$ instead.

Summarizing the formulas developed above gives the Algorithm 7.6.1 derived by Paige and Saunders in 1973. (An independent derivation was later given in [785, 1976].) Mathematically, LSQR generates the same sequence of approximations x_k as CGLS. Both methods require access to A only via subroutines for the matrix-vector products Av_k and $A^T u_k$. However, LSQR is shown in [639, 1982] to be numerically more reliable when many iterations are to be carried out and A is ill-conditioned.

ALGORITHM 7.6.1. LSQR (Paige and Saunders). Initialize

$$x_0 := 0; \qquad \beta_1 u_1 := b; \qquad \alpha_1 v_1 = A^T u_1; \qquad w_1 = v_1;$$
$$\bar{\phi}_1 = \beta_1; \qquad \bar{\rho}_1 = \alpha_1;$$

and for $i = 1, 2, \ldots$ repeat until convergence

$$\beta_{i+1} u_{i+1} := Av_i - \alpha_i u_i;$$
$$\alpha_{i+1} v_{i+1} := A^T u_{i+1} - \beta_{i+1} v_i;$$

7.6. Methods Based on Lanczos Bidiagonalization

$$[c_i, s_i, \rho_i] = \text{givrot}(\bar{\rho}_i, \beta_{i+1});$$
$$\theta_i = s_i \alpha_{i+1}; \quad \bar{\rho}_{i+1} = c_i \alpha_{i+1};$$
$$\phi_i = c_i \bar{\phi}_i; \quad \bar{\phi}_{i+1} = -s_i \bar{\phi}_i;$$
$$x_i = x_{i-1} + (\phi_i/\rho_i) w_i;$$
$$w_{i+1} = v_{i+1} - (\theta_{i+1}/\rho_i) w_i;$$

The algorithm givrot is defined by Algorithm 2.3.1, and the scalars $\alpha_i \geq 0$ and $\beta_i \geq 0$ are chosen to normalize the corresponding vector. LSQR requires $3m + 5n$ multiplications and storage of two m-vectors u, Av, and three n-vectors x, v, w. This can be compared to CGLS, which requires $2m + 3n$ multiplications, two m-vectors, and two n-vectors. Preconditioned versions of LSQR can be derived in the same way as for CGLS.

In Paige and Saunders [639, 1982] reliable stopping criteria are developed for LSQR. Note that LSQR is unusual in not explicitly giving the residual vector $r_k = b - Ax_k$. From $r_k = U_{k+1} t_{k+1}$ (assuming that $U_{k+1}^T U_{k+1} = I$, and using (7.6.20) and (7.6.21)) we get the estimate

$$\|r_k\|_2 = \bar{\phi}_{k+1} s_k s_{k-1} \cdots s_1.$$

A Fortran implementation of LSQR is given in [638, 1982].

7.6.4. Convergence of singular values and vectors. Saad [697] proves the following result on convergence of the Lanczos method applied to a symmetric matrix B.

THEOREM 7.6.1. *Let the eigenvalues of the real symmetric matrix B be ordered decreasingly, $\lambda_1 > \lambda_2 > \cdots > \lambda_n$. Then the angle $\theta(q_i, \mathcal{K}_k)$ between the exact eigenvector q_i associated with λ_i ($i < k$) and the kth Krylov subspace $\mathcal{K}_k = \text{span}\{v, Bb, \ldots, B^{k-1}v\}$ satisfies the inequality*

$$(7.6.22) \qquad \tan \theta(q_i, \mathcal{K}_k) \leq \frac{K_i}{T_{k-i}(1 + 2\gamma_i)} \tan \theta(v, q_i),$$

where T_k is the Chebyshev polynomial, $\kappa_1 = 1$, and

$$K_i = \prod_{j=1}^{i-1} \frac{\lambda_j - \lambda_n}{\lambda_j - \lambda_i}, \quad i > 1, \qquad \gamma_i = \frac{\lambda_i - \lambda_{i+1}}{\lambda_{i+1} - \lambda_n}.$$

In particular, for the eigenvector q_1 corresponding to the largest eigenvalue λ_1,

$$(7.6.23) \qquad \tan \theta(q_1, \mathcal{K}_k) \leq \frac{\tan \theta(v, q_1)}{T_{k-1}(1 + 2\gamma_1)}, \qquad \gamma_1 = \frac{\lambda_1 - \lambda_2}{\lambda_2 - \lambda_n}.$$

This inequality can be shown to be optimal in the sense that, given B and k, v can be chosen so that equality holds.

For the Ritz approximation θ_k to the eigenvalues there are similar bounds. Here we only give the result for $k = 1$.

THEOREM 7.6.2. *Let λ_1 be the largest eigenvalue of B with associated eigenvector q_1, and assume that the Ritz approximation $\lambda_1^{(k)} > \lambda_2$. Let γ_i be defined as in Theorem 7.6.1, and let $K_1 = 1$. Then*

$$(7.6.24) \qquad 0 \leq \lambda_1 - \lambda_1^{(k)} \leq (\lambda_1 - \lambda_n)\left(\frac{\tan\theta(v, q_1)}{T_{k-1}(1 + 2\gamma_1)}\right)^2.$$

The above bounds are invariant under shifts of the spectrum. A similar bound for the Ritz approximation to the smallest eigenvalue follows simply from applying the above result to $-B$.

Saad's result can be used to bound the error in the Lanczos bidiagonalization applied to A. In particular, we can obtain bounds for the errors in the largest and smallest singular values and the associated left singular vectors $\tilde{u}_1$ and $\tilde{u}_n$ by applying Saad's results to AA^T and $-AA^T$, respectively. For example, with the starting vector u_1 we obtain from (7.6.23)

$$(7.6.25) \qquad \tan\theta(u_n, \mathcal{K}_k) \leq \frac{\tan\theta(u_1, \tilde{u}_n)}{T_{k-1}(1 + 2\gamma_1)},$$

where

$$(7.6.26) \qquad \gamma_1 = \frac{\sigma_{n-1}^2 - \sigma_n^2}{\sigma_1^2 - \sigma_n^2} > \frac{\sigma_{n-1}^2 - \sigma_n^2}{\sigma_1^2},$$

and σ_n is the smallest singular value of A. Hence, if the relative gap between σ_{n-1}^2 and σ_n^2 is large, we can expect rapid convergence. For the error in the kth computed smallest singular value $\sigma_n^{(k)}$ we obtain the bound

$$(7.6.27) \qquad 0 \leq (\sigma_n^{(k)})^2 - \sigma_n^2 \leq (\sigma_1^2 - \sigma_n^2)\left(\frac{\tan\theta(u_1, \tilde{u}_n)}{T_{k-1}(1 + 2\gamma_1)}\right)^2,$$

where γ_1 is defined by (7.6.26).

7.6.5. Bidiagonalization and total least squares. We now consider computing an approximate solution to the total least squares (TLS) problem

$$\min_{E,r} \|(E,\ r)\|_F, \qquad (A + E)x = b + r,$$

using the Lanczos process. If $\hat{\sigma}_n > \sigma_{n+1}$, then by (4.6.8) the TLS solution is determined by the left singular vector v_{n+1} of (A, b). One possibility is to apply the second Lanczos bidiagonalization in Section 7.6.1 to the augmented matrix (A, b). This yields an approximate TLS solution in the Krylov subspace $\mathcal{K}_k$. A slightly different way to proceed is to apply the Lanczos bidiagonalization process to A, and then, as in LSQR, seek an approximate TLS solution

$$x_k = V_k y_k \in \mathcal{K}_k.$$

Then from (7.6.13) and (7.6.14)

$$(A + E_k)x_k = (A + E_k)V_k y_k = (U_{k+1}B_k + E_k V_k)y_k = \beta_1 U_{k+1}e_1 + r_k.$$

7.6. Methods Based on Lanczos Bidiagonalization

Hence the consistency relation becomes

(7.6.28) $\quad (B_k + F_k)y_k = \beta_1 e_1 + s_k, \quad F_k = U_{k+1}^T E_k V_k, \quad s_k = U_{k+1}^T r_k.$

Using the orthogonality of U_{k+1} and V_k it follows that

(7.6.29) $\quad \|(E_k, r_k)\|_F = \|(F_k, s_k)\|_F.$

Hence to minimize $\|(E_k, r_k)\|_F$ we should take y_k to be the solution to the TLS subproblem

(7.6.30) $\quad \min_{F,s} \|(F,\ s)\|_F, \quad (B_k + F)y_k = \beta_1 e_1 + s.$

Note the special form of the right-hand side, which holds because the starting vector was taken as b. To solve this subproblem we need to compute the SVD of the bordered bidiagonal matrix $(B_k, \beta_1 e_1)$. Permuting the last column to first position this matrix already has bidiagonal form,

(7.6.31) $\quad (\beta_1 e_1,\ B_k) = \begin{pmatrix} \beta_1 & \alpha_1 & & & \\ & \beta_2 & \alpha_2 & & \\ & & \beta_3 & \ddots & \\ & & & \ddots & \alpha_k \\ & & & & \beta_{k+1} \end{pmatrix} \in \mathbf{R}^{(k+1)\times(k+1)}.$

The SVD of this matrix

$$B_k = P_{k+1} \Omega_k Q_{k+1}^T$$

can be computed cheaply by the standard implicit QR algorithm. Then we have

$$(b,\ A)\begin{pmatrix} V_k^1 \\ V_k^2 \end{pmatrix}(Q_k e_k) = \omega_k (U_{k+1} P_{k+1} e_k).$$

Hence, with

$$\begin{pmatrix} z_k \\ \gamma_k \end{pmatrix} = \begin{pmatrix} V_k^1 \\ V_k^2 \end{pmatrix} Q_k e_k,$$

the approximate TLS solution is given by $x_k = -z_k/\gamma_k \in \mathcal{K}_k$. Note that we only need the last singular vector $Q_k e_k$ to compute x_k, but the vectors v_k need to be saved or regenerated when x_k is computed.

The SVD of this matrix can be computed cheaply by the standard implicit QR algorithm. If the smallest singular value equals $\sigma_{k+1}^{(k)}$, then it holds that

$$\|(E_k, r_k)\|_F = \sigma_{k+1}^{(k)}.$$

Then the inequality (7.6.27) gives us a bound for the convergence

$$\|(E_k, r_k)\|_F^2 \le \min \|(E, r)\|_F^2 + \sigma_1^2 \Big(\frac{\tan\theta(v, u_1)}{T_{k-1}(1 + 2\gamma_1)}\Big)^2.$$

7.7. Methods for Constrained Problems

7.7.1. Problems with upper and lower bounds.
In Section 5.2 we considered methods for solving least squares problems where the solution is subject to inequality constraints. Here we discuss iterative methods for the simple special case of sparse bounded least squares (BLS) problems,

$$\min_x \|Ax - b\|_2, \quad \text{subject to} \quad l \leq x \leq u.$$

Although slow convergence is often a problem, iterative methods can be advantageous for problems where A is well-conditioned as well as large and sparse. For some applications an efficient preconditioner may be found which enhances the rate of convergence.

Many of the methods proposed are based on solving an equivalent linear complementarity problem (LCP)

$$(7.7.1) \qquad y = c - Bx \geq 0, \quad x \geq 0, \quad y^T x = 0,$$

where $c \in \mathbf{R}^n$ is a given vector and $B \in \mathbf{R}^{n \times n}$ a given symmetric matrix. These relations constitute the Karush–Kuhn–Tucker optimality conditions of the quadratic program

$$(7.7.2) \qquad \min_x \frac{1}{2} x^T B x - c^T x \quad \text{subject to} \quad x \geq 0.$$

If we let $B = A^T A$ and $c = A^T b$, this is equivalent to the nonnegativity constrained least squares NNLS problem of Section 5.2.

A good survey of the early history of iterative methods for LCP problems is given in Cottle [187, 1977]. A more recent survey of iterative methods for large convex quadratic programming problems is given by Lin and Pang [532, 1987]. Cryer [194, 1971] first proposed an SOR method for minimizing a strictly convex quadratic function with nonnegativity constraints. Cottle, Golub, and Sacher [188, 1978] developed a block iterative method for linear complementarity problems, which is an extension of Cryer's method.

Pang [646, 1982] introduced a general scheme for solving LCP problems based on the notion of matrix splitting, and subsequently discussed the specialization of this scheme to strictly convex quadratic programs. If $B = M - N$ is a splitting of the matrix B, then given an approximation x^k, the next iterate x^{k+1} is determined as a solution (assumed to exist) to the LCP

$$(7.7.3) \qquad y = c - Nx^k + Mx \geq 0, \quad x \geq 0, \quad x^T y = 0.$$

As an example, with $B = L + D + L^T$, $D = \text{diag}\,(d_i) > 0$, the splitting for the SOR method gives Cryer's method:

$$(7.7.4) \qquad x_i^{k+1} = \max\left\{0, x_i^k - \frac{\omega}{d_i}\left(c_i - \sum_{j<i} b_{ij} x_j^{k+1} - \sum_{j\geq i} b_{ij} x_j^{k+1}\right)\right\},$$

7.7. Methods for Constrained Problems

$i = 1, \ldots, n$. For $B = A^T A$ and $c = A^T b$ this method can be implemented without forming B and c as shown in Section 7.2.4. Using a block splitting of B leads to block SOR schemes, where each step requires the solution of small LCP subproblems.

General convergence results for the basic iterative method (7.7.3) are given in Lin and Pang [532, 1987]. To state the convergence result for the SOR iteration we first need some definitions. We say that a matrix B is strictly copositive if $x^T B x > 0$ for all $x \geq 0$, $x \neq 0$. Further, we say that a method is weakly convergent if for all vectors c and all initial approximations $x^0 \geq 0$, any sequence $\{x^k\}$ generated by the method contains at least one accumulation point; moreover, any such point is a solution to the LCP.

THEOREM 7.7.1. *Let B be symmetric with positive diagonal. Then the SOR method (7.7.4) is weakly convergent for all $0 < \omega < 2$ if and only if B is strictly copositive.*

Many more references to SOR methods can be found in Lin and Pang [532, 1987]. Lin and Cryer [531, 1985] develop an alternating direction implicit algorithm for problems arising from free boundary problems.

The CG method was used by O'Leary [605, 1980] to solve a quadratic programming problem with lower and upper bounds

$$(7.7.5) \qquad \min_x \frac{1}{2} x^T B x - c^T x \quad \text{subject to} \quad l \leq x \leq u,$$

where B is symmetric and positive definite. The algorithm maintains feasibility of the iterates x^k, while iterating toward the proper sign conditions for $y = c - Bx$, i.e., for $j = 1, \ldots, n$,

$$y_j \leq 0 \text{ if } x_j = l_j, \quad y_j \geq 0 \text{ if } x_j = u_j, \quad y_j = 0 \text{ if } l_j < x_j < u_j.$$

In each step a subsystem is solved by the CG method.

Iterative methods can be particularly attractive for solving a sequence of constrained problems with a slowly changing matrix A; see Lötstedt [546, 1984]. Lötstedt gives an active set algorithm for solving time-dependent simulation of contact problems in mechanical systems. His algorithm is, like O'Leary's algorithm, based on the use of a preconditioned CG method. The preconditioner is kept constant during several time steps. The matrix A is usually not of full column rank. Thus a unique solution does not usually exist, and in a second stage the unique minimum norm solution is computed.

In Lötstedt's algorithm only one variable is allowed to leave its bound at each iteration. As in other active set algorithms this may be an undesirable feature when the number of variables is large and many variables must leave their bounds. Based on recent developments in optimization with bound constraints Bierlaire, Toint, and Tuyttens [74, 1991] have developed three new algorithms. The first is an improvement on the algorithm by Lötstedt, which is more efficient, in particular for large scale problems. Two other methods are of projected gradient type, in which the number of active constraints can change more rapidly between the iterations. Their performance on a wide class of problems is shown to be superior to the first algorithm.

7.7.2. Iterative regularization. In Section 5.3 we considered the solution of discretized ill-posed problems by Tikhonov regularization. When two- and three-dimensional problems of this type are to be solved the direct methods given there become impractical, and iterative algorithms have to be considered. Another application for iterative methods are ill-posed problems coming from convolution-type integral equations. Then the matrix A is a Toeplitz matrix, and then matrix-vector products Ax and $A^T y$ can be computed in $O(n \log_2 n)$ multiplications using the fast Fourier transform; see Section 8.4.2.

In iterative methods for computing a regularized solution to

$$(7.7.6) \qquad \min_x \|Ax - b\|_2$$

regularization is achieved by terminating the iterations before the unwanted irregular part of the solution has converged. Thus the regularization is controlled by the number of iterations carried out.

One of the earliest methods of the first class was proposed by Landweber [514, 1951], who considered the iteration

$$(7.7.7) \qquad x_{k+1} = x_k + \omega A^T(b - Ax_k), \qquad k = 0, 1, 2, \ldots.$$

Here ω is a parameter that should be chosen so that $\omega \approx 1/\sigma_1^2(A)$. The approximations x_k here will seem to converge in the beginning, before they deteriorate and finally diverge. This behavior is often called **semiconvergence**. It can be shown that terminating the iterations with x_k gives behavior similar to truncating the singular value expansion of the solution for $\sigma_i \leq \mu \sim k^{-1/2}$. Thus this method produces a *sequence* of less and less regularized solutions. Note that Landweber's method is equivalent to Richardson's stationary first-order method applied to the normal equations $A^T(Ax - b) = 0$.

Strand [761, 1974] analyzed the more general iteration

$$(7.7.8) \qquad x_{k+1} = x_k + p(A^T A) A^T (b - Ax_k),$$

where $p(\lambda)$ is a polynomial or rational function of λ. We note that a special case is the iteration suggested by Riley [687, 1956]:

$$x_{k+1} = x_k + (A^T A + \mu^2 I)^{-1} A^T (b - Ax_k),$$

which corresponds to taking $p(\lambda) = (\lambda + \mu^2)^{-1}$. Riley's method is sometimes called the **iterated Tikhonov method**; see Section 2.7.2.

Assume that $x_0 = 0$ in the iteration (7.7.8), which is no restriction. Then the kth iterate can be expressed in terms of the SVD of A. With $A = U\Sigma V^T$, $U = (u_1, \ldots, u_m)$, $V = (v_1, \ldots, v_n)$, we have

$$(7.7.9) \qquad x_k = \sum_{i=1}^n \varphi_k(\sigma_i^2) \frac{u_i^T b}{\sigma_i} v_i, \qquad \varphi_k(\lambda) = 1 - (1 - \lambda p(\lambda))^k,$$

where φ is called the **filter factors** after k iterations. From (7.7.9) it follows that the effect of terminating the iteration with x_k is to damp the component of the solution along v_i by the factor $\varphi_k(\sigma_i^2)$. For example, the filter function for the Landweber iteration is $\varphi_k(\lambda) = 1 - (1-\omega\lambda)^k$. From this it is easily deduced that, after k iterations only the components of the solution corresponding to $\sigma_i \geq 1/k^{1/2}$ have converged. This means that a large number of iterations are usually required, and therefore Landweber's method cannot be recommended in practice; see Hanke [425, 1991].

We remark that the iteration (7.7.8) can often be performed more efficiently using the factorized polynomial $1 - \lambda p(\lambda) = \prod_{i=1}^{d}(1 - \alpha_j \lambda)$. One iteration in (7.7.8) can then be performed in d minor steps. This leads to a nonstationary Landweber method applied to the normal equations $A^T A x = A^T b$,

$$(7.7.10) \qquad x_{j+1} = x_j + \gamma_j A^T(b - A x_j), \qquad j = 0, 1, \ldots, d-1.$$

Assume that $\sigma_1 = \beta^{1/2}$, and that our object is to compute an approximation to the truncated singular value solution of (7.7.6), with a cut-off for singular values $\sigma_i \leq \sigma_c = \alpha^{1/2}$. Then it is well known (see, e.g., Rutishauser [693, 1959]) that in a certain sense the optimal choice of the parameters in (7.7.10) are $\gamma_j = 1/\xi_j$, where

$$(7.7.11) \qquad \xi_j = \frac{1}{2}(\alpha + \beta) + \frac{1}{2}(\alpha - \beta)x_k, \qquad x_k = \cos\left(\frac{\pi}{2}\frac{2j+1}{k}\right),$$

are the zeros of the Chebyshev polynomial of degree k on the interval $[\alpha, \beta]$. This choice leads to a filter function $R(t)$ of degree k with $R(0) = 0$, and of least maximum deviation from one on $[\alpha, \beta]$. Thus there is no need to construct $p(\lambda)$ first in order to get the parameters γ_j in (7.7.10). Note that we have to pick α in advance, but it is possible to vary the regularization by using a decreasing sequence $\alpha = \alpha_1 > \alpha_2 > \alpha_3 > \cdots$.

Using standard results for Chebyshev polynomials it can be shown that if $\alpha \ll \beta$, then k steps in the iteration (7.7.10)–(7.7.11) reduce the regular part of the solution by the factor

$$(7.7.12) \qquad \delta_k \approx 2e^{-2k(\alpha/\beta)^{1/2}}.$$

From this it follows that the cut-off σ_c for this method is related to the number of iteration steps k in (7.7.10) by $k \approx 1/\sigma_c$. This is a great improvement over Landweber's method, for which $k \approx (1/\sigma_c)^2$.

It is important to note that as it stands the iteration (7.7.10) with parameters (7.7.11) suffers severely from roundoff errors. This instability can be overcome by a reordering of the parameters ξ_j; see [19, 1972]. Alternatively, (7.7.10)–(7.7.11) can be rewritten as a three-term recursion, as in the Chebyshev semi-iterative method in Section 7.2.5.

The CGLS method (Section 7.4.1) and the mathematically equivalent Lanczos-type method LSQR (Section 7.6.3) also are well suited for computing regularized solutions. It was already remarked by Lanczos [512, 1950] that Krylov subspace methods tend to converge quickly to the solution corresponding to the dominating eigenvalues (singular values).

CGLS and LSQR both generate a sequence of approximations x_k, $k = 0, 1, 2, \ldots$, which minimize the quadratic form $\|Ax - b\|_2^2$ over the Krylov subspace $x_k = x_0 + w_k$, where
$$w_k \in \mathcal{K}_k(A^T A, s_0), \quad s_0 = A^T(b - Ax_0).$$
Usually one starts with the smooth initial solution $x_0 = 0$. These methods often converge much more quickly than competing iterative methods to the optimal solution of an ill-posed problem. Under appropriate conditions it can be dramatically faster; see Louis [547, 1987]. However, after the optimal number of iterations the CG method diverges much more rapidly than other methods. Hence, as demonstrated by Hanke [425, 1991], it is essential to stop the iterations after the optimal number of steps.

A strict proof of the regularizing properties has been given by Nemirovskii [598, 1986]. The CG method and stopping rules are discussed in the excellent survey by Hanke and Hansen [427, 1994], in Hanke [426, 1995], and in Hansen [433, 1995]. Unfortunately, a complete understanding of the regularizing effect of Krylov subspace methods is still lacking.

The difficulty in finding reliable stopping rules for Krylov subspace methods can partly be solved by combining them with an inner regularizing algorithm. This was first suggested by O'Leary and Simmons [609, 1981] and independently by Björck [92, 1988]. For example, the CGLS method can be applied to the regularized problem

(7.7.13)
$$\min_x \left\| \begin{pmatrix} A \\ \mu I_n \end{pmatrix} x - \begin{pmatrix} b \\ 0 \end{pmatrix} \right\|_2,$$

which has to be solved for several values of μ. Björck [92, 1988] gives an efficient implementation based on the Golub–Kahan bidiagonalization (7.6.3)–(7.6.12) in the LSQR method. The kth approximation is taken to be $x_k(\mu) = V_k y_k(\mu)$, where $y_k(\mu)$ is the solution to

$$\min_{y_k} \left\| \begin{pmatrix} B_k \\ \mu I_k \end{pmatrix} y_k - \beta_1 \begin{pmatrix} e_1 \\ 0 \end{pmatrix} \right\|_2,$$

which is a regularized version of (7.6.11). Since B_k is bidiagonal its singular value decomposition

$$B_k = P_k \Omega_k Q_k^T = \sum_{i=1}^k \omega_i p_i q_i^T$$

can be computed cheaply. Then, for any value of μ we have

$$y_k(\mu) = \beta_1 \sum_{i=1}^k \frac{\omega_i p_{1i}}{\omega_i^2 + \mu^2} q_i,$$

where $p_{1i} = (P_k)_{1i}$ are the elements in the first row of P_k. Note that the vectors $v_1, \ldots, v_k$ need to be saved or recomputed to construct $x_k(\mu)$. However, this need not be done except at the last iteration step. The problem of choosing the parameter μ by cross-validation is addressed in Björck, Grimme, and Van Dooren [112, 1994].

Chapter 8
Least Squares Problems with Special Bases

8.1. Least Squares Approximation and Orthogonal Systems

8.1.1. General formalism. Frequently a given function $y = f(t)$ has to be modeled by a linear combination of basis functions,

$$(8.1.1) \qquad \tilde{f} = \sum_{j=0}^{n} c_j \phi_j(t).$$

The $n+1$ basis functions $\phi_j(t)$, $j = 0, \ldots, n$, could, e.g., be chosen to span the set of polynomials of degree $\leq n$. In such cases an important consideration is the choice of a proper basis for the space of approximating functions.

We now introduce a formalism related to geometrical ideas, which is convenient in the study of discrete least squares approximation. The set of values of a function $f(x)$ on a finite grid $G = \{x_i\}_{i=0}^{m}$ of distinct points can be considered as a column vector

$$(f(x_0), f(x_1), \ldots, f(x_m))^T.$$

Thinking about functions as vectors leads us to make the following definitions.

DEFINITION 8.1.1. *The* **inner product** *of two real-valued functions f and g defined on $\{x_i\}_{i=0}^{m}$ is denoted by (f,g) and defined by the relation*

$$(8.1.2) \qquad (f,g) = \sum_{i=0}^{m} f(x_i)g(x_i)w_i,$$

where $\{w_i\}_{i=0}^{m}$ are given positive weights. The norm of a function f is defined to be

$$(8.1.3) \qquad \|f\| = ((f,f))^{1/2}.$$

DEFINITION 8.1.2. *Two functions f and g are said to be* **orthogonal** *if $(f,g) = 0$. A sequence of functions $\phi_0, \phi_1, \ldots, \phi_n$ constitutes an* **orthogonal system**, *if $(\phi_i, \phi_j) = 0$, for $i \neq j$ and $\|\phi_i\| \neq 0$ for all i. If, in addition, $\|\phi_i\| = 1$ for all i, then the sequence is called an* **orthonormal system**.

We shall study the least squares problem to determine the coefficients $c_0, c_1, \ldots, c_n$ in (8.1.1) such that the weighted Euclidean norm of the error function $\tilde{f} - f$ is minimized,

$$\|\tilde{f} - f\|^2 = \sum_{i=0}^{m} |\tilde{f}(x_i) - f(x_i)|^2 w_i.$$

Note that interpolation is a special case ($n = m$) of this problem.

THEOREM 8.1.1. *If $\phi_0, \phi_1, \ldots, \phi_n$ are linearly independent, then the least squares approximation problem has a unique solution,*

$$f^* = \sum_{j=0}^{n} c_j^* \phi_j,$$

which is characterized by the orthogonality property $(f^* - f) \perp \phi_j$, $j = 0, 1, \ldots, n$. The coefficients c_j^*, which are called **orthogonal coefficients** *or occasionally Fourier coefficients, satisfy the* **normal equations**

(8.1.4) $$\sum_{j=0}^{n} (\phi_j, \phi_k) c_j = (f, \phi_k).$$

In the important special case when $\phi_0, \phi_1, \ldots, \phi_n$ form an orthogonal system, the coefficients are computed more simply by

(8.1.5) $$c_j^* = (f, \phi_j)/(\phi_j, \phi_j), \qquad j = 0, 1, \ldots, n.$$

We remark that the notations used are such that the results can be generalized with minor changes to also cover the least squares approximation in the continuous case when f is approximated by an infinite sequence of orthogonal functions $\phi_0, \phi_1, \phi_2, \ldots$.

8.1.2. Statistical aspects of the method of least squares. One of the motivations for the method of least squares is that it effectively reduces the influence of random errors in measurements. Suppose that the values of a function have been measured in the points $x_0, x_1, \ldots, x_m$. Let $f(x_p)$ be the measured value, and let $\bar{f}(x_p)$ be the "true" (unknown) function value, which is assumed to be the same as the *expected value* of the measured value. Thus *no systematic errors* are assumed to be present. Suppose further that *the errors in measurement at the various points are statistically independent*. Then we have $f(x_p) = \bar{f}(x_p) + \epsilon$, where

(8.1.6) $$\mathcal{E}(\epsilon) = 0, \qquad \mathcal{V}(\epsilon) = \operatorname{diag}(\sigma_0^2, \ldots, \sigma_n^2),$$

where $\mathcal{E}$ denotes expected value and $\mathcal{V}$ variance. The problem is to use the measured data to estimate the coefficients in the series

$$f(x) = \sum_{j=0}^{n} c_j \phi(x),$$

where $\phi_0, \phi_1, \ldots, \phi_n$ are known functions, $n \leq m$. According to Theorem 1.1.1, the Gauss–Markoff theorem, the estimates c_j^*, which one gets by minimizing the sum

8.2. Polynomial Approximation

$$\sum_{p=0}^{m} w_p \big(f(x_p) - \sum_{j=0}^{n} c_j \phi_j(x_p)\big)^2, \qquad w_p = \sigma_p^{-2},$$

have a smaller variance than the values one gets by any other linear unbiased estimation method. This minimum property holds for estimates not just of the coefficients c_j, but also for every linear function of the coefficients, for example, the estimate

$$f_n^*(\alpha) = \sum_{j=0}^{n} c_j^* \phi(\alpha)$$

of the value $f(\alpha)$ at an arbitrary point α.

Suppose now that $\sigma_p = \sigma$ for all p and that the functions $\{\phi_j\}_{j=0}^{n}$ form an *orthonormal system* with respect to the inner product

$$(f, g) = \sum_{p=0}^{m} f(x_p) g(x_p), \qquad \|\phi_j\|^2 = (\phi_j, \phi_j) = 1, j = 0, \ldots, n.$$

Then the least squares estimates are $c_j^* = (f, \phi_j)$, $j = 0, \ldots, n$.

By Corollary 1.1.1, the covariance matrix of the estimates c_j^* equals $\sigma^2 I$. Hence c_j^* and c_k^* are *uncorrelated if $j \neq k$*:

$$\mathcal{E}\{(c_j^* - \bar{c}_j)(c_k^* - \bar{c}_k)\} = \begin{cases} 0, & \text{if } j \neq k; \\ \sigma^2 & \text{if } j = k, \end{cases}$$

and the variance of the estimate c_j^* is σ^2. From this it follows that

$$\mathcal{V}\{f_n^*(\alpha)\} = \mathcal{V}\left\{\sum_{j=0}^{n} c_j^* \phi_j(\alpha)\right\} = \sum_{j=0}^{n} \mathcal{V}\{c_j^*\} |\phi_j(\alpha)|^2 = \sigma^2 \sum_{j=0}^{n} |\phi_j(\alpha)|^2.$$

As an average, *taken over the grid of measurement points*, the variance of the smoothed function values is

$$\frac{1}{m+1} \sum_{j=0}^{n} \mathcal{V}\{f_n^*(x_i)\} = \frac{\sigma^2}{m+1} \sum_{j=0}^{n} \sum_{i=0}^{m} |\phi_j(x_i)|^2 = \sigma^2 \frac{n+1}{m+1}.$$

Between the grid points, however, *the variance can in many cases be significantly larger*.

8.2. Polynomial Approximation

8.2.1. Triangle family of polynomials. By a polynomial of degree k we mean a function of the form

$$p_k(x) = \sum_{j=0}^{k} a_{k,j} x^j.$$

If the leading coefficient $a_{k,k} \neq 0$, then the polynomial is called a *genuine kth-degree polynomial*. The class of kth-degree polynomials contains all polynomials of lower degree as a special case. A constant is a polynomial of degree zero.

A sequence of polynomials $p_0, p_1, \ldots, p_n$ where $a_{k,k} \neq 0$, $k = 0, 1, \ldots, n$, is called a triangle family of polynomials. The coefficients of such a family form a nonsingular lower triangular matrix $A = (a_{i,j})$, $0 \leq j \leq i \leq n$. It follows that the monomials $1, x, \ldots, x^n$ of x can be expressed recursively and uniquely as linear combinations of $p_0, p_1, \ldots, p_n$,

$$x^k = b_{k,0} p_0 + b_{k,1} p_1 + \cdots + b_{k,k} p_k,$$

where the matrix of coefficients $B = (b_{i,j}) = A^{-1}$. Hence the polynomials of any triangle family can be used as basis functions in the approximation problem.

8.2.2. General theory of orthogonal polynomials. By a family of orthogonal polynomials we mean here a triangle family of polynomials, which is an orthogonal system with respect to the inner product (8.1.2) for some given weights. Expansions of functions in terms of orthogonal polynomials are very useful. They are easy to manipulate, have good convergence properties, and give a well-conditioned representation of a function (with the exception of weight distributions on certain grids).

We shall now prove some results from the general theory of orthogonal polynomials.

THEOREM 8.2.1. *Let $\{x_i\}_{i=0}^m \in (a,b)$ be distinct points and $\{w_i\}_{i=0}^m$ a set of weights. Then there is an associated triangle family of orthogonal polynomials $\phi_0, \phi_1, \ldots, \phi_m$. The family is uniquely determined apart from the fact that the leading coefficients $A_0, A_1, A_2, \ldots$ can be given arbitrary nonzero values. The orthogonal polynomials satisfy a three-term recursion formula, $\phi_{-1}(x) = 0$, $\phi_0(x) = A_0$,*

$$(8.2.1) \qquad \phi_{n+1}(x) = \alpha_n (x - \beta_n) \phi_n(x) - \gamma_n \phi_{n-1}(x), \quad n \geq 0,$$

where $\alpha_n = A_{n+1}/A_n$, and

$$(8.2.2) \qquad \beta_n = \frac{(x\phi_n, \phi_n)}{\|\phi_n\|^2}, \qquad \gamma_n = \frac{\alpha_n \|\phi_n\|^2}{\alpha_{n-1} \|\phi_{n-1}\|^2}, \quad (n \neq 0).$$

If the weight distribution is symmetric about $x = \beta$, then $\beta_n = \beta$ for all n.

Proof. Suppose that the ϕ_j have been constructed for $0 \leq j \leq n$, $\phi_j \neq 0$ ($n \geq 0$). We now seek a polynomial of degree $n+1$ with leading coefficient A_{n+1}, which is orthogonal to $\phi_0, \phi_1, \ldots, \phi_n$.

Since $\{\phi_j\}_{j=0}^n$ is a triangle family, ϕ_{n+1} can be written in the form

$$(8.2.3) \qquad \phi_{n+1} = \alpha_n x \phi_n - \sum_{i=0}^n c_{n,i} \phi_i.$$

Hence ϕ_{n+1} is orthogonal to ϕ_j, $j \leq n$, if and only if

$$\alpha_n (x\phi_n, \phi_j) - \sum_{i=0}^n c_{n,i} (\phi_i, \phi_j) = 0, \qquad j = 0, 1, \ldots, n.$$

8.2. Polynomial Approximation

But since $(\phi_i, \phi_j) = 0$ for $i \neq j$, we have

$$c_{n,j}\|\phi_j\|^2 = \alpha_n(x\phi_n, \phi_j),$$

which determines the coefficients $c_{n,j}$ uniquely. From the definition of inner product it follows that $(x\phi_n, \phi_j) = (\phi_n, x\phi_j)$. But $x\phi_j$ is a polynomial of degree $j+1$, and is therefore orthogonal to ϕ_n if $j+1 < n$. So $c_{nj} = 0$, $j < n-1$, and thus

$$\phi_{n+1} = \alpha_n x\phi_n - c_{n,n}\phi_n - c_{n,n-1}\phi_{n-1}.$$

This has the same form as (8.2.1) if we set

$$\beta_n = c_{n,n}/\alpha_n, \qquad \gamma_n = c_{n,n-1}.$$

To get the expression in (8.2.2) for γ_n we take the inner product of equation (8.2.3) with ϕ_{n+1}. Using orthogonality we get $(\phi_{n+1}, \phi_{n+1}) = \alpha_n(\phi_{n+1}, x\phi_n)$. If we decrease all indices by one we obtain

$$(\phi_n, x\phi_{n-1}) = \|\phi_n\|^2/\alpha_{n-1}, \qquad n \geq 1.$$

Substituting this in the expression for γ_n gives the desired result. ∎

The proof of the above theorem leads to a way to construct the coefficients β_n, γ_n, and the values of the polynomials ϕ_n at the grid points for $n = 1, 2, 3, \ldots$. This is called the **Stieltjes procedure**. Note that for $n = m$, the constructed polynomial must be equal to

$$A_{m+1}(x - x_0)(x - x_1) \cdots (x - x_m),$$

because this polynomial is zero at all the grid points, and thus orthogonal to all functions on the grid. Since $\|\phi_{m+1}\| = 0$, the construction stops at $n = m$. This is natural, since there cannot be more than $m+1$ orthogonal (or even linearly independent) functions on a grid with $m+1$ points.

8.2.3. Discrete least squares fitting. By Theorem 8.1.1 the best approximating polynomial of degree k is given by

$$(8.2.4) \qquad p_k = \sum_{j=0}^{k} c_j \phi_j, \quad c_j = (f, \phi_j)/(\phi_j, \phi_j), \quad k = 0, 1, \ldots, m.$$

Note the important fact that *the coefficients c_j are independent of k.* Hence approximations of increasing degree can be recursively generated as follows. Suppose that ϕ_i, $i = 1, \ldots, k$, and the least squares approximation p_k of degree k have been computed. In the next step the coefficients β_k, γ_k are computed from (8.2.2) and then ϕ_{k+1} by (8.2.1). The next approximation of f is then given by

$$(8.2.5) \qquad p_{k+1} = p_k + c_{k+1}\phi_{k+1}, \qquad c_{k+1} = (f, \phi_{k+1})/\|\phi_{k+1}\|^2.$$

The coefficients $\{\beta_j, \gamma_j\}$ in the recursion formula (8.2.1), and the orthogonal functions ϕ_j at the grid points can be computed using the Stieltjes procedure together with the orthogonal coefficients $\{c_j\}$ for $j = 1, 2, \ldots, n$. The total work required is about $4mn$ flops, assuming unit weights and that the grid is symmetric. If there are differing weights, then about mn additional operations are needed; similarly, mn additional operations are required if the grid is not symmetric. If the orthogonal coefficients are determined simultaneously for several functions on the same grid, then only about mn additional operations per function are required. (In the above, we assume $m \gg 1$, $n \gg 1$.) Hence the procedure is much more economical than the general methods based on normal equations or QR factorization, which all require $O(mn^2)$ flops. The computational advantage of the Stieltjes approach was pointed out by Forsythe [304, 1956].

Since ϕ_{k+1} is orthogonal to p_k, an alternative expression for the new coefficient is

(8.2.6) $$c_{k+1} = (r_k, \phi_{k+1})/\|\phi_{k+1}\|^2, \qquad r_k = f - p_k.$$

Mathematically this formula is equivalent to the classical formula. However, in practice the computed ϕ_{k+1} will gradually lose orthogonality to the previously computed ϕ_j. As pointed out in Shampine [718, 1975] there is an advantage in using the formula involving the residual r_k instead of the classical, because cancellation will take place already when computing the residual, and the coefficient will be more accurately determined. Indeed, when using the classical formula one sometimes finds that the residual norm increases when the degree of the approximation is increased! Note that the difference between the two variants discussed here is similar to the difference between the classical and modified Gram–Schmidt orthogonalization methods.

The Stieltjes procedure may also be sensitive to propagation of roundoff errors. An alternative procedure for computing the recurrence coefficients in (8.2.1) and the values of the orthogonal polynomials has been given by Gragg and Harrod [395, 1984]; see also Boley and Golub [126, 1987]. In this procedure these quantities are computed from an inverse eigenvalue problem for a certain symmetric tridiagonal matrix. Reichel [676, 1991] has compared this scheme with the Stieltjes procedure and shown that the Gragg–Harrod procedure generally yields better accuracy.

When the coefficients c_j in the orthogonal expansion (8.2.4) are known, then the easiest way to compute values of f is to use the following algorithm.

ALGORITHM 8.2.1. CLENSHAW'S RECURSION FORMULA. Let $p(x)$ denote the polynomial in (8.2.4), where $\phi_k(x)$ are orthogonal polynomials which satisfy the recursion (8.2.1). Then $p(x) = A_0 y_0$, where $y_{n+2} = y_{n+1} = 0$, and

(8.2.7) $$y_k = \alpha_k(x - \beta_k) y_{k+1} - \gamma_{k+1} y_{k+2} + c_k, \qquad k = n, n-1, \ldots, 0.$$

Using expansions in orthogonal polynomials also has the very important advantage of avoiding the difficulties with ill-conditioned systems of equations which occur even for moderate n when the coefficients in a polynomial $\sum_{j=0}^{n} c_j x^j$ and the function values are given on an equidistant grid. For equidistant data,

8.2. Polynomial Approximation

the **Gram polynomials** $\{P_{n,m}\}_{n=0}^{m}$ are of interest. These polynomials are orthogonal with respect to the inner product

$$(f,g) = \sum_{i=0}^{m} f(x_i)g(x_i), \qquad x_i = -1 + \frac{2i}{m}.$$

The recursion formula is

$$P_{-1,m}(x) = 0, \qquad P_{0,m} = (m+1)^{-1/2},$$
$$P_{n+1,m}(x) = \alpha_{n,m} x P_{n,m}(x) - \gamma_{n,m} P_{n-1,m}(x), \qquad n \geq 0,$$

where the coefficients are given by $(n < m)$

$$\alpha_{n,m} = \frac{m}{n+1}\left(\frac{4(n+1)^2 - 1}{(m+1)^2 - (n+1)^2}\right)^{1/2}, \qquad \gamma_{n,m} = \frac{\alpha_{n,m}}{\alpha_{n-1,m}}.$$

When $n \ll m^{1/2}$, these polynomials are well behaved. However, when $n \gg m^{1/2}$, they have very large oscillations between the grid points, and a large maximum norm in $[-1,1]$. Related to this is the fact that when fitting a polynomial to *equidistant* data, one should never choose n larger than about $2m^{1/2}$.

The Gram polynomials can be much larger between the grid points if $j \gg m^{1/2}$. Set

$$\sigma_I^2 = \sigma^2 \sum_{j=0}^{n} \frac{1}{2}\int_{-1}^{1} |\phi_j(\alpha)|^2 d\alpha.$$

Thus σ_I^2 is an average variance for $f_n^*(\alpha)$ *taken over the entire interval* $[-1,1]$. The following values were obtained for the ratio k between σ_I^2 and $\sigma^2(n+1)/(m+1)$ when $m = 41$:

n	5	10	15	20	25	30	35
k	1.0	1.1	1.7	26	$7 \cdot 10^3$	$1.7 \cdot 10^7$	$8 \cdot 10^{11}$

These results are related to the recommendation that one should choose $n < 2m^{1/2}$ when fitting a polynomial to equidistant data. This recommendation seems to contradict the Gauss–Markov theorem, but in fact it only means that one gives up the requirement that the estimate be unbiased. Still it is remarkable that this can lead to such a drastic reduction of the variance of the estimates f_n^*.

8.2.4. Vandermonde-like systems. A **Vandermonde** matrix is a matrix of the form

$$V = V(\alpha_0, \alpha_1, \ldots, \alpha_n) = \begin{pmatrix} 1 & 1 & \cdots & 1 \\ \alpha_0 & \alpha_1 & \cdots & \alpha_n \\ \vdots & \vdots & & \vdots \\ \alpha_0^n & \alpha_1^n & \cdots & \alpha_n^n \end{pmatrix},$$

where $\{\alpha_k\}_{k=0}^n$ is a sequence of $n+1$ distinct real numbers. Vandermonde matrices are related to the polynomial problem of finding a polynomial $p(x) = a_n x^n + a_{n-1} x^{n-1} + \cdots + a_0$ that interpolates the data (α_i, f_i), $i = 0, 1, \ldots, n$. It is easily shown that the coefficient vector satisfies the linear system

$$V^T a = f,$$

which is called a dual Vandermonde system. The primal system

$$Vx = b$$

arises when determining weights x_i in quadrature formulas when moments b_i are given.

Because of the structure present in Vandermonde systems they can be solved by special algorithms in $O(n^2)$ multiplications and $O(n)$ memory. Such an algorithm was developed by Björck and Pereyra [115, 1970]. This algorithm corresponds to the decomposition of the inverse V^{-1} into a product

$$V^{-1} = U_0 \cdots U_{n-1} L_{n-1} \cdots L_0$$

of bidiagonal upper and lower triangular matrices.

Vandermonde matrices can be generalized by allowing confluency of some of the points α_i. The corresponding dual system then corresponds to a Hermite interpolation problem. The fast algorithms for the primal and dual systems can be generalized to this case; see Björck and Elfving [107, 1973].

Vandermonde systems are often extremely ill-conditioned because they are related to the interpolation problem with a monomial basis. An interesting phenomenon is that the Björck–Pereyra algorithm is often able to achieve more accuracy in the computed solution than standard (and more expensive) methods like Gaussian elimination with partial pivoting. It was observed in [115] that "some problems, connected with Vandermonde systems, which traditionally have been considered to be too ill-conditioned to be attacked, actually can be solved with good precision." Higham [455, 1987] gives an error analysis, which identifies a class of Vandermonde systems for which the relative error in the computed solution can be bounded by a quantity *independent of* $\kappa(V)$. This holds if the points $\{\alpha_k\}_{k=0}^n$ are positive and monotonically ordered,

$$0 < \alpha_0 \leq \alpha_1 \leq \cdots \leq \alpha_n,$$

and the components of the right-hand side have interchanging signs, that is, $(-1)^k b_k \geq 0$.

The remarkable numerical stability obtained for Vandermonde systems has counterparts for other classes of structured matrices. Boros, Kailath, and Olshevsky [128] have derived fast algorithms and shown similar stability results for Cauchy matrices of the form

$$C = \begin{pmatrix} \frac{1}{x_0 - y_0} & \cdots & \frac{1}{x_0 - y_n} \\ \vdots & \ddots & \vdots \\ \frac{1}{x_n - y_0} & \cdots & \frac{1}{x_n - y_n} \end{pmatrix}.$$

8.2. Polynomial Approximation

Interestingly, the class of systems which can be solved with their algorithm includes Hilbert linear systems with sign-interchanging right-hand side.

Another generalization of Vandermonde matrices is obtained by replacing the monomials by a family of orthonormal polynomials $\{\phi_i\}_0^n$ that satisfy a three-term recurrence relation of the form (8.2.1), where $\theta_j \neq 0$ for all j. The corresponding matrix V with elements $v_{ij} = \phi_i(\alpha_j)$, $0 \leq i,j \leq n$ is called a Vandermonde-like matrix. The interest in this generalization stems from the fact that these Vandermonde-like matrices generally have much smaller condition numbers than the classical Vandermonde matrices. Fast algorithms for Vandermonde-like systems have been given by Higham [457, 1988] and Reichel [676, 1991].

Consider now a rectangular Vandermonde matrix $V \in \mathbf{R}^{n \times n}$ consisting of the first $n < m$ columns of $V(\alpha_0, \alpha_1, \ldots, \alpha_m)$. It is natural to ask if fast methods exist for solving the primal Vandermonde least squares problem

$$(8.2.8) \qquad \min_x \|Vx - b\|_2.$$

Demeure [214, 1989], [215, 1990] has given an algorithm of complexity $5mn + 7n^2/2 + O(m)$ for computing the QR factorization of V, which can be used to solve problem (8.2.8). However, since this algorithm forms $V^T V$ it is likely to be unstable.

Reichel [676, 1991] gives a fast algorithm based on the Rutishauser–Gragg–Harrod scheme for computing the QR decomposition of transposed Vandermonde-like matrices. This can be used to solve overdetermined dual Vandermonde-like systems in the least squares sense in order $O(mn)$ operations.

8.2.5. Chebyshev polynomials. The Chebyshev polynomials are perhaps the most important family of orthogonal polynomials. Their properties can be derived by rather simple methods.

Consider the easily verified formula

$$\cos(n+1)\phi + \cos(n-1)\phi = 2\cos\phi\cos(n\phi), \qquad n \geq 1.$$

This formula can be used recursively to express $\cos(n\phi)$ as a polynomial in $\cos\phi$. If we set $x = \cos\phi$, and thus $\phi = \arccos x$, then we obtain a triangle family of polynomials, the Chebyshev polynomials, defined for $-1 \leq x \leq 1$, by the formula $T_n(x) = \cos(n \arccos x)$, $n \geq 0$. Hence the Chebyshev polynomials satisfy the recursion formula, $T_0(x) = 1$,

$$(8.2.9) \qquad T_1(x) = x, \quad T_{n+1}(x) = 2xT_n(x) - T_{n-1}(x), \quad n \geq 0.$$

The leading coefficient of $T_n(x)$ is 2^{n-1} for $n \geq 1$ and 1 for $n = 0$. The symmetry property $T_n(-x) = (-1)^n T_n(x)$ also follows from the recurrence formula.

$T_n(x)$ has n zeros in $[-1, 1]$ given by the **Chebyshev abscissae**,

$$(8.2.10) \qquad x_k = \cos\left(\frac{2k+1}{n}\frac{\pi}{2}\right), \qquad k = 0, 1, \ldots, n-1,$$

and $n+1$ **extrema**

$$x'_k = \cos\frac{k}{n}\pi, \qquad T_n(x'_k) = (-1)^k, \qquad k = 0, 1, \ldots, n.$$

These results follow directly from the fact that $|\cos(n\phi)|$ has maxima for $\phi'_k = k\pi/n$ and $\cos(n\phi_k) = 0$ for $\phi_k = (2k+1)\pi/(2n)$.

The Chebyshev polynomials $T_0, T_1, \ldots, T_{n-1}$ are orthogonal with respect to the inner product

$$(f, g) = \sum_{k=0}^{n-1} f(x_k)g(x_k),$$

where $\{x_k\}$ are the Chebyshev abscissae (8.2.10) for T_n. Then for $0 \leq i, j < n$, we have $(T_i, T_j) = 0$, if $i \neq j$, and

(8.2.11) $$(T_i, T_j) = \begin{cases} \frac{1}{2}n, & \text{if } i = j \neq 0; \\ n, & \text{if } i = j = 0. \end{cases}$$

If one intends to approximate a function *in the entire interval* $[-1, 1]$ by a *polynomial* and can choose the points at which the function is computed or measured, then one should choose the Chebyshev abscissae. Using these points, interpolation is a fairly *well-conditioned* problem in the entire interval, and one can conveniently fit a polynomial of lower degree than m, if one wishes to smooth errors in measurement. The risk of disturbing surprises between the grid points is insignificant.

For interpolating a function in the Chebyshev abscissae we get the following method.

ALGORITHM 8.2.2. CHEBYSHEV INTERPOLATION. Let $p(x)$ denote the interpolation polynomial in the Chebyshev points x_k given by (8.2.10). Then by Theorem 8.2.1 we have

$$p(x) = \sum_{j=0}^{n-1} c_j T_j(x), \qquad c_i = \frac{1}{\|T_i\|^2} \sum_{k=0}^{n-1} f(x_k) T_i(x_k),$$

where $\|T_i\|_2$ is given by (8.2.11).

Expansions in the terms of Chebyshev polynomials are an important aid in the study of functions on the interval $[-1, 1]$. If one is working in terms of a parameter t which varies in the interval $[a, b]$, then one should make the substitution, $t = \frac{1}{2}(a+b) + \frac{1}{2}(a-b)x$ ($t \in [a, b] \Leftrightarrow x \in [-1, 1]$) and work with the Chebyshev points

$$t_k = \frac{1}{2}(a+b) + \frac{1}{2}(a-b)x_k, \qquad k = 0, 1, \ldots, n-1.$$

The remainder term in interpolation using the values of the function f at the points x_i, $i = 0, 1, \ldots, n-1$, is equal to

$$\frac{f^{(n)}(\xi)}{(n)!}(x - x_0)(x - x_1) \cdots (x - x_{n-1}).$$

8.2. Polynomial Approximation

Here ξ depends on x, but one can say that the error curve behaves for the most part like a polynomial curve $y = a(x - x_0)(x - x_1) \cdots (x - x_{n-1})$. A similar oscillating curve is also typical for error curves arising from least squares approximation. The zeros of the error are then about the same as the zeros for the first neglected term in the orthogonal expansion. This contrasts sharply with the error curve for Taylor approximation at x_0, whose usual behavior is described approximately by the formula $y = a(x - x_0)^{n-1}$.

From the minimax property of the Chebyshev polynomials it follows that placing the interpolation points $x_0, x_1, \ldots, x_{n-1}$ in the Chebyshev abscissae will minimize the maximum magnitude of $q(x) = (x - x_0)(x - x_1) \cdots (x - x_{n-1})$ in the interval $[-1, 1]$. This corresponds to choosing $q(x) = T_{m+1}(x)/2^m$.

For computing $p(x)$, one can use Clenshaw's recursion formula; see the previous section. (Note that $\alpha_k = 2$ for $k > 0$, but $\alpha_0 = 1$.) Occasionally one is interested in the partial sums of the expansion. For example, in order to smooth errors in measurement it can be advantageous to break off the summation before one has come to the last term. If the values of the function are afflicted with statistically independent errors in measurement with standard deviation σ, then (see the next section) the series can be broken off when for the first time

$$\left\| f - \sum_{j=0}^{k} c_j T_j \right\| < \sigma n^{1/2}.$$

If the measurement points are the Chebyshev abscissae, then no difficulties arise in fitting a polynomial to the data. The Chebyshev polynomials have in this case a magnitude between the grid points which is not much larger than their magnitude at the grid points. The average variance for $f_n^*(\alpha)$ becomes the same on the interval $[-1, 1]$ as on the grid of measurement points, $\sigma^2(n+1)/(m+1)$.

The choice of n, when m is given, is a question of compromising between taking into account the systematic error, i.e., the truncation error which decreases when n increases, and taking into account the random errors which grow as n increases. In the Chebyshev case, $|c_j|$ decreases quickly with j if f is a sufficiently smooth function, while the part of c_j^* which comes from errors in measurement varies randomly with magnitude about $\sigma(2/(m+1))^{1/2}$. The expansion should then be broken off when the coefficients begin to "behave randomly." The coefficients in an expansion in terms of the Chebyshev polynomials can hence be used for filtering away the "noise" from the signal, even when σ is initially unknown.

EXAMPLE 8.2.1. Fifty-one equidistant values of a certain analytic function were rounded to four decimals. In Figure 8.2.1 a semilogarithmic diagram is given which shows how $|c_j^*|$ varies in an expansion in terms of the Chebyshev polynomials of the above data. For $j < 20$, approximately, the contribution due to noise dominates the contribution due to signal. Thus it is sufficient to break off the series at $n = 20$.

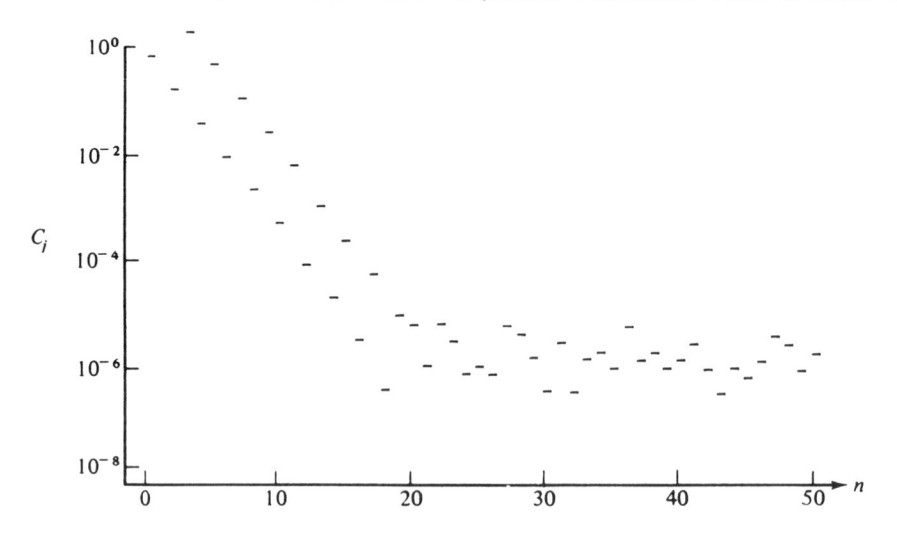

FIG. 8.2.1. *Coefficients in Chebyshev expansion of a rounded analytic function.*

8.3. Discrete Fourier Analysis

8.3.1. Introduction. According to a mathematical theorem first given by Fourier (1758–1830), every function $f(t)$ with period $2\pi/\omega$ can, under certain very general conditions, be expanded into a series of the form

$$(8.3.1) \qquad \sum_{k=0}^{\infty} r_k \sin(k\omega t + v_k).$$

(By a function with period p, we mean here that $f(t+p) = f(t)$, $\forall t$.)

Fourier analysis is one of the most useful and valuable tools in applied mathematics. It has application in the modeling of phenomena which are exactly or approximately periodic in time or space. An important area of application is in digital signal processing, which is used in interpreting radar and sonar signals. Another is statistical time series, which are used in communications theory, control theory, and the study of turbulence.

Frequently the continuous Fourier analysis cannot be directly applied, because the functions to be modeled are known only on a discrete (equidistant) set of sampling points. Then a discrete version of the Fourier analysis has to be used. This also has applications to problems that are not a priori periodic.

An expansion of the form of (8.3.1) can be expressed in many equivalent ways. If we set $a_k = r_k \sin v_k$, $b_k = r_k \cos v_k$, then using the addition theorem for the sine function we can write

$$(8.3.2) \qquad f(t) = \sum_{k=0}^{\infty} (a_k \cos k\omega t + b_k \sin k\omega t),$$

where a_k, b_k are real constants. Another form, which is often the most convenient, can be found with the help of Euler's formulas,

$$\cos x = \frac{1}{2}(e^{ix} + e^{-ix}), \qquad \sin x = \frac{1}{2i}(e^{ix} - e^{-ix})$$

8.3. Discrete Fourier Analysis

$(i = \sqrt{-1})$. Then one gets

(8.3.3) $$f(t) = \sum_{-\infty}^{\infty} c_k e^{ik\omega t},$$

where for $k = 1, 2, 3, \ldots,$

(8.3.4) $$c_0 = a_0, \quad c_k = (a_k - ib_k)/2, \quad c_{-k} = (a_k + ib_k)/2.$$

In the rest of this section we shall study approximations of a function f with period 2π with partial sums of the form of these series. We call these finite sums **trigonometric polynomials**. If a function of t has period p, then the substitution $x = 2\pi t/p$ transforms the function to a function of x with period 2π. We assume that the function can have complex values, since the complex exponential function is convenient for manipulations.

8.3.2. Orthogonality relations. The discrete inner product of two complex-valued functions f and g of period 2π is defined in the following way (the bar over g indicates complex conjugation):

(8.3.5) $$(f, g) = \sum_{\alpha=0}^{M} f(x_\alpha) \bar{g}(x_\alpha), \quad x_\alpha = 2\pi\alpha/(M+1).$$

Note that an equidistant grid $G = \{x_\alpha\}_{\alpha=0}^M$ is used. We have

$$(g, f) = \overline{(f, g)}, \quad \|f\| = (f, f)^{1/2}.$$

THEOREM 8.3.1. *With the inner product (8.3.5) the following orthogonality relations hold for the functions $\phi_j(x) = e^{ijx}$, $j = 0, \pm 1, \pm 2, \ldots$:*

$$(\phi_j, \phi_k) = \begin{cases} M + 1, & \text{if } (j-k)/(M+1) \text{ is an integer,} \\ 0, & \text{otherwise.} \end{cases}$$

Proof. Set $h = 2\pi/(M+1)$, $x_\alpha = h\alpha$,

$$(\phi_j, \phi_k) = \sum_{\alpha=0}^{M} e^{ijx_\alpha} e^{-ikx_\alpha} = \sum_{\alpha=0}^{M} e^{i(j-k)h\alpha}.$$

This is a geometric series with ratio $q = e^{i(j-k)h}$. If $(j-k)/(M+1)$ is an integer, then $q = 1$ and the sum is $M + 1$. Otherwise $q \neq 1$, but $q^{M+1} = e^{i(j-k)2\pi} = 1$. From the summation formula of a geometric series

$$(\phi_j, \phi_k) = (q^{M+1} - 1)/(q - 1) = 0. \quad \blacksquare$$

If one knows that the function f has an expansion of the form $f = \sum_{j=0}^{M} c_j \phi_j$, it follows formally that

$$(f, \phi_k) = \sum_{j=a}^{b} c_j (\phi_j, \phi_k) = c_k (\phi_k, \phi_k), \quad a \leq k \leq b,$$

since $(\phi_j, \phi_k) = 0$ for $j \neq k$. Thus, changing k to j, we have

$$(8.3.6) \qquad c_j = \frac{(f, \phi_j)}{(\phi_j, \phi_j)} = \frac{1}{M+1} \sum_{\alpha=0}^{M} f(x_\alpha) e^{-ijx_\alpha}.$$

THEOREM 8.3.2. *Fourier Analysis, Discrete Case. Every function, defined on the grid* $G = \{x_\alpha\}_{\alpha=0}^{M-1}$, $x_\alpha = 2\pi\alpha/(M+1)$, *can be interpolated by a trigonometric polynomial, which can be written in the form*

$$(8.3.7) \qquad f(x) = \sum_{j=-k}^{k+\theta} c_j e^{ijx}, \qquad \theta = \begin{cases} 0, & \text{if } M \text{ even,} \\ 1, & \text{if } M \text{ odd.} \end{cases}$$

If the sum in (8.3.7) *is terminated when* $j < k + \theta$, *then one obtains the trigonometric polynomial which is the best least squares approximation, among all trigonometric polynomials with the same number of terms, to* f *on the grid.*

Proof. The expression for c_j was formally derived previously (see (8.3.6)), and the derivation is justified by Theorem 8.1.1. The function $f(x)$ coincides *on the grid* with the function $f^*(x) = \sum_{j=0}^{m} c_j e^{ijx}$, because

$$e^{-i(M+1-j)x_\alpha} = e^{ijx_\alpha}, \qquad c_{-j} = c_{M+1-j}. \quad \blacksquare$$

The functions f and f^* are not identical between the grid points. Functions of several variables can be treated analogously by taking one variable at a time.

Notice that the calculations required to compute the coefficients c_j according to (8.3.6), **Fourier analysis**, are of essentially the same type as the calculations needed to tabulate $f^*(x)$ for $x = 2\pi\alpha/(M+1)$, $\alpha = 0, 1, \ldots, M$, when the expansion in (8.3.7) is known, so-called **Fourier synthesis**.

8.3.3. The fast Fourier transform. The application of discrete Fourier analysis to large scale problems became feasible only with the invention of a new family of algorithms called fast Fourier transforms (FFT), which reduced the computational complexity to $O(N \log N)$, where N is the number of points. The FFT was developed in 1965 by Cooley and Tukey [185, 1965], but related ideas can be found in many previous works; see Cooley [183, 1990].

Consider the problem of computing the discrete Fourier coefficients $\{c_j\}_{j=0}^{N-1}$ for a function

$$f(x) = \sum_{j=0}^{N-1} c_j e^{ijx},$$

whose values are known at the points $x_\beta = 2\pi\beta/N$, $\beta = 0, 1, \ldots, N-1$. According to Theorem 8.3.2, with obvious changes in notation, we have

$$(8.3.8) \qquad c_j = \frac{1}{N} \sum_{\beta=0}^{N-1} f(2\pi\beta/N) e^{-ij2\pi\beta/N}, \quad j = 0, 1, \ldots, N-1.$$

Let $\omega = e^{-2\pi i/N}$ be the nth root of unity, $\omega^N = 1$. Then we can rewrite the problem as follows. For $j = 0, 1, \ldots, N-1$ compute

8.3. Discrete Fourier Analysis

$$c_j = \sum_{\beta=0}^{N-1} f_\beta \omega^{j\beta}, \quad f_\beta = \frac{1}{N} f(x_\beta).$$

Here c_j is expressed as a polynomial of degree $N-1$ in ω^j. This can also be written as a matrix-vector multiplication

$$c = F_N f, \qquad (F_N)_{j\beta} = \omega^{j\beta},$$

where $F_N \in \mathbf{R}^{N \times N}$ is the **Fourier matrix**.

If implemented in a naive way this discrete Fourier transform will take N^2 operations. With the FFT, if $N = r_1 r_2 \cdots r_p$, one needs only $N(r_1 + r_2 + \cdots + r_p)$ operations (one operation equals one complex addition and one complex multiplication) to compute all c_j.

Consider the special case $N = 2^k$ and set

$$\beta = \begin{cases} 2\beta_1, & \text{if } \beta \text{ even,} \\ 2\beta_1 + 1, & \text{if } \beta \text{ odd,} \end{cases} \quad 0 \leq \beta_1 \leq \frac{1}{2}N - 1.$$

Then the sum in (8.3.8) can be split into an even and an odd part:

$$(8.3.9) \qquad c_j = \sum_{\beta_1=0}^{\frac{1}{2}N-1} f_{2\beta_1}(\omega^2)^{j\beta_1} + \sum_{\beta_1=0}^{\frac{1}{2}N-1} f_{2\beta_1+1}(\omega^2)^{j\beta_1} \omega^j.$$

Let α be the quotient and j_1 the remainder when j is divided by $\frac{1}{2}N$, i.e., $j = \alpha \frac{1}{2} N + j_1$. Then, since $\omega^N = 1$,

$$(\omega^2)^{j\beta_1} = (\omega^2)^{\alpha(1/2)N\beta_1}(\omega^2)^{j_1\beta_1} = (\omega^N)^{\alpha\beta_1}(\omega^2)^{j_1\beta_1} = (\omega^2)^{j_1\beta_1}.$$

Thus if, for $j_1 = 0, 1, \ldots, \frac{1}{2}N - 1$, we set

$$\phi(j_1) = \sum_{\beta_1=0}^{\frac{1}{2}N-1} f_{2\beta_1}(\omega^2)^{j_1\beta_1}, \quad \psi(j_1) = \sum_{\beta_1=0}^{\frac{1}{2}N-1} f_{2\beta_1+1}(\omega^2)^{j_1\beta_1},$$

where $(\omega^2)^{\frac{1}{2}N} = 1$, then by (8.3.9)

$$c_j = \phi(j_1) + \omega^j \psi(j_1), \qquad j = 0, 1, \ldots, N - 1.$$

The computation of $\phi(j_1)$ and $\psi(j_1)$ means that one does *two* Fourier analyses with $\frac{1}{2}N = 2^{k-1}$ terms instead of one with $N = 2^k$ terms. The number of operations required to compute $\{c_j\}$, when $\{\phi(j_1)\}$ and $\{\psi(j_1)\}$ have been computed, is $N = 2^k$, assuming that the powers of ω are precomputed and stored. The same idea can now be applied to the two Fourier analyses. One then gets *four* Fourier analyses, each of which has $\frac{1}{4}N$ terms; these can be further divided, etc. (This approach is called **divide-and-conquer**.)

Denote by p_k the total number of operations needed to compute the coefficient when $N = 2^k$. Reasoning as above, we have

$$p_k \leq 2p_{k-1} + 2^k, \qquad k = 1, 2, 3, \ldots.$$

Since $p_0 = 0$, it follows by induction that $p_k \leq k \cdot 2^k = N \cdot \log_2 N$. Hence, when N is a power of two, the FFT solves the problem with at most $N \cdot \log_2 N$ operations.

Van Loan [800, 1992] gives a unified treatment of FFT algorithms based on the factorization of the Fourier matrix F_N into a product of sparse matrix factors. There are many excellent surveys of the use of the discrete Fourier transform, e.g., Cooley, Lewis, and Welsh [184, 1969] and Henrici [447, 1979].

8.4. Toeplitz Least Squares Problems

8.4.1. Introduction.
A Toeplitz matrix $T = (t_{ij})$ is a matrix with constant entries along each diagonal, i.e., $t_{ij} = t_{j-i}$. Hence a rectangular Toeplitz $T \in \mathbf{R}^{m \times n}$ matrix has the form

$$T = \begin{pmatrix} t_0 & t_1 & \cdots & t_n \\ t_{-1} & t_0 & \cdots & t_{n-1} \\ \vdots & \vdots & \ddots & \vdots \\ t_{-m} & t_{-m+1} & \cdots & t_0 \end{pmatrix} \in \mathbf{R}^{(m+1) \times (n+1)},$$

and is defined by the $n+m+1$ values of $t_{-m}, \ldots, t_0, \ldots, t_n$. Here we are interested in special methods for solving Toeplitz linear systems $Tx = b$, or the Toeplitz linear least squares problem

(8.4.1) $$\min_x \|Tx - b\|_2,$$

where T has full column rank. Such problems arise in many applications, e.g., digital signal processing and linear prediction problems. They are important in signal restoration, acoustics, and seismic exploration. The dimensions of the Toeplitz matrices in such applications are often large, and the size $10,000 \times 2,000$ not uncommon. Hence there is a great need for special fast methods for solving problem (8.4.1). In large problems storage requirements are also important. The original data in problem (8.4.1) only requires $2m + n + 1$ storage. However, if standard factorization methods are used, at least $n(n+1)/2 + n$ storage is needed.

Many special methods have been devised to solve Toeplitz linear systems and least squares problems. Bunch has [133, 1985] investigated the stability properties of classical and fast algorithms for solving Toeplitz systems. More recent surveys can be found in Brent, [131, 1988], and in Bojanczyk, Brent, and de Hoog [122, 1993]. Several methods of complexity $O(mn)$ exist for solving (8.4.1). There also exist methods which only require $O(m \log n)$ operations. These have been called "superfast," although n may have to be quite large ($n > 256$) for them to be more efficient than the $O(n^2)$ methods. Since the numerical stability properties of superfast methods are generally either bad or unknown, these methods will not be discussed any further here.

A Hankel matrix is a Toeplitz matrix in which the rows have been reversed:

$$H = \begin{pmatrix} h_{-m} & h_{-m+1} & \cdots & h_0 \\ \vdots & \vdots & \ddots & \vdots \\ h_{-1} & h_0 & \cdots & h_{n-1} \\ h_0 & h_1 & \cdots & h_n \end{pmatrix} \in \mathbf{R}^{(m+1)\times(n+1)}.$$

Hence the methods discussed in this section apply also to Hankel linear systems and Hankel least squares problems.

8.4.2. QR decomposition of Toeplitz matrices. In order to find fast and stable methods for solving the Toeplitz least squares problem it is natural to consider computing the QR decomposition of the Toeplitz matrix T. The first such $O(mn)$ method was given by Sweet [768, 1984], and later generalized to the block case [769]. However, this method has been shown to be unstable. Bojanczyk, Brent, and de Hoog [121, 1986] developed an $O(mn)$ algorithm (BBH) for computing Q and R. Here R is implicitly derived from the cross-product of T and its transpose and therefore will fail on ill-conditioned cases. An equivalent algorithm was derived by Chun, Kailath, and Lev-Ari [168, 1987] using a different approach. However, none of these methods perform well on ill-conditioned problems. Park and Eldén [648, 1993] give modifications to improve the accuracy of the BBH algorithm.

Cybenko [199, 1987] has developed an algorithm for computing R^{-1} and $Q \in \mathbf{R}^{m \times n}$ based on the so-called lattice algorithm. This algorithm requires that all submatrices $T_{:,1:i}$, $i = 1, \ldots, n$, be well-conditioned. This requirement is not always satisfied in applications. Note that column pivoting would destroy the Toeplitz structure, and thus is not a possible solution. Hansen and Gesmar [434, 1993] discuss a modification of Cybenko's algorithm where a block step is used to skip over linearly dependent columns.

We now describe the basic idea behind the BBH algorithm. Using the Toeplitz structure T can be partitioned as

$$(8.4.2) \qquad T = \begin{pmatrix} t_0 & u^T \\ v & T_0 \end{pmatrix} = \begin{pmatrix} T_0 & \tilde{u} \\ \tilde{v}^T & t_{m-n} \end{pmatrix},$$

where T_0 is a submatrix of T, u and $\tilde{v}$ are $n-1$ dimensional vectors, v, $\tilde{u}$ are $m-1$ dimensional vectors, and t_0 and t_{m-n} are scalars. Let R be the Cholesky factor of $T^T T$, and partition R as

$$(8.4.3) \qquad R = \begin{pmatrix} r_{11} & z^T \\ 0 & R_b \end{pmatrix} = \begin{pmatrix} R_t & \tilde{z} \\ 0 & r_{nn} \end{pmatrix},$$

where r_{11} and r_{nn} are scalars.

Setting $R^T R = T^T T$ and using the partitioning (8.4.2) and (8.4.3), we get

$$(8.4.4) \qquad \begin{pmatrix} r_{11}^2 & r_{11}z^T \\ r_{11}z & zz^T + R_b^T R_b \end{pmatrix} = \begin{pmatrix} t_0^2 + v^T v & t_0 u^T + v^T T_0 \\ t_0 u + T_0^T v & uu^T + T_0^T T_0 \end{pmatrix}$$

and
(8.4.5) $\begin{pmatrix} R_t^T R_t & R_t^T \tilde{z} \\ \tilde{z}^T R_t & \tilde{z}^T \tilde{z} + r_{nn}^2 \end{pmatrix} = \begin{pmatrix} T_0^T T_0 + \tilde{v}\tilde{v}^T & T_0^T \tilde{u} + t_{m-n}\tilde{v} \\ \tilde{u}^T T_0 + t_{m-n}\tilde{v}^T & \tilde{u}^T \tilde{u} + t_{m-n}^2 \end{pmatrix}.$

From (8.4.4) and (8.4.5) we see that

$$zz^T + R_b^T R_b = uu^T + T_0^T T_0, \qquad R_t^T R_t = T_0^T T_0 + \tilde{v}\tilde{v}^T,$$

and eliminating the term $T_0^T T_0$ we obtain

(8.4.6) $$R_b^T R_b = R_t^T R_t + uu^T - \tilde{v}\tilde{v}^T - zz^T,$$

which is the basic relation used by the BBH algorithm. This shows that if R_t was known, R_b could be computed by one Cholesky updating step and two Cholesky downdating steps; see Section 3.3. Moreover, since updating and downdating can proceed by rows the first k rows of R_b can be obtained from the first k rows of R_b. But the kth row of R_b defines the $(k+1)$th row of R_t, and the first row of R can be obtained from (8.4.4)

$$r_{11} = \sqrt{t_0^2 + v^T v}, \qquad z^T = (t_0 u^T + v^T T_0)/r_{11}.$$

It follows that (8.4.6) provides a method for computing R one row at a time. This algorithm requires $mn^2 + 6n^2$ multiplications.

There are several ways to proceed when computing the least squares solution. Nagy [594, 1993] has modified the BBH algorithm to compute R^{-1}, $Q^T b$ and the solution x in a linear amount of storage and $2mn + 14n^2$ multiplications. Another possibility is to use the seminormal equations $R^T R x = T^T b$ and to compute the solution as $x = R^{-1} R^{-T} (T^T b)$. This can also be implemented in linear storage (see Nagy [594]) and, if $T^T b$ is computed by FFT, reduces the cost to $mn + 9n^2$ multiplications. For ill-conditioned problems the solution can be improved by iterative refinement.

8.4.3. Iterative solvers for Toeplitz systems. The Toeplitz structure implies that the matrix-vector product Tx for a given vector x reduces to a convolution problem, and can be computed via the fast Fourier transform in $O(n \log n)$ operations. This is true also when T is a rectangular Toeplitz matrix; see O'Leary and Simmons [609, 1981].

Let the matrix T be expanded into a square circulant matrix

$$T_C = \begin{pmatrix} t_0 & t_1 & \cdots & t_n & t_{-m} & \cdots & t_{-1} \\ t_{-1} & t_0 & \cdots & t_{n-1} & t_n & \cdots & t_{-m} \\ \vdots & \vdots & \ddots & \vdots & \vdots & \ddots & \vdots \\ t_{-m} & t_{-m+1} & \cdots & t_0 & t_1 & \cdots & t_n \\ t_n & t_{-m} & \cdots & t_{-1} & t_0 & \cdots & t_{n-1} \\ \vdots & \vdots & \ddots & \vdots & \vdots & \ddots & \vdots \\ t_1 & t_2 & \cdots & t_{-m} & t_{-m+1} & \cdots & t_0 \end{pmatrix} \in \mathbf{R}^{p \times p},$$

8.4. TOEPLITZ LEAST SQUARES PROBLEMS

where $p = m + n - 1$ and T corresponds to the $(1,1)$-block. The eigenvectors of the circulant T_C are known to be the Fourier vectors, i.e., the columns of the matrix $F = (f_{jk}) \in \mathbf{R}^{p \times p}$, where

$$f_{jk} = \frac{1}{\sqrt{p}} e^{2\pi i jk/p}, \quad 0 \leq j,k \leq p, \quad i = \sqrt{-1}.$$

The eigenvalues are given by the components of the Fourier transform of its first column
(8.4.7) $\qquad F(t_0, \ldots, t_{-m}, t_n, \ldots, t_1)^T = (\lambda_1 \;\; \ldots \;\; \lambda_p)^T,$

and the matrix T_C can be factorized as
(8.4.8) $\qquad T_C = F \Lambda F^H, \qquad \Lambda = \mathrm{diag}\,(\lambda_1, \ldots, \lambda_p).$

To form the product of $y = Tx$, where $x \in \mathbf{R}^{n+1}$ is an arbitrary vector, we pad x with zeros and calculate

$$y = (\,I_m \;\; 0\,)\,z, \quad z = T_C \begin{pmatrix} x \\ 0 \end{pmatrix} = F \Lambda F^H \begin{pmatrix} x \\ 0 \end{pmatrix}.$$

This can be done with two FFTs, and one multiplication with a diagonal matrix, and hence the cost of this is $O(n \log_2 n)$ operations. Since the transpose of T is also a Toeplitz matrix, a similar scheme can be used for the fast computation of $T^T y$.

Using the fast multiplication described here we obtain fast implementations of several of the iterative methods for least squares problems in Chapter 7. Of particular interest is Algorithm 7.4.1 (CGLS), for which in step k we only need to compute $T p^{(k)}$ and $T^T r^{(k+1)}$.

8.4.4. Preconditioners for Toeplitz systems.
Nagy [593, 1991] has studied circulant preconditioners for the least squares problem

(8.4.9) $\qquad \min \|Tx - b\|_2, \qquad T = \begin{pmatrix} T_1 \\ \vdots \\ T_q \end{pmatrix} \in \mathbf{R}^{m \times n},$

where each block T_j, $j = 1, \ldots, q$, is a square Toeplitz matrix. (Note that if T itself is a rectangular Toeplitz matrix, then each block T_j is necessarily Toeplitz.)

First a circulant approximation C_j is constructed for each block T_j. Each circulant matrix C_j, $j = 1, \ldots, q$, is diagonalized by the Fourier matrix F, $C_j = F \Lambda_j F^H$, where Λ_j is diagonal and F^H denotes the conjugate transpose of the complex Fourier matrix F. The eigenvalues Λ_j can be found from the first column of C_j; cf. (8.4.7). Hence the spectrum of C_j, $j = 1, \ldots, q$, can be computed in $O(m \log n)$ operations by using the FFT.

The preconditioner for T is then defined as a square circulant matrix C, such that

$$C^T C = \sum_{j=1}^{q} C_j^T C_j = F^H \sum_{j=1}^{q} (\Lambda_j^H \Lambda_j) F.$$

Thus $C^T C$ is also circulant, and its spectrum can be computed in $O(m \log n)$ operations. Now C is taken to be the symmetric positive definite matrix defined by

$$(8.4.10) \qquad C \equiv F^H \left(\sum_{j=1}^{q} \Lambda_j^H \Lambda_j \right)^{1/2} F.$$

The preconditioned (PCGLS) method, Algorithm 7.4.3, is then applied with $S = C$ and $A = T$. Notice that to use the preconditioner C we need only know its eigenvalues, since the right-hand side of (8.4.10) can be used to solve linear systems involving C and C^T.

The convergence rate of the conjugate gradient algorithm depends on the distribution of the singular values of the matrix TC^{-1}. It is shown by Chan, Nagy, and Plemmons [150, 1994] that if the generating functions of the blocks T_j are 2π-periodic continuous functions and if one of these functions has no zeros, then the singular values of the preconditioned matrix TC^{-1} are clustered around 1, and the PCGLS converges very quickly. It turns out that the class of 2π-periodic continuous functions contains a class of functions which arises in many signal processing applications.

Similar ideas can be applied to problems where the least squares matrix T has a general Toeplitz block or block Toeplitz structure; see Chan, Nagy, and Plemmons [149, 1993]. Hence the method can be applied also to two-dimensional or multidimensional problems. For the construction of circulant preconditioners for constrained and weighted least squares problems, see Jin [490, 1996].

8.5. Kronecker Product Problems

Sometimes least squares problems occur which have a highly regular block structure. Here we consider least squares problems of the form

$$(8.5.1) \qquad \min_x \|(A \otimes B)x - d\|_2,$$

where the $A \otimes B$ is the **Kronecker product** of $A \in \mathbf{R}^{m \times n}$ and $B \in \mathbf{R}^{p \times q}$. This product is the $mp \times nq$ block matrix,

$$A \otimes B = \begin{pmatrix} a_{11}B & a_{12}B & \cdots & a_{1n}B \\ a_{21}B & a_{22}B & \cdots & a_{2n}B \\ \vdots & \vdots & & \vdots \\ a_{m1}B & a_{m2}B & \cdots & a_{mn}B \end{pmatrix}.$$

Problems of Kronecker structure arise in several application areas including signal and image processing (Eldén and Skoglund [277, 1982]), photogrammetry, and multidimensional approximation; see Fausett and Fulton [289, 1994]. Grosse [402, 1980] describes a tensor factorization algorithm and how it applies to least squares fitting of multivariate data on a rectangular grid. Such problems can be solved with great savings in storage and operations. Since often the size of the matrices A and B is large, resulting in models involving several hundred thousand equations and unknowns, such savings may be essential.

8.5. Kronecker Product Problems

We first state some elementary facts about Kronecker products. From the definition it immediately follows that

$$\begin{aligned}(A + B) \otimes C &= (A \otimes C) + (B \otimes C), \\ A \otimes (B + C) &= (A \otimes B) + (A \otimes C), \\ A \otimes (B \otimes C) &= (A \otimes B) \otimes C, \\ (A \otimes B)^T &= A^T \otimes B^T.\end{aligned}$$

A further important relation, which is not so obvious, is given next.

LEMMA 8.5.1. *If the ordinary multiplications AC and BD are defined, then*

$$(8.5.2) \qquad (A \otimes B)(C \otimes D) = AC \otimes BD.$$

Proof. See Lancaster and Tismenetsky [511, 1985, Chap. 12.1] ∎

As a corollary of this lemma we obtain the identity

$$(A_1 \otimes B_1)(A_2 \otimes B_2) \cdots (A_p \otimes B_p) = (A_1 A_2 \cdots A_n) \otimes (B_1 B_2 \cdots B_n),$$

assuming all the multiplications are defined. We can also conclude that if P and Q are orthogonal $n \times n$ matrices then $P \otimes Q$ is an orthogonal $n^2 \times n^2$ matrix. Further, if A and B are square and nonsingular, then it follows that $A \otimes B$ is nonsingular and

$$(A \otimes B)^{-1} = A^{-1} \otimes B^{-1}.$$

This generalizes to pseudoinverses, as shown in the following lemmas.

LEMMA 8.5.2. *Let $A^\dagger$ and $B^\dagger$ be the pseudoinverses of A and B. Then*

$$(A \otimes B)^\dagger = A^\dagger \otimes B^\dagger.$$

Proof. The theorem follows by verifying that $X = A^\dagger \otimes B^\dagger$ satisfies the four Penrose conditions in Theorem 1.2.11. ∎

We now introduce a function closely related to the Kronecker product, which converts a matrix into a vector. For a matrix $C = (c_1, c_2, \ldots, c_n) \in \mathbf{R}^{m \times n}$ we define

$$(8.5.3) \qquad \operatorname{vec}(C) = \begin{pmatrix} c_1 \\ c_2 \\ \vdots \\ c_n \end{pmatrix}.$$

Hence $\operatorname{vec}(C)$ is the vector formed by stacking the columns of C into one long vector. We now state a result which shows how the vec-function is related to the Kronecker product.

LEMMA 8.5.3. *If* $A \in \mathbf{R}^{m \times n}$, $B \in \mathbf{R}^{p \times q}$, *and* $d = \text{vec}(D)$, *where* $D \in \mathbf{R}^{p \times m}$, *then*
$$(A \otimes B)d = \text{vec}(B^T D A). \tag{8.5.4}$$

If A and B are square and nonsingular the solution of the linear system $(A \otimes B)x = d$ can be written
$$x = (A^{-1} \otimes B^{-1})\text{vec}(D) = \text{vec}(B^{-T} D A^{-1}),$$
where D is the matrix which satisfies $d = \text{vec}(D)$. Using Lemma 8.5.3 the solution to the Kronecker least squares problem (8.5.1) can be written
$$x = (A \otimes B)^\dagger d = (A^\dagger \otimes B^\dagger)d = \text{vec}((B^\dagger)^T D A^\dagger). \tag{8.5.5}$$

This allows a great reduction in the cost of solving (8.5.1). For example, if A and B are both $m \times n$ matrices the cost of computing the least squares solution is reduced from $O(m^2 n^4)$ to $O(mn^2)$.

In some areas the most common approach to computing the least squares solution to (8.5.1) is from normal equations. If we assume that both A and B have full column rank, then we can use the expressions
$$A^\dagger = (A^T A)^{-1} A^T, \qquad B^\dagger = (B^T B)^{-1} B^T.$$

However, because of the instability associated with the explicit formation of $A^T A$ and $B^T B$, an approach based on orthogonal decompositions should generally be preferred. If we have computed the complete QR decompositions of A and B,
$$A\Pi_1 = Q_1 \begin{pmatrix} R_1 & 0 \\ 0 & 0 \end{pmatrix} V_1^T, \qquad B\Pi_2 = Q_2 \begin{pmatrix} R_2 & 0 \\ 0 & 0 \end{pmatrix} V_2^T,$$
with R_1, R_2 upper triangular and nonsingular, then from Section 2.7.3 we have
$$A^\dagger = \Pi_1 V_1 \begin{pmatrix} R_1^{-1} & 0 \\ 0 & 0 \end{pmatrix} Q_1^T, \qquad B^\dagger = \Pi_2 V_2 \begin{pmatrix} R_2^{-1} & 0 \\ 0 & 0 \end{pmatrix} Q_2^T.$$

These expressions can be used in (8.5.5) to compute the pseudoinverse solution of problem (8.5.1) even in the rank deficient case.

We finally note that the singular values and singular vectors of the Kronecker product $A \otimes B$ can be simply expressed in terms of the singular values and singular vectors of A and B.

LEMMA 8.5.4. *Let A and B have the singular value decompositions*
$$A = U_1 \Sigma_1 V_1^T, \qquad B = U_2 \Sigma_2 V_2^T.$$

Then we have
$$A \otimes B = (U_1 \otimes U_2)(\Sigma_1 \otimes \Sigma_2)(V_1 \otimes V_2)^T.$$

Proof. The proof follows from Lemma 8.5.3. ∎

Chapter 9
Nonlinear Least Squares Problems

9.1. The Nonlinear Least Squares Problem

9.1.1. Introduction. In this chapter we discuss the solution of nonlinear least squares problems. Methods for solving such problems are iterative, and each iteration step usually requires the solution of a related linear least squares problem. The nonlinear least squares problem is closely related to the problem of solving a nonlinear system of equations, and is a special case of the general optimization problem in $\mathbf{R}^n$. Here we mainly emphasize those aspects of the nonlinear least squares problem which derive from its special structure. A general treatment of theory and algorithms for solving systems of nonlinear equations is given by Ortega and Rheinboldt [613, 1970]. Methods for unconstrained optimization and nonlinear least squares are discussed in the books by Gill, Murray, and Wright [360, 1981], Dennis and Schnabel [223, 1983], and Fletcher [299, 1987]. Other surveys of methods for the nonlinear least squares problems are given by Dennis [219, 1977] and Fraley [312, 1988]. A very useful guide to software is given by Moré and Wright [591, 1993].

The unconstrained nonlinear least squares problem is to find a global minimizer of the sum of squares of m nonlinear functions,

$$(9.1.1) \qquad \min_{x \in \mathbf{R}^n} f(x), \quad f(x) = \frac{1}{2} \sum_{i=1}^m r_i^2(x), \quad m \geq n.$$

Here each $r_i(x)$, $i = 1, \ldots, m$, is a nonlinear functional defined over $\mathbf{R}^n$. Clearly, if all $r_i(x)$ were linear in x, then (9.1.1) would be a linear least squares problem. For $m = n$ (9.1.1) includes as a special case the solution of a system of nonlinear equations.

One important area in which nonlinear least squares problems arise is in data fitting. Here one attempts to fit given data (y_i, t_i), $i = 1, \ldots, m$, to a model function $g(x, t)$. If we let $r_i(x)$ represent the error in the model prediction for the ith observation,

$$(9.1.2) \qquad r_i(x) = y_i - g(x, t_i), \quad i = 1, \ldots, m,$$

we are led to a problem of the form (9.1.1). The choice of the least squares measure is justified here, as for the linear case, by statistical considerations; see

Bard [35, 1974]. This assumes that only y_i are subject to errors and the values t_i of the independent variable t are exact. The case when there are errors in both y_i and t_i is discussed in Section 9.4.3.

9.1.2. Necessary conditions for local minima. The basic methods for the nonlinear least squares problem require derivative information about the components $r_i(x)$. In the following we assume that $r_i(x)$ are twice continuously differentiable. The Jacobian of the residual vector $r(x) = (r_1(x), \ldots, r_m(x))^T$ is

$$J(x) \in \mathbf{R}^{m \times n}, \qquad J(x)_{ij} = \frac{\partial r_i(x)}{\partial x_j},$$

$i = 1, \ldots, m$, $j = 1, \ldots, n$, and the Hessian matrices of $r_i(x)$ are

$$G_i(x) = \nabla^2 r_i(x) \in \mathbf{R}^{n \times n}, \qquad G_i(x)_{jk} = \frac{\partial^2 r_i(x)}{\partial x_j \partial x_k}, \qquad i = 1, \ldots, m.$$

Then the first and second derivatives of $f(x) = \frac{1}{2} r(x)^T r(x) = \frac{1}{2}\|r(x)\|_2^2$ are given by
(9.1.3) $$\nabla f(x) = J(x)^T r(x),$$

and
(9.1.4) $$\nabla^2 f(x) = J(x)^T J(x) + Q(x), \quad Q(x) = \sum_{i=1}^{m} r_i(x) G_i(x),$$

where $G_i(x) = G_i(x)^T$. The special forms of $\nabla f(x)$ and $\nabla^2 f(x)$ can be exploited by methods for the nonlinear least squares problem.

A necessary condition for x^* to be a **local minimum** of $f(x)$ is that

$$\nabla f(x^*) = J(x^*)^T r(x^*) = 0.$$

Any point which satisfies this condition will be called a **critical point**. We now establish a necessary condition for a critical point x^* to be a local minimum of $f(x)$. We follow a geometrical approach by Wedin [825, 1974], and interpret the problem of minimizing $f(x)$ as the problem of finding a point on the n-dimensional surface $z = r(x)$ in $\mathbf{R}^m$ closest to the origin.

Assuming that $J(x)$ has full column rank we have $J^\dagger(x) J(x) = I_n$, where $J^\dagger(x)$ is the pseudoinverse of $J(x)$. Then we can rewrite (9.1.4) as

(9.1.5) $$\nabla^2 f(x) = J^T J - G_w = J^T \left(I - \gamma (J^\dagger)^T G_w J^\dagger \right) J,$$

where
(9.1.6) $$\gamma = \|r\|_2, \qquad G_w = \sum_{i=1}^{m} w_i G_i, \qquad w = -r/\gamma.$$

The symmetric matrix
(9.1.7) $$K = (J^\dagger)^T G_w J^\dagger$$

9.1. THE NONLINEAR LEAST SQUARES PROBLEM

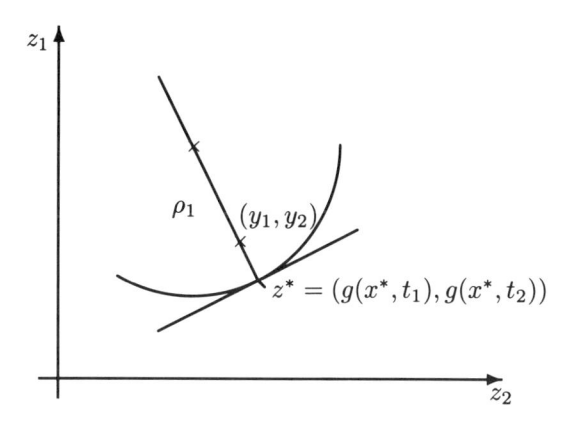

FIG. 9.1.1. *Geometry of the data fitting problem for $m = 2, n = 1$.*

is called the **normal curvature matrix** of the surface $z = r(x)$, with respect to the normal vector w. Let the eigenvalues of K be

$$\kappa_1 \geq \kappa_2 \geq \cdots \geq \kappa_n.$$

The quantities $\rho_i = 1/\kappa_i$, $\kappa_i \neq 0$, are the **principal radii of curvature** of the surface, with respect to the normal w.

If $J(x^*)$ has full column rank, it follows that $\nabla^2 f(x^*) = J^T(I - \gamma K)J$ is positive definite and x^* is a local minimum if and only if $I - \gamma K$ is positive definite at x^*, which is the case when

(9.1.8) $$1 - \gamma \kappa_1 > 0$$

at x^*. If $1 - \gamma \kappa_1 \leq 0$, then $f(x)$ has a saddle point at x^*; if $1 - \gamma \kappa_n < 0$, then $f(x)$ even has a local maximum at x^*.

In the case of data fitting, when $r_i(x)$ is given by (9.1.2), it is more illustrative to consider the surface

$$z = (g(x, t_1), \ldots, g(x, t_m))^T \in \mathbf{R}^m.$$

The problem is then to find the point on this surface closest to the observation vector $y \in \mathbf{R}^m$; cf. Draper and Smith [232, 1981, pp. 500–501]. This is illustrated in Figure 9.1.1 for the simple case of $m = 2$ observations and only a single parameter x. Since in the figure we have $\gamma = \|r\|_2 < \rho$, it follows that $1 - \gamma \kappa_1 > 0$, which is consistent with the fact that x^* is a local minimum.

9.1.3. Basic numerical methods. There are two different ways to view problem (9.1.1). One could think of this problem as arising from an overdetermined system of nonlinear equations $r(x) = 0$. It is then natural to approximate $r(x)$ by a linear model in a neighborhood of a given point x_c,

(9.1.9) $$\tilde{r}_c(x) = r(x_c) + J(x_c)(x - x_c).$$

One can then use the linear least squares problem

(9.1.10) $$\min_x \|r(x_c) + J(x_c)(x - x_c)\|_2$$

to derive an improved approximate solution to (9.1.1). This approach, which only uses first-order derivative information about $r(x)$, leads to the Gauss–Newton and the Levenberg–Marquardt methods, which are discussed in Section 9.2.

In the second approach the problem (9.1.1) is viewed as a special case of an optimization problem, and a quadratic model

(9.1.11) $$\tilde{f}_c(x_c + z) = f(x_c) + \nabla f(x_c)^T z + \frac{1}{2} z^T \nabla^2 f(x_c) z$$

of $f(x)$ is used. This approach uses second derivative information about $r(x)$. The minimizer of $\tilde{f}_c(x)$ is given by

(9.1.12) $$x_N = x_c - \left(J(x_c)^T J(x_c) + Q(x_c)\right)^{-1} J(x_c)^T r(x_c)$$

where $Q(x)$ is given by (9.1.4). This is equivalent to Newton's method applied to problem (9.1.1), for which the local convergence rate usually is quadratic. Methods which explicitly or implicitly take second derivatives into account are discussed in Section 9.3.

Note that the Gauss–Newton method (9.1.10) can be thought of as arising from neglecting the term $Q(x_c)$ in (9.1.12). From (9.1.4) it follows that this term is small if the quantities

$$|r_i(x_c)| \, \|G_i(x_c)\|, \quad i = 1, \ldots, m,$$

are small. This will be the case if either $r_i(x)$ are only mildly nonlinear at x_c or the residuals $r_i(x_c)$, $i = 1, \ldots, m$, are small. In this case the behavior of the Gauss–Newton method can be expected to be similar to that of Newton's method. In particular, for a consistent problem $r(x^*) = 0$ at the solution, hence the local convergence rate will be quadratic for both methods. For moderate-to-large residual problems the local convergence rate for the Gauss–Newton method can be much inferior to that of Newton's method. However, the cost of computing the mn^2 second derivatives $G_i(x)$ can be prohibitively large.

For curve fitting problems the function values $r_i(x) = y_i - g(x, t_i)$ and derivatives can be obtained from the single function $g(x, t)$. If $g(x, t)$ is composed, e.g., of simple exponential and trigonometric functions, then second derivatives can sometimes be computed cheaply. Also, if J is sparse, so often is Q, and the cost for computing $Q(x)$, with analytical derivatives, may in this case not be very large.

9.2. Gauss–Newton-Type Methods

The Gauss–Newton method for problem (9.1.1) is based on a sequence of linear approximations of $r(x)$. If x_k denotes the current approximation, then

9.2. Gauss–Newton-Type Methods

a correction p_k is computed as a solution to the linear least squares problem

(9.2.1) $$\min_p \|r(x_k) + J(x_k)p\|_2, \quad p \in \mathbf{R}^n,$$

and the new approximation is $x_{k+1} = x_k + p_k$. This linear least squares problem can be solved using the QR decomposition of $J(x_k)$.

The Gauss–Newton method as described above has the advantage that it solves linear problems in just one iteration and has fast local convergence on mildly nonlinear and nearly consistent problems. However, it may not even be locally convergent on problems that are very nonlinear or have large residuals. This is illustrated by the following example due to Powell; see also Fletcher [299, 1987].

EXAMPLE 9.2.1. Consider the problem with $m = 2, n = 1$ given by

$$\begin{aligned} r_1(x) &= x + 1, \\ r_2(x) &= \lambda x^2 + x - 1, \end{aligned}$$

where λ is a parameter. The minimizer of $r_1^2(x) + r_2^2(x)$ is $x^* = 0$. It can be shown that for the Gauss–Newton method,

$$x_{k+1} = \lambda x_k + 0(x_k^2),$$

and therefore this method is not locally convergent when $|\lambda| > 1$.

9.2.1. The damped Gauss–Newton method. To get a more useful method we take instead

$$x_{k+1} = x_k + \alpha_k p_k,$$

where p_k is the solution to (9.2.1) and α_k is a step length to be determined. The resulting method, which uses p_k as a **search direction**, is called the **damped Gauss–Newton** method, and the vector p_k the Gauss–Newton direction. When $J(x_k)$ is rank deficient p_k should be chosen as the minimum norm solution of the linear least squares problem (9.2.1)

(9.2.2) $$p_k = -J^\dagger(x_k) r(x_k).$$

The Gauss–Newton direction has the following two important properties.

1. The vector p_k is invariant under linear transformations of the independent variable x.

2. If x_k is not a critical point, then p_k is a **descent direction**, i.e., for sufficiently small $\alpha > 0$, $\|r(x_k + \alpha p_k)\|_2 < \|r(x_k)\|_2$.

The first property is obviously desirable. The second property follows from the relation

(9.2.3) $$\|r(x_k + \alpha p_k)\|_2^2 = \|r(x_k)\|_2^2 - 2\alpha \|P_{J_k} r(x_k)\|_2^2 + 0(|\alpha|^2),$$

where

$$P_{J_k} = J(x_k) J^\dagger(x_k)$$

is the orthogonal projection onto the range of $J(x_k)$; see Section 1.2.5. If x_k is not a critical point, then $J(x_k)^T r(x_k) \neq 0$, and using the singular value decomposition (SVD) of $J(x_k)$, it can be shown that this implies $P_{J_k} r(x_k) \neq 0$. This proves that p_k is a descent direction.

To make the damped Gauss–Newton method into a viable algorithm the step length α_k must be chosen carefully. Two common ways of choosing α_k are:

1. Take α_k to be the largest number in the sequence $1, \frac{1}{2}, \frac{1}{4}, \ldots$ for which the inequality

$$\|r(x_k)\|_2^2 - \|r(x_k + \alpha_k p_k)\|_2^2 \geq \frac{1}{2}\alpha_k \|J(x_k)p_k\|_2^2$$

holds (notice that $-J(x_k)p_k = P_{J_k} r(x_k)$). This is the Armijo–Goldstein step length principle; see Ortega and Rheinboldt [613, 1970, p. 491] and Gill, Murray, and Wright [360, 1981, p. 100].

2. Take α_k to be the solution to the one-dimensional minimization problem

(9.2.4) $$\min_\alpha \|r(x_k + \alpha p_k)\|_2^2.$$

A theoretical analysis of these two step length principles has been given by Ruhe [689, 1979]. Note that the exact line search in (9.2.4) cannot be implemented in a finite number of steps.

Often a step length α_k is chosen to be an approximate solution of (9.2.4). A special line search algorithm for nonlinear least squares has been developed by Lindström and Wedin [536, 1984]. In this, an approximation $p(\alpha)$ of the curve $f(\alpha) = r(x_k + \alpha p_k)$ in $\mathbf{R}^m$ is determined, and then $\|p(\alpha)\|_2$ is minimized as a function of α. One possibility is to choose $p(\alpha)$ to be the unique circle (in the degenerate case, a straight line) determined by the conditions

$$p(0) = f(0), \quad \nabla p(0) = \nabla f(0), \quad p(\alpha_0) = f(\alpha_0),$$

where α_0 is a guess of the step length.

To improve the rate of convergence of the damped Gauss–Newton method one should switch to another search direction, preferably the negative gradient $g_k = -J(x_k)^T r(x_k)$, when either the angle between the Gauss–Newton directions p_k and g_k becomes large, or the reduction achieved in $\|r(x)\|_2^2$ is small.

Since the damped Gauss–Newton method always takes descent steps, this method is locally convergent on almost all nonlinear least squares problems, provided that the line search is carried out appropriately. In fact, it is usually globally convergent. However, the rate of convergence may still be slow on large residual problems and very nonlinear problems.

An implementation of the Gauss–Newton method that uses the minimum norm solution (9.2.2) when $J(x_k)$ is rank deficient must include some strategy for estimating the rank of $J(x_k)$. Such strategies have been discussed for the QR and the SVD in Section 2.7. Usually an underestimate of the rank is preferable except when $f(x)$ is actually close to an ill-conditioned quadratic function. The following example illustrates that the determination of rank may be critical.

9.2. Gauss–Newton-Type Methods

EXAMPLE 9.2.2. (See Gill, Murray, and Wright [360, 1981, p. 136].) Let $J = J(x_k)$ and $r = r(x_k)$ be defined by

$$J = \begin{pmatrix} 1 & 0 \\ 0 & \epsilon \end{pmatrix}, \qquad r = \begin{pmatrix} r_1 \\ r_2 \end{pmatrix},$$

where $\epsilon \ll 1$ and r_1 and r_2 are of order unity. If J is considered to be of rank two then the search direction is $p_k = s_1$, whereas if the assigned rank is one, $p_k = s_2$, where

$$s_1 = -\begin{pmatrix} r_1 \\ r_2/\epsilon \end{pmatrix}, \qquad s_2 = -\begin{pmatrix} r_1 \\ 0 \end{pmatrix}.$$

Clearly the two directions s_1 and s_2 are almost orthogonal, and s_1 is almost orthogonal to the gradient vector $J^T r$. ∎

9.2.2. Local convergence of the Gauss–Newton method. We now discuss the local convergence properties of the Gauss–Newton method. This has been analyzed by Wedin [825, 1974] and Ramsin and Wedin [674, 1977]. One step of the undamped method (i.e., taking $\alpha = 1$) can be written as a fixed point iteration

$$x_{k+1} = F(x_k), \qquad F(x) = x - J(x)^\dagger r(x),$$

where the first derivative of $F(x)$ equals

$$\nabla F = -J^\dagger (J^\dagger)^T \sum_{i=1}^m r_i G_i = \gamma J^\dagger (J^\dagger)^T G_w,$$

using the notations of (9.1.6). The asymptotic rate of convergence ρ is bounded by the spectral radius of the matrix $\nabla F(x^*)$ at the solution x^*. But $\nabla F(x)$ has the same nonzero eigenvalues as the matrix

$$\gamma (J(x)^\dagger)^T G_w J(x)^\dagger = \gamma K,$$

where K is the normal curvature matrix (9.1.7). Hence,

(9.2.5) $$\rho = \rho(\nabla F) = \gamma \max(\kappa_1, -\kappa_n).$$

In general, convergence is linear, but if $\gamma = \|r(x^*)\|_2 = 0$ we have superlinear convergence. From (9.2.5) our earlier conjecture follows, that the local rate of convergence of the undamped Gauss–Newton method is fast when either

1. the residual norm $\gamma = \|r(x^*)\|_2$ is small, or

2. $r(x)$ is mildly nonlinear, i.e., $\|G_i\|$, $i = 1, \ldots, m$, are small.

We showed in Section 9.1.2 that if x^* is a saddle point of $f(x)$, then $\kappa_1 \geq 1$. Hence, using the undamped Gauss–Newton method, one is generally repelled from a saddle point. This is an excellent property, since saddle points are not at all uncommon for nonlinear least squares problems.

The rate of convergence for the undamped Gauss–Newton method can be estimated during the iterations from

$$\|J(x_{k+1})p_{k+1}\|_2 / \|J(x_k)p_k\|_2 \leq \rho + 0(\|x_k - x^*\|_2^2). \tag{9.2.6}$$

When the estimated ρ is greater than 0.5 (say) then one should consider switching to a method using second derivative information, or perhaps evaluate the quality of the underlying model.

The asymptotic rate of convergence for the Gauss–Newton method with exact line search has been shown by Ruhe [689, 1979] to be

$$\tilde{\rho} = \gamma(\kappa_1 - \kappa_n)/(2 - \gamma(\kappa_1 + \kappa_n)). \tag{9.2.7}$$

Hence it holds that $\tilde{\rho} = \rho$ if $\kappa_n = -\kappa_1$ and $\tilde{\rho} < \rho$ otherwise. We also have that $\gamma\kappa_1 < 1$ implies $\tilde{\rho} < 1$, i.e., we always get convergence close to a local minimum. This is in contrast to the undamped Gauss–Newton method, which may fail to converge to a local minimum.

Sometimes one encounters nonlinear least squares problems which are ill behaved in that the radius of curvature at a critical point satisfies $1/\kappa \ll \|r(x^*)\|_2$. Then many insignificant local minima may exist. For such problems it seems reasonable to demand that the surface $r(x)$ is smoothed before one attempts to solve the problem. Wedin [825, 1974] has shown that the estimate (9.2.6) of the rate of convergence of the Gauss–Newton method is often a good confirmation of the quality of the underlying model. Deuflhard and Apostolescu [226, 1980] call problems for which divergence occurs "inadequate problems." Hence it seems important that algorithms for nonlinear least squares problems also attempt to estimate the maximal curvature.

9.2.3. Trust region methods. At an intermediate point x_k, where the Jacobian matrix does not have full column rank, there is still a possibility that the damped Gauss–Newton method can have difficulties proceeding. This can be avoided either by taking second derivatives into account (see Section 9.3) or by further stabilizing the damped Gauss–Newton method to overcome this possibility of failure. Methods using the latter approach were first suggested by Levenberg [526, 1944] and Marquardt [568, 1963], and are now known as trust region methods.

In trust region methods a search direction p_k is computed as the solution to the regularized problem

$$\min_p \{\|r(x_k) + J(x_k)p\|_2^2 + \mu_k\|p\|_2^2\}, \tag{9.2.8}$$

where $\mu_k \geq 0$ is the parameter which limits the size of p_k. Note that p_k is well defined by (9.2.8) also when $J(x_k)$ is rank deficient. The problem (9.2.8) can equivalently be written

$$\min_p \left\| \begin{pmatrix} J(x_k) \\ \mu_k I \end{pmatrix} p - \begin{pmatrix} r(x_k) \\ 0 \end{pmatrix} \right\|_2.$$

9.2. GAUSS–NEWTON-TYPE METHODS

As $\mu_k \to \infty$, $\|p_k\|_2 \to 0$ and for small μ_k the direction p_k becomes parallel to the steepest descent direction $J(x_k)^T r(x_k)$.

It follows from the discussion in Section 5.3 that (9.2.8) is related to the least squares problem with quadratic constraint

$$(9.2.9) \quad \min_p \|r(x_k) + J(x_k)p\|_2, \text{ subject to } \|p\|_2 \leq \delta_k.$$

If the constraint in (9.2.9) is not binding then $\mu_k = 0$ in (9.2.8), and otherwise $\mu_k > 0$. The set of feasible vectors p, $\|p\|_2 \leq \delta_k$ in (9.2.9) can be thought of as a region of trust for the linear model

$$r(x) \approx r(x_k) + J(x_k)p, \qquad p = x - x_k.$$

For a general description of trust region methods for nonlinear optimization, see Moré [587, 1983].

Many different strategies have been used to choose μ_k in (9.2.8). A careful implementation of the Levenberg–Marquardt algorithm as a scaled trust region algorithm has been described by Moré [586, 1978]. Moré considers an iteration of the following form.

ALGORITHM 9.2.1. TRUST REGION ALGORITHM. Let x_0, D_0, and δ_0 be given and set $\beta \in (0,1)$. For $k = 0, 1, 2, \ldots$,

1. Compute $\|r(x_k)\|_2^2$.

2. Determine p_k as a solution to the subproblem

$$\min_p \|r(x_k) + J(x_k)p\|_2, \text{ subject to } \|D_k p\|_2 \leq \delta_k,$$

   where D_k is a diagonal scaling matrix.

3. Compute the model prediction of the decrease in $\|r(x_k)\|_2^2$ as

$$\psi_k(p_k) = \|r(x_k)\|_2^2 - \|r(x_k) + J(x_k)p_k\|_2^2.$$

4. Compute the ratio

$$\rho_k = (\|r(x_k)\|_2^2 - \|r(x_k - p_k)\|_2^2)/\psi_k(p_k).$$

   If $\rho_k > \beta$ then set $x_{k+1} = x_k + p_k$; otherwise set $x_{k+1} = x_k$.

5. Update the scaling matrix D_k and δ_k. ∎

In the algorithm the ratio ρ_k measures the agreement between the linear model and the nonlinear function. An iteration with $\rho_k > \beta$ is successful, and otherwise the iteration is unsuccessful. After an unsuccessful iteration δ_k is reduced. Moré chooses the scaling D_k such that the algorithm is scale invariant, i.e., the algorithm generates the same iterations if applied to $r(Dx)$ for any nonsingular diagonal matrix D.

Moré proves that if $r(x)$ is continuously differentiable, $r'(x)$ uniformly continuous, and $J(x_k)$ bounded, then the algorithm will converge to a critical point. The algorithm has also proven to be very successful in practice and is included in the software package MINPACK-1, which is available from netlib. For a user's guide see Moré, Hillstrom, and Garbow [588, 1980].

Convergence of trust region methods may still be slow for large residual or very nonlinear problems. In the next section we discuss methods using second derivative information which are somewhat more robust but also more complex.

9.3. Newton-Type Methods

9.3.1. Introduction. The analysis in previous sections has shown that methods of Gauss–Newton type may converge slowly for large residual problems and strongly nonlinear problems. These methods can also have problems at points where the Jacobian is rank deficient. When second derivatives of $f(x)$ are available, Newton's method can be used to overcome these problems. Newton's method is based on the quadratic model (9.1.11) of $f(x)$ at the current approximation, and the critical point of this quadratic model is chosen as the next approximation.

It can be shown (see Dennis and Schnabel [223, 1983, p. 229]) that Newton's method is locally quadratically convergent as long as

$$(9.3.1) \qquad \nabla^2 f(x) = J(x)^T J(x) + Q(x), \qquad Q(x) = \sum_{i=1}^{m} r_i(x) G_i(x),$$

is Lipschitz continuous around x^* and $\nabla^2 f(x^*)$ is positive definite.

To get global convergence, Newton's method is used with a line search algorithm, $x_{k+1} = x_k + \alpha_k p_k$, where the search direction p_k is determined from

$$(9.3.2) \qquad \left(J(x_k)^T J(x_k) + Q(x_k)\right) p_k = -J(x_k)^T r(x_k).$$

Note that the matrix $J(x_k)^T J(x_k) + Q(x_k)$ must be positive definite in order for p_k to be a guaranteed descent direction. The linear system (9.3.2) should be solved by a method which is also stable when $J(x_k)$ is ill-conditioned or rank deficient.

9.3.2. A hybrid Newton method. Newton's method is not often used because the second derivative term $Q(x_k)$ is rarely available at a reasonable cost. However, a number of methods have been suggested that partially take the second derivatives into account, either explicitly or implicitly.

Gill and Murray [358, 1978] suggest regarding $J(x_k)^T J(x_k)$ as a good estimate of the Hessian in the invariant subspace corresponding to the large singular values of $J(x_k)$. In the complementary subspace the second derivative term $Q(x_k)$ is taken into account. Let the SVD of $J(x_k)$ be

$$J(x_k) = U \begin{pmatrix} \Sigma \\ 0 \end{pmatrix} V^T, \quad U \in \mathbf{R}^{m \times m}, \quad V \in \mathbf{R}^{n \times n},$$

where $\Sigma = \text{diag}(\sigma_1, \ldots, \sigma_n)$, $\sigma_1 \geq \sigma_2 \geq \cdots \geq \sigma_n$, is the matrix of singular values. The equations (9.3.2) for the Newton direction $p_k = V q_k$ can then be written

$$(9.3.3) \qquad (\Sigma^2 + V^T Q_k V) q_k = -\Sigma \bar{s},$$

where $Q_k = Q(x_k)$, and $\bar{s}$ denotes the first n components of the vector $s_k = U^T r(x_k)$.

The singular values are now split into two groups, $\Sigma = \text{diag}(\Sigma_1, \Sigma_2)$,

$$\Sigma_1 = \text{diag}(\sigma_1, \ldots, \sigma_p), \qquad \Sigma_2 = \text{diag}(\sigma_{p+1}, \ldots, \sigma_n),$$

where Σ_1 contains the "large" singular values. If we partition V, q_k, and $\bar{s}$ conformably, then the first p equations in (9.3.3) can be written

$$(\Sigma_1^2 + V_1^T Q_k V_1) q_1 + V_1^T Q_k V_2 q_2 = -\Sigma_1 \bar{s}_1.$$

If the terms involving Q_k are neglected compared to $\Sigma_1^2 q_1$ we get $q_1 = -\Sigma_1^{-1} \bar{s}_1$. If this is substituted into the last $(n-p)$ equations we can solve for q_2 from

$$(\Sigma_2^2 + V_2^T Q_k V_2) q_2 = -\Sigma_2 \bar{s}_2 - V_2^T Q_k V_1 q_1.$$

The approximate Newton direction is then given by

$$p_k = V q = V_1 q_1 + V_2 q_2.$$

The split of the singular values is updated at each iteration, and the idea is to maintain r close to n as long as adequate progress is made.

There are several alternative ways to implement the method by Gill and Murray [358, 1978]. First, Q_k could be used explicitly, or a finite difference approximation to $Q_k V_2$ could be obtained by differencing the gradient along the columns of V_2. A third option is to use a quasi-Newton approximation to Q_k, similar to the methods described below.

9.3.3. Quasi-Newton methods. In quasi-Newton optimization routines an approximation to the second derivative matrix (9.3.1) is built up successively from evaluations of the gradient. Many of those are known to possess superlinear convergence.

Let S_{k-1} be a symmetric approximation to the Hessian at step k. It is then required that the updated approximation S_k approximate the curvature of f along $x_k - x_{k-1}$, that is,

$$(9.3.4) \qquad S_k(x_k - x_{k-1}) = y_k, \quad y_k = J(x_k)^T r(x_k) - J(x_{k-1})^T r(x_{k-1}),$$

which is called the quasi-Newton relation. Further, S_k should differ from S_{k-1} by a matrix of small rank. The search direction p_k for the next step is then computed from

$$S_k p_k = -J(x_k)^T r(x_k).$$

As a starting approximation, $S_0 = J(x_0)^T J(x_0)$ is usually recommended.

Ramsin and Wedin [674, 1977] gave the following recommendations on the choice between Gauss–Newton and quasi-Newton methods, which can be the basis for a hybrid method with automatic switching between the two methods. The rule is based on the observed rate (9.2.6) of convergence ρ for the Gauss–Newton method.

1. For $\rho \leq 0.5$ Gauss–Newton is better.

2. For globally simple problems quasi-Newton is better for $\rho > 0.5$.

3. For globally difficult problems Gauss–Newton is much faster for $\rho \leq 0.7$, but for larger values of ρ quasi-Newton is safer.

The straightforward application of quasi-Newton methods to the nonlinear least squares problem outlined above has not been very efficient in practice. One reason is that these methods disregard the information in $J(x_k)$, and often $J(x_k)^T J(x_k)$ is the dominant part of $\nabla^2 f(x_k)$. A successful approach has been taken by Dennis, Gay, and Welsch [220, 1981], who approximate $\nabla^2 f(x_k)$ by $S_k = J(x_k)^T J(x_k) + B_k$, where B_k is a quasi-Newton approximation of the term $Q(x_k)$. The quasi-Newton relation now becomes

$$(9.3.5) \quad B_k(x_k - x_{k-1}) = z_k, \quad z_k = J(x_k)^T r(x_k) - J(x_{k-1})^T r(x_{k-1}),$$

where B_k is required to be symmetric. It can be shown (cf. Dennis and Schnabel [223, 1983, pp. 231–232]) that a solution to (9.3.5) which minimizes the change from B_{k-1} is given by the update formula

$$(9.3.6) \quad \begin{aligned} B_k = B_{k-1} &+ \left((z_k - B_{k-1}s_k)y_k^T + y_k(z_k - B_{k-1}s_k)^T\right)/y_k^T s_k \\ &- (z_k - B_{k-1}s_k)^T s_k y_k y_k^T / (y_k^T s_k)^2, \end{aligned}$$

where $s_k = x_k - x_{k-1}$. This update was proposed by Dennis, Gay, and Welsch [220, 1981], and is used in a subroutine NL2SOL. Two new hybrid quasi-Newton methods, which are superlinearly convergent under mild conditions, have been developed by Fletcher and Xu; see [300, 1987].

The NL2SOL code has several interesting features. It maintains the approximation B_k and adaptively decides whether to use it; i.e., it switches between a Gauss–Newton and a quasi-Newton method. For both methods a trust region strategy is used to achieve global convergence. In each iteration NL2SOL computes the reduction predicted by both quadratic models and compares with the actual reduction $f(x_{k+1}) - f(x_k)$. For the next step the model is used whose predicted reduction best approximated the actual reduction. Usually this causes NL2SOL to initially use Gauss–Newton steps until the information in S_k becomes useful.

A different way to obtain second derivative information has been developed by Ruhe [689, 1979], who uses a nonlinear conjugate gradient acceleration of the Gauss–Newton method with exact line searches. This method achieves quadratic convergence and thus gives much faster convergence than the Gauss–Newton method on difficult problems. When exact line search is used, then the conjugate gradient acceleration amounts to a negligible amount of extra work. However, for small residual problems exact line search is a waste of time and then a simpler damped Gauss–Newton method is superior.

Extending the quasi-Newton method to large sparse problems has proved to be difficult. A promising approach has been suggested by Toint [777, 1987] for certain types of large, "partially separable" nonlinear least squares problems. A typical case included is when every function $r_i(x)$ only depends on a small subset of the set of n variables. Then the Jacobian $J(x)$ and the element Hessian matrices $G_i(x)$ will be sparse, and it may not be infeasible to store approximations to all $G_i(x)$, $i = 1, \ldots, m$. An implementation is available as the Fortran subroutine VE10 in the Harwell Subroutine Library; see Toint [778, 1987].

9.4. Separable and Constrained Problems

9.4.1. Separable problems.
A nonlinear least squares problem $\min_x \|r(x)\|_2$ is said to be **separable** if the solution vector x can be partitioned so that a subproblem
$$(9.4.1) \qquad \min_y \|r(y,z)\|_2, \quad x = \begin{pmatrix} y \\ z \end{pmatrix} \begin{matrix} \}p \\ \}q \end{matrix}, \quad p+q=n$$
is easy to solve. Many practical nonlinear least squares problems are separable.

In the following we restrict ourselves to the particular case when $r(y,z)$ is linear in y,
$$(9.4.2) \qquad r(y,z) = F(z)y - g(z), \quad F(z) \in \mathbf{R}^{m \times p}.$$
Then the minimum norm solution to (9.4.1) is
$$y(z) = F^\dagger(z)g(z),$$
where $F^\dagger(z)$ is the pseudoinverse of $F(z)$. It follows that the original separable problem can be written
$$(9.4.3) \qquad \min_z \|g(z) - F(z)y(z)\|_2.$$
Since
$$g(z) - F(z)y(z) = (I - P_{F(z)})g(z),$$
where $P_{F(z)} = F(z)F(z)^\dagger$ is the orthogonal projector onto the range of $F(z)$, algorithms based on (9.4.4) are often called **variable projection algorithms**. A particularly simple case is when $r(y,z)$ is linear in both y and z so that we also have
$$(9.4.4) \qquad r(y,z) = G(y)z - h(y), \quad G(y) \in \mathbf{R}^{m \times q}.$$

EXAMPLE 9.4.1. Consider the exponential fitting problem
$$\min_{y,z} \sum_{i=1}^m (y_1 e^{z_1 t_i} + y_2 e^{z_2 t_i} - g_i)^2.$$

Here the model is nonlinear in the parameters z_1 and z_2, but linear in y_1 and y_2. Given values of z_1 and z_2 the linear subproblem is easily solved. ∎

Special-purpose algorithms for separable nonlinear least squares problems were first considered by Scolnik [716, 1972]. A variable projection algorithm using a Gauss–Newton method applied to the problem (9.4.3) was developed by Golub and Pereyra [378, 1973]. Kaufman [498, 1975] proposed a simplification of this algorithm in which two steps are merged into one.

ALGORITHM 9.4.1. VARIABLE PROJECTION ALGORITHM. Let $(y_k, z_k)^T$ be the current approximation. Then proceed in the following steps.

1. Solve the linear subproblem

$$\min_{\delta y_k} \|F(z_k)\delta y_k - (g(z_k) - F(z_k)y_k)\|_2,$$

and put $y_{k+\frac{1}{2}} = y_k + \delta y_k$, $x_{k+\frac{1}{2}} = (y_{k+\frac{1}{2}}, z_k)^T$.

2. Compute the Gauss–Newton direction p_k at $x_{k+\frac{1}{2}}$, i.e., solve

$$\min_{p_k} \|C(x_{k+\frac{1}{2}})p_k + r(y_{k+\frac{1}{2}}, z_k)\|_2,$$

where $C(x_{k+\frac{1}{2}})$ is the Jacobian matrix

$$C(x_{k+\frac{1}{2}}) = \left(F(z_k), r_z(y_{k+\frac{1}{2}}, z_k)\right).$$

3. Take $x_{k+1} = x_k + \alpha_k p_k$ and go to 1. ∎

In step 2 we have used the fact that by (9.4.2) $r_y(y_{k+\frac{1}{2}}, z_k) = F(z_k)$. Further, we have

$$r_z(y_{k+\frac{1}{2}}, z_k) = B(z_k)y_{k+\frac{1}{2}} - g'(z_k),$$

where

$$B(z) = \left(\frac{\partial F}{\partial z_1}y, \ldots, \frac{\partial F}{\partial z_q}y\right) \in \mathbf{R}^{m \times q}.$$

Note that when $r(y, z)$ is also linear in y, it follows from (9.4.4) that

$$C(x_{k+\frac{1}{2}}) = \left(F(z_k), G(y_{k+\frac{1}{2}})\right).$$

Ruhe and Wedin [691, 1980] have given a general analysis of different algorithms for separable problems. They show that the Gauss–Newton algorithm applied to (9.4.3) and the original problem both give the same asymptotic convergence rate. In particular, both converge quadratically for the zero residual problem. It is important to note that, in contrast, the naive algorithm of alternatively minimizing $\|r(y, z)\|_2$ over y and z *always converges linearly*. They also prove that the simplified algorithm of Kaufman has roughly the same asymptotic convergence rate as the one proposed by Golub and Pereyra.

Golub and LeVeque [372, 1979] have extended the variable projection method for solving problems in which it is desired to fit more than one data vector with

9.4. SEPARABLE AND CONSTRAINED PROBLEMS

the same nonlinear parameter vector, though with different linear parameters for each right-hand side.

The special algorithms for separable problems have the same local rate of convergence as the ordinary Gauss–Newton applied to the full problem. However, one important advantage is that no starting values for the linear parameters have to be provided. For example, in the Kaufman algorithm we can take $y_0 = 0$ and determine $y_1 = \delta y_1$ in the first step. This seems to make a difference in the first steps of the iterations. Krogh [510, 1974] reports that the variable projection algorithm solved several problems which methods not using separability could not solve.

To be robust the algorithms for separable problems must employ stabilizing techniques for the Gauss–Newton steps similar to those described in Sections 9.2.2 and 9.2.3. It is fairly straightforward to implement these techniques for the Kaufman algorithm.

9.4.2. General constrained problems. In a more general setting the solution to nonlinear least squares problems may be subject to constraints. In case of nonlinear equality constraints the problem can be stated as

$$(9.4.5) \qquad \min_x \frac{1}{2}\|r(x)\|_2^2, \text{ subject to } h(x) = 0,$$

where $x \in \mathbf{R}^n$, $r(x) \in \mathbf{R}^m$, $h \in \mathbf{R}^p$, and $p < n$.

The Gauss–Newton method can be generalized to constrained problems by linearizing (9.4.5) at a point x_k. A search direction p_k is then computed as a solution to the linearly constrained problem

$$(9.4.6) \qquad \min_p \|r(x_k) + J(x_k)p\|_2 \text{ subject to } h(x_k) + C(x_k)p = 0,$$

where J and C are the Jacobian matrices for $r(x)$ and $h(x)$, respectively. This problem can be solved by the methods described in Section 5.1. The search direction p_k obtained from (9.4.6) can be shown to be a descent direction to the **merit function**

$$\psi(x, \mu) = \|r(x)\|_2^2 + \mu \|h(x)\|_2^2$$

at the point x_k, provided that μ is large enough. This makes it possible to stabilize the Gauss–Newton method with a line search strategy or a trust region technique; cf. Section 9.2. With a suitable active set strategy such an algorithm can be extended to also handle problems with nonlinear inequality constraints. An algorithm based on this approach has been developed by Lindström [534, 1983].

There are some algorithms specialized to solve the nonlinear least squares problem subject to **linear** inequality constraints. In Holt and Fletcher [472, 1979] the unknowns can be constrained by lower and upper bounds. Lindström [535, 1984] describes two easy-to-use routines, ENLSIP and ELSUNC, for solving the general constrained or the simple bound case. These algorithms are based on the Gauss–Newton method with a specialized line search; see Lindström and Wedin [536, 1984]. Far from the solution the algorithm can be stabilized by a certain subspace minimization. Close to the solution the algorithm switches to a second-order method (Newton's method in the unconstrained case) when the Gauss–Newton method converges slowly. The trust region approach for unconstrained problems is generalized to handle linear inequality constraints in Gay [327, 1984] and Wright and Holt [845, 1985]. Popular general nonlinear optimization algorithms have also been used to solve nonlinear least squares problems with nonlinear inequality constraints; see Schittkowski [707, 1985] and Mahdavi-Amiri [557, 1981].

We mention that implicit curve fitting problems, where a model $h(y, x, t) = 0$ is to be fitted to observations (y_i, t_i), $i = 1, \ldots, m$, can be formulated as a special least squares problem with nonlinear constraints:

$$\min_{x,z} \|z - y\|_2^2, \quad \text{subject to} \quad h(z_i, x, t_i) = 0.$$

This problem is a special case of (9.4.5). It has $n + m$ unknowns x and z, but the Jacobian matrices are sparse, which may be taken advantage of; see Lindström [535, 1984].

We mention also that Kaufman and Pereyra [501, 1978] have extended the Golub–Pereyra method to problems with separable nonlinear constraints. The Kaufman method seems even easier to generalize to constrained problems.

9.4.3. Orthogonal distance regression. A special problem which can be formulated as a constrained nonlinear least squares problem is the problem of **orthogonal distance regression**. This problem arises from the fitting of observations (y_i, t_i), $i = 1, \ldots, m$, to a mathematical model

(9.4.7) $$y = f(x, t),$$

such that the sum of squares of the orthogonal distances are minimized; see Figure 9.4.1. This is appropriate when both the measurements t_i of the independent variable and the observations y_i are subject to random errors. Independent of statistical considerations, the orthogonal distance measure has natural applications in fitting geometrical elements; see Section 9.4.4.

The general orthogonal distance problem did not at first receive the same attention as the standard nonlinear regression problem except for the case when f is linear in x. One reason is that if the errors in the independent variables are small, then ignoring these errors will not seriously degrade the estimates of x. For the special case when

$$y = x^T t \in \mathbf{R}, \qquad x, t \in \mathbf{R}^n,$$

the orthogonal distance problem is a total least squares problem, and an algorithm based on the SVD has been described in Section 4.6.6.

9.4. SEPARABLE AND CONSTRAINED PROBLEMS

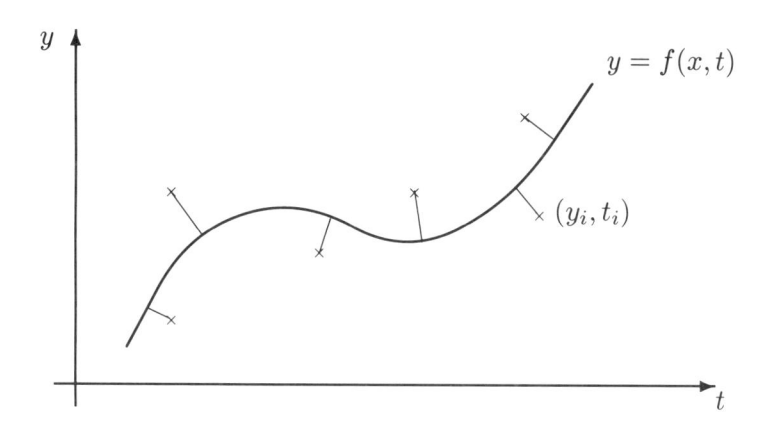

FIG. 9.4.1. *Orthogonal distance fitting.*

We now consider the general problem and assume that y_i and t_i are subject to errors $\bar{\epsilon}_i$ and $\bar{\delta}_i$, respectively, so that

$$(9.4.8) \qquad y_i + \bar{\epsilon}_i = f(x, t_i + \bar{\delta}_i), \qquad i = 1, \ldots, m.$$

If the errors $\bar{\epsilon}_i$ and $\bar{\delta}_i$ are independent random variables with zero mean and variance σ^2, then it seems reasonable to choose the parameters x so that the sum of squares of the orthogonal distances r_i from the observations (y_i, t_i) to the curve (9.4.7) is minimized; cf. Figure 9.4.1. We have that $r_i = (\epsilon_i^2 + \delta_i^2)^{1/2}$, where ϵ_i and δ_i solve

$$\min_{\epsilon_i, \delta_i}(\epsilon_i^2 + \delta_i^2), \quad \text{subject to} \quad y_i + \epsilon_i = f(x, t_i + \delta_i).$$

Hence the parameters x should be chosen as the solution to

$$\min_{x,\epsilon,\delta} \sum_{i=1}^{m}(\epsilon_i^2 + \delta_i^2) \quad \text{subject to}$$

$$y_i + \epsilon_i = f(x, t_i + \delta_i), \quad i = 1, \ldots, m.$$

This is a constrained least squares problem of special form. Eliminating ϵ_i using the constraints, we arrive at the orthogonal distance problem

$$(9.4.9) \qquad \min_{x,\delta} \sum_{i=1}^{m}\left((f(x, t_i + \delta_i) - y_i)^2 + \delta_i^2\right).$$

Note that this is a nonlinear least squares problem even if $f(x,t)$ is linear in x.

So far we have implicitly assumed that y and t are scalar variables. More generally, $y \in \mathbf{R}^{n_y}$ and $t \in \mathbf{R}^{n_t}$, and then we have the problem

$$(9.4.10) \qquad \min_{x,\delta} \sum_{i=1}^{m}\left(\|f(x, t_i + \delta_i) - y_i\|_2^2 + \|\delta_i\|_2^2\right).$$

Finally, if δ_i and ϵ_i do not have constant covariance matrices, then weighted norms should be substituted above.

The problem (9.4.9) has $(m+n)$ unknowns x and δ. In applications usually $m \gg n$ and accounting for the errors in t_i will considerably increase the size of the problem. Therefore, the use of standard software for nonlinear least squares to solve orthogonal distance problems is not efficient or feasible. This is even more accentuated for (9.4.10), which has $mn_t + n$ variables. We now show how the special structure of (9.4.9) can be taken into account to reduce the work. Similar comments apply to the general case (9.4.10).

If we define the residual vector $r^T(\delta, x) = (r_1^T(\delta, x), r_2^T(\delta))$ by

$$r_1(\delta, x)_i = f(x, t_i + \delta_i) - y_i, \quad r_2(\delta)_i = \delta_i, \quad i = 1, \ldots, m,$$

then (9.4.9) is the standard nonlinear least squares problem $\min_{x,\delta} \|r(\delta, x)\|_2^2$.

The Jacobian matrix corresponding to this problem can be written in block form as

(9.4.11) $$\tilde{J} = \begin{pmatrix} D_1 & J \\ I_n & 0 \end{pmatrix} \begin{matrix} \}m \\ \}m \end{matrix} \in \mathbf{R}^{2m \times (m+n)},$$

where

$$\underbrace{}_{m} \underbrace{}_{n}$$

$$D_1 = \operatorname{diag}(d_1, \ldots, d_m), \quad d_i = \left(\frac{\partial f}{\partial t}\right)_{t=t_i},$$

$$J_{ij} = \frac{\partial f(x, t_i + \delta_i)}{\partial x_j}, \quad i = 1, \ldots, m, \quad j = 1, \ldots, n.$$

Note that $\tilde{J}$ is sparse and highly structured. In the Gauss–Newton method we compute corrections $\Delta \delta_k$ and Δx_k to the current approximations which solve the linear least squares problem

(9.4.12) $$\min_{\Delta \delta, \Delta x} \left\| \tilde{J} \begin{pmatrix} \Delta \delta \\ \Delta x \end{pmatrix} - \begin{pmatrix} r_1 \\ r_2 \end{pmatrix} \right\|_2,$$

where $\tilde{J}, r_1$, and r_2 are evaluated at the current estimates of δ and x. To solve this problem we need the QR decomposition of $\tilde{J}$. This can be computed in two steps. First we apply a sequence of Givens rotations $Q_1 = G_m \cdots G_2 G_1$, where $G_i = R_{i,i+m}$, $i = 1, 2, \ldots, m$, to zero the (2,1) block of $\tilde{J}$:

$$Q_1 \tilde{J} = \begin{pmatrix} D_2 & K \\ 0 & L \end{pmatrix}, \quad Q_2 \begin{pmatrix} r_1 \\ r_2 \end{pmatrix} = \begin{pmatrix} s_1 \\ s_2 \end{pmatrix},$$

where D_2 is again a diagonal matrix. The problem (9.4.12) now decouples, and Δx_k is determined as the solution to

$$\min_{\Delta x} \|L \Delta x - s_2\|_2.$$

Here $L \in \mathbf{R}^{m \times n}$, so this is a problem of the same size as that which defines the Gauss–Newton correction in the classical nonlinear least squares problem. We then have

$$\Delta \delta_k = D_2^{-1}(s_2 - K \Delta x_k).$$

9.4. SEPARABLE AND CONSTRAINED PROBLEMS

Algorithms for the nonlinear case, based on regularized Gauss-Newton methods as described in Subsection 9.2.3, have been developed by Schwetlick and Tiller [714, 1985], who use a special Marquardt-type regularization. However, the path

$$p(\mu) := -(\Delta\delta(\mu)^T, \Delta x(\mu)^T)^T$$

used was later shown by Schwetlick and Tiller [715, 1989] to be equivalent to a trust region path defined by a nonstandard scaling matrix D_k the δ-part of which is just the matrix D_2 above wheras the x-part D_3 can be chosen in standard way as diagonal matrix, see Algorithm 9.2.1. In the later paper the step is controlled in trust region style by the condition $\|p(\mu)\|^2 = \|D_2\Delta\delta(\mu)\|^2 + \|D_3\Delta x(\mu)\|^2 \leq \Delta^2$ and not by the Lagrange parameter μ.

The algorithm by Boggs, Byrd, and Schnabel [119, 1987] incorporates a full trust region strategy, and is the basis of a software package ODRPACK for orthogonal distance regression. This ANSI Fortran subroutine library is in the public domain, and is described in the user's guide [118, 1992].

9.4.4. Least squares fit of geometric elements.

A special nonlinear least squares problem that arises in many areas of applications is to fit given data points to a geometrical element, which may be defined in implicit form. We have already discussed fitting data to an affine linear manifold such as a line or a plane. The problem of fitting circles, ellipses, spheres, and cylinders arises in applications such as computer graphics, coordinate meteorology, and statistics.

In the past, algorithms have been given which fit circles and ellipses in *some* least squares sense without minimizing the geometric distance to the given points [671]. More recently the emphasis has been on minimizing the sum of squares of orthogonal distances to the given points. Algorithms for such a **geometric fit** are described, e.g., in Forbes [301, 302, 1989] and Gander, Golub, and Strebel [318, 1994].

EXAMPLE 9.4.2. Consider the problem of fitting a half-circle

$$y = b + (r^2 - (t - a)^2)^{1/2}$$

to a given set of points (y_i, t_i), $i = 1, 2, \ldots, m$. It is obvious (see Figure 9.4.2) that minimizing squares of either horizontal or vertical distances to the circle will normally not be satisfactory. ∎

Circles and ellipses may be represented algebraically by an equation of the form $F(x) = 0$. If a point is on the curve, then its coordinates satisfy $F(x) = 0$. Alternatively, curves may be represented in parametric form, which representation is well suited for minimizing the sum of the squares of the distances. Ellipses, for which the sum of the squares of the distances to the given points is minimal, will be referred to as **geometric fit**. Determining the parameters of the algebraic equation $F(x) = 0$ in the least squares sense will be denoted by **algebraic fit**.

Let $u = (u_1, \ldots, u_n)^T$ be a vector of unknowns and consider the nonlinear

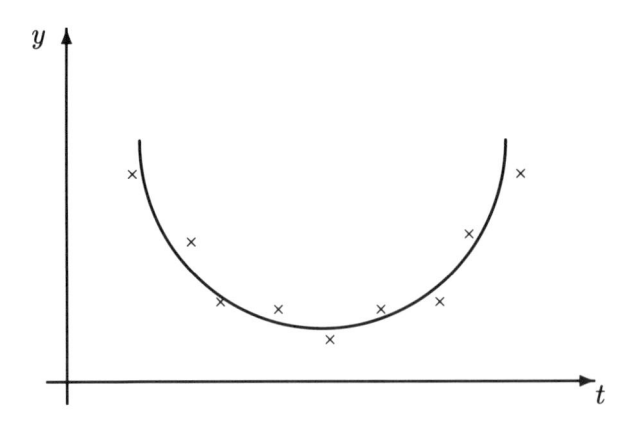

FIG. 9.4.2. *Fitting a half-circle.*

system of m equations $f(u) = 0$. If $m > n$, then we want to minimize

$$\sum_{i=1}^{m} f_i(u)^2 = \min.$$

This is a nonlinear least squares problem, which can be solved by the Gauss–Newton method. We then approximate the solution $\hat{u}$ by $\tilde{u} + h$. Developing $f(u)$ around $\tilde{u}$ in a Taylor series, we obtain the correction vector h from the linear least squares problem
(9.4.13) $$\min \|J(\tilde{u})h - f(\tilde{u})\|_2,$$
where J is the Jacobian of f.

EXAMPLE 9.4.3. One wants to fit a circle with radius r and center (x_0, y_0) to given data (x_i, y_i), $i = 1, 2, \ldots, m$. The orthogonal distance from (x_i, y_i) to the circle
$$d_i(x_0, y_0, r) = r_i - r, \qquad r_i = ((x_i - x_0)^2 + (y_i - y_0)^2)^{1/2},$$
depends nonlinearly on the parameters x_0, y_0. The problem

$$\min_{x_0, y_0, r} \sum_{i=1}^{m} d_i^2(x_0, y_0, r)$$

is thus a nonlinear least squares problem. An approximative linear model is obtained by writing the equation of the circle $(x - x_0)^2 + (y - y_0)^2 = r^2$ in the form
$$\delta(x_0, y_0, c) = 2xx_0 + 2yy_0 + c = x^2 + y^2,$$
which depends linearly on the parameters x_0, y_0 and $c = r^2 - x_0^2 - y_0^2$. If these parameters are known, then the radius of the circle can be determined by $r = (c + x_0^2 + y_0^2)^{1/2}$.

Bibliography

Electronic BibTeX Reference Base

The bibliography in this book is also available in electronic form as a BibTeX reference base, by anonymous ftp at *math.liu.se*. The file is named bjorBOOK.bib, and found in the directory *pub/references*. It has been modeled after a BibTeX reference base created by G. W. Stewart, and uses the same conventions.

The citation key is divided into two fields, <authors>:<date>, separated by a colon. The author field is constructed as follows. If there is only one author, the first four letters of his or her last name are used to form the field (but only one letter each from prefixes, e.g., van de Geign becomes vdge). If there are two authors the first two letters in their last names are used. If there are three authors the first two letters of the first are used with one letter from each of the second and third. If their are four, one letter from each is used.

Names appear in the order in which they appear in the reference, and entries are alphabetized by the first name to appear. To keep BibTeX from becoming confused, only initials of given names are used.

Each entry has a key word field. The program `bibsrch` (see below) will search the file for matches in a user-specified combination of conjunctions, disjunctions, and negations. The user can also specify the author. The program `getkey` dumps the key words on the standard output.

The program `bibsrch`, written by G. W. Stewart, may be generated from the file bibsearch.tar, available by anonymous ftp at *thales.cs.umd.edu* in the directory *pub/references*. (Web surfers may find it easier to download from *http://www.cs.umd.edu/~stewart/*.) Simply place bibsearch.tar in the directory where you want the directory containing the programs. Then do

```
tar xf bibsearch.tar
cd bibsearch
more README
```

and you will be told what to do next.

[1] N. N. ABDELMALEK, *Roundoff error analysis for Gram-Schmidt method and solution of linear least squares problems*, BIT, 11 (1971), pp. 345–368.

[2] R. J. ADCOCK, *A problem in least squares*, The Analyst, 5 (1878), pp. 53–54.

[3] A. V. AHO, J. E. HOPCROFT, AND J. D. ULLMAN, *Data Structures and Algorithms*, Addison-Wesley, Reading, MA, 1983.

[4] A. C. AIKEN, *On least squares and linear combinations of observations*, Proceedings of the Royal Society of Edinburgh, Sec. A, 55 (1934), pp. 42–47.

[5] M. A. AJIZ AND A. JENNINGS, *A robust incomplete Choleski-conjugate gradient algorithm*, Internat. J. Numer. Methods. Engrg., 20 (1984), pp. 949–966.

[6] M. AL-BAALI, *Methods for Nonlinear Least Squares*, Ph. D. thesis, Department of Mathematical Sciences, University of Dundee, Scotland, 1984.

[7] M. AL-BAALI AND R. FLETCHER, *Variational methods for non-linear least squares*, J. Oper. Res. Soc., 36 (1985), pp. 405–421.

[8] ——, *An efficient line search for nonlinear least-squares*, J. Optim. Theory Appl., 48 (1986), pp. 359–377.

[9] S. T. ALEXANDER, C.-T. PAN, AND R. J. PLEMMONS, *Analysis of a recursive least squares hyperbolic rotation algorithm for signal processing*, Linear Algebra Appl., 98 (1988), pp. 3–40.

[10] F. L. ALVARADO, *Manipulating and visualization of sparse matrices*, ORSA J. Comput., 2 (1990), pp. 186–207.

[11] P. R. AMESTOY, *Factorization of Large Sparse Matrices Based on a Multifrontal Approach in a Multiprocessor Environment*, Ph. D. thesis, CERFACS, Toulouse, France, 1991.

[12] P. R. AMESTOY AND I. S. DUFF, *Vectorization of a multiprocessor multifrontal code*, Internat. J. Supercomp. Appl., 3 (1989), pp. 41–59.

[13] ——, *Memory allocation issues in sparse multiprocessor multifrontal methods*, Internat. J. Supercomput. Appl., 7 (1993), pp. 64–82.

[14] A. A. ANDA AND H. PARK, *Fast plane rotations with dynamic scaling*, SIAM J. Matrix Anal. Appl., 15 (1994), pp. 162–174.

[15] ——, *Self-scaling fast rotations for stiff least squares problems*, Linear Algebra Appl., 234 (1996), pp. 137–162.

[16] E. ANDERSON, Z. BAI, C. BISCHOF, J. DEMMEL, J. DONGARRA, J. DU CROZ, A. GREENBAUM, S. HAMMARLING, A. MCKENNEY, S. OSTROUCHOV, AND D. SORENSEN, eds., *LAPACK Users' Guide. Second Edition*, SIAM, Philadelphia, 1995.

[17] N. ANDERSON AND I. KARASALO, *On computing bounds for the least singular value of a triangular matrix*, BIT, 15 (1975), pp. 1–4.

[18] E. ANDERSSEN, Z. BAI, AND J. DONGARRA, *Generalized QR factorization and its applications*, Tech. Report CS-91-131, Computer Science Department, University of Tennessee, Knoxville, TN, 1991.

[19] R. S. ANDERSSEN AND G. GOLUB, *Richardson's non-stationary matrix iterative procedure*, Tech. Report STAN-CS-72-304, Computer Science Department, Stanford University, CA, 1972.

[20] M. ARIOLI, J. DEMMEL, AND I. S. DUFF, *Solving sparse linear systems with sparse backward error*, SIAM J. Matrix Anal. Appl., 10 (1989), pp. 165–190.

[21] M. ARIOLI, I. S. DUFF, AND P. P. M. DE RIJK, *On the augmented system approach to sparse least-squares problems*, Numer. Math., 55 (1989), pp. 667–684.

[22] M. ARIOLI, I. S. DUFF, J. NOAILLES, AND D. RUIZ, *A block projection method for sparse matrices*, SIAM J. Sci. Statist. Comput., 13 (1990), pp. 47–70.

[23] M. ARIOLI, I. S. DUFF, AND D. RUIZ, *Stopping criteria for iterative solvers*,

SIAM J. Matrix Anal. Appl., 13 (1992), pp. 138–144.
[24] M. ARIOLI, I. S. DUFF, D. RUIZ, AND M. SADKANE, *Block Lanczos techniques for accelerating the block Cimmino method*, SIAM J. Sci. Comput., 16 (1995), pp. 1478–1511.
[25] M. ARIOLI AND A. LARATTA, *Error analysis of algorithms for computing the projection of a point onto a linear manifold*, Linear Algebra Appl., 82 (1986), pp. 1–26.
[26] ———, *Error analysis of an algorithm for solving an underdetermined linear system*, Numer. Math., 46 (1986), pp. 255–268.
[27] C. ASHCRAFT, *A vector implementation of the multifrontal method for large sparse positive definite systems*, Tech. Report ETA-TR-51, Engineering Technology Division, Boeing Computer Services, Seattle, WA, 1987.
[28] C. ASHCRAFT AND R. G. GRIMES, *The influence of relaxed supernode partitions on the multifrontal method*, ACM Trans. Math. Software, 15 (1989), pp. 291–309.
[29] V. ASHKENAZI, *Geodetic normal equations*, in Large Sets of Linear Equations, J. K. Reid, ed., Academic Press, New York, 1971, pp. 57–74.
[30] J. K. AVILA AND J. A. TOMLIN, *Solution of very large least squares problems by nested dissection on a parallel processor*, in Proceedings of the Computer Science and Statistics 12th Annual Symposium on the Interface, J. F. Gentleman, ed., University of Waterloo, Canada, 1979.
[31] Z. BAI, *The CSD, GSVD, their applications and computations*, Tech. Report IMA Preprint Series 958, Institute for Mathematics and Its Applications, University of Minnesota, Minneapolis, MN, 1992.
[32] Z. BAI AND J. W. DEMMEL, *Computing the generalized singular value decomposition*, SIAM J. Sci. Comput., 14 (1993), pp. 1464–1486.
[33] Z. BAI AND H. ZHA, *A new preprocessing algorithm for the computation of the generalized singular value decomposition*, SIAM J. Sci. Comput., 14 (1993), pp. 1007–1012.
[34] T. BANACHIEWICZ, *Principes d'une nouvelle technique de la méthode des moindres carrés*, Bull. Internat. Acad. Polon. Sci. A, (1938), pp. 134–135.
[35] Y. BARD, *Nonlinear Parameter Estimation*, Academic Press, New York, 1974.
[36] E. H. BAREISS, *Numerical solution of the weighted linear least squares problems by G-transformations*, Tech. Report 82-03-NAM-03, Department of Electrical Engineering and Computer Science, Northwestern University, Evanston, IL, 1982.
[37] J. L. BARLOW, *Stability analysis of the G-algorithm and a note on its application to sparse least squares problems*, BIT, 25 (1985), pp. 507–520.
[38] ———, *Error analysis and implementation aspects of deferred correction for equality constrained least squares problems*, SIAM J. Numer. Anal., 25 (1988), pp. 1340–1358.
[39] ———, *The accurate solution of sparse weighted and equality constrained least squares problems using a static data structure*, Tech. Report CS-89-03, Dept. of Computer Science, The Pennsylvania State University, State College, PA, 1989.
[40] ———, *On the use of structural zeros in orthogonal factorization*, SIAM J. Sci. Stat. Comput., 11 (1990), pp. 600–601.
[41] J. L. BARLOW AND S. L. HANDY, *The direct solution of weighted and equality constrained least squares problems*, SIAM J. Sci. Statist. Comput., 9 (1988), pp. 704–716.
[42] J. L. BARLOW AND I. C. F. IPSEN, *Scaled Givens rotations for the solution of linear least squares problems on systolic arrays*, SIAM J. Sci. Statist. Comput., 8 (1987), pp. 716–734.

[43] J. L. BARLOW, N. K. NICHOLS, AND R. J. PLEMMONS, *Iterative methods for equality-constrained least squares problems.*, SIAM J. Sci. Statist. Comput., 9 (1988), pp. 892–906.

[44] J. L. BARLOW AND U. B. VEMULAPATI, *Rank detection methods for sparse matrices*, SIAM J. Matrix Anal. Appl., 13 (1992), pp. 1279–1297.

[45] J. L. BARLOW, P. A. YOON, AND H. ZHA, *An algorithm and a stability theory for downdating the ULV decomposition*, BIT, 36 (1996), pp. 14–40.

[46] J. L. BARLOW, H. ZHA, AND P. A. YOON, *Stable chasing algorithms for modifying complete and partial singular value decompositions*, Tech. Report CSE-93-19, Department of Computer Science, The Pennsylvania State University, State College, PA, 1993.

[47] R. BARRET, M. W. BERRY, T. CHAN, J. DEMMEL, J. DONATO, J. DONGARRA, V. EIJKHOUT, R. POZO, C. ROMINE, AND H. VAN DER VORST, eds., *Templates for the Solution of Linear Systems: Building Blocks for Iterative Methods*, SIAM, Philadelphia, 1993.

[48] A. BARRLUND, *Perturbation bounds for the generalized QR factorization*, Linear Algebra Appl., 207 (1994), pp. 251–272.

[49] I. BARRODALE AND C. PHILLIPS, *Algorithm 495: Solution of an overdetermined system of linear equations in the Chebyshev norm*, ACM Trans. Math. Software, 1 (1975), pp. 264–270.

[50] I. BARRODALE AND F. D. K. ROBERTS, *An improved algorithm for discrete ℓ_1 linear approximation*, SIAM J. Numer. Anal., 10 (1973), pp. 839–848.

[51] ———, *An efficient algorithm for discrete ℓ_1 linear approximation with linear constraints*, SIAM J. Numer. Anal., 15 (1978), pp. 603–611.

[52] R. H. BARTELS AND A. R. CONN, *Linearly constrained discrete ℓ_1 problems*, ACM Trans. Math. Software, 6 (1980), pp. 594–608.

[53] R. H. BARTELS, A. R. CONN, AND C. CHARALAMBOUS, *On Cline's direct method for solving overdetermined linear systems in the ℓ_∞ sense*, SIAM J. Numer. Anal., 15 (1980), pp. 255–270.

[54] R. H. BARTELS, A. R. CONN, AND J. W. SINCLAIR, *Minimization techniques for piecewise differentiable functions: The ℓ_1 solution to an overdetermined linear system*, SIAM J. Numer. Anal., 15 (1978), pp. 224–241.

[55] R. H. BARTELS, G. H. GOLUB, AND M. A. SAUNDERS, *Numerical techniques in mathematical programming*, in Nonlinear Programming, J. B. Rosen, O. L. Mangasarian, and K. Ritter, eds., Academic Press, New York, 1970, pp. 123–176.

[56] W. BARTH, R. S. MARTIN, AND J. H. WILKINSON, *Calculation of the eigenvalues of a symmetric tridiagonal matrix by the method of bisection*, Numer. Math., 9 (1967), pp. 386–393.

[57] F. L. BAUER, *Elimination with weighted row combinations for solving linear equations and least squares problems*, Numer. Math., 7 (1965), pp. 338–352.

[58] ———, *Genauigkeitsfragen bei der Lösung linear Gleichungssysteme*, Z. Angew. Math. Mech., 46 (1966), pp. 409–421.

[59] A. BEN-ISRAEL AND S. J. WERSAN, *An elimination method for computing the generalized inverse of an arbitrary matrix*, J. ACM, 10 (1963), pp. 532–537.

[60] A. BEN-TAL AND M. TEBOULLE, *A geometric property of the least squares solution of linear equations*, Linear Algebra Appl., 139 (1990), pp. 165–170.

[61] C. BENOIT, *Sur la méthode de résolution des équationes normales, etc. (procédés du commandant Cholesky)*, Bull. Géodésique, 2 (1924), pp. 67–77.

[62] M. BENZI, *A Direct Row-Projection Method for Sparse Linear Systems*, Ph. D. thesis, North Carolina State University, Raleigh, NC, 1993.

[63] M. BENZI AND C. D. MEYER, *An explicit preconditioner for the conjugate gradient method*, in Proceedings of the Cornelius Lanczos International Centenary Conference, Raleigh, NC, Dec. 1993, J. D. Brown, M. T. Chu, D. C. Ellison, and R. J. Plemmons, eds., SIAM, Philadelphia, 1994, pp. 294–296.

[64] M. BENZI, C. D. MEYER, AND M. TŮMA, *A sparse approximate inverse preconditioner for the conjugate gradient method*, SIAM J. Sci. Comput., 17 (1996), to appear.

[65] A. BERMAN AND R. J. PLEMMONS, *Cones and iterative methods for best least squares solutions of linear systems*, SIAM J. Numer. Anal., 11 (1974), pp. 145–154.

[66] M. W. BERRY, *A Fortran-77 software library for the sparse singular value decomposition*, Tech. Report CS-92-159, University of Tennessee, Knoxville, TN, 1992.

[67] ———, *Large scale sparse singular value computations*, Internat. J. Supercomp. Appl., 6 (1992), pp. 13–49.

[68] ———, *SVDPACKC: Version 1.0 user's guide*, Tech. Report CS-93-194, University of Tennessee, Knoxville, TN, 1993.

[69] ———, *A survey of public-domain Lanczos-based software*, in Proceedings of the Cornelius Lanczos International Centenary Conference, Raleigh, NC, Dec. 1993, J. D. Brown, M. T. Chu, D. C. Ellison, and R. J. Plemmons, eds., SIAM, Philadelphia, 1994, pp. 332–334.

[70] M. W. BERRY AND R. L. AUERBACH, *A block Lanczos SVD method with adaptive reorthogonalization*, in Proceedings of the Cornelius Lanczos International Centenary Conference, Raleigh, NC, Dec. 1993, J. D. Brown, M. T. Chu, D. C. Ellison, and R. J. Plemmons, eds., SIAM, Philadelphia, 1994, pp. 329–331.

[71] M. W. BERRY AND G. H. GOLUB, *Estimating the largest singular values of large sparse matrices via modified moments*, Numer. Algorithms, 1 (1991), pp. 363–374.

[72] J. BERTRAND, *Sur la méthode des moindres carrés*, C. R. Acad. Sci., Paris, 40 (1855), pp. 1190–1192.

[73] I. J. BIENAYMÉ, *Remarques sur les différences qui distinguent l'interpolation de M. Cauchy de la méthode des moindre carrés et qui assurent la supériorité de cette méthode*, C. R. Acad. Sci., Paris, 37 (1853), pp. 5–13.

[74] M. BIERLAIR, P. TOINT, AND D. TUYTTENS, *On iterative algorithms for linear least-squares problems with bound constraints*, Linear Algebra Appl., 143 (1991), pp. 111–143.

[75] C. H. BISCHOF, *A block QR factorization algorithm using restricted pivoting*, in Supercomputing 89, ACM Press, New York, 1989, pp. 248–256.

[76] ———, *Adaptive condition estimation for rank-one downdates of QR factorizations*, Tech. Report ANL/MCS-P166-0790, Argonne National Laboratory, Math. and Computer Sciences Div., Argonne, IL, 1990.

[77] ———, *Incremental condition estimation*, SIAM J. Matrix Anal. Appl., 11 (1990), pp. 312–322.

[78] ———, *A parallel QR factorization algorithm with controlled local pivoting*, SIAM J. Sci. Statist. Comput., 12 (1991), pp. 36–57.

[79] C. H. BISCHOF AND P. C. HANSEN, *Structure preserving and rank-revealing QR factorizations*, SIAM J. Sci. Statist. Comput., 12 (1991), pp. 1332–1350.

[80] ———, *A block algorithm for computing rank-revealing QR factorizations*, Numer. Algorithms, 2 (1992), pp. 371–392.

[81] C. H. BISCHOF, J. G. LEWIS, AND D. J. PIERCE, *Incremental condition estimation for sparse matrices*, SIAM J. Matrix Anal. Appl., 11 (1990), pp. 644–659.

[82] C. H. BISCHOF, C.-T. PAN, AND P. T. P. TANG, *A Cholesky up-and downdating algorithm for systolic and SIMD architectures*, SIAM J. Sci. Comput., 14 (1993), pp. 670–676.

[83] A. BJERHAMMER, *Rectangular reciprocal matrices with special reference to geodetic calculations*, Bull. Géodésique, 52 (1951), pp. 118–220.

[84] A. BJÖRCK, *Iterative refinement of linear least squares solutions I*, BIT, 7 (1967), pp. 257–278.

[85] ———, *Solving linear least squares problems by Gram-Schmidt orthogonalization*, BIT, 7 (1967), pp. 1–21.

[86] ———, *Iterative refinement of linear least squares solutions II*, BIT, 8 (1968), pp. 8–30.

[87] ———, *Methods for sparse least squares problems*, in Sparse Matrix Computations, J. Bunch and D. J. Rose, eds., Academic Press, New York, 1976, pp. 177–199.

[88] ———, *Comment on the iterative refinement of least squares solutions*, J. Amer. Statist. Assoc., 73 (1978), pp. 161–166.

[89] ———, *SSOR preconditioning methods for sparse least squares problems*, in Proceedings of the Computer Science and Statistics 12th Annual Symposium on the Interface, J. F. Gentleman, ed., University of Waterloo, Canada, 1979, pp. 21–25.

[90] ———, *A general updating algorithm for constrained linear least squares problems*, SIAM J. Sci. Statist. Comput., 5 (1984), pp. 394–402.

[91] ———, *Stability analysis of the method of semi-normal equations for least squares problems*, Linear Algebra Appl., 88/89 (1987), pp. 31–48.

[92] ———, *A bidiagonalization algorithm for solving ill-posed systems of linear equations*, BIT, 28 (1988), pp. 659–670.

[93] ———, *A direct method for sparse least squares problems with lower and upper bounds*, Numer. Math., 54 (1988), pp. 19–32.

[94] ———, *Iterative refinement and reliable computing*, in Reliable Numerical Computation, M. G. Cox and S. J. Hammarling, eds., Clarendon Press, Oxford, UK, 1990, pp. 249–266.

[95] ———, *Least squares methods*, in Handbook of Numerical Analysis. I. Solution of Equations in R^n. Part 1, P. G. Ciarlet and J. L. Lions, eds., Elsevier/North-Holland, Amsterdam, 1990, pp. 466–647.

[96] ———, *Algorithms for linear least squares problems*, in Computer Algorithms for Solving Linear Algebraic Equations; The State of the Art, E. Spedicato, ed., vol. 77 of NATO ASI Series F: Computer and Systems Sciences, Springer-Verlag, Berlin, 1991, pp. 57–92.

[97] ———, *Component-wise perturbation analysis and errors bounds for linear least square solutions*, BIT, 31 (1991), pp. 238–244.

[98] ———, *Error analysis of least squares algorithms*, in Numerical Linear Algebra, Digital Signal Processing and Parallel Algorithms, G. H. Golub and P. Van Dooren, eds., vol. 70 of NATO ASI Series, Springer-Verlag, Berlin, 1991, pp. 41–73.

[99] ———, *Pivoting and stability in the augmented system method*, in Numerical Analysis 1991: Proceedings of the 14th Dundee Conference, June 1991, D. F. Griffiths and G. A. Watson, eds., Pitman Research Notes in Mathematics 260, Longman Scientific and Technical, Harlow, UK, 1992, pp. 1–16.

[100] ———, *Generalized and sparse least squares problems*, in Algorithms for Continuous Optimization; The State of the Art, E. Spedicato, ed., vol. 434 of NATO ASI Series C: Mathematical and Physical Sciences, Kluwer Academic Publisher, Dordrecht, 1994, pp. 37–80.

[101] ———, *Numerics of Gram-Schmidt orthogonalization*, Linear Algebra Appl., 197–

198 (1994), pp. 297–316.
[102] ———, *Numerical Methods for Least Squares Problems*, SIAM, Philadelphia, 1996.
[103] Å. BJÖRCK AND C. BOWIE, *An iterative algorithm for computing the best estimate of an orthogonal matrix*, SIAM J. Numer. Anal., 8 (1971), pp. 358–364.
[104] Å. BJÖRCK AND I. S. DUFF, *A direct method for the solution of sparse linear least squares problems*, Linear Algebra Appl., 34 (1980), pp. 43–67.
[105] Å. BJÖRCK AND L. ELDÉN, *Methods in numerical algebra for ill-posed problems*, Tech. Report LiTH-MAT-R-1979-33, Department of Mathematics, Linköping University, Sweden, 1979.
[106] Å. BJÖRCK, L. ELDÉN, AND H. PARK, *Accurate downdating of least squares solutions*, SIAM J. Matrix Anal. Appl., 15 (1994), pp. 549–568.
[107] Å. BJÖRCK AND T. ELFVING, *Algorithms for confluent Vandermonde systems*, Numer. Math., 21 (1973), pp. 130–137.
[108] ———, *Accelerated projection methods for computing pseudoinverse solutions of systems of linear equations*, BIT, 19 (1979), pp. 145–163.
[109] Å. BJÖRCK, T. ELFVING, AND Z. STRAKOS, *Implementing conjugate gradient-type methods for linear least squares problems*, Tech. Report LiTH-MAT-R-95-26, Department of Mathematics, Linköping University, Sweden, 1995.
[110] Å. BJÖRCK AND G. H. GOLUB, *Iterative refinement of linear least squares solution by Householder transformation*, BIT, 7 (1967), pp. 322–337.
[111] ———, *Numerical methods for computing angles between linear subspaces*, Math. Comp., 27 (1973), pp. 579–594.
[112] Å. BJÖRCK, E. GRIMME, AND P. VAN DOOREN, *An implicit bidiagonalization algorithm for ill-posed systems*, BIT, 34 (1994), pp. 510–534.
[113] Å. BJÖRCK AND C. C. PAIGE, *Loss and recapture of orthogonality in the modified Gram-Schmidt algorithm*, SIAM J. Matrix Anal. Appl., 13 (1992), pp. 176–190.
[114] ———, *Solution of augmented linear systems using orthogonal factorizations*, BIT, 34 (1994), pp. 1–26.
[115] Å. BJÖRCK AND V. PEREYRA, *Solution of Vandermonde system of equations*, Math. Comp., 24 (1970), pp. 893–903.
[116] Å. BJÖRCK, R. J. PLEMMONS, AND H. SCHNEIDER, eds., *Large Scale Matrix Problems*, North-Holland, New York, 1981.
[117] J. L. BLUE, *A portable Fortran program to find the Euclidean norm of a vector*, Trans. Math. Software, 4 (1978), pp. 15–23.
[118] P. T. BOGGS, R. H. BYRD, J. E. ROGERS, AND R. B. SCHNABEL, *User's reference guide for ODRPACK version 2.01—Software for weighted orthogonal distance regression*, Tech. Report LiTH-MAT-R-1979-33, National Institute of Standards and Technology, Gaithersburg, MD, 1992.
[119] P. T. BOGGS, R. H. BYRD, AND R. B. SCHNABEL, *A stable and efficient algorithm for nonlinear orthogonal regression*, SIAM J. Sci. Statist. Comput., 8 (1987), pp. 1052–1078.
[120] A. W. BOJANCZYK AND R. P. BRENT, *Parallel solution of certain Toeplitz least-squares problems*, Linear Algebra Appl., 77 (1986), pp. 43–60.
[121] A. W. BOJANCZYK, R. P. BRENT, AND F. R. DE HOOG, *QR factorization of Toeplitz matrices*, Numer. Math., 49 (1986), pp. 81–94.
[122] ———, *A weakly stable algorithm for general Toeplitz matrices*, Tech. Report TR-CS-93-15, Cornell University, Ithaca, NY, 1993.
[123] A. W. BOJANCZYK, R. P. BRENT, P. VAN DOOREN, AND F. DE HOOG, *A note on downdating the Cholesky factorization*, SIAM J. Sci. Statist. Comput., 8 (1987), pp. 210–221.

[124] A. W. BOJANCZYK AND J. M. LEBAK, *Downdating a ULLV decomposition of two matrices*, in Proceedings of the Fifth SIAM Conference on Applied Linear Algebra, J. G. Lewis, ed., SIAM, Philadelphia, 1994, pp. 261–265.

[125] A. W. BOJANCZYK, J. G. NAGY, AND R. J. PLEMMONS, *Block RLS using row Householder reflections*, Linear Algebra Appl., 188/189 (1993), pp. 31–62.

[126] D. BOLEY AND G. H. GOLUB, *A survey of matrix inverse eigenvalue problems*, Inverse Problems, 3 (1987), pp. 595–622.

[127] F. L. BOOKSTEIN, *Fitting conic sections to scattered data*, Computer Graphics and Image Processing, 9 (1979), pp. 56–71.

[128] T. BOROS, T. KAILATH, AND V. OLSHEVSKY, *Error analysis of a fast algorithm for solving Cauchy linear systems*, (1995), submitted.

[129] R. BRAMLEY, *Row-Projection Methods for Linear Systems*, Ph. D. thesis 881, Center for Supercomputing Research and Development, University of Illinois, Urbana, IL, 1989.

[130] R. BRAMLEY AND A. SAMEH, *Row projection methods for large nonsymmetric linear systems*, SIAM J. Sci. Statist. Comput., 13 (1992), pp. 168–193.

[131] R. P. BRENT, *Old and new algorithms for Toeplitz systems*, in Advanced Algorithms and Architectures for Signal Processing III, F. T. Luk, ed., SPIE Proceeding Series, Bellingham, WA, 1988, pp. 2–9.

[132] J. R. BUNCH, *Analysis of the diagonal pivoting method*, SIAM J. Numer. Anal., 8 (1971), pp. 656–680.

[133] ———, *Stability of methods for solving Toeplitz systems of equations*, SIAM J. Sci. Statist. Comput., 6 (1985), pp. 349–364.

[134] ———, *The weak and strong stability of algorithms in numerical linear algebra*, Linear Algebra Appl., 88/89 (1987), pp. 49–66.

[135] J. R. BUNCH, J. W. DEMMEL, AND C. F. VAN LOAN, *The strong stability of algorithms for solving symmetric systems*, SIAM J. Matrix Anal. Appl., 10 (1989), pp. 494–499.

[136] J. R. BUNCH AND L. KAUFMAN, *Some stable methods for calculating inertia and solving symmetric linear systems*, Math. Comp., 31 (1977), pp. 162–179.

[137] J. R. BUNCH, L. KAUFMAN, AND B. N. PARLETT, *Decomposition of a symmetric matrix*, Numer. Math., 27 (1976), pp. 95–109.

[138] J. R. BUNCH AND C. P. NIELSEN, *Updating the singular value decomposition*, Numer. Math., 31 (1978), pp. 111–129.

[139] J. R. BUNCH AND B. N. PARLETT, *Direct methods for solving symmetric indefinite systems of linear systems*, SIAM J. Numer. Anal., 8 (1971), pp. 639–655.

[140] J. R. BUNCH AND D. J. ROSE, eds., *Sparse Matrix Computations*, Academic Press, New York, 1976.

[141] P. BUSINGER, *Updating a singular value decomposition*, BIT, 10 (1970), pp. 376–385.

[142] P. BUSINGER AND G. H. GOLUB, *Linear least squares solutions by Householder transformations*, Numer. Math., 7 (1965), pp. 269–276.

[143] ———, *Algorithm 358: Singular value decomposition of a complex matrix*, Comm. ACM, 12 (1969), pp. 564–565.

[144] D. CALVETTI AND L. REICHEL, *Fast inversion of Vandermonde-like matrices involving orthogonal polynomials*, BIT, 33 (1993), pp. 473–484.

[145] D. CARLSON AND H. SCHNEIDER, *Inertia theorems for matrices: The positive semidefinite case*, J. Math. Anal. Appl., 6 (1963), pp. 430–446.

[146] A. CAUCHY, *Mémoire sur l'interpolation*, J. Math. Pures Appl., 2 (1837), pp. 193–205.

[147] J. M. CHAMBERS, *Regression updating*, J. Amer. Statist. Assoc., 66 (1971), pp. 744–748.
[148] ———, *Computational Methods for Data Analysis*, John Wiley, New York, 1977.
[149] R. H. CHAN, J. G. NAGY, AND R. J. PLEMMONS, *FFT-based preconditioners for Toeplitz-block least squares problems*, SIAM J. Numer. Anal., 30 (1993), pp. 1740–1768.
[150] ———, *Circulant preconditioned Toeplitz least squares iterations*, SIAM J. Matrix Anal. Appl., 15 (1994), pp. 80–97.
[151] T. F. CHAN, *Algorithm 581: An improved algorithm for computing the singular value decomposition*, ACM Trans. Math. Software, 8 (1982), pp. 84–88.
[152] ———, *An improved algorithm for computing the singular value decomposition*, ACM Trans. Math. Software, 8 (1982), pp. 72–83.
[153] ———, *Rank revealing QR-factorizations*, Linear Algebra Appl., 88/89 (1987), pp. 67–82.
[154] T. F. CHAN AND P. C. HANSEN, *Computing truncated SVD least squares solutions by rank revealing QR factorizations*, SIAM J. Sci. Statist. Comput., 11 (1990), pp. 519–530.
[155] ———, *Some applications of the rank revealing QR factorization*, SIAM J. Sci. Statist. Comput., 13 (1992), pp. 727–741.
[156] ———, *Low-rank revealing QR factorizations*, Numer. Linear Algebra Appl., 1 (1994), pp. 33–44.
[157] S. CHANDRASEKARAN AND I. C. F. IPSEN, *Analysis of a QR algorithm for computing singular values*, Tech. Report YALEU/DCS/RR-917, Yale University, New Haven, CT, 1992.
[158] ———, *Backward errors for eigenvalue and singular value decompositions*, Numer. Math., 68 (1994), pp. 215–223.
[159] ———, *On rank-revealing factorizations*, SIAM J. Matrix Anal. Appl., 15 (1994), pp. 592–622.
[160] X.-W. CHANG AND C. C. PAIGE, *A new perturbation analysis for the Cholesky factorization*, IMA J. Numer. Anal., (1996).
[161] J. CHARLIER, M. VANBEGIN, AND P. VAN DOOREN, *On efficient implementations of Kogbetliantz's algorithm for computing the singular value decomposition*, Numer. Math., 52 (1988), pp. 279–300.
[162] S. CHATTERJEE AND G. HELLER, *The numerical effect of measurement error in the explanatory variables on the observed least squares estimate*, SIAM J. Matrix Anal. Appl., 14 (1993), pp. 677–687.
[163] Y. T. CHEN, *Iterative methods for linear least squares problems*, Tech. Report CS-75-04, University of Waterloo, Canada, 1975.
[164] Y. T. CHEN AND R. P. TEWARSON, *On the fill-in when sparse vectors are orthonormalized*, Computing, 9 (1972), pp. 53–56.
[165] E. C. H. CHU, *Orthogonal Decomposition of Dense and Sparse Matrices on Multiprocessors*, Ph. D. thesis, University of Waterloo, Canada, 1988.
[166] E. C. H. CHU AND J. A. GEORGE, *QR factorization of a dense matrix on a hypercube multiprocessor*, Parallel Comput., 11 (1989), pp. 55–71.
[167] E. C. H. CHU, J. A. GEORGE, J. LIU, AND E. NG, *SPARSPAK: Waterloo sparse matrix package user's guide for SPARSPAK-A*, Res. Report CS-84-36, Department of Computer Science, University of Waterloo, Canada, 1984.
[168] J. CHUN, T. KAILATH, AND H. LEV-ARI, *Fast parallel algorithms for QR and triangular factorizations*, SIAM J. Sci. Statist. Comput., 8 (1987), pp. 899–913.
[169] G. CIMMINO, *Calcolo approssimato per le soluzioni dei sistemi di equazioni lineari*,

Ric. Sci. Progr. Tecn. Econom. Naz., 9 (1939), pp. 326–333.

[170] D. I. CLARK AND M. R. OSBORNE, *Finite algorithms for Hubers's M-estimator*, SIAM J. Sci. Statist. Comput., 7 (1986), pp. 72–85.

[171] ——, *On linear restricted and interval least-squares problems*, IMA J. Numer. Anal., 8 (1988), pp. 23–36.

[172] A. K. CLINE, *An elimination method for the solution of linear least squares problems*, SIAM J. Numer. Anal., 10 (1973), pp. 283–289.

[173] ——, *The transformation of a quadratic programming problem into solvable form*, Tech. Report ICASE 75-14, NASA, Langley Research Center, Hampton, VA., 1975.

[174] A. K. CLINE, A. R. CONN, AND C. F. VAN LOAN, *Generalizing the LINPACK condition estimator*, in Numerical Analysis, J. P. Hennart, ed., vol. 909 of Lecture Notes in Mathematics, Springer-Verlag, Berlin, 1982.

[175] A. K. CLINE, C. B. MOLER, G. W. STEWART, AND J. H. WILKINSON, *An estimate for the condition number of a matrix*, SIAM J. Numer. Anal., 16 (1979), pp. 368–375.

[176] R. E. CLINE AND R. J. PLEMMONS, *ℓ_2-solutions to underdetermined linear systems*, SIAM Review, 18 (1976), pp. 92–106.

[177] T. F. COLEMAN, A. EDENBRANDT, AND J. R. GILBERT, *Predicting fill for sparse orthogonal factorization*, J. Assoc. Comput. Mach., 33 (1986), pp. 517–532.

[178] T. F. COLEMAN AND L. A. HULBERT, *A direct active set algorithm for large sparse quadratic programs with simple lower bounds*, Math. Programming, 45 (1989), pp. 373–406.

[179] ——, *A globally and superlinearly convergent algorithm for convex quadratic programs with simple bounds*, SIAM J. Optim., 3 (1993), pp. 298–321.

[180] T. F. COLEMAN AND Y. LI, *A global and quadratically convergent affine scaling method for linear l_1 problems*, Math. Programming, 56 (1992), pp. 189–222.

[181] ——, *A global and quadratically convergent affine scaling method for linear l_∞ problems*, SIAM J. Numer. Anal., 29 (1992), pp. 1166–1186.

[182] ——, *An interior Newton method for quadratic programming*, Tech. Report TR 93-1388, Department of Computer Science, Cornell University, Ithaca, NY, 1993.

[183] J. W. COOLEY, *How the FFT gained acceptance*, in A History of Scientific Computing, S. G. Nash, ed., Addison-Wesley, Reading, MA, 1990, pp. 133–140.

[184] J. W. COOLEY, P. A. W. LEWIS, AND P. D. WELSH, *The fast Fourier transform and its application*, IEEE Trans. Education, E-12 (1969), pp. 27–34.

[185] J. W. COOLEY AND J. W. TUKEY, *An algorithm for the machine calculation of complex Fourier series*, Math. Comp., 19 (1965), pp. 297–301.

[186] R. W. COTTLE, *Manifestations of the Schur complement*, Linear Algebra Appl., 8 (1974), pp. 189–211.

[187] ——, *Numerical methods for complementarity problems in engineering and applied science*, in Computing Methods in Applied Sciences and Engineering, R. Glowinski and J. L. Lions, eds., vol. 704 of Lecture Notes in Mathematics, Springer-Verlag, New York, 1977, pp. 37–52.

[188] R. W. COTTLE, G. H. GOLUB, AND R. S. SACHER, *On the solution of large structured linear complementarity problems: The block partitioned case*, Appl. Math. Optim., 4 (1978), pp. 347–363.

[189] M. G. COX, *The least squares solution of overdetermined linear equations having band or augmented band structure*, IMA J. Numer. Anal., 1 (1981), pp. 3–22.

[190] ——, *The least-squares solution of linear equations with block-angular observation matrix*, in Reliable Numerical Computation, M. G. Cox and S. J. Hammarling,

eds., Oxford University Press, UK, 1990, pp. 227–240.
[191] E. J. CRAIG, *The n-step iteration procedure*, J. Math. Phys., 34 (1955), pp. 65–73.
[192] R. L. CRANE, B. S. GARBOW, K. E. HILLSTROM, AND M. MINKOFF, *LCLSQ: An implementation of an algorithm for linearly constrained linear least squares*, Tech. Report ANL-80-116, Argonne National Laboratory, Argonne, IL, 1980.
[193] P. CRAVEN AND G. WAHBA, *Smoothing noisy data with spline functions*, Numer. Math., 31 (1979), pp. 377–403.
[194] C. CRYER, *The solution of a quadratic programming problem using systematic overrelaxation*, SIAM J. Control Optim., 9 (1971), pp. 385–392.
[195] J. K. CULLUM AND R. A. WILLOUGHBY, *Lanczos Algorithms for Large Symmetric Eigenvalue Computations*, vol. 1 Theory. 2 Programs, Birkhäuser, Stuttgart, 1985.
[196] J. K. CULLUM, R. A. WILLOUGHBY, AND M. LAKE, *A Lanczos algorithm for computing singular values and vectors of large matrices*, SIAM J. Sci. Statist. Comput., 4 (1983), pp. 197–215.
[197] E. CUTHILL, *Several strategies for reducing the bandwidth of matrices*, in Sparse Matrices and Their Applications, D. J. Rose and R. A. Willoughby, eds., Plenum Press, New York, 1972, pp. 157–166.
[198] G. CYBENKO, *The numerical stability of the lattice algorithm for least squares linear prediction problems*, BIT, 24 (1984), pp. 441–455.
[199] ———, *Fast Toeplitz orthogonalization using inner products*, SIAM J. Sci. Statist. Comput., 8 (1987), pp. 734–740.
[200] G. DAHLQUIST, B. SJÖBERG, AND P. SVENSSON, *Comparison of the method of averages with the method of least squares*, Math. Comp., 22 (1968), pp. 833–846.
[201] C. DANIEL AND F. S. WOOD, *Fitting Equations to Data*, 2nd ed., John Wiley, New York, 1980.
[202] J. DANIEL, W. B. GRAGG, L. KAUFMAN, AND G. W. STEWART, *Reorthogonalization and stable algorithms for updating the Gram-Schmidt QR factorization*, Math. Comp., 30 (1976), pp. 772–95.
[203] C. DAVIS AND W. KAHAN, *The rotation of eigenvectors by a perturbation III*, SIAM J. Numer. Anal., 7 (1970), pp. 1–46.
[204] A. DAX, *The ℓ_1 solution of linear equations subject to linear constraints*, SIAM J. Sci. Statist. Comput., 10 (1989), pp. 328–340.
[205] ———, *On row relaxation methods for large constrained least squares problems*, SIAM J. Sci. Comput., 14 (1993), pp. 570–584.
[206] C. DE BOOR, *A Practical Guide to Splines*, Springer-Verlag, Berlin, 1978.
[207] A. DE LA GARZA, *An iterative method for solving systems linear equations*, Tech. Report K-731, Union Carbide, Oak Ridge, TN, 1951.
[208] B. DE MOOR, *Structured total least squares and l_2 approximation problems*, Linear Algebra Appl., 188/189 (1993), pp. 163–206.
[209] ———, *On the structure of generalized singular value and QR decompositions*, SIAM J. Matrix Anal. Appl., 15 (1994), pp. 347–358.
[210] B. DE MOOR AND G. H. GOLUB, *Generalized singular value decomposition: A proposal for a standard nomenclature*, Tech. Report NA-89-05, Numerical Analysis Project, Stanford University, CA, 1989.
[211] ———, *The restricted singular value decomposition: Properties and applications*, SIAM J. Matrix. Anal. Appl., 12 (1991), pp. 401–425.
[212] B. DE MOOR AND P. VAN DOOREN, *Generalizations of the singular value and QR decompositions*, SIAM J. Matrix. Anal. Appl., 13 (1992), pp. 993–1014.
[213] L. M. DELVES AND I. BARRODALE, *A fast direct method for the least squares*

solution of slightly overdetermined sets of linear equations, J. Inst. Maths. Applic., 24 (1979), pp. 149–156.

[214] C. J. DEMEURE, *Fast QR factorization of Vandermonde matrices*, Linear Algebra Appl., 122/3/4 (1989), pp. 165–194.

[215] ———, *QR factorization of confluent Vandermonde matrices*, IEEE Trans. Acoust. Speech Signal Process., 38 (10) (1990), pp. 1799–1802.

[216] J. DEMMEL, *The smallest perturbation of a submatrix which lowers the rank and constrained total least squares problems*, SIAM J. Numer. Anal., 24 (1987), pp. 199–206.

[217] J. DEMMEL AND N. J. HIGHAM, *Improved error bounds for underdetermined system solvers*, SIAM J. Matrix Anal. Appl., 14 (1993), pp. 1–14.

[218] J. DEMMEL AND W. KAHAN, *Accurate singular values of bidiagonal matrices*, SIAM J. Sci. Statist. Comput., 11 (1990), pp. 873–912.

[219] J. E. DENNIS, *Nonlinear least squares and equations*, in The State of the Art in Numerical Analysis, D. A. H. Jacobs, ed., Academic Press, New York, 1977, pp. 269–312.

[220] J. E. DENNIS, D. M. GAY, AND R. E. WELSCH, *An adaptive nonlinear least-squares algorithm*, ACM Trans. Math. Software, 7 (1981), pp. 348–368.

[221] ———, *Algorithm 573 NL2SOL: An adaptive nonlinear least-squares algorithm*, ACM Trans. Math. Software, 7 (1981), pp. 369–383.

[222] ———, *Remark on algorithm 573*, ACM Trans. Math. Software, 9 (1983), p. 139.

[223] J. E. DENNIS AND R. B. SCHNABEL, *Numerical Methods for Unconstrained Optimization and Nonlinear Equations*, Prentice Hall, Englewood Cliffs, NJ, 1983.

[224] J. E. DENNIS AND T. STEIHAUG, *On the successive projection approach to least squares problems*, SIAM J. Numer. Anal., 23 (1986), pp. 717–733.

[225] P. DEUFLHARD AND V. APOSTOLESCU, *An underrelaxed Gauss:s–Newton method for equality constrained nonlinear least squares*, in Proceedings 8th IFIP Conference on Optimization Techniques, J. Stoer, ed., vol. 7 of Lecture Notes in Control and Information Science, Springer-Verlag, Berlin, 1978, pp. 22–32.

[226] ———, *A study of the Gauss–Newton algorithm for the solution of nonlinear least squares problems*, in Special Topics of Applied Mathematics, J. Frehse, D. Pallaschke, and U. Trottenberg, eds., North-Holland, Amsterdam, 1980.

[227] P. DEUFLHARD AND W. SAUTTER, *On rank-deficient pseudoinverses*, Linear Algebra Appl., 29 (1980), pp. 91–111.

[228] J. DONGARRA, J. R. BUNCH, C. B. MOLER, AND G. W. STEWART, *LINPACK Users' Guide*, SIAM, Philadelphia, 1979.

[229] J. J. DONGARRA, J. DU CROZ, I. S. DUFF, AND S. HAMMARLING, *A set of level 3 basic linear algebra subprograms*, ACM Trans. Math. Software, 16 (1990), pp. 1–17.

[230] J. J. DONGARRA, J. DU CROZ, S. HAMMARLING, AND R. J. HANSON, *An extended set of Fortran basic linear algebra subprograms*, ACM Trans. Math. Software, 14 (1988), pp. 1–17.

[231] J. J. DONGARRA AND E. GROSSE, *Distribution of mathematical software via electronic mail*, Comm. ACM, 30 (1987), pp. 403–407.

[232] N. R. DRAPER AND H. SMITH, *Applied Regression Analysis*, 2nd ed., John Wiley, New York, 1981.

[233] Z. DRMAČ, *Implementation of Jacobi rotations for accurate singular value computation in floating point arithmetic*, SIAM J. Sci. Comput., 18 (1997), to appear.

[234] I. S. DUFF, *Pivot selection and row orderings in Givens reduction on sparse*

[235] ——, *On permutations to block triangular form*, J. Inst. Maths. Applic., 19 (1977), pp. 339–342.

[236] ——, *A survey of sparse matrix research*, Proceedings of the IEEE, 65 (1977), pp. 500–535.

[237] ——, *On algorithms for obtaining a maximum transversal*, ACM Trans. Math. Software, 7 (1981), pp. 315–330.

[238] ——, *Parallel implementation of multifrontal schemes*, Parallel Computing, 3 (1986), pp. 193–204.

[239] ——, *The solution of augmented systems*, in Numerical Analysis 1993: Proceedings of the 15th Dundee Conference, June 1993, D. F. Griffiths and G. A. Watson, eds., Pitman Research Notes in Mathematics, Longman Scientific and Technical, Harlow, UK, 1994.

[240] I. S. DUFF, A. M. ERISMAN, AND J. K. REID, *Direct Methods for Sparse Matrices*, Oxford University Press, London, 1986.

[241] I. S. DUFF, N. I. M. GOULD, J. K. REID, J. A. SCOTT, AND K. TURNER, *The factorization of sparse symmetric indefinite matrices*, IMA J. Numer. Anal., 11 (1991), pp. 181–204.

[242] I. S. DUFF, R. G. GRIMES, AND J. G. LEWIS, *Sparse matrix test problems*, ACM Trans. Math. Software, 15 (1989), pp. 1–14.

[243] ——, *User's guide for Harwell-Boeing sparse matrix test problems collection*, Tech. Report RAL-92-086, Computing and Information Systems Department, Rutherford Appleton Laboratory, Didcot, UK, 1992.

[244] I. S. DUFF, M. MARRONE, G. RADICATI, AND C. VITTOLI, *A set of level 3 basic linear algebra subprograms for sparse matrices*, Tech. Report RAL-TR-95-049, Computing and Information Systems Department, Rutherford Appleton Laboratory, Didcot, UK, 1995.

[245] I. S. DUFF AND G. A. MEURANT, *The effect of ordering on preconditioned conjugate gradients*, BIT, 29 (1989), pp. 635–657.

[246] I. S. DUFF, N. MUNKSGAARD, H. B. NIELSEN, AND J. K. REID, *Direct solution of sets of linear equations whose matrix is sparse symmetric and indefinite*, J. Inst. Maths. Applic., 23 (1979), pp. 235–250.

[247] I. S. DUFF AND J. K. REID, *A comparison of some methods for the solution of sparse overdetermined systems of linear equations*, J. Inst. Maths. Applic., 17 (1976), pp. 267–280.

[248] ——, *An implementation of Tarjan's algorithm for the block triangularization of a matrix*, ACM Trans. Math. Software, 4 (1978), pp. 137–147.

[249] ——, *Some design features of a sparse matrix code*, ACM Trans. Math. Software, 5 (1979), pp. 18–35.

[250] ——, *MA27—A set of Fortran subroutines for solving sparse symmetric sets of linear equations*, Tech. Report R.10533, AERE, Harwell, Oxfordshire, UK, 1982.

[251] ——, *The multifrontal solution of indefinite sparse symmetric linear systems*, ACM Trans. Math. Software, 9 (1983), pp. 302–325.

[252] ——, *MA47, A Fortran code for direct solution of indefinite sparse symmetric linear systems*, Tech. Report RAL-95-001, Rutherford Appleton Laboratory, Didcot, UK, 1995.

[253] I. S. DUFF AND G. W. STEWART, eds., *Sparse Matrix Proceedings*, SIAM, Philadelphia, 1978.

[254] H. M. DUFOUR, *Résolution des systemes linéaires par la méthode des résidus conjugués*, Bull. Géodésique, 71 (1964), pp. 65–87.

[255] A. L. DULMAGE AND N. S. MENDELSOHN, *Coverings of bipartite graphs*, Canad. J. Math., 10 (1958), pp. 517–534.

[256] ——, *A structure theory of bipartite graphs of finite exterior dimension*, Trans. Roy. Soc. Canada, 53, III (1959), pp. 1–13.

[257] ——, *Two algorithms for bipartite graphs*, J. Soc. Indust. Appl. Math., 11 (1963), pp. 183–194.

[258] P. S. DWYER, *A matrix presentation of least squares and correlation theory with matrix justification of improved methods of solution*, Ann. Math. Statist., 15 (1944), pp. 82–89.

[259] ——, *The square root method and its use in correlation and regression*, J. Amer. Statist. Assoc., 40 (1945), pp. 493–503.

[260] ——, *Linear Computations*, John Wiley, New York, 1951.

[261] C. ECKHART AND G. YOUNG, *The approximation of one matrix by another of lower rank*, Psychometrica, 1 (1936), pp. 211–218.

[262] E. L. EICHHORN AND C. L. LAWSON, *An ALGOL procedure for solution of constrained least squares problems*, Computing Memorandum No. 374, JPL, Pasadena, CA, 1975.

[263] S. C. EISENSTAT, M. C. GURKY, M. H. SCHULTZ, AND A. H. SHERMAN, *Yale sparse matrix package, 1: The symmetric codes.*, Internat. J. Numer. Methods Engrg., 18 (1982), pp. 235–250.

[264] S. C. EISENSTAT AND I. C. F. IPSEN, *Relative perturbation techniques for singular value problems*, Tech. Report YALEU/DCS/RR-942, Yale University, New Haven, CT, 1993.

[265] S. C. EISENSTAT, M. H. SCHULTZ, AND A. H. SHERMAN, *Algorithms and data structures for sparse symmetric Gaussian elimination*, SIAM J. Sci. Statist. Comput., 2 (1981), pp. 225–237.

[266] H. EKBLOM, *A new algorithm for the Huber estimator in linear models*, BIT, 28 (1988), pp. 60–76.

[267] L. ELDÉN, *Algorithms for the regularization of ill-conditioned least squares problems*, BIT, 17 (1977), pp. 134–145.

[268] ——, *Perturbation theory for the least squares problem with linear equality constraints*, SIAM J. Numer. Anal., 17 (1980), pp. 338–350.

[269] ——, *A weighted pseudoinverse, generalized singular values, and constrained least squares problems*, BIT, 22 (1983), pp. 487–502.

[270] ——, *An algorithm for the regularization of ill-conditioned banded least squares problems*, SIAM J. Sci. Statist. Comput., 5 (1984), pp. 237–254.

[271] ——, *An efficient algorithm for the regularization of ill-conditioned least squares problems with a triangular Toeplitz matrix*, SIAM J. Sci. Statist. Comput., 5 (1984), pp. 229–236.

[272] ——, *A note on the computation of the generalized cross-validation function for ill-conditioned least squares problems*, BIT, 24 (1984), pp. 467–472.

[273] ——, *Parallel QR decomposition of a rectangular matrix*, Tech. Report LiTH-MAT-R-78-3, Linköping University, Sweden, 1988.

[274] L. ELDÉN AND H. PARK, *Block downdating of least squares solutions*, SIAM J. Matrix. Anal. Appl., 15 (1994), pp. 1018–1034.

[275] ——, *Perturbation analysis for block downdating of a Cholesky decomposition*, Numer. Math., 68 (1994), pp. 457–467.

[276] L. ELDÉN AND R. SCHREIBER, *A systolic array for the regularization of ill-conditioned least-squares problems with triangular Toeplitz matrix*, Linear Algebra Appl., 77 (1986), pp. 137–147.

[277] L. ELDÉN AND I. SKOGLUND, *Algorithms for the regularization of ill-conditioned least squares problems with tensor product structure, and application to space-invariant image restoration*, Tech. Report LiTH-MAT-R-82-48, Linköping University, Sweden, 1982.

[278] T. ELFVING, *On the conjugate gradient method for solving linear least squares problems*, Tech. Report LiTH-MAT-R-78-3, Linköping University, Sweden, 1978.

[279] ———, *Block-iterative methods for consistent and inconsistent linear equations*, Numer. Math., 35 (1980), pp. 1–12.

[280] A. M. ERISMAN AND W. F. TINNEY, *On computing certain elements of the inverse of a sparse matrix*, Comm. ACM, 18 (1975), pp. 177–179.

[281] D. J. EVANS AND C. LI, *Gauss–Seidel and SOR methods for least squares problems*, Numer. Math., 53 (1988), pp. 485–498.

[282] ———, *Numerical aspects of the generalized cg-method applied to least squares problems*, Computing, 41 (1989), pp. 171–178.

[283] ———, *The theoretical aspects of the gcg-method applied to least-squares problems*, Internat. J. Comput. Math., 35 (1990), pp. 207–229.

[284] D. K. FADDEEV, V. N. KUBLANOVSKAYA, AND V. N. FADDEEVA, *Solution of linear algebraic systems with rectangular matrices*, Proc. Steklov Inst. Math., 96 (1968).

[285] ———, *Sur les systèmes linéaires algébriques de matrices rectangularies et mal-conditionées*, in Programmation en Mathématiques Numériques, Editions Centre Nat. Recherche Sci., Paris VII, 1968, pp. 161–170.

[286] K. FAN AND A. HOFFMAN, *Some metric inequalities in the space of matrices*, Proc. Amer. Math. Soc., 6 (1955), pp. 111–116.

[287] R. W. FAREBROTHER, *The statistical estimation of the standard linear model, 1756–1853*, in Proc. First Internat. Tampere Seminar on Linear Statistical Models Appl. 1983, T. Pukkila and S. Puntanen, eds., University of Tampere, Tampere, Finland, 1985, pp. 77–99.

[288] ———, *Linear Least Squares Computations*, Marcel Dekker, New York, 1988.

[289] D. W. FAUSETT AND C. T. FULTON, *Large least squares problems involving Kronecker products*, SIAM J. Matrix Anal. Appl., 15 (1994), pp. 219–227.

[290] K. V. FERNANDO, *Linear convergence of the row cyclic Jacobi and Kogbetliantz methods*, Numer. Math., 56 (1989), pp. 73–91.

[291] K. V. FERNANDO AND B. N. PARLETT, *Accurate singular values and differential qd algorithms*, Numer. Math., 67 (1994), pp. 191–229.

[292] R. D. FIERRO, *Perturbation analysis for two-sided (or complete) orthogonal decompositions*, Tech. Report PAM 94-02, Department of Mathematics, California State University, San Marcos, CA, 1994.

[293] R. D. FIERRO AND J. R. BUNCH, *Collinearity and total least squares*, SIAM J. Matrix Anal. Appl., 15 (1994), pp. 1167–1181.

[294] ———, *Orthogonal projection and total least squares*, Numer. Linear Algebra Appl., 2 (1995), pp. 135–154.

[295] ———, *Perturbation theory for orthogonal projection methods with applications to least squares and total least squares*, Linear Algebra Appl., 234 (1996), pp. 71–96.

[296] R. FLETCHER, *Generalized inverse methods for the best least squares solution of systems of non-linear equations*, Comput. J., 10 (1968), pp. 392–399.

[297] ———, *Factorizing symmetric indefinite matrices*, Linear Algebra Appl., 14 (1976), pp. 257–272.

[298] ———, *Expected conditioning*, IMA J. Numer. Anal., 5 (1985), pp. 247–273.

[299] ———, *Practical Methods of Optimization*, 2nd ed., John Wiley, New York, 1987.

[300] R. FLETCHER AND C. XU, *Hybrid methods for nonlinear least squares*, IMA J. Numer. Anal., 7 (1987), pp. 371–389.

[301] A. B. FORBES, *Least-squares best-fit geometric elements*, Tech. Report NPL DITC 140/89, National Physical Laboratory, Teddington, UK, 1989.

[302] ———, *Robust circle and sphere fitting by least squares*, Tech. Report NPL DITC 153/89, National Physical Laboratory, Teddington, UK, 1989.

[303] ———, *Least squares best fit geometrical elements*, in Algorithms for Approximation II, J. C. Mason and M. G. Cox, eds., Chapman & Hall, London, 1990, pp. 311–319.

[304] G. E. FORSYTHE, *Generation and use of orthogonal polynomials for data-fitting with a digital computer*, J. Soc. Indust. Appl. Math., 5 (1956), pp. 74–88.

[305] G. E. FORSYTHE AND G. H. GOLUB, *On the stationary values of a second degree polynomial on the unit sphere*, J. Soc. Indust. Appl. Math., 13 (1965), pp. 1050–1068.

[306] G. E. FORSYTHE, M. A. MALCOLM, AND C. B. MOLER, *Computer Methods for Mathematical Computations*, Prentice-Hall, Englewood Cliffs, NJ, 1977.

[307] G. E. FORSYTHE AND C. B. MOLER, *Computer Solution of Linear Algebraic Systems*, Prentice-Hall, Englewood Cliffs, NJ, 1967.

[308] L. V. FOSTER, *Rank and nullspace calculations using matrix decompositions without column interchanges*, Linear Algebra Appl., 74 (1986), pp. 47–71.

[309] ———, *The probability of large diagonal elements in the QR factorization*, SIAM J. Sci. Statist. Comput., 11 (1990), pp. 531–544.

[310] ———, *Modifications of the normal equations method that are numerically stable*, in Numerical Linear Algebra, Digital Signal Processing and Parallel Algorithms, G. H. Golub and P. Van Dooren, eds., NATO ASI Series, Springer-Verlag, Berlin, 1991, pp. 501–512.

[311] R. FOURER AND S. MEHROTRA, *Solving symmetric indefinite systems in an interior point method for linear programming*, Tech. Report 92-01, Department of Industrial Engineering and Management Science, Northwestern University, Evanston, IL, 1992.

[312] C. FRALEY, *Algorithms for nonlinear least squares*, Tech. Report STAN-CLaSSiC-88-22, Center for Large Scale Scientific Computation, Stanford University, CA, 1988.

[313] ———, *Computational behavior of Gauss–Newton methods*, SIAM J. Sci. Statist. Comput., 10 (1989), pp. 515–532.

[314] R. FREUND, *A note on two block SOR methods for sparse least squares problems*, Linear Algebra Appl., 88/89 (1987), pp. 211–221.

[315] R. W. FREUND, G. H. GOLUB, AND N. NACHTIGAL, *Iterative solution of linear systems*, Acta Numerica, 1 (1991), pp. 57–100.

[316] W. GANDER, *Algorithms for the QR-decomposition*, Tech. Report 80–02, Angewandte Mathematik, ETH, Zürich, Switzerland, 1980.

[317] ———, *Least squares with a quadratic constraint*, Numer. Math., 36 (1981), pp. 291–307.

[318] W. GANDER, G. H. GOLUB, AND R. STREBEL, *Least-squares fitting of circles and ellipses*, BIT, 34 (1994), pp. 558–578.

[319] W. GANDER AND U. VON MATT, *Some least squares problems*, in Solving Problems in Scientific Computing Using Maple and MATLAB, W. Gander and J. Hřbiček, eds., Springer-Verlag, Berlin, 1993, pp. 69–87.

[320] C. F. GAUSS, *Theory of the Motion of the Heavenly Bodies Moving about the Sun in Conic Sections*, C. H. Davis, Trans., Dover, New York, 1963. First published in 1809.

[321] ——, *Disquisitio de elementis ellipticis Palladis*, in Werke, VI, Königlichen Gesellschaft der Wissenschaften zu Göttingen, 1880, pp. 1–24. First published in 1810.

[322] ——, *Theoria combinationis observationum erroribus minimis obnoxiae, pars prior*, in Werke, IV, Königlichen Gesellschaft der Wissenschaften zu Göttingen, 1880, pp. 1–26. First published in 1821.

[323] ——, *Theoria combinationis observationum erroribus minimis obnoxiae, pars posterior*, in Werke, IV, Königlichen Gesellschaft der Wissenschaften zu Göttingen, 1880, pp. 27–53. First published in 1823.

[324] ——, *Theoria combinationis observationum erroribus minimis obnoxiae, supplementum*, in Werke, IV, Königlichen Gesellschaft der Wissenschaften zu Göttingen, 1880, pp. 55–93. First published in 1826.

[325] ——, *Theory of the Combination of Observations Least Subject to Errors. Part 1, Part 2, Supplement*, G. W. Stewart, Trans., SIAM, Philadelphia, 1995.

[326] D. M. GAY, *Algorithm 611. Subroutines for unconstrained minimization using a model/trust-region approach*, ACM Trans. Math. Software, 9 (1983), pp. 503–524.

[327] ——, *A trust-region approach to linearly constrained optimization*, in Proceedings of the 1983 Dundee Conference on Numerical Analysis, D. Griffiths, ed., vol. 1066 of Lecture Notes in Mathematics, Springer-Verlag, Berlin, 1984, pp. 72–105.

[328] ——, *Interval least squares—a diagnostic tool*, in Reliability in Computing, R. E. Moore, ed., Academic Press, London, 1988, pp. 183–206.

[329] W. M. GENTLEMAN, *Least squares computations by Givens transformations without square roots*, J. Inst. Maths. Applic., 12 (1973), pp. 329–336.

[330] ——, *Error analysis of QR decompositions by Givens transformations*, Linear Algebra Appl., 10 (1975), pp. 189–197.

[331] ——, *Row elimination for solving sparse linear systems and least squares problems*, in Proceedings of the Dundee Conference on Numerical Analysis 1975, G. A. Watson, ed., vol. 506 of Lecture Notes in Mathematics, Springer-Verlag, Berlin, 1976, pp. 122–133.

[332] J. A. GEORGE, *Computer Implementation of the Finite-Element Method*, Ph. D. thesis, Stanford University, CA, 1971.

[333] J. A. GEORGE AND M. T. HEATH, *Solution of sparse linear least squares problems using Givens rotations*, Linear Algebra Appl., 34 (1980), pp. 69–83.

[334] J. A. GEORGE, M. T. HEATH, AND E. G. NG, *A comparison of some methods for solving sparse linear least squares problems*, SIAM J. Sci. Statist. Comput., 4 (1983), pp. 177–187.

[335] J. A. GEORGE, M. T. HEATH, AND R. J. PLEMMONS, *Solution of large-scale sparse least squares problems using auxiliary storage*, SIAM J. Sci. Statist. Comput., 2 (1981), pp. 416–429.

[336] J. A. GEORGE AND J. W.-H. LIU, *Computer Solution of Large Sparse Positive Definite Systems*, Prentice-Hall, Englewood Cliffs, NJ, 1981.

[337] ——, *Householder reflections versus Givens rotations in sparse orthogonal decomposition*, Linear Algebra Appl., 88/89 (1987), pp. 223–238.

[338] ——, *The evolution of the minimum degree ordering algorithm*, SIAM Review, 31 (1989), pp. 1–19.

[339] J. A. GEORGE, J. W.-H. LIU, AND E. G. NG, *Row ordering schemes for sparse Givens transformations, I. Bipartite graph model*, Linear Algebra Appl., 61 (1984), pp. 55–81.

[340] ——, *Row ordering schemes for sparse Givens transformations, III. Analysis for a model problem*, Linear Algebra Appl., 75 (1984), pp. 225–240.

[341] ——, *Row ordering schemes for sparse Givens transformations, II. Implicit graph model*, Linear Algebra Appl., 75 (1986), pp. 203–223.

[342] ——, *A data structure for sparse QR and LU factorization*, SIAM J. Sci. Statist. Comput., 9 (1988), pp. 100–121.

[343] J. A. GEORGE AND E. G. NG, *On row and column orderings for sparse least squares problems*, SIAM J. Numer. Anal., 20 (1983), pp. 326–344.

[344] ——, *SPARSPAK: Waterloo sparse matrix package user's guide for SPARSPAK-B*, Res. Report CS-84-37, Department of Computer Science, University of Waterloo, Canada, 1984.

[345] ——, *An implementation of Gaussian elimination with partial pivoting for sparse systems*, SIAM J. Sci. Statist. Comput., 6 (1985), pp. 390–409.

[346] ——, *Symbolic factorization for sparse Gaussian elimination with partial pivoting*, SIAM J. Sci. Statist. Comput., 8 (1987), pp. 877–898.

[347] ——, *On the complexity of sparse QR and LU factorization of finite-element matrices*, SIAM J. Sci. Statist. Comput., 9 (1988), pp. 849–861.

[348] J. A. GEORGE, W. G. POOLE, AND R. G. VOIGT, *Incomplete nested dissection for solving n by n grid problems*, SIAM J. Numer. Anal., 15 (1978), pp. 90–112.

[349] J. R. GILBERT, *Graph Separator Theorems and Sparse Gaussian Elimination*, Ph. D. thesis, Stanford University, CA, 1980.

[350] ——, *Predicting structure in sparse matrix computation*, SIAM J. Matrix. Anal. Appl., 15 (1994), pp. 62–79.

[351] J. R. GILBERT, C. MOLER, AND R. SCHREIBER, *Sparse matrices in MATLAB: Design and implementation*, SIAM J. Matrix. Anal. Appl., 13 (1992), pp. 333–356.

[352] J. R. GILBERT, E. G. NG, AND B. W. PEYTON, *Separators and structure prediction in sparse orthogonal factorization*, Tech. Report CSL-93-15, Xerox Corporation, Palo Alto Research Center, 1993.

[353] J. R. GILBERT AND T. PEIERLS, *Sparse partial pivoting in time proportional to arithmetic operations*, SIAM J. Sci. Statist. Comput., 9 (1988), pp. 862–874.

[354] P. E. GILL, G. H. GOLUB, W. MURRAY, AND M. SAUNDERS, *Methods for modifying matrix factorizations*, Math. Comp, 28 (1974), pp. 505–535.

[355] P. E. GILL, S. J. HAMMARLING, W. MURRAY, M. A. SAUNDERS, AND M. H. WRIGHT, *User's guide for LSSOL (version 1.0): a Fortran package for constrained linear least-squares and convex quadratic programming*, Report SOL, Department of Operations Research, Stanford University, CA, 1986.

[356] P. E. GILL AND W. MURRAY, *Nonlinear least squares and nonlinearly constrained optimization*, in Proceedings Dundee Conference on Numerical Analysis 1975, G. A. Watson, ed., vol. 506 of Lecture Notes in Mathematics, Springer-Verlag, Berlin, 1976, pp. 135–147.

[357] ——, *The orthogonal factorization of a large sparse matrix*, in Sparse Matrix Computations, J. R. Bunch and D. J. Rose, eds., Academic Press, New York, 1976, pp. 201–212.

[358] ——, *Algorithms for the solution of the nonlinear least squares problem*, SIAM J. Numer. Anal., 15 (1978), pp. 977–992.

[359] P. E. GILL, W. MURRAY, M. A. SAUNDERS, AND M. H. WRIGHT, *A Schur-complement method for sparse quadratic programming*, in Reliable Numerical Computation, M. G. Cox and S. Hammarling, eds., Clarendon Press, Oxford, UK, 1990, pp. 113–138.

[360] P. E. GILL, W. MURRAY, AND M. H. WRIGHT, *Practical Optimization*, Academic Press, London and New York, 1981.

[361] W. GIVENS, *Computation of plane unitary rotations transforming a general matrix*

to triangular form, SIAM J. Appl. Math., 6 (1958), pp. 26–50.

[362] J. GLUCHOWSKA AND A. SMOKTUNOWICZ, *Solving the linear least squares problem with very high relative accuracy*, Computing, 45 (1990), pp. 345–354.

[363] H. H. GOLDSTINE, *A History of Numerical Analysis from the 16th through the 19th Century*, Springer-Verlag, New York, 1977.

[364] G. H. GOLUB, *Numerical methods for solving least squares problems*, Numer. Math., 7 (1965), pp. 206–216.

[365] ———, *Least squares, singular values and matrix approximations*, Aplikace Matematiky, 13 (1968), pp. 44–51.

[366] ———, *Matrix decompositions and statistical computation*, in Statistical Computation, R. Milton and J. A. Nelder, eds., Academic Press, New York, 1969, pp. 365–397.

[367] ———, *Some modified matrix eigenvalue problems*, SIAM Review, 15 (1973), pp. 318–344.

[368] G. H. GOLUB, M. T. HEATH, AND G. WAHBA, *Generalized cross-validation as a method for choosing a good ridge parameter*, Technometrics, 21 (1979), pp. 215–223.

[369] G. H. GOLUB, A. HOFFMAN, AND G. W. STEWART, *A generalization of the Eckhard-Young-Mirsky matrix approximation theorem*, Linear Algebra Appl., 88/89 (1987), pp. 317–327.

[370] G. H. GOLUB AND W. KAHAN, *Calculating the singular values and pseudo-inverse of a matrix*, SIAM J. Numer. Anal. Ser. B, 2 (1965), pp. 205–224.

[371] G. H. GOLUB, V. KLEMA, AND G. W. STEWART, *Rank degeneracy and least squares problems*, Tech. Report STAN-CS-76-559, August 1976, Computer Science Department, Stanford University, CA, 1976.

[372] G. H. GOLUB AND R. J. LEVEQUE, *Extensions and uses of the variable projection algorithm for solving nonlinear least squares problems*, in Proceedings of the 1979 Army Numerical Analysis and Computers Conf., White Sands Missile Range, White Sands, NM, ARO Report 79-3, 1979, pp. 1–12.

[373] G. H. GOLUB AND F. T. LUK, *Singular value decomposition: Applications and computations*, in Trans. 22nd Conference of Army Mathematicians, ARO Report 77-1, 1977, pp. 577–605.

[374] G. H. GOLUB, F. T. LUK, AND M. L. OVERTON, *A block Lanczos method for computing the singular values and corresponding singular vectors of a matrix*, ACM Trans. Math. Software, 7 (1981), pp. 149–169.

[375] G. H. GOLUB, F. T. LUK, AND M. PAGANO, *A sparse least squares problem in photogrammetry*, in Proceedings of the Computer Science and Statistics 12th Annual Symposium on the Interface, J. F. Gentleman, ed., University of Waterloo, Canada, 1979, pp. 26–30.

[376] G. H. GOLUB, P. MANNEBACK, AND P. TOINT, *A comparison between some direct and iterative methods for large scale geodetic least squares problems*, SIAM J. Sci. Statist. Comput., 7 (1986), pp. 799–816.

[377] G. H. GOLUB AND S. G. NASH, *Nonorthogonal analysis of variance using a generalized conjugate-gradient algorithm*, J. Amer. Statist. Assoc., 77 (1982), pp. 109–116.

[378] G. H. GOLUB AND V. PEREYRA, *The differentiation of pseudoinverses and nonlinear least squares problems whose variables separate*, SIAM J. Numer. Anal., 10 (1973), pp. 413–432.

[379] ———, *Differentiation of pseudoinverses, separable nonlinear least squares problems and other tales*, in Generalized Inverses and Applications, M. Z. Nashed, ed.,

Academic Press, New York, 1976, pp. 303–324.

[380] G. H. GOLUB AND R. J. PLEMMONS, *Large-scale geodetic least-squares adjustment by dissection and orthogonal decomposition*, Linear Algebra Appl., 34 (1980), pp. 3–28.

[381] G. H. GOLUB, R. J. PLEMMONS, AND A. SAMEH, *Parallel block schemes for large-scale least-squares computations*, in High-Speed Computing, Scientific Applications and Algorithm Design, University of Illinois Press, 1988, pp. 171–179.

[382] G. H. GOLUB AND C. REINSCH, *Singular value decomposition and least squares solution*, Numer. Math., 14 (1970), pp. 403–420.

[383] G. H. GOLUB AND M. A. SAUNDERS, *Linear least squares and quadratic programming*, in Integer and Nonlinear Programming, J. Abadie, ed., North-Holland, Amsterdam, 1970, pp. 229–256.

[384] G. H. GOLUB, K. SOLNA, AND P. VAN DOOREN, *A QR-like SVD algorithm for a product/quotient of several matrices*, in SVD and Signal Processing, III: Algorithms, Architectures and Applications, M. Moonen and B. De Moor, eds., Elsevier Science B.V., Amsterdam, 1995, pp. 139–147.

[385] G. H. GOLUB AND G. P. STYAN, *Numerical computations for univariate linear models*, J. Statist. Comput. Simul., 2 (1973), pp. 253–274.

[386] G. H. GOLUB AND R. UNDERWOOD, *Stationary values of the ratio of quadratic forms subject to linear constraints*, Tech. Report CS-142, Computer Science Department, Stanford University, CA, 1969.

[387] G. H. GOLUB AND C. F. VAN LOAN, *Total least squares*, in Smoothing Techniques for Curve Estimation, T. Gasser and M. Rosenblatt, eds., Springer-Verlag, New York, 1979, pp. 69–76.

[388] ———, *An analysis of the total least squares problem*, SIAM J. Numer. Anal., 17 (1980), pp. 883–893.

[389] ———, *Matrix Computations*, 2nd ed., Johns Hopkins University Press, Baltimore, 1989.

[390] G. H. GOLUB AND R. S. VARGA, *Chebyshev semi-iterative methods, successive overrelaxation iterative methods and second order Richardson iterative methods, Parts I and II*, Numer. Math., 3 (1961), pp. 147–168.

[391] G. H. GOLUB AND U. VON MATT, *Quadratically constrained least squares and quadratic problems*, Numer. Math., 59 (1991), pp. 561–580.

[392] G. H. GOLUB AND J. H. WILKINSON, *Iterative refinement of least squares solutions*, in Proceedings of the IFIP Congress 65, New York, 1965, W. A. Kalenich, ed., Spartan Books, Washington, 1965, pp. 606–607.

[393] ———, *Note on the iterative refinement of least squares solution*, Numer. Math., 9 (1966), pp. 139–148.

[394] G. H. GOLUB AND H. ZHA, *Perturbation analysis of the canonical correlation of matrix pairs*, Linear Algebra Appl., 210 (1994), pp. 3–28.

[395] W. H. GRAGG AND W. J. HARROD, *The numerically stable reconstruction of Jacobi matrices from spectral data*, Numer. Math., 44 (1984), pp. 317–335.

[396] J. P. GRAM, *Über die Entwickelung reeller Funktionen in Reihen mittelst der Methode der kleinsten Quadrate*, J. Reine Angew. Math., 94 (1883), pp. 41–73.

[397] A. GREENBAUM, *Behavior of slightly perturbed Lanczos and conjugate-gradient recurrences*, Linear Algebra Appl., 113 (1989), pp. 7–63.

[398] A. GREENBAUM AND Z. STRAKOS, *Predicting the behavior of finite precision Lanczos and conjugate gradient computations*, SIAM J. Matrix Anal. Appl., 13 (1992), pp. 121–137.

[399] T. E. GREVILLE, *Note on the generalized inverse of a matrix product*, SIAM Review, 8 (1966), pp. 518–521.

[400] P. GRIFFITHS AND I. D. HILL, *Applied Statistical Algorithms*, Ellis Horwood Ltd., Chichester, UK, 1985.

[401] R. G. GRIMES AND J. G. LEWIS, *Condition number estimation for sparse matrices*, SIAM J. Sci. Statist. Comput., 2 (1981), pp. 384–388.

[402] E. GROSSE, *Tensor spline approximations*, Linear Algebra Appl., 34 (1980), pp. 29–41.

[403] M. GU AND S. C. EISENSTAT, *A divide-and-conquer algorithm for the bidiagonal SVD*, Tech. Report YALEU/DCS/RR-933, Department of Computer Science, Yale University, New Haven, CT, 1992.

[404] ———, *A stable and fast algorithm for updating the singular value decomposition*, Tech. Report YALEU/DCS/RR-966, Department of Computer Science, Yale University, New Haven, CT, 1993.

[405] ———, *An efficient algorithm for computing a strong rank-revealing QR decomposition*, Tech. Report YALEU/DCS/RR-967, Department of Computer Science, Yale University, New Haven, CT, 1994.

[406] ———, *A divide-and-conquer algorithm for the bidiagonal SVD*, SIAM J. Matrix. Anal. Appl., 16 (1995), pp. 79–92.

[407] ———, *Downdating the singular value decomposition*, SIAM J. Matrix. Anal. Appl., 16 (1995), pp. 793–810.

[408] M. GULLIKSSON, *Algorithms for Overdetermined Systems of Equations*, Ph. D. thesis, Institute of Information Processing, University of Umeå, Umeå, Sweden, 1993.

[409] ———, *Iterative refinement for constrained and weighted linear least squares*, BIT, 34 (1994), pp. 239–253.

[410] ———, *Backward error analysis for the constrained and weighted linear least squares problem when using the weighted QR decomposition*, SIAM J. Matrix. Anal. Appl., 16 (1995), pp. 675–687.

[411] ———, *On modified Gram-Schmidt for weighted and constrained linear least squares*, BIT, 35 (1995), pp. 458–473.

[412] M. GULLIKSSON, I. SÖDERQUIST, AND P.-Å. WEDIN, *Algorithms for weighted and constrained nonlinear least squares using the modified QR decomposition*, Tech. Report UMINF-93-05, Institute of Information Processing, University of Umeå, Umeå, Sweden, 1993.

[413] M. GULLIKSSON AND P.-Å. WEDIN, *Modifying the QR decomposition to constrained and weighted linear least squares*, SIAM J. Matrix. Anal. Appl., 13:4 (1992), pp. 1298–1313.

[414] ———, *Numerical aspects on algorithms for overdetermined linear systems in l_p norm*, Tech. Report UMINF-93-11, Institute of Information Processing, University of Umeå, Umeå, Sweden, 1993.

[415] F. G. GUSTAVSON, *Finding the block lower triangular form of a matrix*, in Sparse Matrix Computations, J. R. Bunch and D. J. Rose, eds., Academic Press, New York, 1976, pp. 275–289.

[416] G. D. HACHTEL, *Extended applications of the sparse tableau approach—finite elements and least squares*, in Basic questions in design theory, W. Spillers, ed., North-Holland, Amsterdam, 1974.

[417] ———, *The sparse tableau approach to finite element assembly*, in Sparse Matrix Computations, J. Bunch and D. J. Rose, eds., Academic Press, New York, 1976.

[418] L. A. HAGEMAN, F. T. LUK, AND D. M. YOUNG, *On the equivalence of certain*

iterative acceleration methods, SIAM J. Numer. Anal., 17 (1980), pp. 852–873.

[419] L. A. HAGEMAN AND D. M. YOUNG, *Applied Iterative Methods*, Academic Press, New York, London, 1981.

[420] W. W. HAGER, *Condition estimates*, SIAM J. Sci. Statist. Comput., 5 (1984), pp. 311–316.

[421] ———, *Updating the inverse of a matrix*, SIAM Review, 31 (1989), pp. 221–239.

[422] S. HAMMARLING, *A note on modifications to the Givens plane rotation*, J. Inst. Maths. Applic., 13 (1974), pp. 215–218.

[423] ———, *The numerical solution of the general Gauss-Markov linear model*, NAG Tech. Report TR2/85, Numerical Algorithms Group Ltd., Oxford, UK, 1985.

[424] C. G. HAN, P. M. PARDALOS, AND Y. YE, *Computational aspects of an interior point algorithm for quadratic programming problems with box constraints*, in Large Scale Numerical Optimization, Proceedings, Cornell University, Ithaca, October 19–20, 1989, T. F. Coleman and Y. Li, eds., SIAM, Philadelphia, 1990, pp. 92–112.

[425] M. HANKE, *Accelerated Landweber iterations for the solution of ill-posed equations*, Numer. Math., 60 (1991), pp. 341–373.

[426] ———, *Conjugate Gradient Type Methods for Ill-posed Problems*, Pitman Research Notes in Mathematics, Longman Scientific and Technical, Harlow, UK, 1995.

[427] M. HANKE AND P. C. HANSEN, *Regularization methods for large-scale problems*, Surveys Math. Indust., 3 (1994), pp. 253–315.

[428] M. HANKE AND M. NEUMANN, *The geometry of the set of scaled projections*, Linear Algebra Appl., 190 (1992), pp. 137–148.

[429] P. C. HANSEN, *The truncated SVD as a method for regularization*, BIT, 27 (1987), pp. 534–553.

[430] ———, *Relations between SVD and GSVD of discrete regularization problems in standard and general form*, Linear Algebra Appl., 141 (1990), pp. 165–176.

[431] ———, *Truncated singular value decomposition solutions to discrete ill-posed problems with ill-determined numerical rank*, SIAM J. Sci. Statist. Comput., 11 (1990), pp. 503–518.

[432] ———, *Analysis of ill-posed problems by means of the L-curve*, SIAM Review, 34 (1992), pp. 561–580.

[433] ———, *Rank-Deficient and Discrete Ill-Posed Problems*, Doctoral thesis, UNI•C, Technical University of Denmark, Lyngby, Denmark, 1995.

[434] P. C. HANSEN AND H. GESMAR, *Fast orthogonal decomposition of rank deficient toeplitz matrices*, Numer. Algorithms, 4 (1993), pp. 151–166.

[435] P. C. HANSEN AND D. O'LEARY, *The use of the L-curve in the regularization of discrete ill-posed problems*, SIAM J. Sci. Comput., 14 (1993), pp. 1487–1503.

[436] R. J. HANSON, *Linear least squares with bounds and linear constraints*, SIAM J. Sci. Statist. Comput., 7 (1986), pp. 826–834.

[437] R. J. HANSON AND C. L. LAWSON, *Extensions and applications of the Householder algorithm for solving linear least squares problems*, Math. Comp., 23 (1969), pp. 787–812.

[438] R. J. HANSON AND J. L. PHILLIPS, *An adaptive numerical method for solving linear Fredholm equations of the first kind*, Numer. Math., 24 (1975), pp. 291–307.

[439] D. R. G. HARE, C. R. JOHNSON, D. D. OLESKY, AND P. V. D. DRIESSCHE, *Sparsity analysis of the QR factorization*, SIAM J. Matrix. Anal. Appl., 14 (1993), pp. 655–669.

[440] K. H. HASKELL AND R. J. HANSON, *Selected algorithms for the linearly constrained least squares problem: A user's guide*, Tech. Report SAND78-1290, Sandia National Laboratories, Albuquerque, NM, 1979.

[441] ———, *An algorithm for linear least squares problems with equality and nonnegativity constraints*, Math Programming, 21 (1981), pp. 98–118.

[442] M. T. HEATH, *Some extensions of an algorithm for sparse linear least squares problems*, SIAM J. Sci. Statist. Comput., 3 (1982), pp. 223–237.

[443] ———, *Numerical methods for large sparse linear least squares problems*, SIAM J. Sci. Statist. Comput., 5 (1984), pp. 497–513.

[444] M. T. HEATH, A. J. LAUB, C. C. PAIGE, AND R. C. WARD, *Computing the SVD of a product of two matrices*, SIAM J. Sci. Statist. Comput., 7 (1986), pp. 1147–1149.

[445] F. R. HELMERT, *Die Mathematischen und Physikalischen Theorien der höheren Geodäsie*, 1 *Teil*, Teubner, Leipzig, 1880.

[446] H. V. HENDERSON AND S. R. SEARLE, *The vec-permutation matrix, the vec operator and Kronecker products: A review*, Linear and Multilinear Algebra, 9 (1981), pp. 271–188.

[447] P. HENRICI, *Fast Fourier methods in computational complex analysis*, SIAM Review, 21 (1979), pp. 481–527.

[448] G. T. HERMAN, A. LENT, AND S. W. ROWLAND, *ART: Mathematics and applications*, J. Theoret. Biol., 42 (1973), pp. 1–32.

[449] M. R. HESTENES, *Inversion of matrices by biorthogonalization and related results*, J. Soc. Indust. Appl. Math., 6 (1958), pp. 51–90.

[450] ———, *Pseudoinverses and conjugate gradients*, Comm. ACM, 18 (1975), pp. 40–43.

[451] M. R. HESTENES AND E. STIEFEL, *Methods of conjugate gradients for solving linear system*, J. Res. Nat. Bur. Standards., B49 (1952), pp. 409–436.

[452] K. L. HIEBERT, *An evaluation of mathematical software that solves nonlinear least squares problems*, ACM Trans. Math. Software, 7 (1986), pp. 1–16.

[453] N. J. HIGHAM, *Computing the polar decomposition–with applications*, SIAM J. Sci. Statist. Comput., 7 (1986), pp. 1160–1174.

[454] ———, *Error analysis of the Björck-Pereyra algorithms for solving Vandermonde systems*, Numer. Math., 50 (1987), pp. 613–632.

[455] ———, *A survey of condition number estimation for triangular matrices*, SIAM Review, 29 (1987), pp. 575–596.

[456] ———, *Computing a nearest symmetric positive semidefinite matrix*, Linear Algebra Appl., 103 (1988), pp. 103–118.

[457] ———, *Fast solution of Vandermonde-like systems involving orthogonal polynomials*, IMA J. Numer. Anal., 8 (1988), pp. 473–486.

[458] ———, *Fortran codes for estimating the one-norm of a real or complex matrix, with applications to condition estimation*, ACM Trans. Math. Software, 14 (1988), pp. 381–396.

[459] ———, *The accuracy of solutions to triangular systems*, SIAM J. Numer. Anal., 26(5) (1989), pp. 1252–1265.

[460] ———, *Analysis of the Cholesky decomposition of a semi-definite matrix*, in Reliable Numerical Computation, M. G. Cox and S. J. Hammarling, eds., Clarendon Press, Oxford, 1990, pp. 161–185.

[461] ———, *Computing error bounds for regression problems*, in Contemporary Mathematics 112: Statistical Analysis of Measurement Error Models and Applications, P. J. Brown and W. A. Fuller, eds., American Mathematical Society, Providence, RI, 1990, pp. 195–210.

[462] ———, *Experience with a matrix norm estimator*, SIAM J. Sci. Statist. Comput., 11 (1990), pp. 804–809.

[463] ———, *How accurate is Gaussian elimination?*, in Numerical Analysis 1989: Proceedings of the 13th Dundee Conference, D. F. Griffiths and G. A. Watson, eds., Pitman Research Notes in Mathematics 228, Longman Scientific and Technical, Harlow, UK, 1990, pp. 137–154.

[464] ———, *Iterative refinement enhances the stability of QR factorization methods for solving linear equations*, BIT, 31 (1991), pp. 447–468.

[465] ———, *Estimating the matrix p-norm*, Numer. Math., 62 (1992), pp. 539–556.

[466] ———, *A survey of componentwise perturbation theory in numerical linear algebra*, in Mathematics of Computation 1943–1993. A Half Century of Computational Mathematics, W. Gautschi, ed., vol. 48 of Proceedings of Symposia in Applied Mathematics, American Mathematical Society, Providence, RI, 1994, pp. 49–77.

[467] ———, *Accuracy and Stability of Numerical Algorithms*, SIAM, Philadelphia, 1995.

[468] ———, *Stability of the diagonal pivoting method with partial pivoting*, Tech. Report 265, July 1995, Department of Mathematics, University of Manchester, UK, 1995.

[469] N. J. HIGHAM AND G. W. STEWART, *Numerical linear algebra in statistical computing*, in The State of the Art in Numerical Analysis, A. Iserles and M. J. D. Powell, eds., Oxford University Press, London, UK, 1987, pp. 41–57.

[470] W. HOFFMAN, *Basic Transformations in Linear Algebra for Vector Computing*, Ph. D. thesis, University of Amsterdam, the Netherlands, 1989.

[471] ———, *Iterative algorithms for Gram-Schmidt orthogonalization*, Computing, 41 (1989), pp. 335–348.

[472] J. N. HOLT AND R. FLETCHER, *An algorithm for constrained non-linear least-squares*, J. Inst. Maths. Applic., 23 (1979), pp. 449–463.

[473] H. P. HONG AND C. T. PAN, *Rank-revealing QR factorization and SVD*, Math. Comp., 58 (1992), pp. 213–232.

[474] P. D. HOUGH AND S. A. VAVASIS, *Complete orthogonal decomposition for weighted least squares*, Tech. Report CTC93TR203, 12/94, Advanced Computing Research Institute, Cornell University, Ithaca, NY, 1994.

[475] A. S. HOUSEHOLDER, *Unitary triangularization of a nonsymmetric matrix*, J. Assoc. Comput. Mach., 5 (1958), pp. 339–342.

[476] ———, *The Theory of Matrices in Numerical Analysis*, Dover, New York, 1974.

[477] A. S. HOUSEHOLDER AND F. L. BAUER, *On certain iterative methods for solving linear systems*, Numer. Math., 2 (1960), pp. 55–59.

[478] P. J. HUBER, *Robust Statistics*, John Wiley, New York, 1981.

[479] M. F. HUTCHINSON AND F. R. DE HOOG, *Smoothing noisy data with spline functions*, Numer. Math., 47 (1985), pp. 99–106.

[480] T.-M. HWANG, W.-W. LIN, AND D. L. PIERCE, *An alternate column selection criterion for a rank revealing QR factorization*, Tech. Report BCSTECH-93-021, Boeing Computer Services, Research and Technology, Seattle, WA, 1993.

[481] T.-M. HWANG, W.-W. LIN, AND E. K. YANG, *Rank revealing LU factorizations*, Linear Algebra Appl., 175 (1992), pp. 115–141.

[482] H. D. IKRAMOV, *Sparse linear least-squares problems*, in Advances in Sciences and Technology: Mathematical Analysis, R. V. Gamkredlidze, ed., vol. 23, Academy of Sciences, Moscow, 1985, pp. 219–285. In Russian.

[483] C. J. G. JACOBI, *Über eine neue Auflösungsart der bei der Methode der kleinsten Quadrate vorkommenden linearen Gleichungen*, Astronomische Nachrichten, 22 (1845), pp. 297–306.

[484] W. JALBY AND B. PHILIPPE, *Stability analysis and improvement of the block Gram-Schmidt algorithm*, SIAM J. Sci. Statist. Comput., 12 (1991), pp. 1058–1073.

[485] D. JAMES, *Implicit nullspace iterative methods for constrained least squares problems*, SIAM J. Matrix. Anal. Appl., 13 (1992), pp. 962–978.

[486] M. JANKOWSKI AND H. WOZNIAKOWSKI, *Iterative refinement implies numerical stability*, BIT, 17 (1977), pp. 303–311.

[487] A. JENNINGS AND M. A. AJIZ, *Incomplete methods for solving $A^T A x = b$*, SIAM J. Sci. Statist. Comput., 5 (1984), pp. 978–987.

[488] L. S. JENNINGS AND M. R. OSBORNE, *A direct error analysis for least squares*, Numer. Math., 22 (1974), pp. 322–332.

[489] E. JESSUP AND D. C. SORENSON, *A parallel algorithm for computing the singular value decomposition of a matrix*, Tech. Report CU-CS-623-92, Department of Computer Science, University of Colorado, Boulder, CO, 1992.

[490] X.-Q. JIN, *A preconditioner for constrained and weighted least squares problems with Toeplitz structure*, BIT, 36 (1996), pp. 101–109.

[491] D. M. JOHNSON, A. L. DULMAGE, AND N. S. MENDELSOHN, *Connectivity and reducibility of graphs*, Canad. J. Math., 14 (1963), pp. 529–539.

[492] T. L. JORDAN, *Experiments on error growth associated with some linear least-squares procedures*, Math. Comp., 22 (1968), pp. 579–588.

[493] S. KACZMARZ, *Angenäherte Auflösung von Systemen linearer Gleichungen*, Bulletin Internat. Acad. Polon. Sciences et Lettres, (1937), pp. 355–357.

[494] W. KAHAN, *Accurate eigenvalues of a symmetric tridiagonal matrix*, Tech. Report CS-41, Computer Science Department, Stanford University, CA, 1966. Revised June 1968.

[495] ———, *Numerical linear algebra*, Canad. Math. Bull., 9 (1966), pp. 757–801.

[496] C. KAMATH AND A. SAMEH, *A projection method for solving nonsymmetric linear systems on multiprocessors*, Parallel Computing, 9 (1988/89), pp. 291–312.

[497] I. KARASALO, *A criterion for truncation of the QR decomposition algorithm for the singular linear least squares problem*, BIT, 14 (1974), pp. 156–166.

[498] L. KAUFMAN, *Variable projection methods for solving separable nonlinear least squares problems*, BIT, 15 (1975), pp. 49–57.

[499] ———, *Application of dense Householder transformation to a sparse matrix*, ACM Trans. Math. Software, 5 (1979), pp. 442–450.

[500] ———, *The generalized Householder transformation and sparse matrices*, Linear Algebra Appl., 90 (1987), pp. 221–234.

[501] L. KAUFMAN AND V. PEREYRA, *A method for separable nonlinear least squares problems with separable nonlinear equality constraints*, SIAM J. Numer. Anal., 15 (1978), pp. 12–20.

[502] L. KAUFMAN AND G. SYLVESTER, *Separable nonlinear least squares problems with multiple right hand sides*, SIAM J. Matrix Anal. Appl., 13 (1992), pp. 68–89.

[503] J. KELLER, *On the solution of singular and semidefinite linear systems by iterations*, J. Soc. Indust. Appl. Math. Ser. B, 2 (1965), pp. 281–290.

[504] A. KIEŁBASIŃSKI AND J. JANKOWSKA, *Fehleranalyse der Schmidtschen und Powellschen Orthonormalisierungsverfahren*, ZAMM, 54 (1974), p. 223.

[505] A. KIEŁBASIŃSKI AND H. SCHWETLICK, *Numerische Lineare Algebra*, VEB Deutscher Verlag der Wissenschaften, Berlin, 1988.

[506] E. G. KOGBETLIANTZ, *Solution of linear equations by diagonalization of coefficients matrix*, Quart. Appl. Math., 13 (1955), pp. 123–132.

[507] G. B. KOLATA, *Geodesy: Dealing with an enormous computer task*, Science, 200 (1978), pp. 421–422.

[508] S. KOUROUKLIS AND C. C. PAIGE, *A constrained approach to the general Gauss-Markov linear model*, J. Amer. Statist. Assoc., 76 (1981), pp. 620–625.

[509] Z. V. KOVARIK, *Some iterative methods for improving orthonormality*, SIAM J. Numer. Anal., 7 (1970), pp. 386–389.

[510] F. T. KROGH, *Efficient implementation of a variable projection algorithm for nonlinear least squares*, Comm. ACM, 17 (1974), pp. 167–169.

[511] P. LANCASTER AND M. TISMENETSKY, *The Theory of Matrices*, Academic Press, New York, 1985.

[512] C. LANCZOS, *An iteration method for the solution of the eigenvalue problem of linear differential and integral operators*, J. Res. Nat. Bur. Standards, 45 (1950), pp. 255–282.

[513] ———, *Linear Differential Operators*, D. Van Nostrand, London, UK, 1961.

[514] L. LANDWEBER, *An iterative formula for Fredholm integral equations of the first kind*, Amer. J. Math., 73 (1951), pp. 615–624.

[515] P. S. LAPLACE, *Théorie analytique des probabilités. Premier supplément*, 3rd ed., Courcier, Paris, 1816.

[516] P. LÄUCHLI, *Iterative Lösung und Fehlerabschätzung in der Ausgleichrechnung*, ZAMP, 10 (1959), pp. 245–280.

[517] ———, *Jordan-Elimination und Ausgleichung nach kleinsten Quadraten*, Numer. Math., 3 (1961), pp. 226–240.

[518] C. L. LAWSON, *Contributions to the Theory of Linear Least Maximum Approximation*, Ph. D. thesis, UCLA, Los Angeles, CA, 1961.

[519] ———, *Sparse matrix methods based on orthogonality and conjugacy*, Tech. Mem. 33-627, June 1973, Jet Propulsion Lab., Cal. Inst. of Tech., Pasadena, CA, 1973.

[520] C. L. LAWSON AND R. J. HANSON, *Solving Least Squares Problems*, Prentice Hall, Englewood Cliffs, NJ, 1974.

[521] ———, *Solving Least Squares Problems*, Classics in Applied Mathematics, SIAM, Philadelphia, 1995.

[522] C. L. LAWSON, R. J. HANSON, D. R. KINCAID, AND F. T. KROGH, *Basic linear algebra subprograms for Fortran usage*, ACM Trans. Math. Software, 5 (1979), pp. 308–323.

[523] A. M. LEGENDRE, *Nouvelles méthodes pour la détermination des orbites des comètes*, Courcier, Paris, 1805.

[524] F. LEMEIRE, *Bounds for condition numbers of triangular and trapezoid matrices*, BIT, 15 (1975), pp. 58–64.

[525] Ö. LERINGE AND P.-Å. WEDIN, *A comparison between different methods to compute a vector x which minimizes $\|Ax - b\|_2$ when $Gx = h$*, Tech. Report, Department of Computer Science, Lund University, Lund, Sweden, 1970.

[526] K. LEVENBERG, *A method for the solution of certain non-linear problems in least squares*, Quart. Appl. Math., 2 (1944), pp. 164–168.

[527] J. G. LEWIS, D. J. PIERCE, AND D. K. WAH, *Multifrontal Householder QR factorization*, Tech. Report ECA-TR-127-Revised, Boeing Computer Services, Seattle, WA, 1989.

[528] T. Y. LI, N. H. RHEE, AND Z. ZENG, *An efficient and accurate parallel algorithm for the singular value problem of bidiagonal matrices*, Tech. Report, Department of Mathematics, University of Missouri–Kansas City, 1995.

[529] Y. LI, *A globally convergent method for l_p problems*, SIAM J. Optim., 3 (1993), pp. 609–629.

[530] ———, *Solving l_p problems and applications*, Tech. Report CTC93TR122, 03/93, Advanced Computing Research Institute, Cornell University, Ithaca, NY, 1993.

[531] Y. LIN AND C. W. CRYER, *An alternating direction implicit algorithm for the solution of linear complementarity problems arising from free boundary problems,*

J. Appl. Math. Optim., 13 (1985), pp. 1–17.

[532] Y. LIN AND J. S. PANG, *Iterative methods for large convex quadratic programs: A survey*, SIAM J. Control Optim., 25 (1987), pp. 383–411.

[533] P. LINDSTRÖM, *A stabilized Gauss-Newton algorithm for unconstrained nonlinear least squares problems*, Tech. Report UMINF–102.82, Institute of Information Processing, University of Umeå, Sweden, 1982.

[534] ———, *A general purpose algorithm for nonlinear least squares problems with nonlinear constraints*, Tech. Report UMINF–102.83, Institute of Information Processing, University of Umeå, Sweden, 1983.

[535] ———, *Two user guides, one (ENLSIP) for constrained — one (ELSUNC) for unconstrained nonlinear least squares problems*, Tech. Report UMINF–109.82 and 110.84, Institute of Information Processing, University of Umeå, Sweden, 1984.

[536] P. LINDSTRÖM AND P.-Å. WEDIN, *A new linesearch algorithm for unconstrained nonlinear least squares problems*, Math. Programming, 29 (1984), pp. 268–296.

[537] ———, *Methods and software for nonlinear least squares problems*, Tech. Report UMINF–133.87, Institute of Information Processing, University of Umeå, Sweden, 1986.

[538] I. LINNIK, *Method of Least Squares and Principles of the Theory of Observations*, Pergamon Press, New York, 1961.

[539] R. J. LIPTON, D. J. ROSE, AND R. E. TARJAN, *Generalized nested dissection*, SIAM J. Numer. Anal., 16 (1979), pp. 346–358.

[540] J. W.-H. LIU, *On general row merging schemes for sparse Givens transformations*, SIAM J. Sci. Statist. Comput., 7 (1986), pp. 1190–1211.

[541] ———, *The role of elimination trees in sparse factorization*, SIAM J. Matrix Anal. Appl., 11 (1990), pp. 134–172.

[542] ———, *The multifrontal method for sparse matrix solution: Theory and practice*, SIAM Review, 34 (1992), pp. 82–109.

[543] J. W. LONGLEY, *Modified Gram-Schmidt process vs. classical Gram-Schmidt*, Comm. Statist. Simulation Comput., B10, 5 (1981), pp. 517–527.

[544] ———, *Least Squares Computations Using Orthogonal Methods*, Marcel Dekker, New York, 1984.

[545] P. LÖTSTEDT, *Perturbation bounds for the linear least squares problem subject to linear inequality constraints*, BIT, 23 (1983), pp. 500–519.

[546] ———, *Solving the minimal least squares problem subject to bounds on the variables*, BIT, 24 (1984), pp. 206–224.

[547] A. K. LOUIS, *Convergence of the conjugate gradient method for compact operators*, in Inverse and Ill-posed Problems, H. W. Engl and C. W. Groetsch, eds., Academic Press, New York, 1987, pp. 177–183.

[548] S. M. LU AND J. L. BARLOW, *Computation of orthogonal factors of sparse matrices*, Tech. Report CSE-93-014, Department of Computer Science, The Pennsylvania State University, State College, PA, 1993.

[549] F. T. LUK, *Computing the singular value decomposition on the ILLIAC IV*, ACM Trans. Math. Software, 6 (1980), pp. 524–539.

[550] ———, *A parallel method for computing the generalized singular value decomposition*, J. Parallel Distributed Comput., 2 (1985), pp. 250–260.

[551] ———, *A rotation method for computing the QR-decomposition*, SIAM J. Sci. Statist. Comput., 7 (1986), pp. 452–459.

[552] F. T. LUK AND S. QIAO, *A fast but unstable orthogonal triangularization technique for Toeplitz matrices*, Linear Algebra Appl., 88/89 (1987), pp. 495–506.

[553] ———, *A new matrix decomposition for signal processing*, Automatica, 30 (1994),

pp. 39–43.

[554] I. LUSTIG, R. MARSTEN, AND D. SHANNO, *Computational experience with a primal-dual interior point method for linear programming*, Linear Algebra Appl., 152 (1991), pp. 191–222.

[555] K. MADSEN AND H. B. NIELSEN, *Finite algorithms for robust regression*, BIT, 30 (1990), pp. 682–699.

[556] ———, *A finite smoothing algorithm for linear l_1 estimation*, SIAM J. Optim., 3 (1993), pp. 68–80.

[557] N. MAHDAVI-AMIRI, *Generally Constrained Nonlinear Least Squares and Generating Test Problems: Algorithmic Approach*, Ph. D. thesis, The John Hopkins University, Baltimore, MD, 1981.

[558] J. H. MAINDONALD, *Statistical Computation*, John Wiley, New York, 1984.

[559] J. MALARD, *Block Solvers for Dense Linear Systems On Local Memory Multiprocessors*, Ph. D. thesis, School of Computer Science, McGill University, Montréal, Canada, 1992.

[560] P. MANNEBACK, *On Some Numerical Methods for Solving Large Sparse Linear Least Squares Problems*, Ph. D. thesis, Facultés Universitaires Notre-Dame de la Paix, Namur, Belgium, 1985.

[561] P. MANNEBACK, C. MURIGANDE, AND P. L. TOINT, *A modification of an algorithm by Golub and Plemmons for large linear least squares in the context of Doppler positioning*, IMA J. Numer. Anal., 5 (1985), pp. 221–234.

[562] T. A. MANTEUFFEL, *Numerical rank determination in linear least squares problems*, Tech. Report SAND 79-8243, Sandia Laboratories, Albuquerque, NM, 1979.

[563] ———, *An incomplete factorization technique for positive definite linear systems*, Math. Comp., 34 (1980), pp. 473–497.

[564] ———, *An interval analysis approach to rank determination in linear least squares problems*, SIAM J. Sci. Statist. Comp., 2 (1981), pp. 335–348.

[565] T. L. MARKHAM, M. NEUMANN, AND R. J. PLEMMONS, *Convergence of a direct-iterative method for large-scale least-squares problems*, Linear Algebra Appl., 69 (1985), pp. 155–167.

[566] A. A. MARKOFF, *Wahrscheinlichkeitsrechnung*, 2nd. ed., H. Liebmann, Trans., Leipzig, 1912.

[567] H. M. MARKOWITZ, *The elimination form of the inverse and its application to linear programming*, Management Sci., 3 (1957), pp. 255–269.

[568] D. W. MARQUARDT, *An algorithm for least-squares estimation of nonlinear parameters*, J. Soc. Indust. Appl. Math., 11 (1963), pp. 431–441.

[569] J. M. MARTINEZ, *An algorithm for solving sparse nonlinear least squares problems*, Computing, 39 (1987), pp. 307–325.

[570] P. MATSTOMS, *The Multifrontal Solution of Sparse Linear Least Squares Problems*, Licentiat thesis, Department of Mathematics, Linköping University, Sweden, 1991.

[571] ———, *QR27—Specification sheet*, Tech. Report March 1992, Department of Mathematics, Linköping University, Sweden, 1992.

[572] ———, *Sparse QR factorization in MATLAB*, ACM Trans. Math. Software, 20 (1994), pp. 136–159.

[573] ———, *Sparse QR Factorization with Applications to Linear Least Squares Problems*, Ph. D. thesis, Department of Mathematics, Linköping University, Sweden, 1994.

[574] ———, *Parallel sparse QR factorization on shared memory architectures*, Parallel

Comput., 21 (1995), pp. 473–486.

[575] J. A. MEIJERINK AND H. VAN DER VORST, *An iterative method for linear systems of which the coefficient matrix is a symmetric M-matrix*, Math. Comp., 31 (1977), pp. 148–162.

[576] P. MEISSL, *A priori prediction of roundoff error accumulation in the solution of a super-large geodetic normal equation system*, Professional Paper 12, National Oceanic and Atmospheric Administration, 1980.

[577] C. D. MEYER *Generalized inversion of modified matrices*, SIAM. J. Appl. Math., 24 (1973), pp. 315–323.

[578] L. MIRSKY, *Symmetric gauge functions and unitarily invariant norms*, Quart. J. Math. Oxford, 11 (1960), pp. 50–59.

[579] J. J. MODI AND M. R. B. CLARKE, *An alternate Givens ordering*, Numer. Math., 43 (1984), pp. 83–90.

[580] C. B. MOLER, *Iterative refinement in floating point*, J. Assoc. Comput. Mach., 14 (1967), pp. 316–321.

[581] L. MOLINARI, *Gram-Schmidt'sches Orthogonalisierungsverfahren*, in Numerische Prozeduren aus Nachlass und Lehre von Prof. Heinz Rutishauser, W. Gander, L. Molinari, and H. Svecová, eds., Birkhäuser, Stuttgart, 1977, pp. 77–93.

[582] M. MOONEN AND P. VAN DOOREN, *On the QR algorithm and updating the SVD and URV decompositions in parallel*, Linear Algebra Appl., 188/189 (1993), pp. 549–568.

[583] M. MOONEN, P. VAN DOOREN, AND J. VANDEWALLE, *An SVD updating algorithm for subspace tracking*, SIAM J. Matrix Anal. Appl., 13 (1992), pp. 1015–1038.

[584] E. H. MOORE, *General Analysis. Part I*, American Philosophical Society, Philadelphia, 1935.

[585] R. MORANDI AND F. SGALLARI, *Parallel algorithms for the iterative solution of sparse least-squares problems*, Parallel Comput., 13 (1990), pp. 271–280.

[586] J. J. MORÉ, *The Levenberg-Marquardt algorithm: Implementation and theory*, in Numerical Analysis. Proceedings Biennial Conference Dundee 1977, G. A. Watson, ed., vol. 630 of Lecture Notes in Mathematics, Springer-Verlag, Berlin, 1978, pp. 105–116.

[587] ———, *Recent developments in algorithms and software for trust region-methods*, in Mathematical Programming. The State of the Art, Proceedings Bonn 1982, A. Bachem, M. Grötchel, and B. Korte, eds., Springer-Verlag, Berlin, 1983, pp. 258–287.

[588] J. J. MORÉ, B. S. GARBOW, AND K. E. HILLSTROM, *Users' guide for MINPACK-1*, Tech. Report ANL-80-74, Applied Math. Div., Argonne National Laboratory, Argonne, IL, 1980.

[589] J. J. MORÉ AND D. C. SORENSON, *Computing a trust region step*, SIAM J. Sci. Statist. Comput., 4 (1981), pp. 553–572.

[590] J. J. MORÉ AND G. TORALDO, *Algorithms for bound constrained quadratic programming problems*, Numer. Math., 55 (1989), pp. 377–400.

[591] J. J. MORÉ AND S. J. WRIGHT, *Optimization Software Guide*, SIAM, Philadelphia, 1993.

[592] N. MUNKSGAARD, *Solving sparse symmetric sets of linear equations by preconditioned conjugate gradients*, ACM Trans. Math. Software, 6 (1980), pp. 206–219.

[593] J. G. NAGY, *Toeplitz Least Squares Computations*, Ph. D. thesis, North Carolina State University, Raleigh, NC, 1991.

[594] ———, *Fast inverse QR factorization for Toeplitz matrices*, SIAM J. Sci. Comput.,

14 (1993), pp. 1174–1193.

[595] J. C. NASH, *A one-sided transformation method for the singular value decomposition and algebraic eigenproblem*, Comput. J., 18 (1975), pp. 74–76.

[596] M. Z. NASHED, ed., *Generalized Inverses and Applications*, Academic Press, New York, 1976.

[597] L. NAZARETH, *Some recent approaches to solving large residual nonlinear least squares problems*, SIAM Review, 22 (1980), pp. 1–11.

[598] A. S. NEMIROVSKII, *The regularization properties of the adjoint gradient method in ill-posed problems*, USSR Comput. Math. and Math. Phys., 26:2 (1986), pp. 7–16.

[599] A. NEUMAIER, *Hybrid norms and bounds for overdetermined linear systems*, Linear Algebra Appl., 216 (1995), pp. 257–265.

[600] E. G. NG, *A scheme for handling rank deficiency in the solution of sparse linear least squares problems*, SIAM J. Sci. Statist. Comput., 12 (1991), pp. 1173–1183.

[601] W. NIETHAMMER, J. DE PILLIS, AND R. S. VARGA, *Convergence of block iterative methods applied to sparse least-squares problems*, Linear Algebra Appl., 58 (1984), pp. 327–341.

[602] B. NOBLE, *Methods for computing the Moore-Penrose generalized inverse and related matters*, in Generalized Inverses and Applications. Proceedings of an Advanced Seminar, The University of Wisconsin-Madison, Oct. 1973, M. Z. Nashed, ed., Academic Press, New York, 1976, pp. 245–301.

[603] W. OETTLI AND W. PRAGER, *Compatibility of approximate solution of linear equations with given error bounds for coefficients and right-hand sides*, Numer. Math., 6 (1964), pp. 405–409.

[604] D. P. O'LEARY, *Estimating matrix condition numbers*, SIAM J. Sci. Statist. Comput., 1 (1980), pp. 205–209.

[605] ———, *A generalized conjugate gradient algorithm for solving class of quadratic programming problems*, Linear Algebra Appl., 34 (1980), pp. 371–399.

[606] ———, *On bounds for scaled projections and pseudo-inverses*, Linear Algebra Appl., 132 (1990), pp. 115–117.

[607] ———, *Robust regression computation using iteratively reweighted least squares*, SIAM J. Matrix Anal. Appl., 11 (1990), pp. 466–480.

[608] D. P. O'LEARY AND B. W. RUST, *Confidence intervals for inequality-constrained least squares problems, with applications to ill-posed problems*, SIAM J. Sci. Statist. Comput., 7 (1986), pp. 473–489.

[609] D. P. O'LEARY AND J. A. SIMMONS, *A bidiagonalization-regularization procedure for large scale discretizations of ill-posed problems*, SIAM J. Sci. Statist. Comput., 2 (1981), pp. 474–489.

[610] D. P. O'LEARY AND P. WHITMAN, *Parallel QR factorization by Householder and modified Gram-Schmidt algorithms*, Parallel Comput., 16 (1990), pp. 99–112.

[611] S. J. OLSZANSKYJ, J. M. LEBAK, AND A. BOJANCZYK, *Rank-k modification methods for recursive least squares problems*, Numer. Algorithms, 7 (1994), pp. 325–354.

[612] U. OREBORN, *A Direct Method for Sparse Nonnegative Least Squares Problems*, Licentiat thesis, Department of Mathematics, Linköping University, Sweden, 1986.

[613] J. M. ORTEGA AND W. C. RHEINBOLDT, *Iterative Solution of Nonlinear Equations in Several Variables*, Academic Press, New York, 1970.

[614] E. E. OSBORNE, *On least squares solutions of linear equations*, J. Assoc. Comput. Mach., 8 (1961), pp. 628–636.

[615] ———, *Nonlinear least squares—the Levenberg algorithm revisited*, J. Austral.

Math. Soc., Ser. B, 19 (1976), pp. 343–357.

[616] M. R. OSBORNE, *On the computation of stepwise regressions*, Austral. Comput. J., 8 (1976), pp. 61–63.

[617] ———, *Some aspects of nonlinear least squares calculations*, in Numerical Methods for Nonlinear Optimization, F. A. Lootsma, ed., Academic Press, New York, 1976, pp. 171–189.

[618] ———, *Finite Algorithms in Optimization and Data Analysis*, John Wiley, New York, 1985.

[619] ———, *Solving least squares problems on parallel vector processors*, Tech. Report, School of Mathematical Sciences, Australian National University, Canberra, 1994.

[620] O. ØSTERBY AND Z. ZLATEV, *Direct Methods for Sparse Matrices*, vol. 157 of Lecture Notes in Computer Science, Springer-Verlag, Berlin, 1983.

[621] G. OSTROUCHOV, *Symbolic Givens reduction and row-ordering in large sparse least squares problems*, SIAM J. Sci. Statist. Comput., 8 (1987), pp. 248–264.

[622] C. C. PAIGE, *The Computation of Eigenvalues and Eigenvectors of Very Large Sparse Matrices*, Ph. D. thesis, University of London, UK, 1971.

[623] ———, *An error analysis of a method for solving matrix equations*, Math. Comp., 27 (1973), pp. 355–359.

[624] ———, *Bidiagonalization of matrices and solution of linear eqautions*, SIAM J. Numer. Anal., 11 (1974), pp. 197–209.

[625] ———, *Error analysis of the Lanczos algorithm for tridiagonalizing a symmetric matrix*, J. Inst. Math. Applic., 18 (1976), pp. 341–349.

[626] ———, *Numerically stable computations for general univariate linear models*, Comm. Statist., B7 (1978), pp. 437–453.

[627] ———, *Computer solution and perturbation analysis of generalized linear least squares problems*, Math. Comp., 33 (1979), pp. 171–184.

[628] ———, *Fast numerically stable computations for generalized least squares problems*, SIAM J. Numer. Anal., 16 (1979), pp. 165–171.

[629] ———, *Error analysis of some techniques for updating orthogonal decompositions*, Math. Comp., 34 (1980), pp. 465–471.

[630] ———, *Properties of numerical algorithms related to computing controllability*, IEEE Trans. Automat. Control, 26 (1981), pp. 130–138.

[631] ———, *A note on a result of Sun Ji-guang: Sensitivity of the CS and GSV decomposition*, SIAM J. Numer. Anal., 21 (1984), pp. 186–191.

[632] ———, *The general linear model and the generalized singular value decomposition*, Linear Algebra Appl., 70 (1985), pp. 269–284.

[633] ———, *Computing the generalized singular value decomposition*, SIAM J. Sci. Statist. Comput., 7 (1986), pp. 1126–1146.

[634] ———, *Some aspects of generalized QR factorizations*, in Reliable Numerical Computation, M. G. Cox and S. J. Hammarling, eds., Clarendon Press, Oxford, UK, 1990, pp. 71–91.

[635] ———, *History and generality of the CS-decomposition*, Linear Algebra Appl., 208/209 (1994), pp. 303–326.

[636] C. C. PAIGE AND M. A. SAUNDERS, *Least squares estimation of discrete linear dynamic systems using orthogonal transformations*, SIAM J. Numer. Anal., 14 (1977), pp. 180–193.

[637] ———, *Toward a generalized singular value decomposition*, SIAM J. Numer. Anal., 18 (1981), pp. 398–405.

[638] ———, *Algorithm 583 LSQR: Sparse linear equations and sparse least squares*, ACM Trans. Math. Software, 8 (1982), pp. 195–209.

[639] ———, *LSQR. An algorithm for sparse linear equations and sparse least squares*, ACM Trans. Math. Software, 8 (1982), pp. 43–71.

[640] C. C. PAIGE AND P. VAN DOOREN, *On the quadratic convergence of Kogbetliantz's algorithm for computing the singular value decomposition*, Linear Algebra Appl., 77 (1986), pp. 301–313.

[641] C. C. PAIGE AND M. WEI, *Analysis of the generalized total least squares problem $AX \approx B$ when some columns are free of errors*, Numer. Math., 65 (1993), pp. 177–202.

[642] ———, *History and generality of the CS decomposition*, Linear Algebra Appl., 108/109 (1994), pp. 303–326.

[643] C. T. PAN, *A modification to the LINPACK downdating algorithm*, BIT, 30 (1990), pp. 707–722.

[644] ———, *A perturbation analysis on the problem of downdating a Cholesky factorization*, Linear Algebra Appl., 183 (1993), pp. 103–116.

[645] C. T. PAN AND R. J. PLEMMONS, *Least squares modifications with inverse factorizations: Parallel implementations*, J. Comput. Appl. Math., 27 (1989), pp. 109–127.

[646] J. S. PANG, *On the convergence of a basic iterative method for the implicit complementarity problem*, J. Optim. Theory Appl., 37 (1982), pp. 149–162.

[647] E. P. PAPADOPOULOU, Y. G. SARIDAKIS, AND T. S. PAPATHEODOROU, *Block AOR iterative schemes for large-scale least-squares problems*, SIAM J. Numer. Anal., 26 (1989), pp. 637–660.

[648] H. PARK AND L. ELDÉN, *Fast and accurate triangularization of Toeplitz matrices*, Tech. Report LiTH-MAT-R-1993-17, Department of Mathematics, Linköping University, Sweden, 1993.

[649] ———, *Downdating the rank-revealing URV decomposition*, SIAM J. Matrix. Anal. Appl., 16 (1995), pp. 138–155.

[650] H. PARK AND S. VAN HUFFEL, *Two-way bidiagonalization scheme for downdating the singular value decomposition*, Linear Algebra Appl., 222 (1995), pp. 23–40.

[651] B. N. PARLETT, *The Symmetric Eigenvalue Problem*, Prentice-Hall, Englewood Cliffs, NJ, 1980.

[652] B. N. PARLETT AND J. K. REID, *On the solution of a system of linear equations whose matrix is symmetric but not definite*, BIT, 10 (1970), pp. 386–397.

[653] S. V. PARTER, *The use of linear graphs in Gauss elimination*, SIAM Review, 3 (1961), pp. 119–130.

[654] P. PAZELT, *Ein Algoritmus zur Lösung von Ausgleichproblemen mit Ungleichungen als Nebenbedingungen*, diplomarbeit, University of Würtzburg, Germany, 1973.

[655] R. PENROSE, *A generalized inverse for matrices*, Proc. Cambridge Philos. Soc., 51 (1955), pp. 406–413.

[656] ———, *On best approximate solutions of linear matrix equations*, Proc. Cambridge Philos. Soc., 52 (1956), pp. 17–19.

[657] V. PEREYRA, *Iterative methods for solving nonlinear least squares problems*, SIAM J. Numer. Anal., 4 (1967), pp. 27–36.

[658] G. PETERS AND J. H. WILKINSON, *The least squares problem and pseudo-inverses*, Comput. J., 13 (1970), pp. 309–316.

[659] D. J. PIERCE AND J. G. LEWIS, *Sparse multifrontal rank revealing QR factorization*, Tech. Report MEA-TR-193-Revised, Boeing Information and Support Services, Seattle, WA, 1995.

[660] R. L. PLACKET, *The discovery of the method of least squares*, Biometrika, 59 (1972), pp. 239–251.

[661] R. J. PLEMMONS, *Monotonicity and iterative approximations involving rectangular matrices*, Math. Comp., 26 (1972), pp. 853–858.

[662] ———, *Linear least squares by elimination and MGS*, J. Assoc. Comput. Mach., 21 (1974), pp. 581–585.

[663] ———, *Adjustment by least squares in geodesy using block iterative methods for sparse matrices*, in Proceedings of the 1979 Army Numerical Analysis and Computer Conference, El Paso, TX, 1979, pp. 151–186.

[664] ———, *A proposal for FFT based fast recursive least squares*, Tech. Report #982, Institute for Mathematics and Its Applications, University of Minnesota, Minneapolis, MN, 1992.

[665] L. F. PORTUGAL, J. J. JÚDICE, AND L. N. VICENTE, *A comparison of block pivoting and interior-point algorithms for linear least squares problems with nonnegative variables*, Math. Comp., 63 (1994), pp. 625–643.

[666] A. POTHEN, *Sparse Null Bases and Marriage Theorems*, Ph. D. thesis, Cornell University, Ithaca, NY, 1984.

[667] ———, *Predicting the structure of sparse orthogonal factors*, Linear Algebra Appl., 194 (1993), pp. 183–204.

[668] A. POTHEN AND C. J. FAN, *Computing the block triangular form of a sparse matrix*, ACM Trans. Math. Software, 16 (1990), pp. 303–324.

[669] A. POTHEN AND P. RAGHAVAN, *Distributed orthogonal factorization: Givens and Householder algorithms*, SIAM J. Sci. Statist. Comput., 10 (1989), pp. 1113–1134.

[670] M. J. D. POWELL AND J. K. REID, *On applying Householder's method to linear least squares problems*, in Proceedings of the IFIP Congress 68, A. J. M. Morell, ed., North-Holland, Amsterdam, 1969, pp. 122–126.

[671] V. PRATT, *Direct least squares fitting of algebraic surfaces*, ACM J. Comput. Graphics, 21:4 (1987).

[672] C. PUGLISI, *QR Factorization of Large Sparse Overdetermined and Square Matrices with the Multifrontal Method in a Multiprocessing Environment*, Ph. D. thesis, Institut National Polytechnique de Toulouse, Toulouse, France, 1993.

[673] C. M. RADER AND A. O. STEINHARDT, *Hyperbolic Householder transforms*, SIAM J. Matrix Anal. Appl., 9 (1988), pp. 269–290.

[674] H. RAMSIN AND P.-Å. WEDIN, *A comparison of some algorithms for the nonlinear least squares problem*, BIT, 17 (1977), pp. 72–90.

[675] R. C. RAO, *Linear Statistical Inference and Its Applications*, 2nd ed., John Wiley, New York, 1973.

[676] L. REICHEL, *Fast QR decomposition of Vandermonde-like matrices and polynomial least squares approximation*, SIAM J. Matrix Anal. Appl., 12 (1991), pp. 552–564.

[677] L. REICHEL, G. S. AMMAR, AND W. B. GRAGG, *Discrete least squares approximation by trigonometric polynomials*, Math. Comp., 57 (1991), pp. 273–289.

[678] L. REICHEL AND W. B. GRAGG, *FORTRAN subroutines for updating the QR decomposition*, ACM Trans. Math. Software, 16 (1990), pp. 369–377.

[679] J. K. REID, *A note on the least squares solution of a band system of linear equations by Householder reductions*, Comput J., 10 (1967), pp. 188–189.

[680] ———, *On the method of conjugate gradients for the solution of large systems of linear equations*, in Large Sparse Sets of Linear Equations, J. K. Reid, ed., Academic Press, New York, 1971, pp. 231–254.

[681] ———, *On the use of conjugate gradients for systems of linear equations possessing "Property A"*, SIAM J. Numer. Anal., 9 (1972), pp. 325–332.

[682] ———, *TREESOLVE, a Fortran package for solving large sets of linear finite*

element solutions, Tech. Report CSS 155, 1984, Computer Science and Systems Division, AERE Harwell, Oxfordshire, UK, 1984.

[683] C. H. REINSCH, *Smoothing by spline functions*, Numer. Math., 16 (1971), pp. 451–454.

[684] J. R. RICE, *Experiments on Gram-Schmidt orthogonalization*, Math. Comp., 20 (1966), pp. 325–328.

[685] ———, *PARVEC workshop on very large least squares problems and supercomputers*, Tech. Report CSD-TR 464, Purdue University, West Lafayette, IN, 1983.

[686] J. L. RIGAL AND J. GACHES, *On the compatability of a given solution with the data of a linear system*, J. Assoc. Comput. Mach., 14 (1967), pp. 543–548.

[687] J. D. RILEY, *Solving systems of linear equations with a positive definite symmetric but possibly ill-conditioned matrix*, Math. Tables Aids. Comput., 9 (1956), pp. 96–101.

[688] D. J. ROSE, *A graph-theoretic study of the numerical solution of sparse positive definite systems of linear equations*, in Graph Theory and Computing, R. C. Read, ed., Academic Press, New York, 1972, pp. 183–217.

[689] A. RUHE, *Accelerated Gauss-Newton algorithms for nonlinear least squares problems*, BIT, 19 (1979), pp. 356–367.

[690] ———, *Numerical aspects of Gram-Schmidt orthogonalization of vectors*, Linear Algebra Appl., 52/53 (1983), pp. 591–601.

[691] A. RUHE AND P.-Å. WEDIN, *Algorithms for separable nonlinear least squares problems*, SIAM Review, 22 (1980), pp. 318–337.

[692] H. RUTISHAUSER, *Der Quotienten-Differenzen-Algorithmus*, ZAMP, 5 (1954), pp. 233–251.

[693] ———, *Theory of gradient methods*, in Refined Methods for Computation of the Solution and the Eigenvalues of Self-Adjoint Boundary Value Problems, M. Engeli, T. Ginsburg, H. Rutishauser, and E. Stiefel, eds., Birkhäuser, Basel/Stuttgart, 1959, pp. 24–50.

[694] ———, *Description of Algol 60*, Handbook for Automatic Computation, Vol. 1a, Springer-Verlag, Berlin, 1967.

[695] ———, *Once again: The least squares problem*, Linear Algebra Appl., 1 (1968), pp. 479–488.

[696] ———, *Vorlesungen über Numerische Mathematik. Band 1*, Birkhäuser, Basel, 1976.

[697] Y. SAAD, *On the rates of convergence of the Lanczos and block-Lanczos methods*, SIAM J. Numer. Anal., 17 (1980), pp. 687–706.

[698] ———, *Preconditioning techniques for nonsymmetric and indefinite linear systems*, J. Comput. Appl. Math., 24 (1988), pp. 89–105.

[699] ———, *Numerical Methods for Large Eigenvalue Problems*, Manchester University Press, Manchester, UK, 1992.

[700] D. E. SALANE, *A continuation approach for solving large residual least squares problems*, SIAM J. Sci. Statist. Comput., 8 (1987), pp. 655–671.

[701] M. A. SAUNDERS, *Large-scale linear programming using the Cholesky factorization*, Tech. Report CS252, Computer Science Department, Stanford University, CA, 1972.

[702] ———, *Sparse least squares by conjugate gradients: A comparison of preconditioning methods*, in Proceedings of Computer Science and Statistics: Twelfth Annual Conference on the Interface, Waterloo, Canada, 1979.

[703] ———, *Solution of sparse rectangular systems using LSQR and CRAIG*, BIT, 35 (1995), pp. 588–604.

[704] W. SAUTTER, *Fehleranalyse für die Gauss-Elimination zur Berechnung der Lösung minimaler Länge*, Numer. Math., 30 (1978), pp. 165–184.

[705] R. SCHABACK, *Convergence analysis of the general Gauss-Newton algorithm*, Numer. Math., 46 (1985), pp. 281–309.

[706] K. SCHITTKOWSKI, *The numerical solution of constrained linear least-squares problems*, IMA J. Numer. Anal., 3 (1983), pp. 11–36.

[707] ———, *Solving constrained nonlinear least squares problems by a general purpose SQP-method*, in Trends in Mathematical Optimization, K.-H. Hoffmann, J. B. Hiriart-Urruty, C. Lemaréchal, and J. Zowe, eds., vol. 84 of International Series of Numerical Mathematics, Birkhäuser-Verlag, Basel, Switzerland, 1985, pp. 49–83.

[708] K. SCHITTKOWSKI AND J. STOER, *A factorization method for the solution of constrained linear least squares problems allowing subsequent data changes*, Numer. Math., 31 (1979), pp. 431–463.

[709] K. SCHITTKOWSKI AND P. ZIMMERMANN, *A factorization method for constrained least squares problems with data changes. Part 2: Numerical tests, comparisons, and ALGOL codes*, Tech. Report Preprint No. 30, Institut für Angewandte Mathematik und Statistik, Universität Würzburg, Germany, 1985.

[710] E. SCHMIDT, *Zur Theorie der linearen und nichtlinearen Integralgleichungen. I. Teil: Entwicklung willkürlicher Funktionen nach System vorgeschriebener*, Math. Ann., 63 (1907), pp. 433–476.

[711] ———, *Über die Auflösung linearer Gleichungen mit unendlich vielen Unbekannten*, Rend. Circ. Mat. Palermo. Ser. 1, 25 (1908), pp. 53–77.

[712] R. SCHREIBER AND W.-P. TANG, *On systolic arrays for updating the Cholesky factorization*, BIT, 26 (1986), pp. 451–466.

[713] R. SCHREIBER AND C. F. VAN LOAN, *A storage efficient WY representation for products of Householder transformations*, SIAM J. Sci. Statist. Comput., 10 (1989), pp. 53–57.

[714] H. SCHWETLICK AND V. TILLER, *Numerical methods for estimating parameters in nonlinear models with errors in the variables*, Technometrics, 27 (1985), pp. 17–24.

[715] ———, *Nonstandard scaling matrices for trust region Gauss-Newton methods*, SIAM J. Sci Statist. Comput., 10 (1989), pp. 654–670.

[716] H. D. SCOLNIK, *On the solution of non-linear least squares problems*, in Proceedings of the IFIP Congress 71, H. Freeman, ed., North-Holland, Amsterdam, 1972, pp. 1258–1265.

[717] P. L. SEIDEL, *Über ein Verfahren, die Gleichungen, auf welche die Methode der kleinsten Quadrate führt, sowie lineare Gleichungen überhaupt, durch successive Annäherung aufzulösen*, Abh. Bayer Akad. Wiss., 11:81 (1874).

[718] L. F. SHAMPINE, *Discrete least squares polynomial fits*, Comm. ACM, 18 (1975), pp. 179–180.

[719] A. H. SHERMAN, *Algorithm 533, NSPIV, a FORTRAN subroutine for sparse Gaussian elimination with partial pivoting*, ACM Trans. Math. Software, 4 (1978), pp. 391–398.

[720] I. H. SIEGEL, *Deferment of computation in the method of least squares*, Math. Comp., 19 (1965), pp. 329–331.

[721] R. D. SKEEL, *Scaling for numerical stability in Gaussian elimination*, J. Assoc. Comput. Mach., 26 (1979), pp. 494–526.

[722] ———, *Iterative refinement implies numerical stability for Gaussian elimination*, Math. Comp., 35 (1980), pp. 817–832.

[723] I. SÖDERQUIST, *Computing Parameters in Nonlinear Least Squares Models*, Ph. D. thesis, University of Umeå, Sweden, 1993.

[724] H. SPÄTH, *On discrete linear orthogonal l_p approximation*, Z. Angew. Math. Mech., 62 (1982), pp. 354–355.

[725] ———, *Orthogonal least squares fitting with linear manifolds*, Numer. Math., 48 (1986), pp. 441–445.

[726] ———, *Modified Gram-Schmidt for solving linear least squares problems is equivalent to Gaussian elimination for the normal equations*, Appl. Math., 20 (1990), pp. 587–589.

[727] ———, *Mathematical Algorithms for Linear Regression*, Academic Press, New York, 1992.

[728] G. W. STEWART, *On the continuity of the generalized inverse*, SIAM J. Appl. Math., 17 (1969), pp. 33–45.

[729] ———, *Introduction to Matrix Computations*, Academic Press, New York, 1973.

[730] ———, *The economical storage of plane rotations*, Numer. Math., 25 (1976), pp. 137–138.

[731] ———, *On the perturbation of pseudo-inverses, projections, and linear least squares problems*, SIAM Review, 19 (1977), pp. 634–662.

[732] ———, *Perturbation bounds for the QR factorization of a matrix*, SIAM J. Numer. Anal., 14 (1977), pp. 509–518.

[733] ———, *Research development and LINPACK*, in Mathematical Software III, J. R. Rice, ed., Academic Press, New York, 1977, pp. 1–14.

[734] ———, *The effects of rounding error on an algorithm for downdating a Cholesky factorization*, J. Inst. Maths. Applic., 23 (1979), pp. 203–213.

[735] ———, *A note on the perturbation of singular values*, Linear Algebra Appl., 28 (1979), pp. 213–216.

[736] ———, *The efficient generation of random orthogonal matrices with an application to condition estimators*, SIAM J. Numer. Anal., 17 (1980), pp. 403–404.

[737] ———, *On the implicit deflation of nearly singular systems of linear equations*, SIAM J. Sci. Statist. Comput., 2 (1981), pp. 136–140.

[738] ———, *Computing the CS decomposition of a partitioned orthogonal matrix*, Numer. Math., 40 (1982), pp. 297–306.

[739] ———, *A generalization of the Eckart-Young approximation theorem*, Tech. Report TR-1325, Department of Computer Science, University of Maryland, College Park, MD, 1983.

[740] ———, *A method for computing the generalized singular value decomposition*, in Matrix Pencils. Proceedings, Pite Havsbad, 1982, B. Kågström and A. Ruhe, eds., vol. 973 of Lecture Notes in Mathematics, Springer-Verlag, Berlin, 1983, pp. 207–220.

[741] ———, *A nonlinear version of Gauss's minimum variance theorem with applications to an errors-in-the-variables model*, Tech. Report TR-1263, Department of Computer Science, University of Maryland, College Park, MD, 1983.

[742] ———, *On the asymptotic behavior of scaled singular and QR decompositions*, Math. Comp., 43(168) (1984), pp. 168–489.

[743] ———, *On the invariance of perturbed null vectors under column scaling*, Numer. Math., 44 (1984), pp. 61–65.

[744] ———, *Rank degeneracy*, SIAM J. Sci. Statist. Comput., 5 (1984), pp. 403–413.

[745] ———, *Collinearity and least squares regression*, Statistical Science, 2 (1987), pp. 68–100.

[746] ———, *On scaled projections and pseudoinverses*, Linear Algebra Appl., 132 (1990), pp. 115–117.

[747] ———, *Perturbation theory and least squares with errors in the variables*, in Con-

temporary Mathematics 112: Statistical Analysis of Measurement Error Models and Applications, P. J. Brown and W. A. Fuller, eds., American Mathematical Society, Providence, RI, 1990, pp. 171–181.

[748] ——, *On an algorithm for refining a rank-revealing URV factorization and a perturbation theorem for singular values*, Tech. Report UMIACS-TR-91-38, CS-TR-2626, Department of Computer Science, University of Maryland, College Park, MD, 1991.

[749] ——, *An updating algorithm for subspace tracking*, IEEE Trans. Signal Processing, 40 (1992), pp. 1535–1541.

[750] ——, *On the early history of the singular value decomposition*, SIAM Review, 35 (1993), pp. 551–566.

[751] ——, *On the perturbation of LU, Cholesky, and QR factorizations*, SIAM J. Matrix Anal. Appl., 14 (1993), pp. 1141–1145.

[752] ——, *Updating a rank-revealing ULV decomposition*, SIAM J. Matrix Anal. Appl., 14 (1993), pp. 494–499.

[753] ——, *On the stability of sequential updates and downdates*, Tech. Report CS-TR-3238, Department of Computer Science, University of Maryland, College Park, MD, 1994.

[754] G. W. STEWART AND J. SUN, *Matrix Perturbation Theory*, Academic Press, Boston, 1990.

[755] E. STIEFEL, *Ausgleichung ohne Aufstellung der Gausschen Normalgleichungen*, Wiss. Z. Tech. Hochsch. Dresden, 2 (1952/53), pp. 441–442.

[756] T. J. STIELTJES, *Sur les racines de l'équation $x_n = 0$*, Acta. Math., 9 (1887), pp. 385–400.

[757] S. M. STIGLER, *An attack on Gauss, published by Legendre in 1820*, Hist. Math., 4 (1977), pp. 31–35.

[758] ——, *Gauss and the invention of least squares*, Ann. Statist., 9 (1981), pp. 465–474.

[759] J. STOER, *On the numerical solution of constrained least squares problems*, SIAM J. Numer. Anal., 8 (1971), pp. 382–411.

[760] Z. STRAKOS, *On the real convergence of the conjugate gradient method*, Linear Algebra Appl., 154/156 (1991), pp. 535–549.

[761] C. N. STRAND, *Theory and methods related to the singular function expansion and Landweber's iteration for integral equations of the first kind*, SIAM J. Numer. Anal., 11 (1974), pp. 798–825.

[762] C. SUN, *Parallel solution of sparse linear least squares problems on distributed-memory multiprocessors*, Tech. Report CTC95TR212, 05/95, Advanced Computing Research Institute, Cornell University, Ithaca, NY, 1995.

[763] ——, *Parallel sparse orthogonal factorization on distributed-memory multiprocessors*, SIAM J. Sci. Comput., (1996), to appear.

[764] J.-G. SUN, *Perturbation bounds for the Cholesky and QR factorizations*, BIT, 31 (1991), pp. 341–352.

[765] ——, *On perturbation bounds for the QR factorization*, Linear Algebra Appl., 215 (1995), pp. 95–111.

[766] ——, *Perturbation analysis of the Cholesky downdating and QR updating problems*, SIAM J. Matrix Anal. Appl., 16 (1995), pp. 760–775.

[767] ——, *Optimal backward perturbation bounds for the linear LS problem with multiple right-hand sides*, IMA J. Numer. Anal., 16 (1996), pp. 1–11.

[768] D. R. SWEET, *Fast Toeplitz orthogonalization*, Numer. Math., 43 (1984), pp. 1–21.

[769] ——, *Fast block Toeplitz orthogonalization*, Numer. Math., 58 (1991), pp. 613–

629.

[770] K. TANABE, *Projection method for solving a singular system of linear equations and its applications*, Numer. Math., 17 (1971), pp. 203–214.

[771] P. T. P. TANG, *Dynamic condition estimation and Rayleigh-Ritz approximation*, SIAM J. Matrix Anal. Appl., 15 (1994), pp. 331–346.

[772] W.-P. TANG AND G. H. GOLUB, *The block decomposition of a Vandermonde matrix and its applications*, BIT, 21 (1987), pp. 505–517.

[773] R. E. TARJAN, *Depth-first search and linear graph algorithms*, SIAM J. Comput., 1 (1972), pp. 146–159.

[774] R. P. TEWARSON, *A computational method for evaluating generalized inverses*, Comput. J., 10 (1968), pp. 411–413.

[775] A. N. TIKHONOV, *Regularization of incorrectly posed problems*, Soviet Math., 4 (1963), pp. 1624–1627.

[776] W. F. TINNEY AND J. W. WALKER, *Direct solution of sparse network equations by optimally ordered triangular factorization*, Proc. IEEE, 55 (1967), pp. 1801–1809.

[777] P. L. TOINT, *On large scale nonlinear least squares calculations*, SIAM J. Sci. Statist. Comput., 8 (1987), pp. 416–435.

[778] ——, *VE10AD a routine for large-scale nonlinear least squares*, Harwell subroutine library, AERE Harwell, Oxfordshire, UK, 1987.

[779] N.-K. TSAO, *A note on implementing the Householder transformation*, SIAM J. Numer. Anal., 12 (1975), pp. 53–58.

[780] A. VAN DER SLUIS, *Condition numbers and equilibration of matrices*, Numer. Math., 14 (1969), pp. 14–23.

[781] ——, *Stability of the solutions of linear least squares problems*, Numer. Math., 23 (1975), pp. 241–254.

[782] A. VAN DER SLUIS AND H. VAN DER VORST, *Numerical solution of large sparse linear equations and least squares problems*, in Seismic Tomography, G. Nolet, ed., Reidel, Dordrecht, the Netherlands, 1987, pp. 49–83.

[783] ——, *SIRT- and CG-type methods for iterative solution of sparse linear least-squares problems*, Linear Algebra Appl., 130 (1990), pp. 257–302.

[784] A. VAN DER SLUIS AND G. VELTKAMP, *Restoring rank and consistency by orthogonal projection*, Linear Algebra Appl., 28 (1979), pp. 257–278.

[785] J. VAN HEIJST, J. JACOBS, AND J. SCHERDERS, *Kleinste-Kvadraten Problemen.*, Tech. Report, Department of Mathematics, Eindhoven University of Technology, Eindhoven, the Netherlands, 1976.

[786] S. VAN HUFFEL, *Analysis of the Total Least Squares Problem and Its Use in Parameter Estimation*, Ph. D. thesis, Katholieke Universiteit Leuven, Belgium, 1987.

[787] ——, *Iterative algorithms for computing the singular subspace of a matrix associated with its smallest singular values*, Linear Algebra Appl., 154/156 (1991), pp. 675–709.

[788] S. VAN HUFFEL AND H. PARK, *Efficient reduction algorithms for bordered band matrices*, Numer. Linear Algebra Appl., 2 (1995), pp. 95–113.

[789] S. VAN HUFFEL AND J. VANDEWALLE, *Algebraic relationships between classical regression and total least-squares estimation*, Linear Algebra Appl., 93 (1987), pp. 149–162.

[790] ——, *Analysis and solution of the nongeneric total least squares problem*, SIAM J. Matrix Anal. Appl., 9 (1988), pp. 360–372.

[791] ——, *Analysis and properties of the generalized total least squares problem*

$AX \approx B$ when some or all columns in A are subject to error, SIAM J. Matrix. Anal. Appl., 10 (1989), pp. 294–315.

[792] ———, *The Total Least Squares Problem: Computational Aspects and Analysis*, vol. 9 of Frontiers in Applied Mathematics, SIAM, Philadelphia, 1991.

[793] S. VAN HUFFEL, J. VANDEWALLE, AND A. HAEGEMANS, *An efficient and reliable algorithm for computing the singular subspace of a matrix, associated with its smallest singular values*, J. Comp. Appl. Math., 19 (1987), pp. 313–330.

[794] S. VAN HUFFEL AND H. ZHA, *The restricted total least squares problem: Formulation, algorithm, and properties*, SIAM J. Matrix Anal. Appl., 12 (1991), pp. 292–309.

[795] ———, *An efficient total least squares algorithm based on a rank revealing two-sided orthogonal decomposition*, Numer. Algorithms, 4 (1993), pp. 101–133.

[796] C. F. VAN LOAN, *Generalizing the singular value decomposition*, SIAM J. Numer. Anal., 13 (1976), pp. 76–83.

[797] ———, *A generalized SVD analysis of some weighting methods for equality constrained least squares*, in Matrix Pencils. Proceedings, Pite Havsbad, 1982, B. Kågström and A. Ruhe, eds., vol. 973 of Lecture Notes in Mathematics, Springer-Verlag, Berlin, 1983, pp. 245–262.

[798] ———, *Computing the CS and the generalized singular value decomposition*, Numer. Math., 46 (1985), pp. 479–492.

[799] ———, *On the method of weighting for equality constrained least squares*, SIAM J. Numer. Anal., 22 (1985), pp. 851–864.

[800] ———, *Computational Frameworks for the Fourier Transform*, vol. 10 of Frontiers in Applied Mathematics, SIAM, Philadelphia, 1992.

[801] J. M. VARAH, *On the numerical solution of ill-conditioned linear systems with applications to ill-posed problems*, SIAM J. Numer. Anal., 10 (1973), pp. 257–267.

[802] ———, *A lower bound for the smallest singular value of a matrix*, Linear Algebra Appl., 11 (1975), pp. 1–2.

[803] ———, *A practical examination of some numerical methods for linear discrete ill-posed problems*, SIAM Review, 21 (1979), pp. 100–111.

[804] ———, *Backward error estimates for Toeplitz systems*, SIAM J. Matrix Anal. Appl., 15 (1994), pp. 408–417.

[805] R. S. VARGA, *Matrix Iterative Analysis*, Prentice-Hall, Englewood Cliffs, NJ, 1962.

[806] S. A. VAVASIS, *Stable numerical algorithms for equilibrium systems*, SIAM J. Matrix Anal. Appl., 15 (1994), pp. 1108–1131.

[807] N. VEMPATI, I. W. SLUTSKER, AND W. F. TINNEY, *Enhancements to Givens rotations for power system state estimation*, IEEE Trans. Power Systems, 6 (1991), pp. 842–849.

[808] U. VON MATT, *Large Constrained Quadratic Problems*, Ph. D. thesis, Institute for Scientific Computing, ETH, Zürich, Switzerland, 1993.

[809] ———, *The orthogonal QD algorithm*, in SVD and Signal Processing, III: Algorithms, Architectures and Applications, M. Moonen and B. De Moor, eds., Elsevier, Amsterdam, 1995, pp. 99–106.

[810] B. WALDÉN, *Least Squares Methods and Application in Robotics*, Ph. D. thesis, Department of Mathematics, Linköping University, Sweden, 1994.

[811] B. WALDÉN, R. KARLSSON, AND J.-G. SUN, *Optimal backward perturbation bounds for the linear least squares problem*, Numer. Linear Algebra Appl., 2 (1995), pp. 271–286.

[812] R. H. WAMPLER, *An evaluation of linear least squares computer programs*, J. Res.

Nat. Bur. Standards, 73B (1969), pp. 59–90.

[813] ———, *A report on the accuracy of some widely used least squares computer programs*, J. Amer. Statist. Assoc., 65 (1970), pp. 549–565.

[814] ———, *Some recent developments in linear least squares computations*, in Proceedings of the Computer Science and Statistics: Sixth Annual Symposium on the Interface, M. Tarter, ed., Academic Press, New York, 1972, pp. 94–110.

[815] ———, *L2A and L2B, weighted least squares solutions by modified Gram–Schmidt with iterative refinement*, ACM Trans. Math. Software, 5 (1979), pp. 494–99.

[816] ———, *Solutions to weighted least squares problems by modified Gram–Schmidt with iterative refinement*, ACM Trans. Math. Software, 5 (1979), pp. 457–465.

[817] ———, *Test procedures and test problems for least squares algorithms*, J. Econom., 12 (1980), pp. 3–22.

[818] X. WANG, *Incomplete Factorization Preconditioning for Least Squares Problems*, Ph. D. thesis, Department of Mathematics, University of Illinois at Urbana-Champaign, Urbana, IL, 1993.

[819] X. WANG, K. A. GALLIVAN, AND R. BRAMLEY, *CIMGS: An incomplete orthogonal factorization preconditionier*, SIAM J. Sci. Comput., 17 (1996), to appear.

[820] G. A. WATSON, *The numerical solution of total ℓ_p approximation problems*, in Numerical Analysis. Proceedings of the 10th Biennial Conference Dundee, Scotland 1983, D. Griffiths, ed., vol. 630 of Lecture Notes in Mathematics, Springer-Verlag, Berlin, 1984, pp. 72–105.

[821] P.-Å. WEDIN, *The non-linear least squares problem from a numerical point of view. I Geometrical properties*, Tech. Report, Department of Computer Sciences, Lund University, Lund, Sweden, 1972.

[822] ———, *Perturbation bounds in connection with the singular value decomposition*, BIT, 12 (1972), pp. 99–111.

[823] ———, *On the almost rank deficient case of the least squares problem*, BIT, 13 (1973), pp. 344–354.

[824] ———, *Perturbation theory for pseudo-inverses*, BIT, 13 (1973), pp. 217–232.

[825] ———, *On the Gauss-Newton method for the nonlinear least squares problems*, Working Paper 24, Institute for Applied Mathematics, Stockholm, Sweden, 1974.

[826] ———, *Perturbations of the pseudo-inverse and the linear least squares problem analysed with suitable matrix decompositions*, Tech. Report UMINF–69.78, Institute of Information Processing, University of Umeå, Sweden, 1978.

[827] ———, *Notes on the constrained linear least squares problem. A new approach based on generalized inverses*, Tech. Report UMINF–75.79, Institute of Information Processing, University of Umeå, Sweden, 1979.

[828] ———, *Perturbation theory and condition numbers for generalized and constrained linear least squares problems*, Tech. Report UMINF–125.85, Institute of Information Processing, University of Umeå, Sweden, 1985.

[829] M. WEI, *Perturbation of the least squares problem*, Linear Algebra Appl., 141 (1990), pp. 177–182.

[830] ———, *Algebraic properties of the rank-deficient equality constrained and weighted least squares problem*, Linear Algebra Appl., 161 (1992), pp. 27–43.

[831] ———, *Algebraic relations between the total least squares and least squares problems with more than one solution*, Numer. Math., 62 (1992), pp. 123–148.

[832] ———, *The analysis for the total least squares problems with more than one solution*, SIAM J. Matrix. Anal. Appl., 13 (1992), pp. 746–763.

[833] ———, *Perturbation theory for the rank-deficient equality constrained and weighted*

least squares problem, SIAM J. Numer. Anal., 29 (1992), pp. 1462–1481.
[834] P. R. WEIL AND P. C. KETTLER, *Rearranging matrices to block-angular form for decomposition (and other) algorithms*, Management Sci., 18 (1971), pp. 98–108.
[835] J. H. WILKINSON, *Rounding Errors in Algebraic Processes*, Prentice-Hall, Englewood Cliffs, NJ, 1963.
[836] ——, *The Algebraic Eigenvalue Problem*, Clarendon Press, Oxford, UK, 1965.
[837] ——, *Error analysis of transformations based on the use of matrices of the form $I - 2xx^H$*, in Error in Digital Computation, L. B. Rall, ed., John Wiley, New York, 1965, pp. 77–101.
[838] ——, *A priori error analysis of algebraic processes*, in Proceedings International Congress Math., Izdat. Mir, Moscow, 1968, pp. 629–639.
[839] ——, *Modern error analysis*, SIAM Review, 13 (1971), pp. 548–568.
[840] ——, *Some recent advances in numerical linear algebra*, in The State of the Art in Numerical Analysis, D. Jacobs, ed., Academic Press, New York, 1977, pp. 1–53.
[841] ——, *The use of single precision-residuals in the solution of linear systems*, unpublished report, National Physical Laboratory, Teddington, UK, 1977.
[842] ——, *Error analysis revisited*, IMA Bulletin, 22 (1986), pp. 192–200.
[843] J. H. WILKINSON AND C. REINSCH, *Handbook for Automatic Computation Volume II. Linear Algebra*, Springer-Verlag, New York, 1971.
[844] H. WITTMEYER, *Einfluss der Änderung einer Matrix auf der Lösung des zugehörigen Gleichungssystems, sowie auf die charakterischen Zahlen und die Eigenvektoren*, Z. Angew. Math. Mech., 16 (1936), pp. 189–199.
[845] S. J. WRIGHT AND J. N. HOLT, *Algorithms for nonlinear least squares with linear inequality constraints*, SIAM J. Sci. Statist. Comput., 6 (1985), pp. 1033–1048.
[846] K. YOO AND H. PARK, *Accurate downdating of a modified Gram-Schmidt QR decomposition*, BIT, 36 (1996), pp. 166–181.
[847] D. M. YOUNG, *Iterative Solution of Large Linear Systems*, Academic Press, New York, 1971.
[848] J.-Y. YUAN, *The convergence of the 2-block SAOR method for the least-squares problem*, Appl. Numer. Math., 11 (1993), pp. 429–441.
[849] ——, *Iterative Methods for the Generalized Least-Squares Problem*, Ph. D. thesis, Instituto de Matemática Pura e Aplicada, Rio de Janeiro, Brazil, 1993.
[850] M. ZELEN, *Linear estimation and related topics*, in Survey of Numerical Analysis, J. Todd, ed., McGraw-Hill, New York, 1962, pp. 558–584.
[851] H. ZHA, *Implicit QR factorization of a product of three matrices*, BIT, 31 (1991), pp. 375–379.
[852] ——, *Restricted singular value decomposition of matrix triples*, SIAM J. Matrix Anal. Appl., 12 (1991), pp. 172–194.
[853] ——, *A two-way chasing scheme for reducing a symmetric arrowhead matrix to tridiagonal form*, J. Numerical Linear Algebra Appl., 1 (1992), pp. 49–57.
[854] ——, *A componentwise perturbation analysis of the QR decomposition*, SIAM J. Matrix Anal. Appl., 14 (1993), pp. 1124–1131.
[855] G. ZIELKE, *Report on test matrices for generalized inverses*, Computing, 36 (1986), pp. 105–162.
[856] P. ZIMMERMANN, *Ein Algorithmus zur Lösung linearer Least Squares Probleme mit unteren und oberen Schranken als Nebenbedingungen*, Diplomarbeit, Universität Würzburg, Germany, 1977.
[857] Z. ZLATEV, *Comparison of two pivotal strategies in sparse plane rotations*, Comput. Math. Appl., 8 (1982), pp. 119–135.
[858] ——, *Computational Methods for General Sparse Matrices*, Kluwer Academic

Publishers, Norwell, MA, 1992.

[859] Z. ZLATEV AND H. NIELSEN, *LLSS01 - a Fortran subroutine for solving least squares problems (User's guide)*, Tech. Report 79-07, Institute of Numerical Analysis, Technical University of Denmark, Lyngby, Denmark, 1979.

[860] ———, *Solving large and sparse linear least-squares problems by conjugate gradient algorithms*, Comput. Math. Appl., 15 (1988), pp. 185–202.

Index

active set method, 257–262
acute perturbation, 26
adjacency set, 230
algebraic fit, 358
angle
    between subspaces, 29, 286
    principal, 18
approximation
    of lower rank, 12, 101
    unitary, 13
augmented band structure, 224
augmented system, 8, 216, 265
    in Matlab, 264
    optimal scaling, 80
    solution by LDL^T, 77–80
    solution by QR, 20, 64

backward error
    analysis, 38
    componentwise, 35
    normwise, 34
    Oettli–Prager bound, 36
banded matrix
    in standard form, 217
    QR decomposition, 221–224
    storage scheme, 218–219
banded problem
    Cholesky factorization, 219
    Givens QR, 221
    Householder QR, 223
    normal equations, 219–221
bandwidth, 217
BBH algorithm, 333, 334
bidiagonal decomposition, 81–83
    Chan's algorithm, 82
    Lanczos process, 303–306
bidiagonal matrix
    graded, 89

    splitting of, 88
block angular form, 224–227
    covariance matrix, 227
    doubly bordered, 225
block angular problem
    QR algorithm, 225–227
block triangular form, 234–235
    algorithm, 235
    coarse decomposition, 234
    fine decomposition, 235
BLS, *see* bounded least squares
BLUE, *see* unbiased estimated, best linear
bounded least squares
    sparse, 257–262
Bunch–Kaufman pivoting, 77

cancellation, 233
Cauchy matrix, 324
CGLS, 288–293
CGNR, *see* CGLS
Chebyshev abscissae, 325, 327
    interpolation, 326
Chebyshev expansion, 327
Chebyshev polynomials, 325–327
Cholesky factorization, 8, 44–48
    block, 48
    columnwise, 46
    incomplete, 294–297
    of banded matrix, 219
    of bordered matrix, 45
    outer product, 47
    row-wise, 46
    sparse, 242
Clenshaw's recursion formula, 322
clique, 231, 240
column ordering
    Cuthill–McKee, 237–238

minimum degree, 237–240
nested dissection, 237, 240–241
column scaling
optimal, 31
complete orthogonal decomposition, 23, 110–112
from RRQR, 111
condition estimation
by QR, 114
Hager's, 116–118
LINPACK, 114–116
condition number, 38
Bauer–Skeel, 33
estimation of, 114–118
of least squares problem, 31
of matrix, 27
conjugate gradient method, 288–294
convergence, 290–292
convergence rate, 289
effective convergence, 292
in finite precision, 292–293
preconditioned, 293–294
pseudoinverse solution, 291
constrained problem
bounded, 201–203, 312–314
Cryer's method for, 312
iterative methods, 312–314
linear equality, 187–194
linear inequality, 194–203
nonlinear, 353–354
quadratic, 203–213
quadratic inequality, 205–206
sparse, 257–264
sparse BLS, 257–262
Courant–Fischer theorem, 14
covariance matrix, 3, 164
block angular problem, 227
computing, 118–120
selected elements, 256–257
method, 131
cross-validation, 211–213, 316
CS decomposition, 155–157
CSNE, *see* seminormal equations, corrected
curvature matrix, 341
curvature radius, 341

decomposition
bidiagonal, 23, 81–83

complete orthogonal, 23, 110–112
CS, 155–157
GQR, 153
GSVD, 155
LDL^T, 77
PQR, 153
QR, 19–22
singular value, 9–18, 80–98
ULV, 112
URV, 111
defect matrix, 294
derivatives, 32
distance
between subspaces, 18
to set of singular matrices, 12
downdating
by hyperbolic rotation, 143–144
by seminormal equations, 142–143
Cholesky factorization, 140–144
Gram–Schmidt decomposition, 138–140
QR decomposition, 137
Saunders algorithm, 141–142

Eckhart–Young theorem, 12
elimination tree, 241, 247–249
post ordering of, 249
topological ordering of, 249
transitive reduction, 248
envelope, 218
equality constrained problem
by direct elimination, 188–189
by nullspace method, 189–191
error analysis, 38
error estimation
a posteriori, 34
backward, 34–36
forward, 27–32
of inner product, 39
estimator
unbiased, 4
expected value, 2

fast Fourier transform, 330–332
FFT, *see* fast Fourier transform
fill in sparse matrix, 216
filter factor, 101, 315
floating point arithmetic, 37

standard model, 37
flop, 43
Fourier analysis
    discrete, 328–332
Fourier coefficients, 318
Fourier matrix, 331
Fourier synthesis
    discrete, 330
fundamental matrix, 163
fundamental subspaces, 17

Gauss–Markoff
    general linear model, 160–162
    theorem, 3
Gauss–Newton direction, 343
Gauss–Newton method, 342–348
    damped, 343 345
    local convergence, 345–346
    search direction, 343
    trust region, 346–348
Gauss–Seidel's method, 277
generalized SVD, 155–160
    computation of, 159–160
geometric fit, 357
Givens rotation, 53
Givens transformation, 53–57
    algorithm, 54
    fast, 56–57
    in QR, 60
    self-scaling, 57
    storage of, 55
    unitary, 55
Golub–Kahan bidiagonalization, see
    Lanczos bidiagonalization
Golub's method, 64
Gragg–Harrod procedure, 322
Gram–Schmidt decomposition, 60–63
    block algorithm, 71–73
    classical, 63
    downdating of, 138–140
    modified, 61
    modifying, 138–140
    square root-free, 62
    with reorthogonalization, 66–69
graph
    bipartite, 236
    clique in, 231, 240
    connected, 230
    directed, 231
    edges, 230
    elimination, 232, 238–240
    filled, 232, 248
    labeled, 230
    nodes, 230
    ordered, 230
    path in, 230
    planar, 241
    representation of sparse matrix, 230–231
    separator, 231
    strongly connected, 231
    subgraph of, 230
    undirected, 230

H-matrix, 295
Hall property, 233
Harwell–Boeing collection, 237, 264, 266
Harwell subroutine library, 235, 265
Hessian matrix, 340
Hestenes method, 92–94
    parallel implementation, 94
Hölder inequality, 24
Householder reflector, see Householder transformation
Householder transformation, 51–53
    algorithm, 52
    in QR, 59
    unitary, 53
hybrid algorithms, 69–70
hyperbolic rotation, 143–144
    Chambers modification, 144

ill-posed problems, 203–204
IMGS, see incomplete MGS
incomplete Cholesky
    algorithm, 294
    correction, 295
    existence, 295
    higher level, 294
    level zero, 294
    threshold, 296
incomplete factorizations, 294–299
incomplete MGS, 297–298
    by Cholesky, 298
incomplete QR decomposition, 297–299
    drop-tolerance, 299
interior point method

sparse constrained problem, 262–264
interlacing property, 14
IRLS, *see* iteratively reweighted least squares
iterative method
   block, 284–287
   block SOR, 285
   CGLS, 288–293
   classical, 276–279
   convergence
      asymptotic rate, 275
      average rate, 275
      conditions for, 274
   error reducing, 278
   polynomial acceleration of, 281
   preconditioned, 283–284
   residual reducing, 277
   SOR, 279–280
   splitting, 276
   SSOR, 280, 283
   stationary, 274–275
   symmetrizable, 275
   Toeplitz system, 334–335
   two-block, 286–287
iterative refinement, 120–126
   extended precision, 121–124
   fixed precision, 124–126
   for linear systems, 120–121
   for sparse problem, 250–252
iterative regularization, 314–316
iteratively reweighted least squares, 173–175

Jacobian matrix, 340
Jacobi methods
   for SVD, 92–96
Jacobi's iterative method, 277

Kalman gain vector, 131
Karush–Kuhn–Tucker conditions, 262, 312
Kogbetliantz's method, 92, 95–96
Kronecker
   least squares problem, 336–338
   product, 337
      pseudoinverse, 337
      QR decomposition, 338
      singular value decomposition, 338
Krylov subspace, 288, 293, 303

best approximation in, 306–307
orthogonal basis for, 304

l_1 and l_∞ approximation, 5, 175–176
l_p approximation, 5, 172–176
Lagrange multipliers, 8
Lanczos bidiagonalization, 303–306
   convergence of singular values, 309–310
   for total least squares, 310–311
   in finite precision, 306
Landweber's method, 314
LBD, *see* Lanczos bidiagonalization
least squares fitting
   discrete, 321–323
   of geometric elements, 357–358
least squares problem
   damped, 101
   dual, 8
   generalized, 162–163
   geometric interpretation, 7
   Kronecker, 336–338
   nonlinear, 339–342
   primal, 8
   sequential, 188
   slightly overdetermined, 74
   statistical aspects, 318–319
   Toeplitz, 332–336
   weighted, 165–171
least squares solution
   basic, 106
   derivative of, 32
   minimum norm, 7
linear complementarity problem, 312
linear equality constraints
   by GSVD, 191–192
   by updating, 194
   by weighting, 192–194
linear inequality constraints
   active set algorithms, 198–203
   basic transformations, 196–197
   by GSVD, 206–208
   by QR, 208–211
   classification, 194–195
linear model
   errors-in-variables, 176–177
   general univariate, 4
   standard, 3
   total least squares, 176–177

linear regression, 50
linear system
    homogeneous, 10
    overdetermined, 1, 15
    underdetermined, 7, 15
LINPACK algorithm, *see* Saunders algorithm
LSQI, *see* quadratic inequality constraints
LSQR, 307–309
LU factorization, 73
    of rectangular matrix, 76–77
    partial, 75
LU preconditioner, 299–303
    for CGLS, 300
    rate of convergence, 301

M-matrix, 295
matrix
    consistently ordered, 280
    reducible, 231
    sparse, 215
mean, 5
median, 5
merit function, 353
MGS, *see* modified Gram–Schmidt
midrange, 5
modification
    of low rank, 128
modified Gram–Schmidt
    as a Householder method, 65
    least squares solution, 65
    minimum norm solution, 66
modified linear systems, 128
multifrontal method, 245–250
    for QR decomposition
        data management, 249
        update matrix, 247

Newton-type method, 348–351
    hybrid, 348–349
NNLS, *see* nonnegative least squares
no-cancellation assumption, 231–233, 254
node(s)
    adjacent, 230
    amalgamation of, 250
    connected, 230
    degree, 230
    indistinguishable, 239
    supernode, 239, 250

nonlinear problem, 339–342
    constrained, 353–354
    Gauss–Newton method, 342–348
    local minima, 340
    Newton-type methods, 348–351
    separable, 351–353
nonnegative least squares, 195
    sparse, 262–264
norm
    Euclidian, 5
    Frobenius, 12, 25
    Hölder, 5, 24
normal equations, 6, 42
    factored form, 269
    forming of, 42
    loss of information in, 44
    method of, 42–51
    of second kind, 7, 42, 45, 269, 278
    scaling of, 49–51
normalized residuals, 118
nullspace, 6
    method, 189–191
    numerical, 101
        from RRQR, 110
        from SVD, 101
        from ULV, 112
        from URV, 111
numerical cancellation, 233
numerical rank, 99

ODR, *see* orthogonal distance regression
ODRPACK, 357
Oettli–Prager bound, 36
orthogonal bases problem, 68
orthogonal coefficients, 318
orthogonal distance
    fitting circles and ellipses, 357
    regression, 354–357
    linear, 184–186
orthogonal polynomials
    Chebyshev, 325–327
    general theory, 320–321
    Gram, 323
    trigonometric, 329–330
orthogonal projection, 7, 17
    derivative of, 32
orthogonal systems, 317–319

orthogonal transformation
 elementary, 51–57
 Givens, 53–57
 Householder, 51–53

Paige's method, 164–165
PCCGLS, *see* conjugate gradient method, preconditioned
Penrose's conditions, 16
perturbation analysis
 asymptotic form, 32
 componentwise, 32–34
 least squares solutions, 27–34
 pseudoinverse, 26–27
Peters–Wilkinson method, 73–76
pivoting
 row, 169
 standard column, 103–106
  failure of, 105
polar decomposition, 13
polynomial
 approximation, 319–327
 triangle family, 319–320
positive definite, 6
preconditioner
 block column, 284–286
 block SSOR, 286
 cyclic Jacobi, 286
 diagonal scaling, 283
 for Toeplitz systems, 335–336
 incomplete Cholesky, 294–297
 incomplete MGS, 297
 incomplete QR decompositions, 297–299
 LU factorization, 299–303
 reduced system, 287
 SSOR, 283–285
predicting
 structure of $A^T A$, 231
 structure of R, 232–234
principal
 angle, 18, 286
 vector, 18
property A, 279, 286
pseudoinverse, 4, 15–17
 characterization of, 16
 derivative of, 32
 from QR decomposition, 106–107
 from SVD, 15

 Kronecker product, 337
 Moore–Penrose, 16
 solution, 15
pseudoinverse solution
 by LU factorization, 76–77

QR algorithm
 convergence criteria, 85, 88
 Demmel–Kahan, 91
 for SVD, 85–92
 implicit, 83
 operation count, 90
 perfect shifts, 89
 real symmetric matrices, 83–85
 zero shift, 90–92
QR decomposition, 19–22, 58–73
 and Cholesky factorization, 19
 appending a column, 135–136
 appending a row, 136
 block algorithm, 71–73
 column pivoting, 103–106
 deleting a column, 133–134
 deleting a row, 137
 for weighted problem, 168–170
 full rank, 19–21
 generalized, 153–155
 Kronecker product, 338
 modifying, 132–137
 multifrontal, 245–250
 partial, 69
 rank one change, 132–133
 rank revealing, 21–22, 108–110, 252–254
  Chan's algorithm, 109
 row ordering for, 221, 244–245
 row pivoting, 169
 row sequential, 242–244
 Toeplitz matrix, 333–334
 Vandermonde matrix, 325
quadratic inequality constraints, 203–213
quasi-Newton method, 349–351

random errors, 3, 165
 uncorrelated, 3, 176
random variable, 2
rank
 numerical, 99
 structural, 236
rank revealing QR, 21–22, 108–110, 252–254

modifying, 149–152
recursive least squares, 131
regression
    linear, 50
    orthogonal distance, 354–357
    robust, 175
regularization, 100–102
    filter factor, 101, 315
    iterated, 314
    Krylov subspace methods, 315
    Landweber, 314
    methods, 204
    semiconvergence, 314
    Tikhonov, 204
relaxation parameter, 279
    optimal, 286, 303
    optimal for SOR, 280
reorthogonalization
    Kahan–Parlett algorithm, 68
    superorthogonalization, 69
residual polynomial, 290
Richardson's method
    first order, 276
    second order, 282
ridge estimate, 204
Riley's method, 102, 314
rounding error analysis, 37–40
    running, 39
RRQR, *see* rank revealing QR

Saunders algorithm, 141–142
semi-iterative method, 280–283
    Chebyshev, 282
seminormal equations, 70, 250–252
    corrected, 70, 250, 261
    for downdating, 142–143
Sherman–Morrison formula, 129
singular value decomposition, 9–10
    and pseudoinverse, 15–17
    computation, 80–98
    generalized, 155–160
    Kronecker product, 338
    modifying, 145–149
    numerical rank, 99–100
    of 2×2 matrix, 93
    partial, 182
    related eigenvalue problems, 11
    subset selection, 113
    truncated solution, 100–102
singular values, 10
    absolute gap, 14
    by spectrum slicing, 96–98
    minmax property, 13, 15
    relative gap, 90
    sensitivity, 13
    of bidiagonal matrix, 90
singular vectors, 10
    sensitivity, 14
    of bidiagonal matrix, 90
    uniqueness, 10
software
    LLSS01, 266
    MA27, 265, 266
    MA45, 265
    MA47, 265
    QR27, 266
    SMMS, 266
    SPARSPAK, 266
SOR, 279–280, 286
    block, 285
    symmetric, 283
    three-block, 302
    two-block, 303
sparse least squares problem
    MATLAB solver, 267
    banded, 217–224
    constrained, 257–264
    general, 227–252
    Harwell–Boeing collection, 237, 264, 266
    out-of-core solution, 255–256
    software, 264–268
    sources, 215
    test results, 266–268
    updating, 254–255
sparse matrix
    block angular form, 224–227
    block triangular form, 234–237
    column ordering, 237–241
    definition, 215
    irreducible, 235
    reducible, 235
sparse product
    matrix-vector, 270–273
spectral radius, 274
splitting, 274
    proper, 276
    standard, 277
square root of matrix, 13
SSOR, *see* SOR, symmetric

stability of algorithm
  acceptable error, 42
  backward, 40
  forward, 40
  strong backward, 41
  weak, 42
standard form
  of banded matrix, 217
  of LSQI, 205
    transformation to, 210
Stieltjes procedure, 321, 322
storage scheme
  compressed column, 272
  compressed diagonals, 273
  compressed form, 228
  compressed matrix, 272
  compressed row, 218, 229, 271
  coordinate scheme, 228
  dynamic, 229
  for banded matrix, 218–219
  general sparse, 227–230
  static, 229
structural cancellation, 233
structure
  of Cholesky factor, 232–234
Sturm sequence, 98
subspaces
  fundamental, 6
SVD, see singular value decomposition
Sylvester's law of inertia, 96

Tikhonov regularization, 101, 204
  iterated, 102, 314
TLS, see total least squares
Toeplitz
  least squares problem, 332–336
  matrix
    fast multiplication, 335
    QR decomposition, 333–334
    upper triangular, 211
  systems
    circulant preconditioner, 335
    iterative solvers, 334–335
    preconditioner, 335–336
total least squares, 176–184
  restricted, 183
  algorithm, 183
  by SVD, 177–179
  generalized, 182–184
  mixed, 182
  multidimensional, 181–182
  relationship to least squares, 180
trigonometric polynomials, 329
truncated SVD, 100–102
  solution, 101
trust region algorithm, 347
TSVD, see truncated SVD

ULV decomposition, 112
unbiased estimate, 4, 100
  best linear, 4, 153, 162
  of σ^2, 118
unit roundoff, 37
updating
  QR decomposition, 136
URV decomposition, 111

Vandermonde
  matrix
    QR decomposition, 325
  systems, 323–325
    fast algorithm, 324
variable projection algorithm, 352
variance-covariance matrix, see covariance matrix
Volterra integral equation, 211

weighted problem, 165–171
  by updating, 171
  condition number, 166
  Gaussian elimination, 166–168
  QR decomposition, 168–170
  stiff, 166
Wielandt–Hoffman theorem, 14
Wilkinson shift, 84, 87
Woodbury formula, 129